D0026022

Mechanical Engineering

License Review

Mechanical Engineering

License Review

Fifth Edition

John D. Constance, P.E.

Engineering Registration Consultant; Registered Professional Engineer, New York and New Jersey; National Engineering Certification (National Council of Engineering Examiners)

Engineering Press **Austin, Texas**

WISSER MEMORIAL LIBRARY

REF
TJ159
.C75
1996
Copy 1

© Copyright 1996, Engineering Press.
All rights reserved. Reproduction or translation of any part of this work beyond that permitted by section 107 or 108 of the 1976 United States Copyright Act without the permission of the copyright owner is unlawful.

Printed in the United States of America.

ISBN 1-57645-003-1

Engineering Press P.O. Box 200129 Austin, Texas 78720-0129

Mechanical Engineering

License Review

Exam Files

Professors around the country have opened their exam files and revealed their examination problems and solutions. These are actual exam problems with the complete solutions prepared by the same professors who wrote the problems. Exam Files are currently available for these topics:

Calculus I
Calculus II
Calculus III
Circuit Analysis
College Algebra
Differential Equations
Dynamics
Engineering Economic Analysis
Fluid Mechanics
Linear Algebra
Materials Science
Mechanics of Materials
Organic Chemistry
Physics I Mechanics
Physics III Electricity and Magnetism
Probability and Statistics
Statics
Thermodynamics

For a description of all available **Exam Files**, or to order them, ask at your college or technical bookstore, or call **1-800-800-1651** or write to:

Engineering Press
P.O. Box 200129
Austin, TX 78720-0129

TO OUR CHILDREN
Thomas, John, Anita-Marie, Joseph

About the Author

John D. Constance is an Engineering Registration Consultant holding a Certificate of Verification from the Committee on Verification, NCEE. During his 35 years of experience, Mr. Constance has assisted corporate management in the professional development of their engineering and technical personnel and has conducted P.E. refresher courses for the ASME, ASCE, IEEE, and a large number of private engineering firms. A graduate chemical engineer, Constance is a contributor to the *Encyclopedia of Chemical Processing and Design* and the *Encyclopedia of Environmental Science and Engineering*. The author of more than 100 professional articles, Mr. Constance has also written four engineering books.

CONTENTS

FOREWORD

Professional Engineering Examination

The main purpose of this book is to help prepare the candidate for licensure in Mechanical Engineering. The examination is also known as the Principles and Practice—Part B—of the examination and consists of two 4-hour sessions, one in the morning and one in the afternoon of the same day, for a total of 8 hours of examination.

Here is a breakdown of subject areas for the examination relating to the subjects included in this book: Mechanical Design—eight problem situations that a mechanical engineer may encounter on the job, in the office, or in the field; Production Engineering—one problem situation representing a professional situation that a mechanical engineer may encounter in practice; Energy Systems—six problem situations representing professional situations; Control Systems—one problem situation; Thermal and Fluid Processes—three problem situations. The number of problems listed above are those included in the examination; broader coverage is included in this book.

The examination is given open-book from two 4-hour examination booklets. Each 4-hour examination presents 10 problem situations in Mechanical Engineering as indicated above, with one of the problem situations in the afternoon 4-hour session being a problem in engineering economics common to all disciplines. Thus, there will be 19 problem situations in Mechanical Engineering for the full 8-hour examination (10 in the morning and 9 in the afternoon). Each candidate is asked to work only four problem situations from each 4-hour examination booklet. Budget your time to work one problem situation per hour. Textbook problems, proofs, and derivations are avoided as well as esoteric problems. The examination is intended to show your ability to apply sound engineering principles and judgment to the solutions of problem situations normally encountered in practice.

Only the most general engineering principles are required for the solutions to problem situations. Considered important are the application of good engineering judgment to the selection and evaluation of pertinent information, and the ability to make reasonable assumptions when necessary.

The solutions are hand-graded *not* machine-graded. Your score will be based on the number of problem situations you solve correctly through eight. Partial credit is given for partially correct solutions. *Each question is presented in a problem situation/requirement format with step-by-step solutions required.*

The examination is professional in character—composed and graded by licensed professional engineers. You are expected to demonstrate that you think, work, and write in a professional manner.

How to prepare for the examination and what you are permitted to take into the examination room are covered in the introduction to this book.

For information regarding specific requirements in your state and for application forms for registration together with the guidelines to your state's administration of the NCEE uniform examinations, contact your state board office. Their addresses are available from the National Council of Engineering Examiners, Box 1686, Clemson, SC 29633-1686.

John D. Constance

PREFACE

Surely, there are many well-qualified practicing mechanical engineers who have entertained the thought at one time or other of seeking licensure but who have not taken the necessary steps to become registered. Perhaps they thought the need for licensure would "go away" but were still left with the burning feeling they would have to do something about it some day. They argue, "who needs it?" that licensure is solely for civil and sanitary engineers.

Suddenly, these same mechanical engineers find themselves in a "bind." The promotional ladder is closed off to them; they may miss a lifetime opportunity to become a department head or chief mechanical engineer; they may be affected by a reduction in the work force; they may be close to retirement and be considering private consulting practice. Finally, that impulse to become a licensed professional engineer is triggered into action.

Thus, the aim and purpose of this book is to provide the needed encouragement for these engineers and to implement the impulse and urge to become licensed.

Every state board examines in the major field of mechanical engineering. The National Council of Engineering Examiners (NCEE), providing an organization through which state boards may act and counsel together, is ever in search of ways to improve examination procedures and their conduct, so that most states offer the mechanical engineering examination in the NCEE format. While the first day's examination, Fundamentals of Engineering (FE), is designed to test the candidate's facility in mathematics and engineering theory, the second day's examination in Principles and Practice (PP) of Mechanical Engineering requires evidence of proper judgment in the practical and economical approaches to the more advanced field of mechanical engineering.

This part of the examination is professional in character and is designed for persons at the entry level with several years of experience. It encompasses the application of scientific principles to the problems the mechanical engineer meets in everyday practice in the design engineering office, field, and operations. Equally, the book provides the mechanical engineering student just completing undergraduate studies an insight into the type and style of problems required at the beginning of internship in engineering.

In the true practice of their profession, mechanical engineers must selectively be qualified to design power plants, including economical evaluation of sites and original investment costs with relationship to source of fuel, water, and transportation and the mechanical equipment by which the power accomplishes useful results; design and building of heat engines and hydraulic engines, engines for railroads, steamships, airplanes, jet engines, spacecraft, missiles, automobiles, trucks, and various other industrial machines; or design of devices that control the direction, force, and nature of energy, and machines for gearing, belting, and shafting. In addition, the mechanical engineer must be able to design and construct units for environmental control, including heating, ventilating, refrigeration, and air conditioning. When the specialty requires it, the trainee must make broad use of mechanics, mechanical design, physics, graphics, thermodynamics, strength of materials, mathematics, and related engineering subjects.

The examination in mechanical engineering is not comprehensive in each of these areas. It is intended to be a test of whether or not the candidate for licensure has the necessary technical knowledge and can apply it to the solution of several well-sampled problem situations in the major field.

More recently, as the result of a *Task Analysis Survey of Licensed Engineers*,* the contents of the third edition of this book have been reviewed and revised to reflect the findings of the survey as it applies to mechanical engineering examinations.

While the written examination is not the complete test of candidates' qualifications, it is one of those devices in universal use by all the professions to assist a state board of registration in determining whether or not they have the necessary technical knowl-

* Copies of the survey may be obtained by writing the National Council of Engineering Examiners, box 1686, Clemson, SC 29633-1686.

edge. Boards realize that replies from the applicant's references do not and cannot tell the full story about capabilities, and the board of record is required to determine, as far as possible, an applicant's qualifications to practice engineering.

The examination is a test of the applicant's experience. Has it been broad, of high quality and character, and diverse? Although there is a close relationship between job performance and experience, a distinction should be made between the two. Job performance relates to the quality and quantity of work performed, whereas experience is the knowledge or skill gained by performing that work. The quality of experience that one gains on the job is directly proportional to the quality of the job performance. Boards evaluate the quality and quantity of work performed as documented on the application; they test for the knowledge and skills obtained from performing that work through the written examination. The written examination through the efforts of NCEES and its various committees has been found to be a reliable means of supplying a good part of the required information needed to grant licensure. It will continue to be used until a better way is found.

In this fifth edition, the specific purpose and original intent are retained: to help prepare the license candidate to pass the mechanical engineering examination for professional engineers with the latest updated material. Many candidates who have used the book in its earlier editions report that it has served this purpose in a uniquely useful way. This fifth edition has been revised, modified, and amplified to reflect closely the NCEES examination trend in mechanical engineering. Improvements have been incorporated to reflect greater accuracy and notation. Deletions have been made to make way for more up-to-date material that reflects survey findings, and new material has been added throughout the new edition. The more fundamental material has been deleted and has been replaced by new problem/situation/requirement approaches. Those chapters most affected in this way include machine design, hydraulics and fluid mechanics, thermodynamics, heat and power, fuels and combustion products, the steam power plant, steam turbines and cycles, gas turbines and cycles, internal-combustion engines and cycles, pumps and pumping, fans, blowers and compressors, heat transmission, refrigeration, heating and ventilating, air conditioning, environmental control. Also in keeping with survey findings, a new chapter on Production Engineering has been added to provide a sampling of situations in tool design, maintenance, manufacturing, materials handling, plant design, fire protection, and metal fabrication. In the forepart of the new edition, discussion of basics has been minimized and several nonessential areas have been trimmed. The aim is to provide greater concentration on more advanced areas in examination preparation reflecting job-related problem situations.

The question/answer approach to the solutions has been retained as in the previous edition, although the NCEE examination format presentations use the situations/requirement/solution nomenclature. The author feels from long experience preparing engineers for licensure over many years that a firm understanding of the material is more important than outer appearances.

The book continues to be more than a mere compilation of questions and answers. To provide that "at home" feeling in the examination room, and for quick reference, a problem index and a suggested reference list have been included. Such a presentation, it is felt, provides the examinee with a sense of quick mastery of the entire range of the testing materials. The reader will take note that the reference list has been not only updated but enlarged and sectionalized for better-organized examination preparation and exam-room use.

In reviewing the material before making the necessary revisions and updating, a number of guidelines were used: (1) What are the most job-related tasks which licensed engineers perform? (2) What are the most accurate as well as most convenient methods for solving these problems? (3) What other question material can be expected over the long haul for examinations of the future?

When the answers to these questions and other related questions were obtained and evaluated, the discussion material, procedures, and worked-out questions were selected. Thus, the end result is a distillation of theory and basic formulas with maximum emphasis on situations and requirements, their analyses, and their solutions.

There are those who claim that the question-and-answer (situation/requirement/solution) approach to examination preparation makes engineering "too easy." The author wishes to point out that for many years engineering educators have recognized the importance of problem solving as an excellent focusing medium in the development of engineering judgment and experience. Problem courses have been popular in many engineering schools for many years and are still presented.

An important point to remember in preparing for the examination is that a previously presented problem may be rewritten and presented again in such a style that it may not be easily recognized accordingly. However, in anticipation of such occurrences, sufficient background material is presented to make it possible for the examinee to augment this with suggested references and texts. The material covered should cause the examinee to recall previous studies and job-related experiences or to pursue further study of unfamiliar material. Textbooks and bound personal notes should be referred to for detailed information and for background that may have been forgotten with time. Intimate familiarity with texts and references is almost a necessity on open-book tests which have time constraints.

Studying the contents of this book carefully and critically and working many practice problems will:

- Provide you with the opportunity to practice with the kind of material you know will be used.
- Uncover weaknesses when you can still do something about them.
- Enable you to ask yourself—are there tough subjects I can avoid, and others where I can pick up valuable time more easily . . . and still get the answers?
- Help speed up your technique to cope.
- Help guard against exam-room emotional block.
- An important feature of this book is that the solution is worked out directly below the question, a timesaver whose value has been repeatedly demonstrated in over 35 years of providing guidance and preparation of candidates for licensure. Another important feature of the book is presentation of shortcut methods to solutions to help speed up exam performance to provide the examinee a boost over the exam hurdle the first time, thus avoiding time-consuming and costly reexaminations.

This book can also serve you:

- In testing for the Mechanical Engineer Group
- In civil service examinations for engineers

- In a reference capacity—at your desk, in the plant, in the field

Technical school graduates and engineering technicians will find the book helpful as they seek licensure for professional and legal status. The author acknowledges with gratitude his many readers who have commented constructively to formulate new ideas for the continued improvement of this work, truly a labor of love.

For his several contributions to the chapters on mechanics and machine design, strength of machine elements, and environmental control, the author acknowledges the able assistance of his son, John A. Constance, P.E.

John D. Constance

Introduction

How You Can Pass the First Time

BECOMING A PROFESSIONAL ENGINEER

To achieve registration as a Professional Engineer there are four distinct steps: education, fundamentals of engineering (engineer-in-training) exam, professional experience, and finally, the professional engineer exam. These steps are described in the following sections.

Education

The obvious appropriate education is a B.S. degree in mechanical engineering from an accredited college or university. This is not an absolute requirement. Alternative, but less acceptable, education is a B.S. degree in something other than mechanical engineering, or from a non-accredited institution, or four years of education but no degree.

Fundamentals of Engineering (FE/EIT) Exam

Most people are required to take and pass this eight-hour multiple-choice examination. Different states call it by different names (Fundamentals of Engineering, E.I.T., or Intern Engineer) but the exam is the same in all states. It is prepared and graded by the National Council of Examiners for Engineering and Surveying (NCEES).

Review materials for this exam are found in other books like Newnan: *Engineer-In-Training License Review*.

Experience

Typically one must have four years of acceptable experience before being permitted to take the Professional Engineer exam, but this requirement may vary from state to state. Both the length and character of the experience will be examined. It may, of course, take more than four years to acquire four years of acceptable experience.

Professional Engineer Exam

The second national exam is called Principles and Practice of Engineering by NCEES, but probably everyone else calls it the Professional Engineer or P.E. exam. All states, plus Guam, the District of Columbia, and Puerto Rico use the same NCEES exam.

MECHANICAL ENGINEERING PROFESSIONAL ENGINEER EXAM

Background

The reason for passing laws regulating the practice of mechanical engineering is to protect the public from incompetent practioners. Beginning about 1907 the individual states began passing *title* acts regulating who could call themselves a mechanical engineer. As the laws were strengthened, the *practice* of certain aspects of mechanical engineering was limited to those who were registered mechanical engineers, or working under the supervision of a registered mechanical engineer. There is no national registration law; registration is based on individual state laws and is administered by boards of registration in each of the states.

Examination Development

Initially the states wrote their own examinations, but beginning in 1966 the NCEES took over the task for some of the states. Now the NCEES exams are used by all states. This greatly eases the ability of a mechanical engineer to move from one state to another and achieve registration in the new state. About 6000 mechanical engineers take the exam each year. As a result about 23% of all mechanical engineers are registered professional engineers.

The development of the mechanical engineering exam is the responsibility of the NCEES Committee on Examinations for Professional Engineers. The committee is composed of people from industry, consulting, and education, plus consultants and subject matter experts. The starting point for the exam is a mechanical engineering task analysis survey that NCEES does at roughly five to ten year intervals. People in industry, consulting and education are surveyed to determine what mechanical engineers do and what knowledge is needed. From this NCEES develops what they call a "matrix of knowledge" that form the basis for the mechanical engineering exam structure described in the next section.

The actual exam questions are prepared by the NCEES committee members, subject matter experts, and other volunteers. All people participating must hold professional registration. Using workshop meetings and correspondence by mail, the questions are written and circulated for review. The problems relate to current professional situations. They are structured to quickly orient one to the requirements, so the examinee can judge whether he or she can successfully solve it. While based on an understanding of engineering fundamentals, the problems require the application of practical professional judgement and insight. Although four hours are allowed for four problems, probably any problem can be solved in 20 minutes by a specialist in the field. A professionally competent applicant can solve the problem in no more than 45 minutes. Multi-part questions are arranged so the solution of each succeeding part does not depend on the correct solution of a prior part. Each part will have a single answer that is reasonable.

Examination Structure

The ten problems in the morning four-hour session are regular computation "essay" problems. In the afternoon four-hour session all ten problems are multiple choice. In each category (Machine Design, Stress Analysis, Power Plant Systems, and so on) about half of the problems will be in the morning session and half in the afternoon. Engineering economics may appear as a component within one or two of the problems.

1. **MACHINE DESIGN** - Two Problems.
   Fasteners, gears, brakes, belts, clutches, wire rope, bearings, conveyors

2. **STRESS ANALYSIS/STRUCTURAL DESIGN** - Two Problems.
   Stress analysis of machines, tools or structures including theories of failure, static and fatigue life

3. **KINEMATICS AND DYNAMICS** - One Problem.
   Motion of machines and vehicles and associated forces/energy

4. **POWER PLANT SYSTEMS** - Two Problems
   Power plant cycles, compressors and turbines, thermal and mechanical efficiency, co-generation

5. **POWER PLANT PROCESSES** - One Problem.
   Fuels and combustion, combustion stoichiometry, products of combustion; efficiency

6. **POWER PLANT COMPONENTS** - One Problem.
   Boilers and pressure vessels, expansion tanks, piping for gases and liquids

7. **HVAC/R SYSTEMS** - Two Problems.
   Heating/cooling/ventilation load calculations, psychrometrics, evaporative cooling, refrigeration specifications

8. **HVAC/R COMPONENTS** - One Problem.
   Components used in the operation and control of heating, ventilating, cooling and refrigeration equipment

9. **CONTROL SYSTEMS** - One Problem
   Analysis and performance of general mechanical systems (fluid, thermal, etc.) May include (but is not limited to) root locus, stability and/or control diagram components, first and second order systems

10. **INSTRUMENTATION/MEASUREMENTS** - One Problem
    Specifications of measuring systems, static and dynamic measurement of temperature, pressure flow, etc.

11. **VIBRATIONS** - One Problem.
    One and two degree of freedom systems, forced vibrations, transmissibility, isolation

12. **HEAT TRANSFER** - One Problem.
    Conduction, convection and radiation in practical applications

13. **THERMODYNAMICS** - One Problem
    Work, energy, compressible gases in various processes such as compressors, engines and nozzles

14. **HYDRAULICS/PNEUMATICS** - One Problem.
    Hydraulic equipment, hydraulic power and control diagrams, pumps and piping head loss calculations

15. **MANAGEMENT** - One Problem.
    Estimation of production for various products such as sheet metal parts, selection of alternative manufacturing methods based on economics and life cycle studies

16. FIRE PROTECTION - One Problem.
System specifications, sprinkler systems, mobile systems, control valves, flow measurement and control. NFPA 13, 14, 20 and 291.

Note: The examination is developed with problems that will require a variety of approaches and methodologies including design, analysis, application, economic aspects, and operations.

Taking The Exam

Exam Dates

The National Council of Examiners for Engineering and Surveying (NCEES) prepares Mechanical Engineering Professional Engineer exams for use on a Friday in April and October each year. Some state boards administer the exam twice a year in their state, while others offer the exam once a year. The scheduled exam dates are:

	April	October
1996	19	25
1997	18	31
1998	24	30
1999	23	29
2000	14	27

People seeking to take a particular exam must apply to their state board several months in advance.

Exam Procedure

Before the morning four-hour session begins, the proctors will pass out an exam booklet and solutions pamphlet to each examinee. There are likely to be civil, chemical, and electrical engineers taking their own exams at the same time. You must solve four of the ten mechanical engineering problems.

The solution pamphlet contains grid sheets on right-hand pages. Only work on these grid sheets will be graded. The left-hand pages are blank and are for scratch paper. The scratchwork will not be considered in the scoring.

If you finish more than 30 minutes early, you may turn in the booklets and leave. In the last 30 minutes, however, you must remain to the end to insure a quiet environment for all those still working, and to insure an orderly collection of materials.

The afternoon session will begin following a one-hour lunch break. The afternoon exam booklet will be distributed along with an answer sheet. The booklet will have ten 10-part multiple choice questions. You must select and solve four of them. An HB or #2 pencil is to be used to record your answers on the scoring sheet.

Exam-Taking Suggestions

People familiar with the psychology of exam-taking have several suggestions for people as they prepare to take an exam.

1. Exam taking is really two skills. One is the skill of illustrating knowledge that you know. The other is the skill of exam-taking. The first may be enhanced by a systematic review of the technical material. Exam-taking skills, on the other hand, may be improved by practice with similar problems presented in the exam format.

2. Since there is no deduction for guessing on the multiple choice problems, an answer should be given for all ten parts of the four selected problems. Even when one is going to guess, a logical approach is to attempt to first eliminate one or two of the four alternatives. If this can be done, the chance of selecting a correct answer obviously improves from 1 in 4 to, say, 1 in 3.

3. Plan ahead with a strategy. Which is your strongest area? Can you expect to see one or two problems in this area? What about your second strongest area? What will you do if you still must find problems in other areas?

4. Have a time plan. How much time are you going to allow yourself to initially go through the entire ten problems and grade them in difficulty *for you to solve them*? Consider assigning a letter, like A, B, C and D, to each problem. If you allow 15 minutes for grading the problems, you might divide the remaining time into *five* parts of 45 minutes each. Thus 45 minutes would be scheduled for the first - and easiest - problem to be solved. Three additional 45 minute periods could be planned for the remaining three problems. Finally, the last 45 minutes would be in reserve. It could be used to switch to a substitute problem in case one of the selected problems proves too difficult. If that is unnecessary, the time can be used to check over the solutions of the four selected problems. A time plan is very important. It gives you the confidence of being in control, and at the same time keeps you from making the serious mistake of misallocation of time in the exam.

5. Read all four multiple choice answers before making a selection. The first answer in a multiple choice question is sometimes a plausible decoy - not the best answer.

6. Do not change an answer unless you are absolutely certain you have made a mistake. Your first reaction is likely to be correct.

7. Do not sit next to a friend, a window, or other potential distractions.

Exam Day Preparations

There is no doubt that the exam will be a stressful and tiring day. This will be no day to have unpleasant surprises. For this reason we suggest that an advance visit be made to the examination site. Try to determine such items as:

1. How much time should I allow for travel to the exam on that day? Plan to arrive about 15 minutes early. That way you will have ample time, but not too much time. Arriving too early, and mingling with others who also are anxious, will increase your anxiety and nervousness.

2. Where will I park?

3. How does the exam site look? Will I have ample workspace? Where will I stack my reference materials? Will it be overly bright (sunglasses) or cold (sweater), or noisy (earplugs)? Would a cushion make the chair more comfortable?

4. Where is the drinking fountain, lavatory facilities, payphone?

5. What about food? Should I take something along for energy in the exam? A bag lunch during the break probably makes sense.

What To Take To The Exam

The NCEES guidelines say you may bring the following reference materials and aids into the examination room for your personal use only:

1. Handbooks and textbooks

2. Bound reference materials, provided the materials are and remain bound during the entire examination. The NCEES defines "bound" as books or materials fastened securely in its cover by fasteners which penetrate all papers. Examples are ring binders, spiral binders and notebooks, plastic snap binders, brads, screw posts, and so on.

3. Battery operated, silent non-printing calculators.

At one time NCEES had a rule that did not permit "review publications directed principally toward sample questions and their solutions" in the exam room. This set the stage for restricting some kinds of publications from the exam. State boards may adopt the NCEES guidelines, or adopt either more or less restrictive rules. Thus an important step in preparing for the exam is to know what will - and will not - be permitted. We suggest that if possible you obtain a written copy of your state's policy for the specific exam you will be taking. Recently there has been considerable confusion at individual examination sites, so a copy of the exact applicable policy will not only allow you to carefully and correctly prepare your materials, but also will insure that the exam proctors will allow all proper materials that you bring to the exam.

As a general rule we recommend that you plan well in advance what books and materials you want to take to the exam. Then they should be obtained promptly so you use the same materials in your review that you will have in the exam.

License Review Books

We suggest you select license review books that have a 1995 or more recent copyright. The exam used to be 17 essay questions and 3 multiple-choice problems, including a full scale engineering economics problem. The engineering economics problem is gone (at least as a separate question) and half the remaining 20 problems are now multiple choice.

Textbooks

If you still have your university textbooks, we think they are the ones you should use in the exam, unless they are too out of date. To a great extent the books will be like old friends with familiar notation.

Bound Reference Materials

The NCEES guidelines suggest that you can take any reference materials you wish, so long as you prepare them properly. You could, for example, prepare several volumes of bound reference materials with each volume intended to cover a particular category of problem. Maybe the most efficient way to use this book would be to cut it up and insert portions of it in your individually prepared bound materials. Use tabs so specific material can be located quickly. If you do a careful and systematic review of mechanical engineering, and prepare a lot of well organized materials, you just may find that you are so well prepared that you will not have left anything of value at home.

Other Items

Calculator - NCEES says you may bring a battery operated, silent, non-printing calculator. You need to determine whether or not your state permits pre-programmed calculators. Extra batteries for your calculator are essential, and many people feel that a second calculator is also a very good idea.

Clock - You must have a time plan and a clock or wristwatch.

Pencils - You should consider mechanical pencils that you twist to advance the lead. This is no place to go running around to sharpen a pencil, and you surely do not want to drag along a pencil sharpener.

Eraser - Try a couple to decide what to bring along. You must be able to change answers on the multiple choice answer sheet, and that means a good eraser. Similarly you will want to make corrections in the essay problem calculations.

Exam Assignment Paperwork - Take along the letter assigning you to the exam at the specified location. To prove you are the correct person, also bring something with your name and picture.

Items Suggested By Advance Visit - If you visit the exam site you probably will discover an item or two that you need to add to your list.

Clothes - Plan to wear comfortable clothes. You probably will do better if you are slightly cool.

Box For Everything - You need to be able to carry all your materials to the exam and have them conveniently organized at your side. Probably a cardboard box is the answer.

Exam Scoring

Essay Questions

The exam booklets are returned to Clemson, SC. There the four essay question solutions are removed from the morning workbook. Each problem is sent to one of many scorers throughout the country.

For each question an item specific scoring plan is created with six possible scores: 0, 2, 4, 6, 8, and 10 points. For each score the scoring plan defines the level of knowledge exhibited by the applicant. An applicant who is minimally qualified in the topic is assigned a score of 6 points. The scoring plan shows exactly what is required to achieve the 6 point score. Similar detailed scoring criteria are developed for the two levels of superior performance (8 and 10 points) and the three levels of inferior performance (0, 2, and 4 points). Every essay problem submitted for grading receives one of these six scores. The scoring criteria may be based on positive factors, like identifying the correct computation approach, or negative factors, like improper assumptions or calculation errors, or a mixture of both positive and negative factors. After scoring, the graded materials are returned to NCEES, which reassembles the applicants work and tabulates the scores.

Multiple Choice Questions

Each of the four multiple choice problems is 10 points, with each of the ten questions of the problem worth one point. The questions are machine scored by scanning. The input data are evaluated by computer programs to do error checking. Marking two answers to a question, for example, would be detected and no credit given. In addition, the programs identify those questions with statistically unlikely results. There is, of course, a possibility that one or more of the questions is in some way faulty. In that case a decision will be made by subject matter experts on how the situation should be handled.

Passing The Exam

In the exam you must answer eight problems, each worth 10 points, for a total raw score of 80 points. Since the minimally qualified applicant is assumed to average six points per problem, a raw score of 48 points is set equal to a converted passing score of 70. Stated bluntly, you must get 48 of the 80 possible points to pass. The converted scores are reported to the individual state boards in about two months, along with the recommended pass or fail status of each applicant. The state board is the final authority of whether an applicant has passed or failed the exam.

Although there is some variation from exam to exam, the following gives the approximate passing rates:

Applicant's Degree	Passing Exam
Engineering from accredited school	62%
Engineering from non-accredited school	50
Engineering Technology from accredited school	42
Engineering Technology from non-accredited school	33
Non-Graduates	36
All Applicants	56

Although you want to pass the exam on your first attempt, you should recognize that if necessary you can always apply and take it again.

THIS BOOK

The task of preparing for the day-long mechanical engineer examination is a formidable one. The logical approach is to undertake a systematic review of the categories within the mechanical engineering exam, and to do it at about the level of understanding required to successfully pass the exam. This book is designed to help with that task. The problems have been selected to represent an appropriate topic and level of difficulty. Many are actual exam problems, but are not from the NCEES exams, as they do not release their problems for publication.

Thanks to people who have studied the prior editions of this book, a number of errors have been identified and removed. If you note any errors in this edition, a letter, written to the Engineering Press address, would be greatly appreciated.

Chapter 1

STRENGTH OF MACHINE ELEMENTS

Questions and Answers

Q1-1. An air compressor has a crank arrangement similar to that shown in Fig. 1-1. The stroke is 4 in., and the length of the rod is 7 in. The crank cheek is circular in cross section and is 1 in. in diameter. Distance *AC* is equal to $1\frac{1}{8}$ in. If the torsional shearing stress at the midlength of the cheek is equal to 10,000 psi for a cranking angle of 30°, find the value of the gas force on the piston.

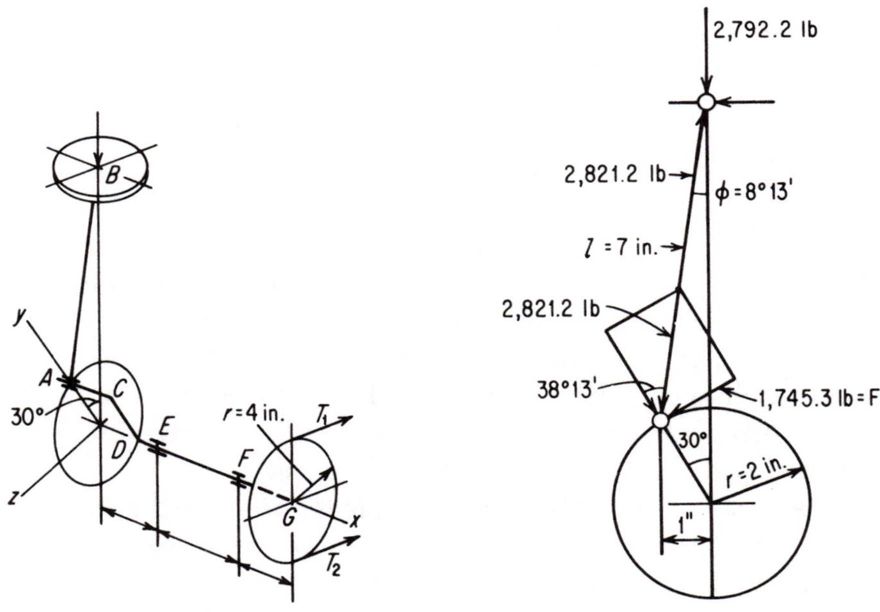

Fig. 1-1. Belt-driven air compressor.

Fig. 1-2

ANSWER. Refer to Fig. 1-2 for the solution. Then $\sin \varphi = 1/7 = 0.14286$ and $\varphi = 8°13'$.

$$T = \frac{\pi d^3 s_s}{16} = \frac{\pi \times 1^3 \times 10{,}000}{16} = 1{,}963.5 \text{ in.-lb}$$

$$F = \frac{1{,}963.5}{1.125} = 1{,}745.3 \text{ lb}$$

$$\text{Force in rod} = \frac{1{,}745.3}{\sin 38°13'} = 2{,}821.2 \text{ lb}$$

$$\text{Force on piston} = 2{,}821.2 \times \cos 8°13' = \mathbf{2{,}792.2 \text{ lb}}$$

Q1-2. A partial single-throw crankshaft drive is shown in Fig. 1-3. The shaft portion that appears has been isolated by

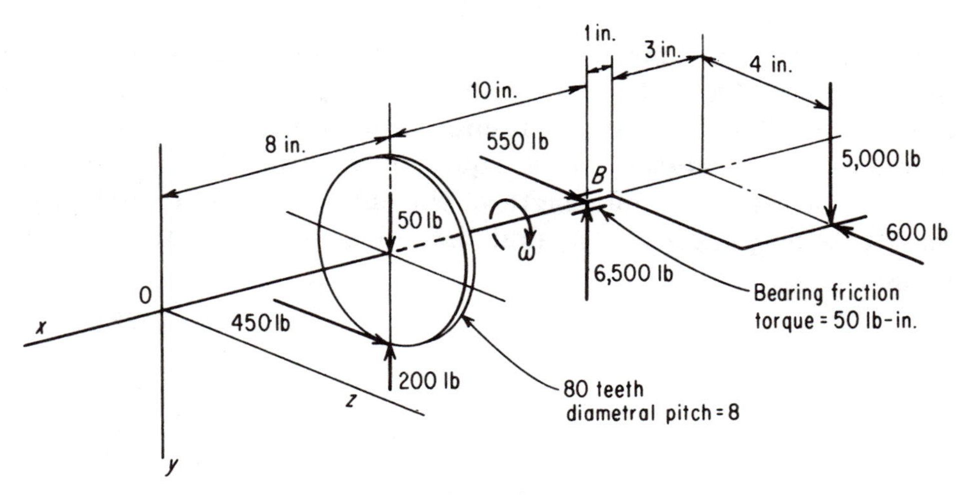

Fig. 1-3

passing a cutting plane at section O between the outboard and inboard bearings (the latter is not shown). Assume constant angular speeds; account for the weight of the gear; otherwise neglect gravitational and inertial forces. Draw the stress-resultant vector components at section O which are required for equilibrium of the system, and report the magnitude of each.

Note. In solving this problem, please observe the following conventions: A vector representing a couple is perpendicular to the plane of the couple. A double arrowhead (⟶») is to be used to indicate the sense of the vector, which, in turn, may be established by use of the right-hand rule. This rule states that when the fingers surround the member being analyzed and point in the sense of the couple, the thumb will indicate the sense of the couple vector.

ANSWER. Refer to Fig. 1-4 and note the direction of bearing friction force.

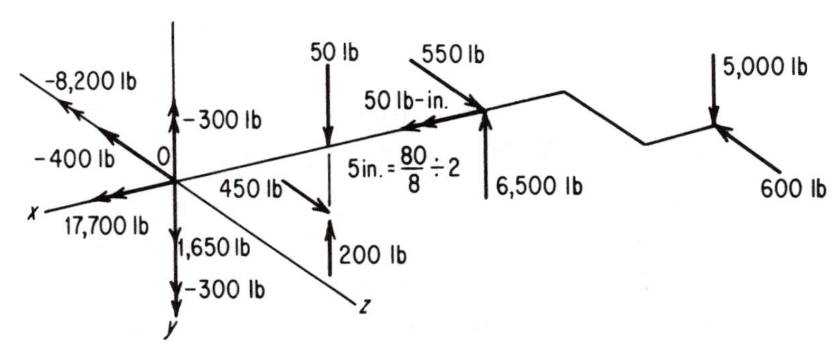

Fig. 1-4

$\Sigma F = 0$

$\Sigma F_x = 0 \qquad 0 + R_x = 0$

$R_x = \mathbf{0}$

$\Sigma F_y = 0 \qquad 5{,}000 - 6{,}500 + 50 - 200 + R_y = 0$

$R_y = \mathbf{+1{,}650\ lb}$

$\Sigma F_z = 0 \qquad -600 + 550 + 450 + R_z = 0$

$R_z = \mathbf{-400\ lb}$

$\Sigma M = 0$

$\Sigma M_x = 0 \qquad -4(5{,}000) + 50 + 5(450) + M_x = 0$

$M_x = \mathbf{17{,}700\ lb\text{-}in.}$

$\Delta M_y = 0 \qquad -22(600) + 18(550) + 8(450) + M_y = 0$

$M_y = \mathbf{-300\ lb\text{-}in.}$

$\Sigma M_z = 0 \qquad -22(5{,}000) + 18(6{,}500) + 8(200 - 50) + M_z = 0$

$M_z = \mathbf{-8{,}200\ lb\text{-}in.}$

Q1-3. The pulley shown in Fig. 1-5 is 16 in. in diameter with a 3½ in.-OD hub. The web plate has a ¼-in. fillet weld on each side. The belt transmits 35 hp to the shaft, which turns at 1,200 rpm. What is the value of the torsional shearing stress in the welds?

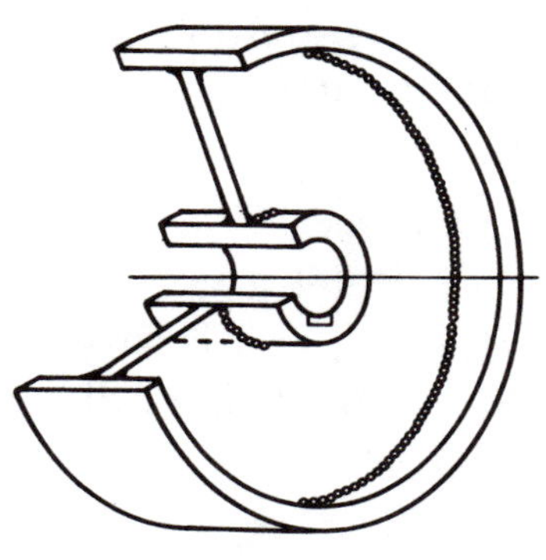

Fig. 1-5

ANSWER. The torque developed = 63,000 × hp/n = (63,000 × 35)/1,200 = 1,840 in.-lb. The radius to center of the throat = r_1 = (3.5/2) + (0.25/4) =

1.81 in. Force carried by welds = P = (1,840/1.81) = 1,020 lb. Length of both welds = $l = 2 \times 2\pi r_1 = 4\pi \times 1.81 = 22.8$ in. Relation between working load P and the torsional shearing stress s is given by

$$s = \frac{P}{0.707hl} = \frac{1{,}020}{0.707 \times 0.25 \times 22.8} = \mathbf{255\ psi}$$

Q1-4. A short gray cast iron (Class 35) post is subjected to an eccentrically applied vertical load P. Load P acts vertically, intersecting the plane at the triangular cross section below and to the left of its centroid (see Fig. 1-6). Given: Ultimate stress in tension

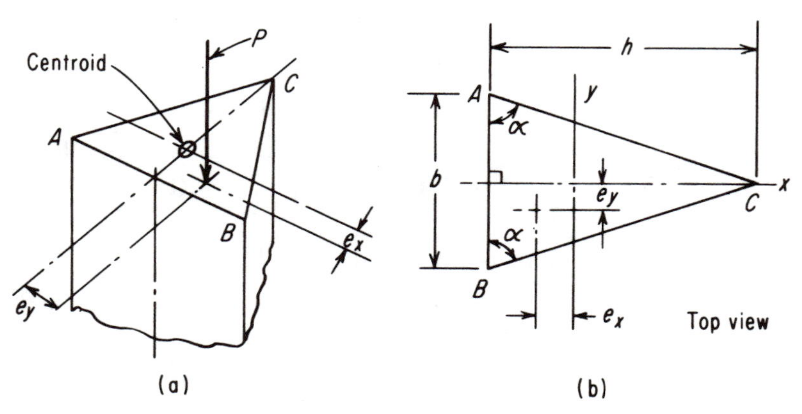

Fig. 1-6

S_{ut} = 35,000 psi. Ultimate stress in compression S_{uc} = 125,000 psi. Factor of safety = 1.75. Base b = 2 in. Height h = 4 in. (*a*) Find the maximum allowable force P, and its eccentricity e, which will induce maximum allowable tensile and compressive stresses simultaneously. (*b*) Using the letters noted in Fig. 1-6*b*, indicate at which points in the cross section these maximum allowable tensile and compressive stresses act.

ANSWER.

(*a*) Refer to Fig. 1-7. Properties of the section are: Area = 4 sq in., $I_y = bh^3/36 = 3.56$ in.^{4}, $I_x = hb^5/48 = 0.67$ in.^{4} The general relationship of direct and bending stress about principal axes X and Y is

$$S = \pm\frac{P}{A} \pm \frac{M_xC}{I_x} \pm \frac{M_yC}{I_y}$$

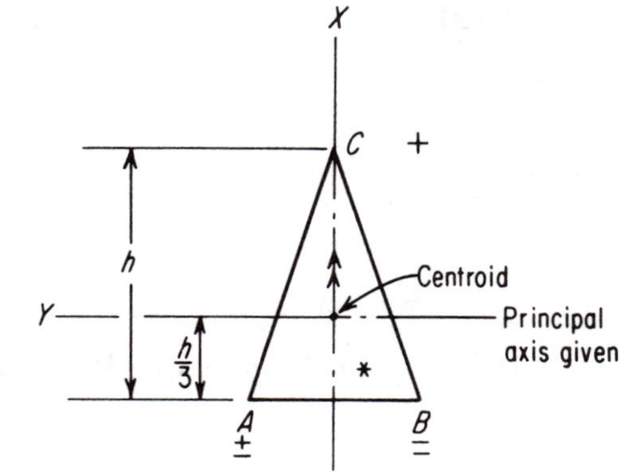

Fig. 1-7

Applying this relationship to points A, B, and C ($+$ = tension, $-$ = compression),

$$S_A = -\frac{P}{A} - \frac{Pe_x(h/3)}{I_y} + \frac{Pe_y(b/2)}{I_x}$$

$$S_B = -\frac{P}{A} - \frac{Pe_x(h/3)}{I_y} - \frac{Pe_y(b/2)}{I_x}$$

$$S_C = -\frac{P}{A} + \frac{Pe_x(2/3h)}{I_y}$$

Now

$$S_A = -\frac{P}{4} - \frac{Pe_x(1.33)}{3.56} + \frac{Pe_y(1)}{0.67} \quad (1)$$

$$S_B = -\frac{P}{4} - \frac{Pe_x(1.33)}{3.56} - \frac{Pe_y(1)}{0.67} \quad (2)$$

$$S_C = -\frac{P}{4} + \frac{Pe_x(2.67)}{3.56} \quad (3)$$

Obviously, maximum compression occurs at B, while maximum tension may occur at A or C.

Solve equations (1), (2), and (3) simultaneously, assuming allowable compressive stress at point B and maximum allowable tensile stress at point A and C simultaneously. Then allowable compressive stress is

$$S_{c\ \text{allowable}} = -125{,}000/1.75 = -71{,}500 \text{ psi}$$
$$S_{t\ \text{allowable}} = 35{,}000/1.75 = 20{,}000 \text{ psi}$$

Solve equations (1) and (2) simultaneously as follows:

$$\frac{20{,}000}{P} = -0.25 - 0.375e_x + 1.5e_y$$

$$\frac{-71{,}500}{P} = -0.25 - 0.375e_x - 1.5e_y$$

From equation (3),

$$\frac{-31{,}500}{P} = -0.75 \qquad \text{and therefore } \mathbf{P = 42{,}000\ lb}$$

(*b*) From (3),

$$e_x = \left(\frac{20{,}000}{42{,}000 + 0.25}\right)^{4/3} = 0.97 \text{ in.}$$

From (1),

$$e_y = \left(\frac{20{,}000}{42{,}000 + 0.25 + 0.375 + 0.97}\right)^{2/3}$$

$$e_y = 0.727$$

Finally,

Location **A,** stress = **20,000 psi**
Location **B,** stress = **−71,500 psi**
Location **C**, stress = **20,000 psi**

Q1-5. An elevator weighs 8,000 lb and moves downward at a constant velocity of 300 ft per min. At the instant when the length of cable from elevator to cable drum is 50 ft, an accident occurs that causes the drum to stop instantaneously. Assume a hemp-filled steel cable of six strands, with each strand made up of 19 wires having a metal area of 2 sq in. Neglecting the weight of the cable, find: (*a*) total stretch of the cable, (*b*) maximum stretch induced in the cable, and (*c*) frequency of oscillation.

ANSWER.

(*a*) Refer to Fig. 1-8. δ_d = dynamic deflection due to dynamic force.

$$\text{Total stretch of cable after accident} = \delta_{\text{static}} + \delta_{\text{dynamic}} = \delta_s + \delta_d$$

$$\delta_s = \frac{P_wL}{AE} = \frac{WL^*}{AE} = \frac{8{,}000(50 \times 12)}{2 \times 12 \times 10^6} = 0.2 \text{ in.}$$

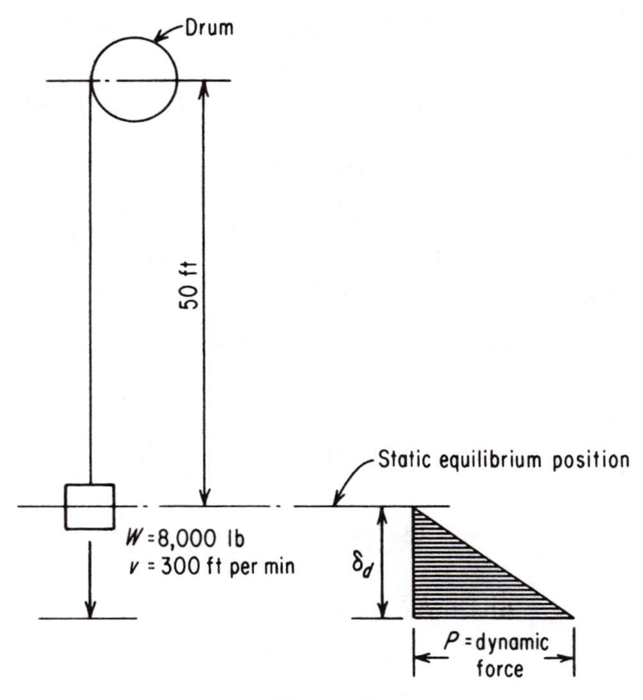

Fig. 1-8

Strain energy u in wire rope due to dynamic load:

$$u = 1/2 \times \delta_d \times P$$

From elastic behavior of wire rope,

$$\delta_d = \frac{PL}{AE}$$

Therefore,
$$u_d = \frac{AE\delta_d}{2L}$$

Kinetic energy at accident $= \dfrac{W}{g} \times \dfrac{v^2}{2}$

Because of conservation of energy, $E_{\text{total}} = 0 = \Delta u$. Then strain energy = kinetic energy. Thus

$$\frac{AE\delta_d^2}{2L} = \frac{Wv^2}{2g}$$

By rearrangement,
$$\delta_d^2 = \frac{Wv^2L}{AEg}$$

$$\delta_d = v\sqrt{\frac{W}{g} \times \frac{L}{AE}} = \frac{300 \times 12}{60}\sqrt{\frac{8{,}000}{386} \times \frac{50 \times 12}{2 \times 12 \times 10^6}} = 1.366 \text{ in.}$$

$$\text{Total stretch} = \delta_s + \delta_d = 0.2 + 1.366 = \mathbf{1.566 \text{ in.}}$$

(*b*) Max stress $= S_s + S_d$. Also, $S = P/A = E\delta/L$.

$$S = \frac{E}{L}(\delta_s + \delta_d) = \frac{12 \times 10^6}{50 \times 12} \times 0.2 + \frac{12 \times 10^6}{50 \times 12} \times 1.366$$

$$S = 4{,}000 + 27{,}300 = \mathbf{31{,}300 \text{ psi}}$$

(*c*) Oscillation frequency $= f$, and t (time of oscillation) $= 1/f$. Refer to Fig. 1-9. Then

$$f = \frac{1}{2\pi}\sqrt{\frac{K}{M}} \text{ cycles per sec (Hz)} = \frac{1}{2\pi}\sqrt{\frac{W}{\delta_s}\Big/\frac{W}{g}} = \frac{1}{2\pi}\sqrt{\frac{g}{\delta_s}}$$

$$f = \frac{1}{2\pi}\sqrt{\frac{386}{0.2}} = \mathbf{7 \text{ cycles per sec}}$$

Note that

$$W/\delta_s = AE/L = K$$

Also, 386 = in. per sec per sec.*

Q1-6. The bars in Fig. 1-10 have the same cross-sectional area.

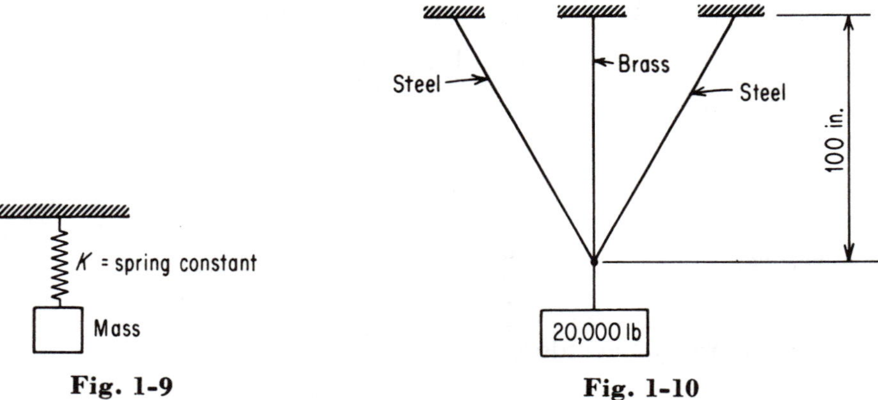

Fig. 1-9 **Fig. 1-10**

There is no stress in the bars before the load is applied. Each bar is 0.5 in. square. Outer bars hang at 60° from horizontal.

(*a*) Find the force in each bar.
(*b*) Find the force in each bar if the temperature drops 100°F.

* See Marks Handbook for wire rope.

ANSWER.

(*a*) $\delta = Pl/AE$, and by rearrangement, $P = \delta AE/l$. Now for the steel,

$$P_s = \frac{0.866\delta A \times 30{,}000{,}000}{115.47} = 225{,}000\delta A$$

See Fig. 1-11.

Vertical component.

$$(P_s)_v = 0.866\delta \times 225{,}000A = 194{,}850\delta A$$

For brass: $P_b = \dfrac{(\delta A \times 15 \times 10^6)}{(100)} = 150{,}000\delta A$

Total: $\delta A(2 \times 194{,}850 + 150{,}000) = 20{,}000$

$\delta A = \dfrac{20{,}000}{539{,}700}$ and $P_s = 225{,}000 \times \dfrac{20{,}000}{539{,}700} = \mathbf{8{,}338\ lb}$

$P_b = 150{,}000 \times \dfrac{20{,}000}{539{,}700} = \mathbf{5{,}559\ lb}$

(*b*) *Due to temperature effect.*

For steel, $\delta = 115.47 \times 0.0000065 \times 100 = 0.075056$ in. shortening
For brass, $\delta = 100 \times 0.0000102 \times 100 = 0.102000$ in. shortening

Stretch of brass due to load. See Fig. 1-12.

$$\delta_b = \frac{P_b \times 100 \times 4}{15 \times 10^6}$$

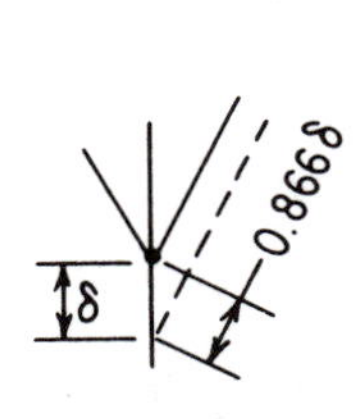

Fig. 1-11

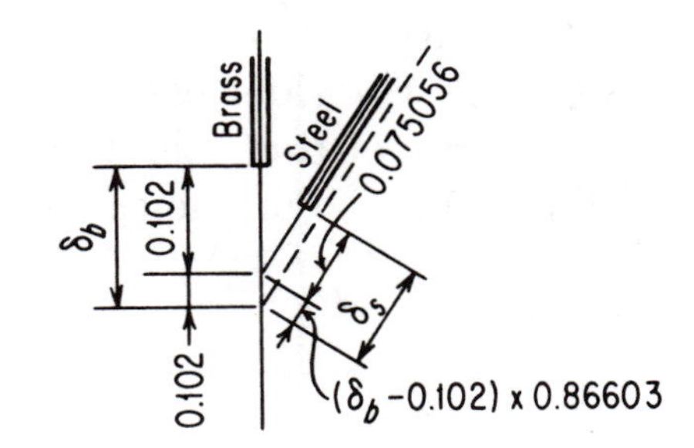

Fig. 1-12

Stretch of steel due to load.

$$\delta_s = (\delta_b - 0.102) \times 0.86603 + 0.075056 = 0.86603\delta_b - 0.013279$$
$$2(P_s)_v = 20{,}000P_b$$

Force in steel.

$$P_s = \frac{20{,}000 - P_b}{2 \times 0.86603} = 11{,}547 - 0.57735P_b$$
$$\delta_s = 0.86603\delta_b - 0.013278 = \frac{P_s \times 115.47 \times 4}{30 \times 10^6}$$
$$0.86603 \times \frac{P_b \times 400}{15 \times 10^6} - 0.013278 = \frac{(11{,}547 - 0.57735P_b)(115.47)}{0.25 \times 30 \times 10^6}$$
$$0.000023094P_b - 0.013278 = 0.177778 - 0.000008889P_b$$
$$0.000031983P_b = 0.191057$$
$$P_b = \mathbf{5{,}974\ lb}$$
$$P_s = 11{,}547 - 0.57735 \times 5{,}974 = 11{,}547 - 3{,}449 = \mathbf{8{,}098\ lb}$$

Q1-7. Figure 1-13 is a view of two beams, both of which have the same EI of 33,333 lb-in.². The end of the 10-in. cantilever

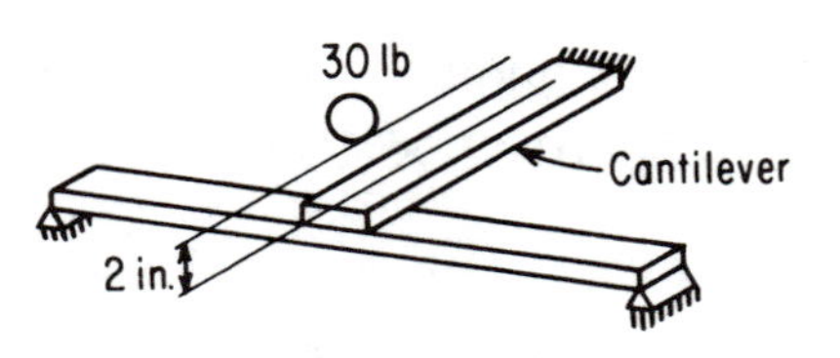

Fig. 1-13

beam rests on the midpoint of the 20-in. simply supported beam. A 30-lb weight falls freely through a distance of ½ in. and strikes the free end of the cantilever beam. Find the maximum deflection that the cantilever beam undergoes.

ANSWER. Consider each beam as a spring that can store the potential energy of the weight $W = 30$ lb in the form of strain energy (see Fig. 1-14). For the cantilever beam alone, shown in Fig. 1-15,

$$\Delta = \frac{PL^3}{3EI} \qquad K_1 = \frac{P}{\Delta} = \text{lb/in.} = \frac{3EI}{L^3}$$
$$= \frac{3(33{,}333)}{10^3} = 100 \text{ lb per in.}$$

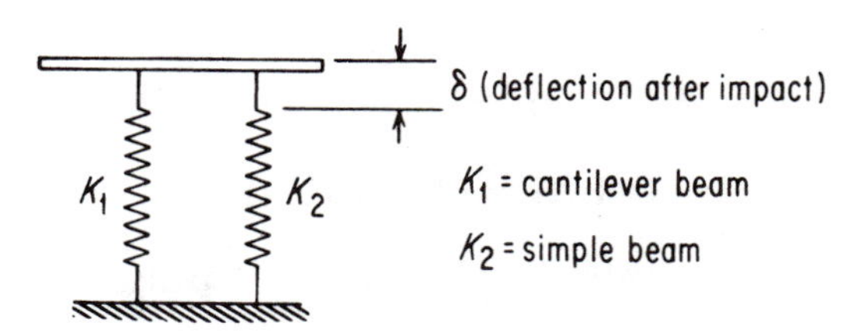

Fig. 1-14

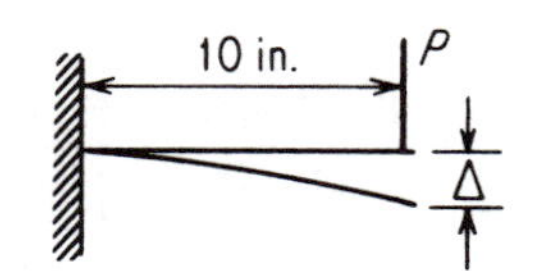

Fig. 1-15. Cantilever beam.

For the simple beam alone, shown in Fig. 1-16,

$$\Delta = \frac{PL^3}{48EI} \qquad K_2 = \frac{P}{\Delta} = \frac{48EI}{L^3} = 200 \text{ lb per in.}$$

$$K \text{ (effective)} = K_1 + K_2 = 100 + 200 = 300 \text{ lb per in.}$$

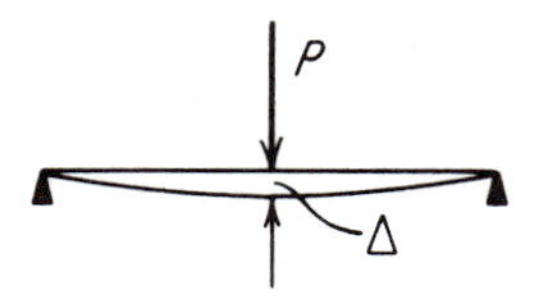

Fig. 1-16. Simple beam.

The strain energy stored in the beams for deflection δ is

$$u = \tfrac{1}{2}K_e\delta^2 = \tfrac{1}{2}(300)(\delta^2) = 150\delta^2$$

$$\text{Potential energy of weight} = E = Wy = 30(0.5 + \delta)$$

Conservation of energy principle:

$$u = E \qquad \text{and} \qquad 150\delta^2 = 30(0.5 + \delta)$$

Then by rearrangement, $150\delta^2 - 30\delta - 15 = 0$, and by application of the binomial theorem,

$$\delta = (30 \pm 100)/300 = \mathbf{0.43 \text{ in.}}$$

Q1-8. A punch punches a 1-in.-diameter hole in a steel plate ¾ in. thick every 10 sec. The actual punching takes 1 sec. The ultimate shear strength of the plate is 60,000 psi. The flywheel of the punch press has a mass moment of inertia of 500 in.-lb-sec² and rotates at a mean speed of 150 rpm.

(*a*) What is the horsepower required for the punch operation?

(*b*) What is the total speed fluctuation of the flywheel in revolutions per minute?

ANSWER.

(*a*) The force required for the actual punching of the metal is

$$F = \pi dSt = \pi \times 1 \times 60{,}000 \times 0.75 = 141{,}372 \text{ lb}$$

Assume that the force required during half the thickness of punch is

$$\text{Energy} = \tfrac{1}{2} \times 141{,}372 \times 0.75 \times 1/12 = 4{,}419 \text{ ft-lb} \approx 4{,}420$$

Then the horsepower required to punch is $4{,}420/(1 \times 550) =$ **8.04 hp.**

(*b*) The speed fluctuation may be found from

$$\left(\frac{I_0}{2}\right)(W_2 - W_1)(W_2 + W_1)$$

where $W_2 - W_1$ is the total fluctuation in angular velocity and $W_2 + W_1$ is equal to twice the average angular velocity $= 2(150/60)(2\pi) = 2 \times 15.70$.

$$W_2 - W_1 = (4{,}420 \times 12)/(500 \times 15.7) = 6.76 \text{ radians per sec}$$

Expressed as revolutions per minute,

$$(6.76)(60/2\pi) = \mathbf{64.55\ rpm}$$

Q1-9. The valve push rod for an overhead valve engine is ¼ in. in diameter and 14 in. long. Find the critical load when the rod is considered as a column with round ends.

ANSWER.

$$I = \frac{\pi d^4}{64} = \frac{\pi}{4^4 \times 64} = \frac{\pi}{256 \times 64}$$

$$P_{cr} = \frac{\pi^2 EI}{l^2} = \frac{\pi^2 \times 30 \times 10^6 \pi}{14^2 \times 256 \times 64} = \mathbf{290\ lb}$$

Q1-10. A structural steel rod extends through an aluminum tube as shown in Fig. 1-17. The cross-sectional area of the rod is 0.8 sq in., and its upper end has 20 threads per inch. The aluminum tube is 20 in. long and has a cross-sectional area of 1.8 sq in. Determine the stresses in the rod and in the tube due to a quarter turn of the nut on the bolt. Assume $E_s = 30 \times 10^6$ psi and $E_a = 10 \times 10^6$ psi.

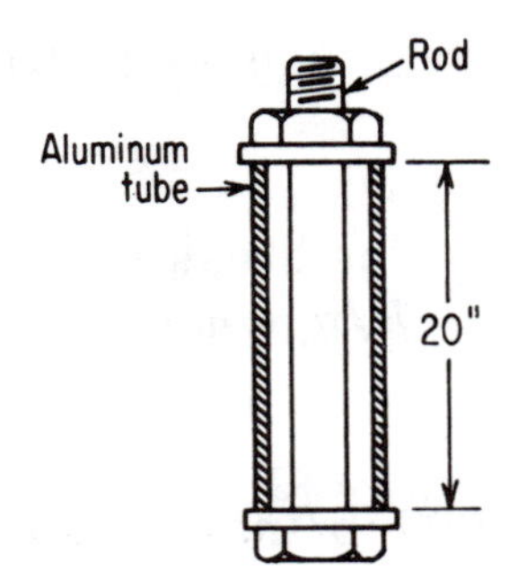

Fig. 1-17

ANSWER. The same force P must exist in both the tube and the bolt. Also, the one-quarter-turn deformation will be taken up by a shortening of the tube and a lengthening of the bolt. Accordingly,

$$\Delta = \Delta_{\text{bolt}} + \Delta_{\text{tube}} = \left(\frac{P \times L}{A \times E}\right)_{\text{bolt}} + \left(\frac{P \times L}{A \times E}\right)_{\text{tube}}$$

$$\Delta = \frac{(P)(20)}{(0.8)(30 \times 10^6)} + \frac{(P)(20)}{(1.8)(10 \times 10^6)} = \left(\frac{1}{4}\right)\left(\frac{1}{20}\right)$$

$$(8.33 \times 10^{-7})P + (1.111 \times 10^{-6})P = 0.0125$$

From which P is found to equal 6,430 lb.

$$\sigma_s = \frac{P}{A} = \frac{6{,}430}{0.8} = \textbf{8,040 psi} \text{ for the steel}$$

$$\sigma_a = \frac{P}{A} = \frac{6{,}430}{1.8} = \textbf{3,570 psi} \text{ for the aluminum}$$

Q1-11. A steel specimen is undergoing destructive testing. Data supplied on the specimen: total length is 19 in; sections at each end of the specimen are 1 by 1 in by 6 in long; center section is $\frac{7}{16}$ in in diameter by 7 in long with center section concentric with the square sections at each end. Center 7-in section has a 2-in gauge length marked. The specimen under test broke when the load reached a maximum of 11,274.75 lb, and the break occurred through the original $\frac{7}{16}$-in diameter. But the diameter was now $\frac{1}{4}$ in. The cross-sectional area of a $\frac{7}{16}$-in-diameter bar is 0.15033 in^2. The length of the 2-in gauge section had stretched to 2.55 in.

(*a*) Determine the ultimate strength of the material.

(*b*) What is the ultimate strength in pounds per square inch of the square section at each end?

(*c*) Determine the percentage elongation of the 2-in gauge section.

ANSWER.

(*a*) Refer to Fig. 1-18. The break occurred in the 7/16-in section, so that the stress S is $S = F/a$, where a is original cross-sectional area.

$$S = \frac{11{,}274.75}{0.15033} = 75{,}000 \text{ lb/in}^2 \text{ ultimate strength} \qquad \textbf{Ans.}$$

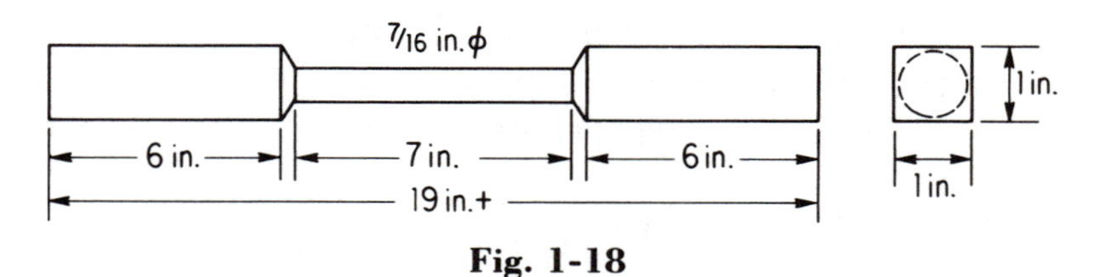

Fig. 1-18

(*b*) The ultimate strength at the square section at each end is the same (75,000 lb/in²) because it is still the same material. It did not break at this section because the cross-sectional area is larger than at the 7/16-in-diameter section.

(*c*)

$$\begin{aligned}\text{Percentage elongation} &= \frac{(\text{final length} - \text{original length}) \times 100}{\text{original length}}\\ &= (2.55 - 2)/2 \times 100\\ &= 27.5 \text{ percent} \qquad \textbf{Ans.}\end{aligned}$$

Q1-12. A steel ring having an internal diameter of 8.99 in and a thickness of ¼ in is heated and allowed to shrink over an aluminum cylinder having an external diameter of 9.00 in and a thickness of ½ in. After the steel cools, the cylinder is subjected to an internal pressure of 800 lb/in². Find the stresses in the two materials. For aluminum, $E = 10 \times 10^6$ lb/in².

ANSWER.

(*a*) Compute the radial pressure caused by prestressing.

$$p = \frac{2\,\Delta D}{D^2(1/t_a E_a + 1/t_s E_s)}$$

where p = radial pressure resulting from prestressing, lb/in^2
D = internal diameter of cylinder, in
t_a = cylinder wall thickness for aluminum, in
t_s = cylinder wall thickness for steel, in

$$p = \frac{2(0.01)}{9^2[1/0.5 \times 10 \times 10^6) + 1/0.25 \times 30 \times 10^6]} = 741 \text{ lb/in}^2$$

(b) Compute the corresponding prestresses. Using the subscripts 1 and 2 denote the stresses caused by prestressing and internal pressure, respectively, $s_{a1} = pD/2t_a$. Thus,

$$s_{a1} = \frac{741(9)}{2(0.5)} = 6670 \text{ lb/in}^2 \text{ compression}$$

$$s_{s1} = \frac{741(9)}{2(0.25)} = 13{,}340 \text{ lb/in}^2 \text{ tension}$$

(c) Compute the stresses caused by internal pressure. From the relation

$$\frac{s_{s2}}{s_{a2}} = \frac{E_s}{E_a} = 30 \times \frac{10^6}{10} \times 10^6 = 3$$

Now compute s_{a2} from the relation $t_a s_{a2} + t_s s_{s2} = pD/2$, from which we obtain $s_{a2} = 2800$ lb/in^2 tension. Also, $s_{s2} = 3 \times 2880 = 8640$ lb/in^2 tension.

(d) Compute the final stresses. Summate the final results from the above to obtain the final stresses as follows:

$$s_{a3} = 6670 - 2880 = 3790 \text{ lb/in}^2 \text{ compression} \qquad \textbf{Ans.}$$

$$s_{s3} = 13{,}340 + 8640 = 21{,}980 \text{ lb/in}^2 \text{ tension} \qquad \textbf{Ans.}$$

Chapter 2

MACHINE DESIGN

Questions and Answers

Q2-1. It is desired to check the design of a 2-in. medium steel shaft subjected to a turning moment of 40,000 in.-lb. Ultimate stress is 50,000.

ANSWER.

$$T = S_s Z_p = 40{,}000. \quad \text{Solve for } S_s.$$

$$Z_p = \pi \frac{d_0^3}{16} = \pi \frac{2^3}{16} = 0.5\pi \text{ in.}^3$$

$$S_s = \frac{T}{Z_p} = \frac{40{,}000}{0.5\pi} = 25{,}400 \text{ psi}$$

$$F = \frac{U_s}{S_s} = \frac{50{,}000}{25{,}400} = 1.98. \quad \text{Too low.} \quad \textbf{Unsafe}$$

Q2-2. A hydraulic turbine in a water-power plant is rated at 12,000 hp. The steel vertical shaft connecting the turbine and generator is 24 in. in diameter and rotates at 60 rpm. Calculate the maximum shearing stress developed in the shaft at full load.

ANSWER. Since the shaft is vertical, there will be no stresses caused by bending and the maximum shearing stress will be really the maximum shaft stress. Now, since horsepower is

$$\frac{2\pi NT}{33{,}000}$$

and torque is found by rearrangement to be

$$T = \frac{12{,}000 \times 33{,}000}{6.28 \times 60} = 1.05 \times 10^6 \text{ lb-ft in shaft}$$

the polar moment of inertia I_p is

$$\frac{\pi d^4}{32} = 3.14 \times \frac{24^4}{32} = 32{,}600 \text{ in.}^4$$

For strength, $T = S_s I_p / r$. By rearrangement and substitution in this equation

$$S_s = 1.05 \times 10^6 \times (24/2) \times (1/32{,}600) \times 12 = \mathbf{4{,}640 \text{ psi}}$$

Q2-3. An automobile weighing 3,000 lb is slowed uniformly from 40 to 10 mph in a distance of 100 ft by a brake on the drive-shaft. Rear wheels are 28 in. in diameter. Drive-shaft diameter is 1 in. Ratio of differential is 40:11. Calculate extreme fiber stress in drive shaft, neglecting tire, bearing, and gear losses.

ANSWER. The following relation for deceleration holds.

$$-a = \frac{v^2 - v_0^2}{2s} = \frac{14.7^2 - 58.7^2}{2 \times 100} = 16.1 \text{ ft per sec}^2$$

where $(-a)$ is the deceleration, v is final velocity, and v_0 is initial velocity expressed in fps. Now, the braking force required is

$$F = \frac{W}{g} a = \frac{3{,}000}{32.2} (-16.1) = -1{,}500 \text{ lb}$$

Total torque on driving wheels is now determined:

$$T = 1{,}500 \times (28/2) = 21{,}000 \text{ in.-lb}$$

Torque on drive shaft is

$$T = 21{,}000 \times (11/40) = 5{,}775 \text{ in.-lb}$$

Unit stress in shaft is

$$S_s = \frac{16}{\pi d^3} T = \frac{16}{\pi 1^3} (5{,}775) = \mathbf{29{,}400 \text{ psi}}$$

Q2-4. A circular bar of solid cast iron 60 in. long carries a solid circular head 60 in. in diameter. The bar is subjected to a torsional moment of 60,000 in.-lb which is supplied at one end. It is desired to keep the torsional deflection of the circular head below $\frac{1}{32}$ in. when the bar is transmitting power over its entire length in order to prevent the chattering of the piece. What would be the diameter of the bar if the working stress is taken as 3,000 psi and the transverse modulus of elasticity is 6 million psi?

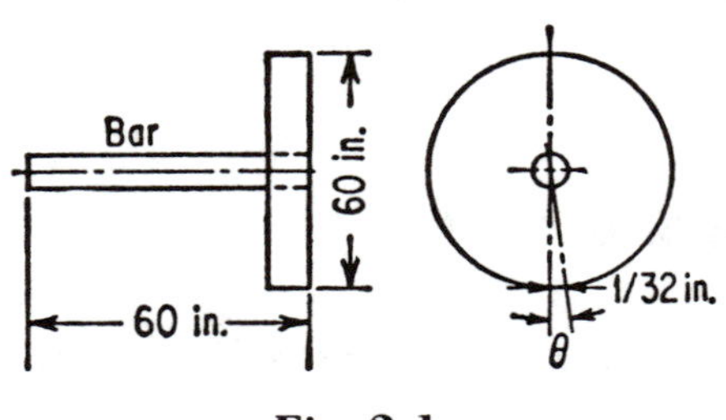

Fig. 2-1

ANSWER. Refer to Fig. 2-1.

$$\text{Torque} = S_s \frac{I_p}{C} = 60{,}000 = 3{,}000\pi \frac{d_0^3}{16}$$

Solving for d_0,

$$d_0^3 = (60{,}000 \times 16)/(3{,}000 \times 3.14)$$
$$d_0 = 4.6 \text{ in.}$$

For torsional stiffness, $\theta = (\frac{1}{32})/r$. This is equal to $\frac{1}{960}$ radian. Since arc length along head is θr, where r is radius of head in inches, therefore $\frac{1}{32} = \theta r = (\theta)(30)$. Note S equals 60 in.

$$d_0^4 = \frac{32 \times T \times S}{\pi \times E_t \times \theta}$$

Now, by direct substitution in this formula $d_0^4 = 5{,}870$ and $d_0 = $ **8.8 in.** Since 8.8 in. is greater than 4.6 in., the shaft must be designed for torsional stiffness.

Q2-5. A $1\frac{13}{16}$-in.-diameter steel shaft is supported on bearings 6 ft apart. A 24-in.-diameter pulley weighing 50 lb is attached to the center of the span. The pulley runs 400 rpm and delivers 15 hp to the shaft. The shaft weight is 8.77 lb per ft. A belt pulls on the pulley

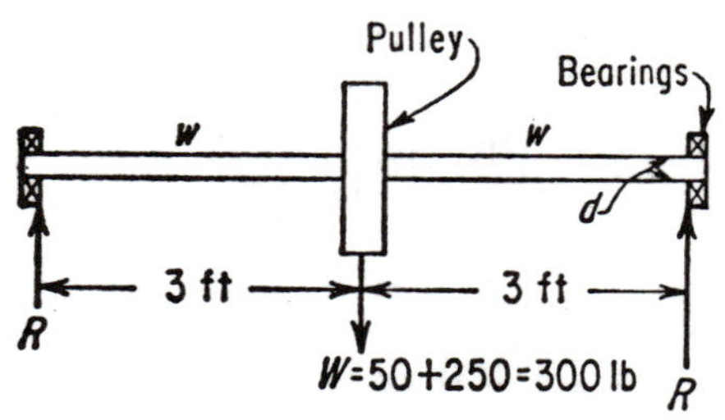

Fig. 2-2

with a force of 250 lb in a vertically downward direction. Calculate the maximum stress in the shaft due to the combination of bending and torsional (twisting) stresses.

ANSWER. Consider the shaft to be a beam with fixed ends (see Fig. 2-2). The maximum bending moment due to loads occurring at bearings is

$$\frac{wl^2}{8} + \frac{Pl}{4} = \frac{8.77 \times 6^2 \times 12}{8} + \frac{300 \times 6 \times 12}{4}$$
$$= 474 + 5{,}400 = 5{,}874 \text{ in.-lb}$$

Torque delivered by power:

$$T = \frac{33{,}000 \times 15 \times 12}{2\pi 400} \equiv 2{,}360 \text{ in.-lb}$$

Maximum shearing stress due to combined loads:

$$S_s = \frac{16}{\pi(1^{13}/_{16})^3} \times \sqrt{5{,}874^2 + 2{,}360^2} = 5{,}415 \text{ psi}$$

Maximum normal stress due to combined loads:

$$S_n = \frac{16}{\pi(1^{13}/_{16})^3} \times [5{,}874 + \sqrt{5{,}874^2 + 2{,}360^2}] = \mathbf{10{,}439 \text{ psi}}$$

Q2-6. A solid circular shaft is used to transmit 200 hp at 1,000 rpm. (*a*) What diameter shaft is required if the allowable maximum shearing stress is 20,000 psi? (*b*) If a hollow shaft is used having an inside diameter equal to the diameter of the solid shaft determined in part (*a*), what must be the outside diameter of this shaft if the angular twist of the two shafts is to be equal?

ANSWER. The torque required to transmit 200 hp at 1,000 rpm is

$$T = (33{,}000 \times 12 \times 200)/2\pi 1{,}000 = 12{,}600 \text{ in.-lb}$$

(*a*) The torsional shearing stress is first determined by use of formula

$$S_s = \frac{T \times d}{2 \times I_p}$$

where I_p is the polar moment of inertia. Now, since $I_p = \pi d^4/32$,

$$S_s = \frac{16T}{\pi d^3}$$

$d^3 = (16 \times 12{,}600)/(\pi \times 20{,}000) = 3.2$ and $d = \mathbf{1.475\ in.}$

(*b*) Now, if a hollow shaft is used with an inside diameter of 1.475 in. and an outside diameter of D is to be determined such that the angular twist θ_a of the solid shaft and that of the hollow shaft θ_b are equal,

$$\theta_a = T_a \frac{L_a}{G_s \times I_{pa}}$$

$$\theta_b = T_b \frac{L_b}{G_s \times I_{pb}}$$

where G_s is the modulus of elasticity in shear for the material. Now, since torque and length are the same for both shafts and are made of the same material, by equating angular twists to each other, $I_{pa} = I_{pb}$, or

$$\frac{\pi d^4}{32} = \frac{\pi}{32}(D^4 - d^4)$$

We find that D (outside diameter of the hollow shaft) is **1.755 in.** and d (inside diameter of the hollow shaft) is equal to **1.475 in.**

Q2-7. A cylinder head of a steam engine is held by 14 bolts. The diameter of the cylinder is 14 in. and the steam pressure is 125 psi. What size of bolts is required if tensile stress is 3,000 psi?

ANSWER. The force due to the steam pressure is pressure × area.

$$F = 125 \times \frac{\pi \times 14^2}{4} = 19{,}250 \text{ lb}$$

Since the pressure is tending to push the cylinder head away from the block, the load per bolt is simply 19,250/14 = 1,375 lb. Then the root diameter of each bolt is found by

$$F = A \times S_t = \frac{\pi d^2}{4} S_t$$

$$1{,}375 = \frac{\pi d^2}{4} \times 3{,}000$$

Rearranging and solving for d,

$$d = \sqrt{\frac{1{,}375 \times 4}{3{,}000 \times \pi}} = \mathbf{0.76\ in.}$$

The standard root diameter that is just above the 0.76 in. is 0.8376 in. Therefore, the outside diameter is **1 in.**

Q2-8. The load on a helical spring is 1,600 lb and the corresponding deflection is to be 4 in. Rigidity modulus is 11 million psi and the maximum intensity of safe torsional stress is 60,000 psi. Design the spring for the total number of turns if the wire is circular in cross section with a diameter of $\frac{5}{8}$ in. and a center-line radius of $1\frac{1}{2}$ in.

ANSWER. In order to find the total deflection, given the number of active coils, the following may be used:

$$y = \frac{N \times 64 \times P \times r^3}{Gd^4}$$

where y = total deflection, in.
r = radius as axis to center line of wires, in.
N = number of active coils
G = rigidity modulus of steel in shear

$$N = \frac{y \times G \times d^4}{64 \times P \times r^3} = \frac{4 \times 11 \times 10^6 \times 0.625^4}{64 \times 1{,}600 \times 1.5^3} = 19.4 \text{ active coils}$$

Now let us check for safe limit of torsional stress using

$$S = \frac{16 \times 1{,}600 \times 1.5}{\pi \times 0.625^3} = 50{,}066 \text{ psi}$$

We see that 50,066 is less than the limit of 60,000 set. Therefore, the spring is acceptable. The total number of turns for two inactive coils (one for each end) is $N + 2 = 19.4 + 2 = 21.4$, or **22 coils.**

Q2-9. A coiled spring with $1\frac{3}{4}$-in outside diameter (OD) is required to work under load of 140 lb. Wire diameter used is 0.192 in., spring is to have seven active coils, and the ends are to be closed and ground. Determine unit deflection, total number of coils, and length of spring when under load. Assume G equal to 12 million and mean radius to be 0.779 in.

ANSWER. Safe shearing stress is found from the formula below to be

$$S = \frac{8 \times 140 \times 0.779 \times 2}{\pi \times 0.192^3} = 78{,}475 \text{ psi}$$

Deflection may also be found in another way than previously shown by

$$y = \frac{4\pi N r^2 S}{Gd}$$

$$y = \frac{4\pi 7 \times 0.779^2 \times 78{,}475}{12{,}000{,}000 \times 0.192} = 1.817, \text{ or about } 1^{13}\!/_{16} \textbf{ in.}$$

For coiled springs with circular wire, the formula for the relation between fiber stress and load is

$$S = \frac{8PD}{\pi d^3}$$

where S = fiber stress in shear, psi
P = axial load, lb
D = mean diameter of spring (OD minus wire diameter)
d = diameter of wire, in.

For ends to be closed and ground smooth, 1½ coils should be taken as inactive. For compression springs the number of active coils depends on the style of ends as follows:

Open ends, not ground—all coils active
Open ends, ground—½ coil inactive
Closed ends, not ground—1 coil inactive
Closed ends, ground—1½ coils inactive
Squared ends, ground—2 coils inactive

Total number of coils = 7 + 1½ = 8½, say **9**
The solid height of the spring, when it is entirely compressed, is

$$9 \text{ coils} \times 0.192 = 1.728 \text{ in., or about } 1\tfrac{3}{4} \text{ in.}$$

We determined above that the total deflection under load of 140 lb was $1^{13}\!/_{16}$ in. And assuming that a total free space between coils to be 1 in., the free length of the spring would be

$$1\tfrac{3}{4} + 1\tfrac{13}{16} + 1 = 4\tfrac{9}{16} \text{ in.}$$

Length of spring, when under load of 140 lb, is

$$4\tfrac{9}{16} - 1\tfrac{13}{16} = 2\tfrac{3}{4} \text{ in.}$$

Q2-10. A cylindrical helical spring of circular cross-section wire is to be designed to carry safely an axial compressive load of 1,200 lb at a maximum stress of 110,000 psi. The spring is to have a deflection scale of about 150 lb per in. Proportions are to be as follows:

$$\frac{\text{Mean diameter of coil}}{\text{Diameter of wire}} = 6 \text{ to } 8$$

$$\frac{\text{Length closed}}{\text{Mean diameter of coil}} = 1.7 \text{ to } 2.3$$

Determine (*a*) mean diameter of coil, (*b*) diameter of wire, (*c*) length of coil when closed, and (*d*) length of coil before application of load.

ANSWER. Assume G to be 11,500,000. Now find trial wire size. In this trial design we can assume the ratio of D/d to be 6:8, as given in problem. Then $d = D/7$, and we can substitute the value of d in a rearrangement of the equation on page 22

$$1{,}200 = \frac{\pi \times (D/7)^3 \times 110{,}000}{8 \times D}$$

When we solve for D, it is found to be **3.09 in.** Then wire diameter d is 3.09/7, or 0.441 in. The nearest commercial wire size to this is **0.437 in.** Then $D/d = 3.09/0.437 = 7.07$. This lies within the limits set down in problem. The spring scale is given by

$$\frac{P}{y} = \frac{Gd^4}{N \times 64 \times r^3}$$

In this problem P/y is equal to 150. Now let us substitute values of d, D, G, and P/y in the above equation

$$150 = \frac{11.5 \times 10^6 \times 0.437^4}{N \times 64 \times 1.545^3}$$

Solving for N, the number of active coils, we find it to be equal to 11.9. Total turns, assuming closed ends and ground, are equal to $11.9 + 1.5 = 13.4$. We must now check the second limiting ratio given in the problem: 1.7 to 2.3. Use the following relation:

$$\frac{\text{Total turns} \times d}{D} = \frac{13.4 \times 0.437}{3.09} = 1.9 \text{ (within limits)}$$

Length of the closed coil is $13.4 \times 0.437 =$ **5.86 plus.**
In order to obtain length of coil before application of load, we add closed length to total deflection. Now, total deflection is given by

$$y = \frac{11.9 \times 64 \times 1{,}200 \times 1.545^3}{11.5 \times 10^6 \times 0.437^4} = 8.04 \text{ in.}$$

Then, length of coil before application of load is $5.86 + 8.04 = 13.9$, say **14.**

Q2-11. Determine the width and thickness of the leaves of a six-leaf steel cantilever spring 13 in. long to carry a load of 375 lb with a deflection of 1¼ in. The maximum stress in this spring is limited to 50,000 psi.

ANSWER. We can set up an equation relating deflection to other factors involved.

$$F = \frac{S \times l^2}{Et} = 1.25 = \frac{50{,}000 \times 13^2}{30 \times 10^6 \times t}$$

Rearranging and solving for t, this is found to be equal to **0.225 in.** Now to solve for b, the width of each leaf. The following likewise holds:

$$W = \frac{S \times N \times bt^2}{6l} = 375 = \frac{50{,}000 \times 6 \times b \times 0.225^2}{6 \times 13}$$

Rearranging and solving for b, this is found to be equal to **1.93 in.**

Q2-12. A solid steel machine shaft with a safe shearing stress of 7,000 psi transmits a torque of 10,500 in.-lb. (*a*) Find the shaft diameter. (*b*) A square key is used whose width is equal to one-fourth the shaft diameter and whose length is equal to 1½ times the shaft diameter. Find key dimensions and check the key for its induced shearing and compressive stresses. (*c*) Obtain the factors of safety of the key in shear and in crushing, allowing an ultimate shearing stress of 50,000 psi and a stress for compression of 60,000 psi.

ANSWER.

(*a*)
$$d^3 = \frac{T \times 16}{\pi \times S} = \frac{10{,}500 \times 16}{\pi \times 7{,}000}$$

Solving for d the shaft diameter, this is found to be 1.96, say, **2 in.**

(*b*) Width of key is one-fourth the shaft diameter, or 2/4, i.e., **1/2 in.** Length of key is L equal to 1.5 × shaft diameter, or 1.5 × 2 = **3 in.** Now check key for its induced shearing and compressive stresses. The tangential force set up at the outside of the shaft is P_t lb. Thus

$$P_t = \frac{\text{torque, in.-lb}}{\text{radius of shaft, in.}}$$
$$= \frac{10{,}500}{1} = 10{,}500 \text{ lb}$$

The shearing stress of the key is given by the following relation:

$$S_s = \frac{P_t}{bL} = \frac{10{,}500}{0.5 \times 3} = 7{,}000 \text{ psi}$$

The crushing stress is found by

$$S_c = \frac{2P_t}{bL} = 2 \times \frac{10{,}500}{(0.5 \times 3)} = 14{,}000 \text{ psi}$$

(*c*) Factor of safety F_s = 50,000/7,000 = **7.1 for shear**
(*d*) Factor of safety F_c = 60,000/14,000 = **4.3 for crushing**

Q2-13. What is the minimum length of key ⅝ in. wide you would use with a gear driving shaft 3 7/16 in. in diameter designed to operate at a torsional working stress of 11,350 psi?

ANSWER. The relation between shaft diameter d, torque T, working stress S_s is given by

$$d^3 = 5.1 \frac{T}{S_s}$$

Then after rearrangement and substitution in the above formula

$$T = \frac{d^3 S_s}{5.1} = \frac{3.4375^3 \times 11{,}350}{5.1} = 90{,}397 \text{ in.-lb.}$$

The tangential force on the shaft is $P = T/r$, where r is radius of shaft.

$$P = 90{,}397/1.71875 = 52{,}595$$

Length of key is then determined by use of $L = P/(W \times S_p)$

$$L = 52{,}595/(0.875 \times 11{,}350) = 5.3 \text{ in., say } \mathbf{5\ in.}$$

Check.

$$L = 1.5D = 1.5 \times 3.4375 = 5.15 \text{ in.} \qquad \text{(Checks)}$$

Q2-14. A double-ply leather belt transmits 10 hp from a motor with a pulley 8 in. in diameter, running at 1,700 rpm, to a 24-in.-diameter pulley. The difference in tension may be taken as 20 lb per inch of belt width. Belt thickness is 0.2 in. Calculate width of belt.

Horsepower transmitted by a belt. The power-transmitting capacity of a belt should be less than the maximum possible capacity to ensure a reasonable length of belt life and to avoid unnecessary expenses and repairs. The turning or tangential force on the rim of a pulley driven by a flat belt is equal to $T_1 - T_2$, where T_1 and T_2 are, respectively, the tension in the driving or tight side and the driven or slack side of the belt. Then, the horsepower transmitted is given by

$$\text{hp} = \frac{(T_1 - T_2)v}{33{,}000}$$

where v is belt speed in feet per minute.

ANSWER. Refer to equation above. Then let F equal effective pull in pounds, w equal belt width in inches, and N equal to motor speed in rpm. Then, $F = T_1 - T_2 = 20w$. Substituting in the equation above, we get

$$10 = 20w\pi(8/12)1{,}700/33{,}000$$

Rearranging and solving for w,

$$w = \frac{10 \times 33{,}000}{20\pi(8/12)1{,}700} = 4.12 \text{ in.}$$

Q2-15. An 80-in. flywheel on an air compressor is connected by a flat belt to a 20-in. pulley on the shaft of a 50-hp motor. The distance between centers of pulleys is 8 ft. Motor speed is 800 rpm. If the coefficient of friction between belt and pulleys is 0.30 and the safe tensile stress of the belt is taken as 300 psi, find the width of a 9/32-in. medium double belt to be used. Given belt speed is 4,188.8 fpm; angle of contact between the belt and smaller pulley is 144°.

ANSWER. *Centrifugal tension on a belt.* Since the belt surrounds the pulley, it is subjected to centrifugal force. The effect of this force is to create an additional tensile stress upon the belt and, therefore, increase the cross-sectional requirements.

When belt speed is less than 3,000 fpm, the effect of centrifugal force may be neglected. At velocities greater than 3,000 fpm, it must be considered in the design. Table 2-1* gives loss in horsepower transmitted due to centrifugal tension for belt speeds from 1,000 to 10,000 fpm. Centrifugal tension may be found by use of the following

TABLE 2-1

Belt speed, fpm	*Percentage of rated horsepower lost*
1,000	1
2,000	4
3,000	8
4,000	15
5,000	23
6,000	34
7,000	46
8,000	60
9,000	76
10,000	95

$$C_t = 0.013v^2 \quad \text{psi}$$

where v is expressed in feet per second.

Since belt speed is greater than the 3,000 fpm indicated above, centrifugal belt tension must be taken into consideration.

$$C_t = 0.013 \times (4{,}188.8/60)^2 = 63.3 \text{ psi of belt section}$$

* This table was developed by William Staniar, consulting engineer, and appeared in *Industry and Power* for April, 1947.

In order to determine effective belt pull ($T_1 - T_2 = P_t$), we must first determine torque in the motor shaft, using the basic horsepower equation and rearranging to solve for torque.

$$T = \frac{33{,}000 \times 50 \times 12}{2 \times \pi \times 800} = 3{,}940 \text{ in.-lb}$$

$$\text{Effective pull } (P_t) = \frac{\text{torque}}{\text{radius of driver pulley, in.}}$$

$$P_t = \frac{3{,}940}{10} = 394 \text{ lb}$$

Also, $P_t = T_1 - T_2$. Therefore find belt width w. To do this, use the equation $T_1 = wT(S_t - C_t)$ lb, where w is in inches, t is belt thickness in inches and S_t is safe tensile stress in psi. In order to find T_1 we must first find T_2.

There is a relationship

$$\frac{T_1}{T_2} = e^{\mu\alpha}$$

where μ is coefficient of friction for circular surfaces, α is angle of contact between belt and pulley in radians, and e is the natural log base, 2.718. Thus, as we apply it here

$$\frac{T_1}{T_2} = 2.718^{0.3 \times 2.5} = 2.13$$

From this relation we can find T_1 to be 743 lb. Then using the above equation involving S_t we determine w:

$$w = \frac{743}{\frac{9}{32} \times (300 - 63.3)} = 11.2 \text{ in., say } \mathbf{12\ in.}$$

Q2-16. A double leather belt 8 in. wide and $\frac{5}{16}$ in. thick runs crossed on cast-iron pulleys 15 and 30 in. in diameter. Centers are 8 ft apart. The belt speed is 3,500 fpm and the allowable unit tensile pressure is 425 psi. Determine arc of contact, the centrifugal tension, and the horsepower which may be transmitted.

ANSWER. Draw diagram showing crossed-belt arrangement. Obtain arc of contact directly from diagram. Arc of contact is found to be equal to **207°**. Centrifugal tension is next determined using the equation previously shown.

$$C_t = 0.013 \times (3{,}500/60)^2 = \mathbf{44.3\ psi}$$

Assume, or find from tables, that μ (coefficient of kinetic friction for circular surfaces) is 0.505 for leather on cast iron. Now determine the horsepower transmitted, using the following equation:

$$\text{hp} = \frac{(S_t - C_t)CvA}{550}$$

The constant C is determined by the following equation:

$$C = \frac{10^{0.0076\mu\alpha} - 1}{10^{0.0076\mu\alpha}}$$

Note here α is in degrees, v is in feet per second, A is cross-sectional area of belt in square inches. Now, when we solve for C, we find it to be 0.84.

Then

$$\text{hp} = \frac{(425 - 44.3) \times 0.84 \times (3{,}500/60) \times (8 \times 5/16)}{550} = \mathbf{84.7\ hp}$$

Q2-17. (*a*) What horsepower is to be taken off the driven shaft of an open-belt drive operated under the following conditions: driving pulley 84 in. in diameter, driven pulley 48 in. in diameter, speed of driving pulley 150 fpm, total tension on tight side of belt 300 lb, total tension on slack side of belt 50 lb? (*b*) What belt material and what size of belt would you use for this drive?

ANSWER.

(*a*) $$\text{hp} = \frac{(300 - 50)(\pi \times 84/12 \times 150)}{33{,}000} = \mathbf{25\ hp}$$

(*b*) The belt should be made of **leather**, and allowing 50 lb per in. of belt width of double-thickness belts for $T_1 - T_2$, the belt should be **5 in.** wide.

Q2-18. Two shafts 20 ft apart are to be connected by belting. The driving shaft runs at 150 rpm, the driven shaft is to run in the opposite direction at 450 rpm. Twenty horsepower is to go through the drive. Sketch the drive. Assume a proper belt speed. Compute the belt tensions, the pulley diameters, and the belt size and thickness.

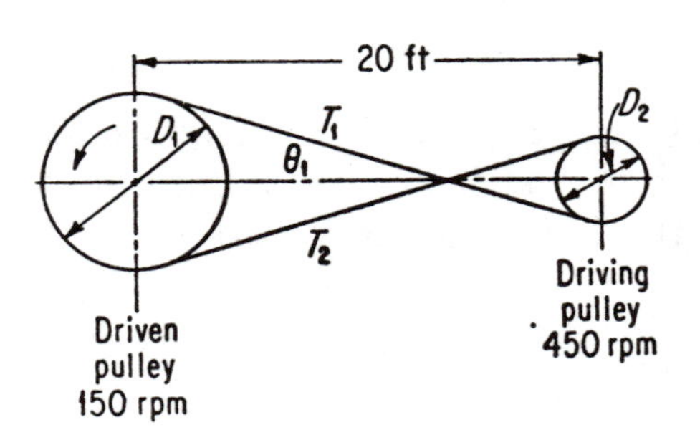

Fig. 2-3

ANSWER. See Fig. 2-3 for drive sketch. Assume a belt speed of 4,500 fpm.

$$T_1 - T_2 = \frac{20 \times 33{,}000}{4{,}500} = 147 \text{ lb}$$

The following relations hold: $D_1\pi 150 = 4{,}500$ fpm; D_1 is found from this to be 9 ft, 6 in. Likewise, $D_2\pi 450 = 4{,}500$, and D_2 is found to be equal to 3 ft, 2 in. Now, allowing $T_1 - T_2$ at 50 lb per in. of belt width for double-thickness belts, the belt for this system should be 3 in. wide. Next find T_1/T_2. The relation given in question Q2-15 has another form and may be used also.

$$\log \frac{T_1}{T_2} = 0.0076\mu(1 - x)\alpha$$

Also, $\sin \theta = 4.75/15.00 = 0.316$. From which we obtain $\theta = 18°$. Angle of contact $\alpha = 180 + (2 \times 18) = 216°$. For a belt velocity of 4,500 fpm or 75 fps, $x = 0.246$, and for leather belts $\mu = 0.3$.

$$\log \frac{T_1}{T_2} = 0.0076 \times 0.30(1 - 0.246)216 = 0.372$$

From the above we obtain the ratio of the tensions

$$\frac{T_1}{T_2} = 2.35$$

Since $T_1 - T_2 = 147$ lb, we can substitute this in the above equation and calculate that $T_1 =$ **256 lb** and $T_2 =$ **109 lb.**

Q2-19. A condensing steam engine with a bore and a stroke of 24 in. cuts off at one-third stroke and has a mean effective pressure of 50 psi. The flywheel is to be 18 ft in mean diameter and it makes 75 rpm with a variation of 1 per cent. Determine the weight of the rim.

ANSWER. The work done on the piston is equal to the mean effective pressure times the distance traveled (twice the stroke in feet) in one revolution. Piston area may be found to be 452.4 sq in., and the distance of travel is 4 ft. Thus, work done is found to be

$$452.4 \times 50 \times 4 = 90{,}480 \text{ ft-lb}$$

From handbooks the factor of energy excess for steam engines at one-third steam cutoff is 0.163. Then the average work done by the flywheel is

$$E = 90{,}480 \times 0.163 = 14{,}748 \text{ ft-lb}$$

The weight of the flywheel is then found to be

$$W = \frac{11{,}745nE}{D^2N^2} = \frac{11{,}745 \times 100 \times 14{,}748}{18^2 \times 75^2} = \mathbf{9{,}504\ lb}$$

Q2-20. Neglecting spoke effect, calculate the energy stored in the rim of a flywheel made of cast iron 24 in. in diameter, having a rim 5 in. wide by 4 in. deep when running at 1,000 rpm.

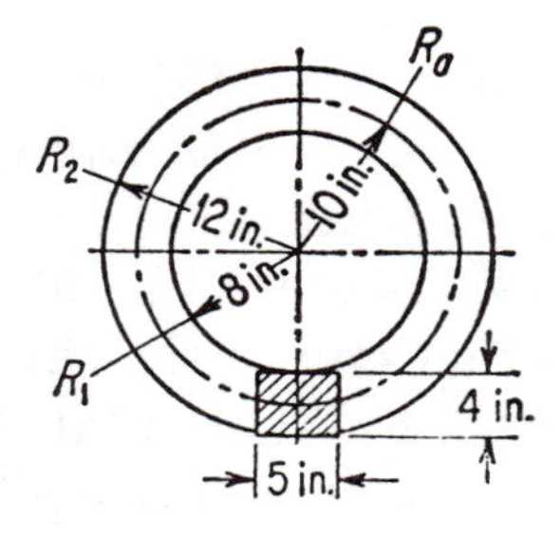

Fig. 2-4

ANSWER. Note this is running above the safe speed listed in design books. Refer to Fig. 2-4. Let W be weight of flywheel in pounds; ρ is radius of gyration in feet; R_1 is internal radius of rim in feet; R_2 is external radius of rim in feet; ω is angular velocity in radians per second; R_a is average radius in feet; density of cast iron is 450 lb per cu ft.

Weight of flywheel is

$$W = 5 \times 4 \times 2 \times \pi \times 10 \times 450/1{,}728 = 327 \text{ lb}$$
$$\rho = \sqrt{\tfrac{1}{2}(R_1^2 + R_2^2)} = \sqrt{\tfrac{1}{2}[(8/12)^2 + (12/12)^2]} = 0.85 \text{ ft}$$

Angular velocity ω = rpm $\times\ 2\pi/60$ = 104.8 radians per second as determined.

$$\omega = 1{,}000 \times 6.28/60 = 104.8 \text{ radians per sec}$$

Energy stored in the rim is next found in accordance with

$$\frac{W\rho^2\omega^2}{2g} = 327 \times 0.85^2 \times 104.8^2/64.4 = \mathbf{40{,}200\ ft\text{-}lb}$$

NEW YORK INSTITUTE
OF TECHNOLOGY LIBRARY

Q2-21. A 100-lb wheel 18 in. in diameter which is turning at 150 rpm in stationary bearings is brought to rest by pressing a brake shoe radially against a rim with a force of 20 lb. If the radius of gyration of the wheel is 7 in. and if the coefficient of friction between shoe and rim has the steady value of 0.25, how many revolutions will the wheel make in coming to rest?

ANSWER. Energy stored in the wheel must first be determined in the usual manner by use of equation from previous problem. The angular velocity may be found to be 15.7 radians per second. And the energy stored in wheel is

$$E = \frac{100 \times (7/12)^2}{64.4} \times (5\pi)^2 = 130 \text{ ft lb}$$

Frictional force to stop car wheel is $20 \times 0.25 = 5$ lb on perimeter. Wheel perimeter is $\pi D = \pi 18/12 = 4.7$ ft. Distance force must travel is $130/5 = 26$ linear feet. And the number of turns is $26/4.7 =$ **5.53 turns.**

Q2-22. A flywheel whose weight is 200 lb and whose radius of gyration is 15 in. is secured to one end of a 6-in. diameter shaft; the other end of the shaft is connected through a chain and sprocket to a motor that rotates at 1,800 rpm. The motor sprocket is 6 in. in diameter and the shaft sprocket is 36 in. in diameter. The total shaft length between flywheel and sprocket is 72 in. Determine the maximum stress in the shaft resulting from instantaneous stopping of the motor drive, assuming that sprocket and chain have no ability to absorb impact loading. Assume shear modulus equal to 12,000,000 psi. Neglect effect of shaft kinetic energy.

ANSWER. When the flywheel is stopped short, the kinetic energy stored is converted to torsional impact. The magnitude of this energy is found by

$$E = \frac{W\rho^2\omega^2}{2g} = \frac{200 \times 1.25^2 \times (2\pi 300/60)^2}{64.4 \times 1/12} = 57{,}600 \text{ in.-lb}$$

The shaft offers resilience to torsional twist. Resilience U (in.-lb) is the potential energy stored up in the deformed body. The amount of resilience is equal to the work required to deform the body (here the shaft) from zero stress to stress S. The modulus of resilience U_p (in.-lb per cu in.), or unit resilience, is the elastic energy stored up in a cubic inch of material at the elastic limit. The unit resilience used in the solution of this problem (for a solid shaft) is

$$U_p = \frac{1}{4}\frac{S_t^2}{G}$$

For the full volume of the shaft, where V is $0.785 \times 6^2 \times 72 = 2{,}035$ cu in.

$$U_p \text{ total} = \tfrac{1}{4} \times S_t^2 \times 2{,}035 \times 1/12{,}000{,}000 = 57{,}600 \text{ in.-lb}$$

Solving for $S_t = \sqrt{4 \times 12{,}000{,}000 \times 57{,}600/2{,}035} =$ **37,000 psi.**

Q2-23. A journal bearing with a diameter of 2.25 in. is subjected to a load of 1,000 lb while rotating at 200 rpm. If the coefficient of friction is taken as 0.02 and L/D is 3.0, find (*a*) projected area, (*b*) pressure on bearing, (*c*) total work of friction, (*d*) work of friction, (*e*) total heat generated, and (*f*) heat generated per minute.

ANSWER. Since L/D is 3, then $L = 3D = 3 \times 2.25 = 6.75$ in.

(*a*) Projected area $= L \times D = 6.75 \times 2.25 =$ **15.19 sq in.**

(*b*) Pressure on bearing $= P/(L \times D) = 1{,}000/(15.19) =$ **65.8 psi**

(*c*) Total work $= W = 0.02 \times 1000 \times (\pi)(2.25/12)(200) =$ **2,356 ft-lb/min**

(*d*) Work $= w = W/LD = 2{,}356/15.19 =$ **155.1 ft-lb per min per sq in.**

(*e*) Total heat $= Q = W/778 = 2{,}356/778 =$ **3.03 Btu per min**

(*f*) $q = w/778 = 155.1/778 =$ **0.2 Btu/(min)(sq in. projected area)**

P = total load on bearing, lb
p = pressure or load per sq in. of the projected area, psi
L = length of bearing, in.
D = diameter of bearing, in.
N = rpm of journal
μ = coefficient of friction
W = total work of friction, ft-lb per min
w = work per sq in. of projected area, ft-lb per sq in. per min
Q = total heat generated, Btu per min
q = heat generated per sq in. of projected area, Btu per min
V = rubbing velocity, fpm

Q2-24. An 8-in. nominal diameter journal is designed for 140° optimum bearing when bearing length is 9 in., speed is 1,800 rpm, and total load is 20,000 lb. Calculate the frictional horsepower loss when this journal operates under stated conditions with oil of optimum viscosity.

ANSWER. The frictional loss in horsepower is given by the equation below, where V_r is rubbing speed and other factors are known.

$$\frac{\mu P V_r}{33{,}000}$$

The rubbing speed is found to be

$$\pi D \text{ rpm} = 3.1416 \times 8/12 \times 1{,}800 = 3{,}770 \text{ fpm}$$

Projected surface of rubbing area is

$$\frac{D}{2} \times \alpha \times L = 8/2 \times \frac{140°}{180°} \times \pi \times 9$$

or 87.5 sq in. Bearing pressure is 20,000/87.5 = 228 psi. The coefficient of friction,* using pressure and rubbing speed as criteria, may be taken as 0.002.

$$\frac{0.002 \times 20{,}000 \times 3{,}770}{33{,}000} = \mathbf{4.56\ hp}$$

Q2-25. Calculate the torque in pound feet produced by a pressure of 200 psi acting on the piston of an automobile engine when the crank is 30° past top center. Cylinder bore is 3.25 in., stroke is 4 in., and connecting rod is 10 in. long.

ANSWER. Refer to Fig. 2-5, and let d be cylinder bore in inches.

$$P_p = 0.785 \times d^2 \times 200 = 1{,}660 \text{ lb}$$

* "Machinery's Handbook," 14th ed., pp. 518–520, The Industrial Press, New York, 1960.

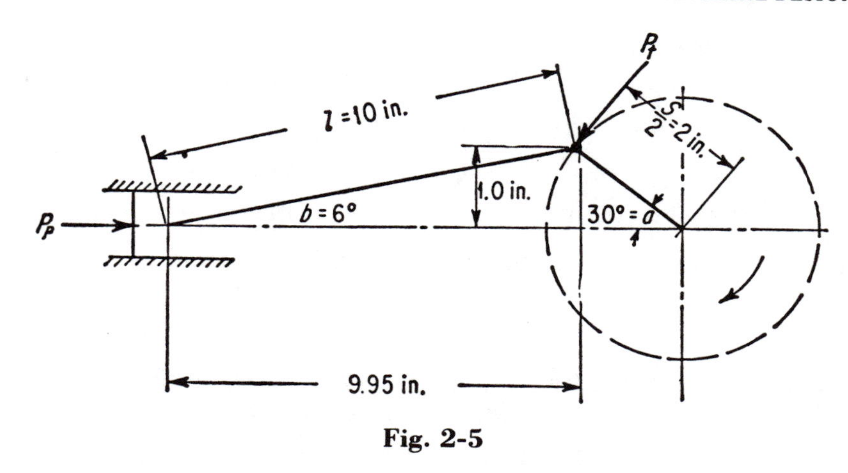

Fig. 2-5

The above holds since we inserted 3.25 for d. Now determine connecting-rod pressure P_c by drawing free-body diagram showing forces on piston pin (see Fig. 2-6). From this we observe that

$$\cos b = \frac{P_p}{P_c}$$

$$P_c = 1{,}660/9.95 \times 10 = 1{,}670 \text{ lb}$$

Having determined P_c, now draw free-body diagram of forces on

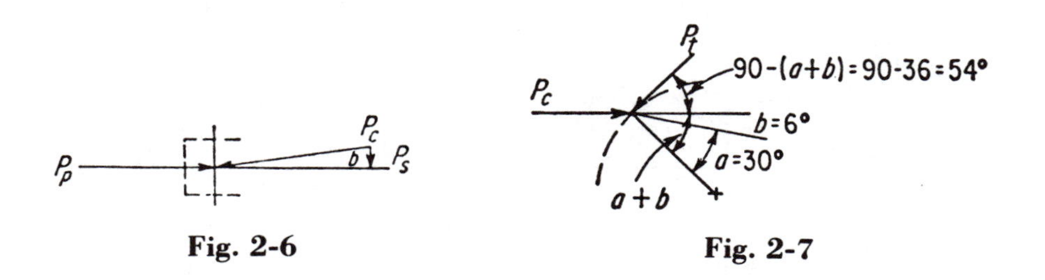

Fig. 2-6 Fig. 2-7

crankpin (Fig. 2-7), and we see that

$$P_t = P_c \times \cos 54° = 1{,}670 \times 0.588 = 982 \text{ lb}$$

$$\text{Torque} = P_t \frac{s}{2} = 982 \times \frac{2}{12} = \mathbf{163.7 \text{ lb-ft}}$$

ANSWER. Refer to Fig. 2-8. At the beginning of the stroke length A is equal to $60 + 12 = 72$ in. At 60° past turning

$$x = 12 \sin 60° = 12 \times \sqrt{3/2} = 10.4 \text{ in.}$$
$$Z = \sqrt{60^2 - x^2} = \sqrt{60^2 - 10.4^2} = 59.1 \text{ in.}$$
$$y = 12 \cos 60° = 12 \times 1/2 = 6 \text{ in.}$$

Therefore, the distance through which piston moved is determined by

$$A - (Z + y) = 72 - (59.1 + 6) = \mathbf{6.9\ in.}$$

Q2-26. Calculate the side thrust against the cylinder walls of an engine of an automobile when the connecting rod is at an angle of 90° to the crank, the connecting rod being 15 in. long and the

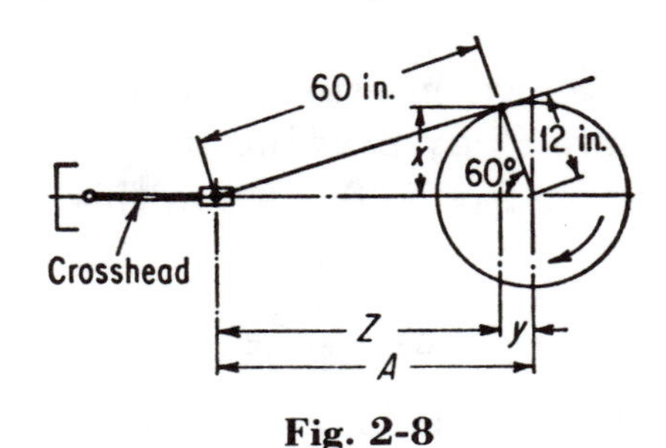

Fig. 2-8

crank 3 in. The piston is 4 in. in diameter and the pressure on it is 200 psi.

ANSWER. Refer to Fig. 2-9, and let the following considerations

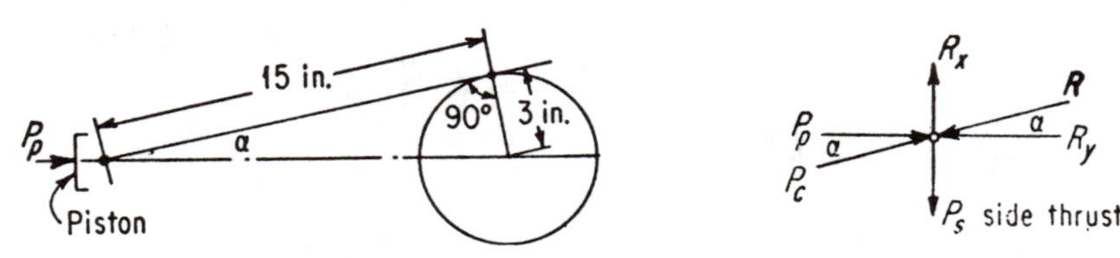

Fig. 2-9 **Fig. 2-10**

hold: P_p is total piston force in pounds; R is rod reaction on piston point in pounds; P_s is side thrust, in pounds.

$$P_p = 0.785 \times 4^2 \times 200 = 2{,}510 \text{ lb}$$

From force diagram (Fig. 2-10): $P_s = R_x$, $R_y = P_p$, $R_x = R_y \tan \alpha$. Therefore,

$$P_s = P_p \tan \alpha = 2{,}510 \times 3/15 = \mathbf{502\ lb}$$

Q2-27. (*a*) **A navy ship has a De Laval steam turbine whose** rotating parts weigh 10 tons, with a radius of gyration of 1 ft about its axis of rotation, and which rotates at 3,000 rpm. Under the worst sea conditions the ship will pitch so that its "fore and aft" angular velocity has a maximum value of 2° per sec. Find the maximum gyroscopic couple to which the ship is subjected.

(*b*) In order to maintain its time schedule, an express train must run at a steady speed of 55 mph over a certain section of its journey. However, a mile of track on this section is under repair, and a distant signal will limit the train to 15 mph over it. Tests have shown that the train takes 2 m from rest to attain a speed of 55 mph and ¼ mi to come to a stop. Determine how late the train will be as a result of this track condition, assuming uniform acceleration and deceleration.

ANSWER.

(*a*) **Refer to Fig. 2-11 and let** n **= 3,000 rpm,** W **= 20,000 lb,** and $K = 12$ in. radius of gyration. Then the vector equation is

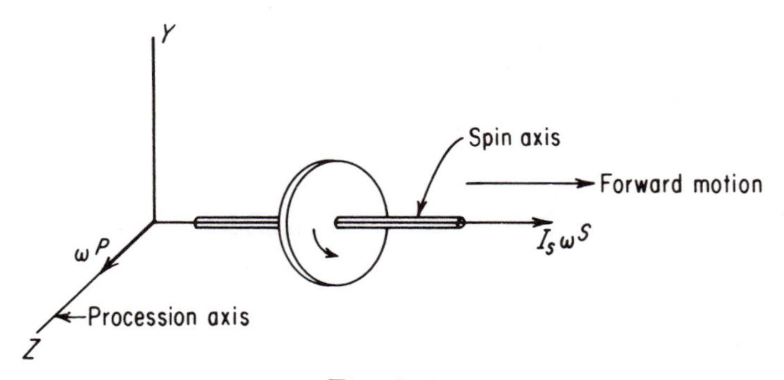

Fig. 2-11

$T = \alpha^p \times I_s\omega^s = I_s\omega^p\omega^s$. The scalar equation is $I_s = W/gK^2 = (20{,}000 \times 12^2)/(32.2 \times 12 \text{ in. per ft}) = 7{,}460$ lb-in.-sec.

$$\omega^p = 2°/\text{sec}(2 \text{ radians}/360°) = 0.0349 \text{ radians per sec}$$
$$\omega_s = 3{,}000 \text{ rpm } (2\pi/60) = 314 \text{ radians per sec}$$
$$T = I_s\omega^p\omega^s = 7{,}460 \times 0.0340 \times 314 = \mathbf{81{,}800 \text{ lb-in.}}$$

(*b*) Refer to Fig. 2-12.

$$\text{Acceleration characteristic} = 55 \text{ mph}/2 \text{ mi} = 27.5 \text{ mph per mi}$$

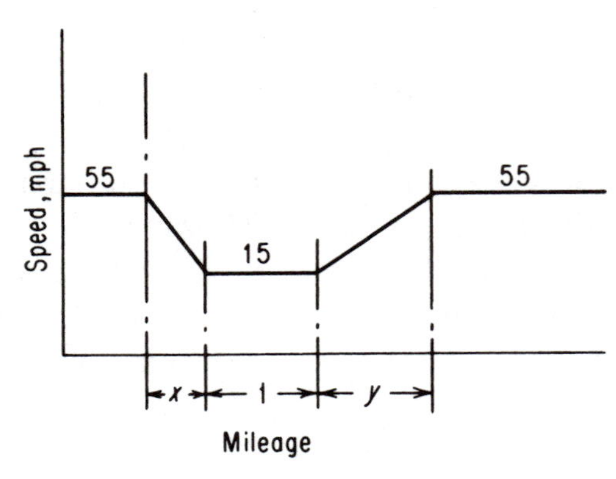

Fig. 2-12

$$\text{Deceleration characteristic} = 55 \text{ mph}/\tfrac{1}{4} \text{ mi} = 220 \text{ mph per mi}$$

Therefore,

$$x = (55 - 15)/220 = 0.1818 \text{ mi}$$
$$y = (55 - 15)/27.5 = 1.454 \text{ mi}$$

Therefore, time for train to travel $(1 + x + y)$ mi at reduced speed is

$$t_1 = \frac{0.1818}{(55 + 15)/2} + \frac{1}{15} + \frac{1.454}{(55 + 15)/2} = 0.113 \text{ hr}$$

Time for train to travel $(1 + x + y)$ mi at regular speed is t_0.

$$t_0 = \frac{1 + x + y}{55} = \frac{2.6358}{55} = 0.048 \text{ hr}$$

Therefore, lost time $\Delta t = t_1 - t_0 = 0.113 - 0.048 =$ **0.065 hr.** Say 4 min.

Q2-28. A pivoted motor drive has forces of 300 lb when running under no-load conditions. The small pulley is of wood, 9 in. in diameter, at the same elevation as the large pulley of 34 in. in diameter. The center distance is 32½ in. The belt is oak-tanned leather; $m = 12$ in., $f = 14$ in. Find the horsepower capacity for **dry conditions. Refer to Fig. 2-13. The motor turns at 1,150 rpm**

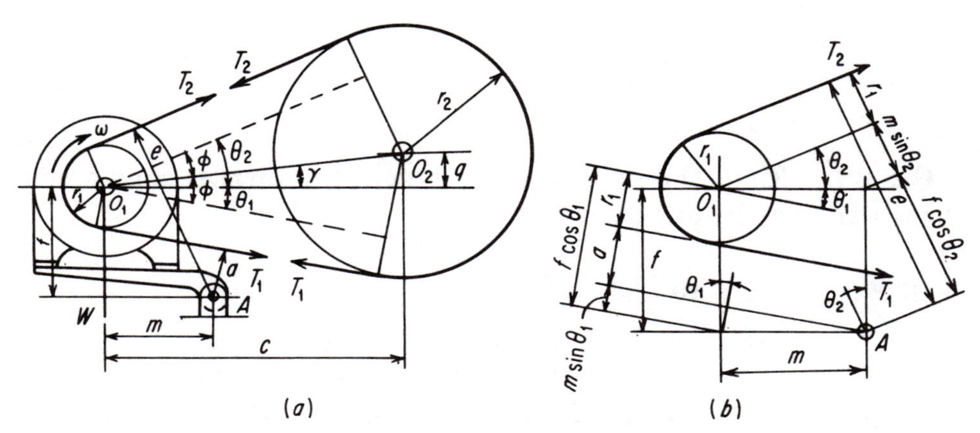

Fig. 2-13

ANSWER.

$$\theta_1 = \theta_2 = \phi \qquad \cot\theta_1 = {}^{12}\!/_5 = 2.4 \qquad \theta_1 = 22°37'$$

$$a = f\cos\theta_1 - m\sin\theta_1 - r_1 = 14 \times 0.92310 - 12 \times 0.38456 - 4.5$$
$$= 12.9234 - 4.6147 - 4.5 = 3.809 \text{ in.}$$

$$e = f\cos\theta_2 + m\sin\theta_2 + r_1 = 12.9234 + 4.6147 + 4.5 = 22.038 \text{ in.}$$
$$(22.038 + 3.808)300 = 12W$$

From which

$$W = 646.2 \text{ lb}$$
$$\beta = 180° - 2(22°37') = 134°46' = 2.35212 \text{ radians}$$
$$e^{\mu\beta} = e^{0.3\times2.35212} = e^{0.7056} = 2.025 = \frac{T_1}{T_2} \qquad \text{and} \qquad T_1 = 2.025T_2$$
$$12W = 3.808T_1 + 22.038T_2$$
$$12 \times 646.2 = 3.808 \times 2.025T_2 + 22.038T_2$$

From which

$$T_2 = 260.7 \text{ lb}$$
$$T_1 = 627.8 \text{ lb}$$
$$V = \frac{\pi dn}{12} = \frac{\pi \times 9 \times 1{,}150}{12} = 2{,}709.6 \text{ fpm}$$
$$\text{Hp} = \frac{(T_1 - T_2)V}{33{,}000} = \frac{267.5 \times 2{,}709.6}{33{,}000} = \mathbf{22.0\ hp}$$

where n = revolutions per-minute
r = radius
T = torque, in.-lb
T_1 = force in tight side
T_2 = force in slack side
V = belt velocity, fpm .

Q2-29. Find the horsepower which the brake shown can absorb and the length of arm a.

ANSWER. Refer to Fig. 2-15.

$$\alpha = 5/4 \times \pi \qquad \mu\alpha = 0.2 \times 1.25\pi = 0.7854$$

$$e^{\mu\alpha} = e^{0.7854} = 2.193 \qquad F_1 = e^{\mu\alpha}F_2 = 2.193F_2$$

$$300 \times 12 = 3F_1 - \mathbf{1.464}F_2 = 3 \times 2.193F_2 - \mathbf{1.464}F_2 = 5.115F_2$$

$$F_2 = 703.8 \text{ lb} \qquad F_1 = 1{,}543.4 \text{ lb}$$

$$T = (1{,}543.4 - 703.8)(5) = 4{,}198 \text{ in.-lb}$$

$$\text{Hp} = \frac{Tn}{63{,}000} = \frac{4{,}198 \times 200}{63{,}000} = \mathbf{13.33\ hp}$$

The problem is shown in Fig. 2-14.

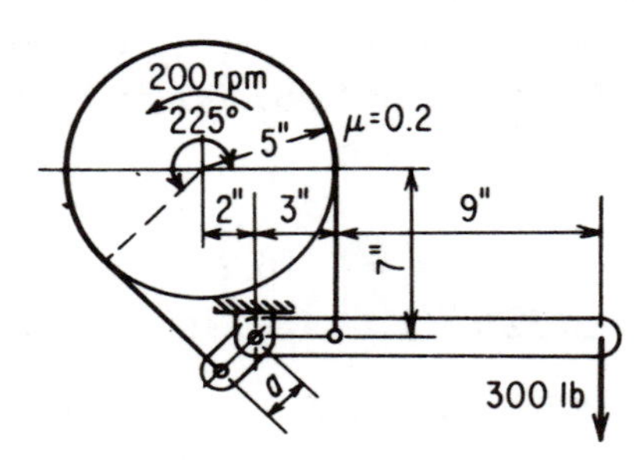

Fig. 2-14

Q2-30. Find the lowest critical speed for the steel shaft shown in Fig. 2-16.

ANSWER.

$$y = \frac{Pbx}{6lEI} \times (l^2 - b^2 - x^2) \qquad I = \frac{\pi d^4}{64} = \frac{\pi}{64}$$

$$EI = \frac{30 \times 10^6\pi}{64} = 1{,}472{,}600$$

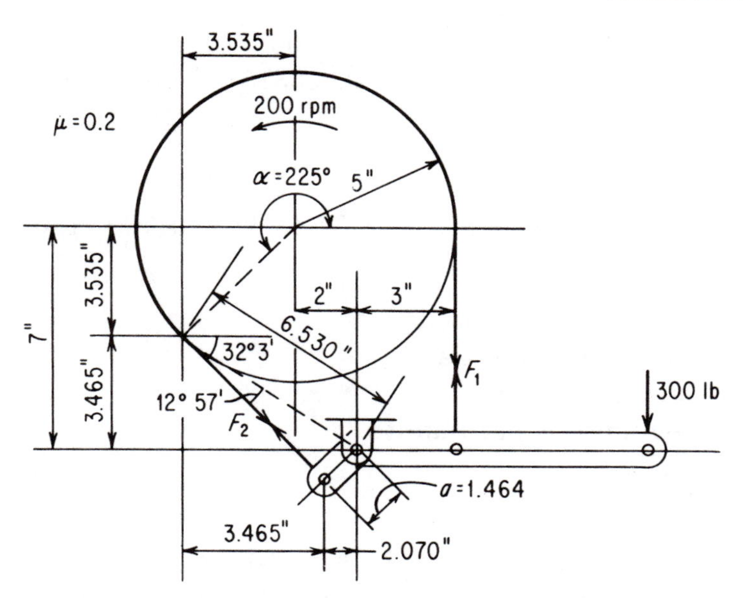

Fig. 2-15

At W_1: Due to 30 lb; $b = 20$ in.; $x = 5$ in.

$$y_1' = \frac{30 \times 20 \times 5}{6 \times 25 \times 1{,}472{,}600} \times (25^2 - 20^2 - 5^2) = 0.002716$$

Due to 50 lb; $b = 10$ in.; $x = 5$ in.

$$y_1'' = \frac{50 \times 10 \times 5}{6 \times 25 \times 1{,}472{,}600} \times (25^2 - 10^2 - 5^2) = 0.005659$$

$$\text{Total } y_1 = 0.008375$$

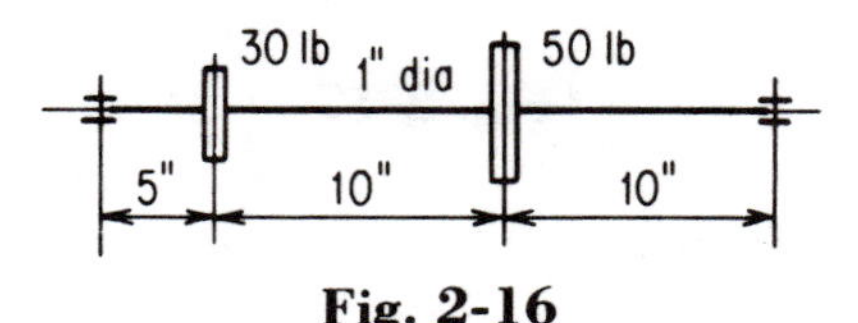

Fig. 2-16

At W_2: Due to 30 lb; $b = 5$ in.; $x = 10$ in.

$$y_2' = \frac{30 \times 5 \times 10}{6 \times 25 \times 1{,}472{,}600} \times (25^2 - 5^2 - 10^2) = 0.003395$$

Due to 50 lb.; $b = 10$ in.; $x = 15$ in.

$$y_2'' = \frac{50 \times 10 \times 15}{6 \times 25 \times 1{,}472{,}600} \times (25^2 - 10^2 - 15^2) = 0.010186$$

$$\text{Total } y_2 = 0.013582$$

$$f = \frac{1}{2\pi}\sqrt{\frac{g\Sigma Wy}{\Sigma Wy^2}} = \frac{1}{2\pi}\sqrt{\frac{386(30 \times 0.008375 + 50 \times 0.013582)}{30 \times 0.008375^2 + 50 \times 0.013582^2}}$$
$$= 28.34 \text{ cycles per sec}$$

Or

$$\text{Critical speed} = n_{cr} = 28.34 \times 60 = \mathbf{1{,}700\ rpm}$$

where d = diameter
E = modulus of elasticity
I = moment of inertia
l = length
n = revolutions per minute
r = radius
y = deflection

Q2-31. The weight shown in Fig. 2-17 must be raised at the rate of 10 fps. The motor pulley is 14 in. in diameter and turns

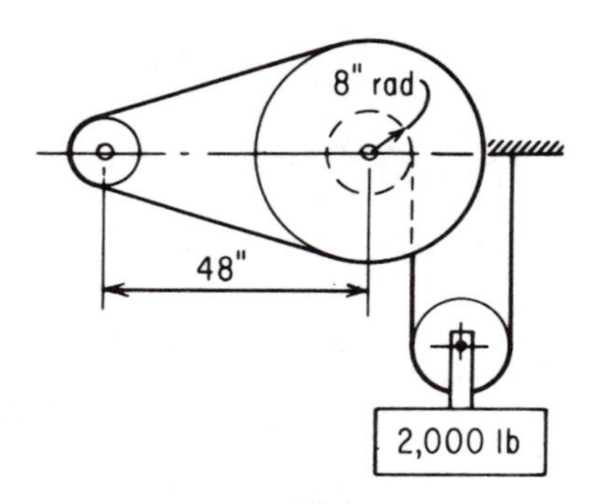

Fig. 2-17

at 850 rpm. Ignore friction, but use a service factor of 1.4. Find the diameter of the large pulley and the number of standard D-section V belts required for an expected life of 10,000 hr.

ANSWER.

$$\text{Cable drum } n = \frac{12V}{\pi d} = \frac{12 \times 20 \times 60}{\pi \times 16} = 286.5 \text{ rpm}$$

$$r_2 = \frac{n_1}{n_2} r_1 = \frac{7 \times 850}{286.5} = 20.769 \text{ in.} \qquad \text{or} \qquad d_2 = \mathbf{41.538\ in.}$$

$$\cos\psi = \frac{13.769}{48} = 0.28685 \qquad \text{and} \qquad \psi = 73.331°$$

$$l = \pi d_2 + (d_2 - d_1) \text{ inv } \psi$$
$$= \pi \times 41.538 + 27.538 \times 2.06035 = 187.2 \text{ in.}$$
$$\text{Belt speed } V = \frac{\pi \times 14 \times 850}{12} = 3{,}115 \text{ fpm}$$
$$\text{Belt passes per min} = \frac{3{,}115 \times 12}{187.2} = 199.7$$
$$N \text{ (peaks)} = 199.7 \times 60 \times 10{,}000 = 119{,}800{,}000$$

From the curves in Fig. 2-18, peak force = 570 lb. Angle of contact = 2 × 73.331° = 146.662°.

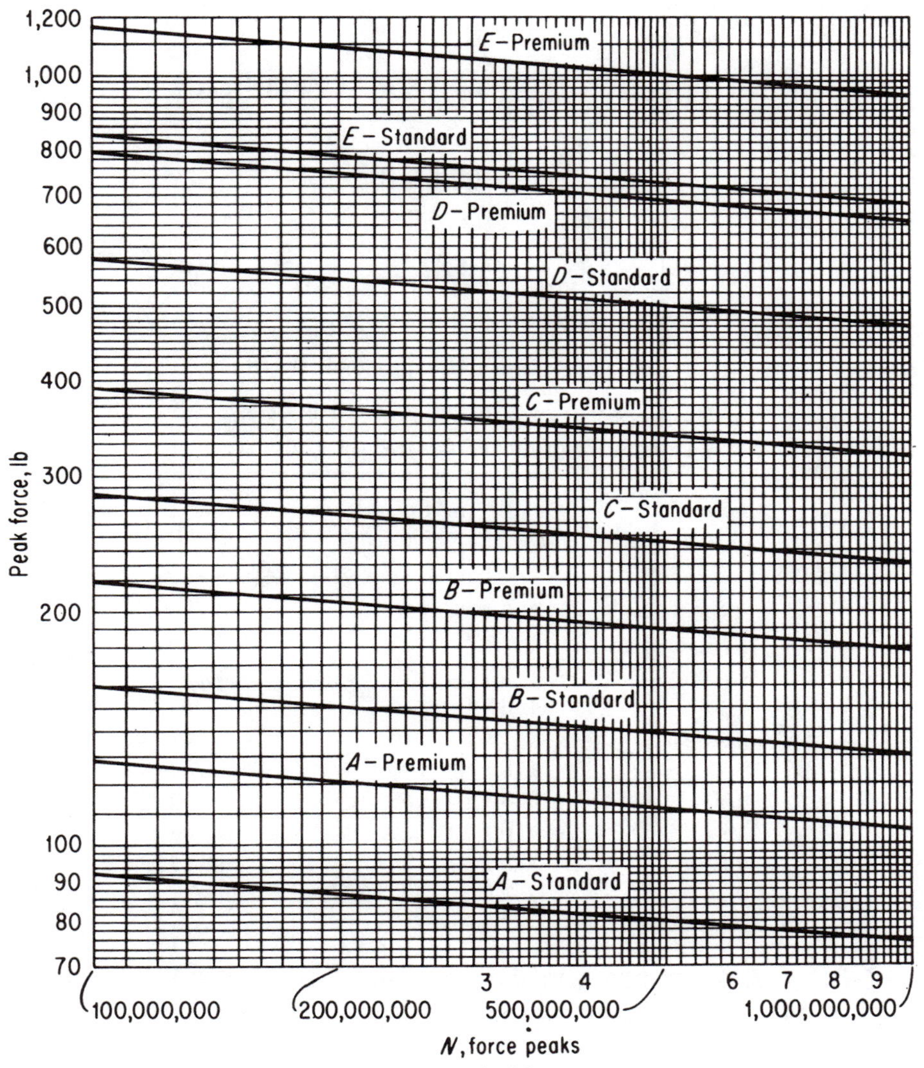

Fig. 2-18

From Table 2-2, $T_1/T_2 = 3.71$ and $T_2 = 0.2695T_1$.

Small Pulley.

$$T_b = \frac{K_b}{d} = \frac{3{,}873}{14} = 276.6 \text{ lb}$$

$$T_c = K_c\left(\frac{V}{1{,}000}\right)^2 = 3.498 \times 3.115^2 = 33.9 \text{ lb}$$

$$T_1 = 570 - 276.6 - 33.9 = 259.5 \text{ lb}$$

$$\text{hp} = \frac{T_1(1 - 0.269)V}{33{,}000} = \frac{259.5 \times 0.731 \times 3{,}115}{33{,}000} = 17.9 \text{ for one belt}$$

$$\text{Required input hp} = \frac{2{,}000 \times 10 \times 60}{33{,}000} \times 1.4 = 50.9$$

$$\text{No. of belts} = \frac{50.9}{17.9} = 2.85.$$ **Use 3 belts.**

Large Pulley.

$$T_1 = 259.5 \text{ lb}$$

$$T_b = \frac{K_b}{d_2} = \frac{3{,}873}{41.538} = 93.2$$

$T_c = 33.9$		33.9
$T_1 = 259.5$		259.5
	Total	386.6 lb

TABLE 2-2 RATIO T_1/T_2 FOR V BELTS FOR VARIOUS VALUES OF ANGLE OF CONTACTS

Angle of contact	T_1/T_2	*Angle of contact*	T_1/T_2
180°	5.00	130°	3.20
175	4.78	125	3.06
170	4.57	120	2.92
165	4.37	115	2.80
160	4.18	110	2.67
155	4.00	105	2.56
150	3.82	100	2.44
145	3.66	95	2.34
140	3.50	90	2.24
135	3.34		

Q2-32. A premium-quality C-section V belt carries a net horsepower of 11, but a service factor of 1.4 must be used. Both pulleys are 12-in. pitch diameter and turn at 1,160 rpm. The belt length is 197.9 in. Find the expected life in hours for this belt.

ANSWER. Lineal velocity of the belt is $V = \pi dn/12 = \pi \times 12 \times 1{,}160/12 = 3{,}644$ fpm. $T_1 = 33{,}000 \times \text{hp}/(0.8 \times V) = 33{,}000 \times 15.4/(0.8 \times 3{,}644) = 174.3$ lb.

$$T_b = \frac{K_b}{d} = \frac{1{,}600}{12} = 133.4$$

$$T_c = K_c\left(\frac{V}{1{,}000}\right)^2 = 1.716 \times (3.644)^2 = 22.8 \text{ lb}$$

Peak force $= T_1 + T_b + T_c = 174.3 + 133.4 + 22.8 = 330.5$ lb From the curve in Fig. 2-19, $N = 69{,}000$ cycles. Cycles per min $= (3{,}644 \times 12/197.9) \times 2 = 442$, where $197.9 = T_1 + T_c = 174.3 + 22.8$.

$$\text{Life of belt} = (690{,}000{,}000)/(442 \times 60) = \mathbf{26{,}000\ hr}$$

Q2-33. **A solid steel machine shaft with a safe shearing stress of 7000 lb/in² transmits a torque of 10,500 in-lb.**

(*a*) Determine the diameter of the shaft.

(*b*) A square key is used whose width is equal to ¼ the shaft diameter and whose length is equal to 1½ times the shaft diameter. Determine the dimensions of the key and check the key for its induced shearing and compressive stresses.

(*c*) Obtain the factors of safety of the key in shear and crushing, allowing an ultimate stress in shear of 50,000 lb/in² and an ultimate stress in crushing of 60,000 lb/in².

ANSWER.

(*a*) Here $S_s = 7000$ lb/in², $T = 10{,}500$ in-lb, and $Z_p = \pi d^3/16$. Apply following formula: $T = S_s Z_p = 10{,}500 = 7000 \times (\pi d^3/16)$, from which

$$d^3 = \frac{10{,}500 \times 16}{7000\pi} = 7.63 \text{ and } d = 1.96, \text{ say } \mathbf{2\ in}$$

(*b*) $b = t = d/4 = 2/4 =$ **½ in**

$L = 1½ \times 2 =$ **3 in**

Key check: $P_t = T/R = 10{,}500/1 = 10{,}500$ lb

The shearing stress of the key is obtained from $S_s = P_t/bL$.

$$S_s = \frac{10{,}500}{\frac{1}{2} \times 3} = \mathbf{7000\ lb/in^2}$$

The crushing stress of the key is obtained from $S_c = 2P_t/tL$.

$$S_c = \frac{2 \times 10{,}500}{\frac{1}{2} \times 3} = \mathbf{14{,}000\ lb/in^2}$$

(*c*) Here U_s = 50,000 lb/in². Then $F = U_s/S_s$ = 50,000/7000 = **7.2**

With U_c = 60,000 lb/in². Then $F = U_c/S_c$ = 60,000/14,000 = **4.3**

where S_s = shearing stress, lb/in²
T = torque, in-lb
Z_p = polar moment of inertia, in³
d = shaft diameter, in
t = key width, in
S_s = shearing stress, lb/in²
S_c = crushing stress, lb/in²
L = key length, in
U_s = ultimate shearing stress, lb/in²
U_c = ultimate crushing stress, lb/in²
F = factor of safety

Q2-34. In our plant we want to connect two 4-in shafts by means of a cast-iron flange coupling which employs six bolts. Our stress lab people tell us the allowable shearing stress of the bolts is 6000 lb/in² while that of the shafting is 8000 lb/in².

(*a*) Determine for us the diameter of the bolts to be used.

(*b*) Determine the induced crushing stress S_c if the thickness of the flange is ⅝ in.

(*c*) Is it a safe stress?

ANSWER.

(*a*) Torque $T = S_s Z_p = S_s(\pi d^3/16) = 8000(\pi 4^3/16)$ = 100,500 in-lb. Tangential force P_t occurring at the bolt circle

$$B = 3d = 3 \times 4 = 12 \text{ in}$$

$$P_t = \frac{T}{(B/2)} = \frac{100{,}500}{12/2} = 16{,}750 \text{ lb}$$

To find the bolt diameter d_1, where $P_t = 16{,}750$ lb, $n = 6$ bolts, and $S_s = 6000$ lb/in².

$$\frac{P_t}{n} = S_s \frac{r\, d_1^2}{4}$$

$$\frac{16{,}750}{6} = 6000(0.7854 \times d_1^2) \qquad \text{from which } d_1^2 = 0.59$$

$d_1 = 0.77$ in, say **⅞ in**, which is the next higher standard diameter

(*b*) $$\frac{P_t}{n} = d_1 t \times S_c \qquad \frac{16{,}750}{6} = \frac{7}{8} \times \frac{5}{8} \times S_c$$

$$S_c = \mathbf{5100\ lb/in^2}$$

This value of S_c is the induced unit crushing stress that is set up by a crushing load of (16,750/6) lb on the projected area or bearing area of each bolt. If we assume that U_c (ultimate crushing stress) for cast iron is 80,000 lb/in² and for steel is 60,000 lb/in², it is evident there will be a greater tendency to crush the steel than to crush the cast iron. Hence we shall determine the factor of safety F for steel. $F = 60{,}000/5100 = 11.8$. This is proof that the induced stress S_c is also a safe stress. **Ans.**

Q2-35. The advent of energy-conservation programs is making it more attractive to switch to synchronous belts as V-belt loss of efficiency becomes more expensive. Synchronous belts, commonly called timing belts, were developed for timing applications requiring synchronization, or "timing" between shafts. However, they have made their way into many other power-transmission applications. Timing belts are nearly 100 percent efficient, and the loss of about 0.5 percent is due to bending. V belts are usually only 95 to 96 percent efficient.

The 40-hp pump drive is to be converted to synchronous belts. The drive is belted with five C-120 belts running on two 10-ft, five-groove C-section sheaves. What is the saving for the first year of intallation? Assume the following and solve accordingly:

Cost of V-belt system: sheaves and belts
Cost of synchronous system: sprockets and belts
Efficiency of V-belt system—handbooks

Efficiency of synchronous system—handbooks
Electricity costs equal 6 cents per kilowatthour

ANSWER.

Cost of V-belt system = \$250
Cost of synchronous system = \$375
V-belt transmission efficiency loss = 5 percent
Synchronous system efficiency loss = 0.5 percent

Operating the 40-hp pump motor for 40 h per week at 6 cents per kilowatthour costs $40 \times 746/1000 \times 40 \times 0.06 =$ \$71.62. With V belts, 5 percent of this cost, or \$3.58 per week, is burned up in slippage. This loss represents \$186.16 per year.

The synchronous system will cost \$125 more, but, at 40 hp for 40 h/week, it will save

$$71.62 \times 0.005 = \$0.36 \text{ burned up in 1 week in slippage}$$

And for 1 year, $0.36 \times 52 =$ \$18.72.

Then the savings of synchronous over V belt per year is

$$186.16 - 18.72 = \$167.44$$

Thus, the change to the synchronous system will save \$167.44 in electricity in the first year, not to mention the savings in reduced maintenance. An additional advantage of timing belts is that the recommended installation belt tension is 30 to 50 percent less than that for V belts, allowing motor and sheave bearings to run under less load and pressure with considerably longer life.

Savings that can be realized by converting to timing belts can be quite considerable. For example, a plant whose V belts carry 1500 hp annually is burning up about $167.44 \times 1500/40 =$ \$6279.

Q2-36. Consider a clamping system operated at 15 in vacuum. The system consists of six clamps, each having an internal free volume of 0.15 ft^3. The clamps cycle four times per minute. Associated tubing and valves between the vacuum pump and clamps have a total free volume of 0.5 ft^3. If a receiver is added to the system having a volume of 20 ft^3 and is controlled by a vacuum switch for 14 and 18 in Hg and the system must be evacuated in 2 min:

What must be the vacuum pump capacity?

ANSWER.

Total system volume is $(6 \times 0.15) + 0.5 = 1.4$ ft³

and the required pump flow rating is given by

$$w = (\Sigma v_d)n \frac{v_s}{29.92} C$$

where w = required "free" air capacity at required vacuum, ft³/min

v_d = volume of each device in system, ft³

n = number of system cycles per minute

v_s = required system vacuum, in Hg

C = safety factor of 1.1 to 1.25

Assuming a safety factor of 1.25

$$w = 1.4 \times 4 \times \frac{15}{29.92} \times 1.25 = 3.51 \text{ ft}^3\text{/min}$$

The air to be removed depends on the cut-in and cutout vacuum switch settings. The required "free" air flow is given by

$$w = \frac{c_o - c_i}{29.92} \times \frac{v_r}{t}$$

where c_o = cutout switch setting, in Hg

c_i = cut-in switch setting, in Hg

v_r = receiver volume, ft³

t = time required to evacuate, min

Then the required vacuum pump capacity is

$$w = \frac{18 - 14}{29.92} \times \frac{20}{2} = 1.34 \text{ ft}^3\text{/min} \qquad \textbf{Ans.}$$

If the vacuum switch on the receiver controls the pump motor rather than a solenoid valve, careful coordination with both motor and pump manufacturers is needed to ensure that both units can withstand the expected intermittent duty cycle.

Chapter 3

GEARING

Questions and Answers

Q3-1. With spur gears having twisted teeth (cycloidal), what is the form of the line of contact along the tooth surface? What is the form of pitch surface in spur gears, bevel gears, hypoid gears? What conditions favor the use of each?

ANSWER. The form is a curved line.

Form of pitch surfaces:

Spur gears—cylindrical
Bevel gears—conical
Hypoid gears—approximately conical

Spur gears are used for the transmission of power between two parallel shafts. Bevel gears are used for the transmission of power and motion between two axes making an angle with each other. Hypoid gears use same as for bevel gears except two axes are past each other.

Q3-2. What is meant by pressure angle in gears? What are the usual angles in involute gears? What conditions are necessary to make involute gears interchangeable?

ANSWER. The pressure angle is that made by the line of action of the tooth with a line tangent to both the pitch circles where these come together.

The following angles are commonly employed for involute gears:

Brown and Sharpe	14.5° (most common)
Sellers	20.0°
Hunt stub teeth	14.5°
Logue stub teeth	20.0°
Fellows	20.0°

Interchangeability of gears is obtained by limiting both dimensions and form of the engaging teeth.

Involute gears will mesh properly, provided the height and thickness of the tooth will allow it. Pressure angle must be constant for gears of same series, 14.5° being the angle most commonly used. Least number of teeth found desirable is 12. The tooth is made long enough to obtain smooth running but not so long as to be too weak under load. Wherever the addendum, dedendum, and pressure angle are not properly related, interchangeability between gears with different numbers of teeth is secured by modifying the tooth outline so as to prevent interference.

Q3-3. A mild-steel spur-gear (15-tooth, 20°, full depth involute) pinion, rotating at 750 rpm, is to transmit 30 hp to a mild-steel gear rotating at 200 rpm. Determine (*a*) number of teeth in gear, (*b*) diametral pitch of teeth required, (*c*) face width of teeth, and (*d*) required shaft center-to-center distance.

ANSWER. Refer to Foote Bros. Gear and Machine Co., Catalogue No. 200, pp. 416–420. Now let the following nomenclature apply:

t_p = number of teeth on pinion
t_g = number of teeth on gear
N_p = pinion rpm
N_g = gear rpm
P = diametral pitch
S = allowable unit stress
A = factor of 4 for cut gears
Y = Lewis formula outline factor
V = pitch line velocity, fpm

(*a*) $N_p/N_g = t_g/t_p = 750/200 = t_g/15$, from which we find t_g to be 56.25. Thus, number of teeth should be **56** for the gear.

(*b*) Gears which have to work together must have the same diametral pitch, i.e., number of teeth divided by diameter of pitch circle. Then

$$P = \sqrt{\pi SAYV/(33{,}000 \times \text{hp})}$$
$$= \sqrt{\pi \times 5000 \times 4 \times 0.289 \times 1200/(33{,}000 \times 30)} = 4.64, \text{ say } \mathbf{5}$$

Diametral pitch is generally expressed as a whole number.

(*c*) General practice indicates gear face width measured along axis is to be four times circular pitch. Then

$$\text{Circular pitch} = \frac{\pi}{P} = \frac{3.1416}{5} = 0.63$$

The face width is, therefore, $4 \times 0.63 =$ **2.52 in.**

(*d*) Center distance is $\dfrac{t_p + t_g}{2 \times P} = \dfrac{15 + 56}{2 \times 5} =$ **7.1 in.**

Note. V for average conditions is taken as 1,200 fpm maximum. For carefully cut gears V should not be greater than 1,800 fpm.

Q3-4. Find the dimensions of the teeth of a cast spur gear whose circular pitch is 1.25 in.

ANSWER.

$$\text{Addendum} = 0.3183 \times 1.25 = 0.40 \text{ in.}$$
$$\text{Dedendum} = 0.39 \times 1.25 = 0.4875 \text{ in.}$$
$$\text{Working depth} = 0.64 \times 1.25 = 0.80 \text{ in.}$$
$$\text{Clearance} = 0.05 \times 1.25 = 0.0625 \text{ in.}$$
$$\text{Whole depth} = 0.6866 \times 1.25 = 0.8583 \text{ in.}$$
$$\text{Tooth thickness} = 0.5 \times 1.25 = 0.625 \text{ in.}$$
$$\text{Width of space} = 0.52 \times 1.25 = 0.65 \text{ in.}$$
$$\text{Backlash} = 0.65 - 0.625 = 0.025 \text{ in.}$$

Q3-5. Some values dealt with in gear design are the following: (*a*) number of teeth, (*b*) pitch of teeth, (*c*) depth of teeth, (*d*) thickness of teeth, (*e*) pressure angle, (*f*) gear diameters, (*g*) face width, (*h*) transmitted torque, and (*i*) radii of curvature of contacting profiles.

Indicate by letter which of the above-mentioned values are involved in calculating the tooth load in pounds per square inch of face width, assuming all the load to be carried on one pair of teeth.

What additional values are used in estimating the breaking strength of a tooth, i.e., tensile stress?

In addition to those given to answer the first question, what value or values are involved most directly in estimating the surface durability of the teeth, i.e., compressive stress?

ANSWER. Refer to Boston Gear Catalogue in which Barth's revision of the Lewis formula is recommended. It reads as follows:

$$W = S \times p' \times f \times y \times \frac{600}{600 + V}$$

$$\text{Horsepower} = W \frac{V}{33{,}000}$$

(*a*) In calculating the tooth load in pounds per inch of face width (W/f above) the following values are used:

Number of teeth, to obtain y above
Pitch of teeth, to obtain p' above
Gear diameters, to obtain V above
Face width, to obtain f above
Transmitted torque, to obtain horsepower above
Radii of curvature of contacting profiles, to obtain y above

(*b*) Additional values used in estimating the breaking strength of a tooth, i.e., tensile stress or S above are:

Depth of tooth
Thickness of tooth
Pressure angle and tooth load W

(*c*) Other values involved most directly in estimating the surface durability of the teeth, i.e., compressive stress, are the kind of material and static stress S.

Q3-6. Design a pair of spur gears for a motor drive as follows: motor delivers 50 hp at 1,200 rpm; gear reduction, 6 to 1. Use cut-steel gears and make bore of drive gear as small as good practice allows. Indicate outside diameter, pitch diameter, diametral pitch, number of teeth, type of teeth, width of face, key size, and distance between shafts.

ANSWER. First determine bore of drive gear in accordance with

$$d = 68.5 \sqrt[3]{\frac{\text{hp}}{N \times S_s}} = 68.5 \sqrt[3]{\frac{50}{1{,}200 \times 600}} = 1.375 \text{ in.}$$

Now assume pinion pitch diameter to be 5 in.; number of teeth, 15; diametral pitch, 3. The velocity of the driver is easily determined.

$$V = \frac{5\pi}{12} \times 1{,}200 = 1{,}570 \text{ fpm}$$

Find the tooth load in accordance with the previous problem, simply by rearrangement:

$$W = \frac{50 \times 33{,}000}{1{,}570} = 1{,}050 \text{ lb}$$

The face width may be found by use of the formula for tooth load.

$$W = S \times p' \times f \times y \times \frac{600}{600 + V}$$

$$1{,}050 = 12{,}000 \times 1.0472 \times f \times 0.075 \times \frac{600}{600 + 1{,}570}$$

From which face width f is found to be **4 in.** Thus, for the **pinion gear** the following design criteria apply: **5-in. pitch diameter, 15 teeth, diametral pitch of 3, 1,200 rpm, face width of 4 in.** The driven gear, because of the gear reduction of 6 to 1, has $6 \times 15 = 90$ teeth, a $6 \times 5 = 30$ pitch diameter, diametral pitch of 3 (same as pinion), rpm of $1{,}200/6 = 200$. In the same manner as the above, face width is found to be 4½ in.

From the expression that the total number of teeth equals twice the center distance multiplied by the diametral pitch we see that

$$15 + 90 = 2 \times C \times 3 \qquad \text{or} \qquad C = 105/(2 \times 3) = \mathbf{17\tfrac{1}{2}\ in.}$$

The key size may be obtained by reference to a standard handbook as **7/16 × 5/16 in.**

Q3-7. Design a pair of gears to transmit 20 hp. Pinion is to rotate at 800 rpm and is to have 17 teeth and width is to be three times its circular pitch. Ratio of pinion speed to driven gear speed is 5:1. Material is cast iron with cut teeth. Assume proper values for allowable stress. Both gears are to be keyed to their shafts.

ANSWER. Try 4 pitch, 14½-in. involute, pitch diameter equal to 17/4, or 4.25 in. Pinion velocity is found as before in previous problem to be

$$V = \frac{4.25 \times \pi}{12} \times 800 = 890 \text{ fpm}$$

$$W = 13{,}500 \times 0.7854 \times 3(0.7854) \times 0.080 \times \frac{600}{600 + 890} = 810 \textbf{ lb}$$

$$\text{hp} = \frac{W \times V}{33{,}000} = \frac{810 \times 890}{33{,}000} = 21.5 \qquad \text{(Satisfactory)}$$

Summarizing. Pinion: 800 rpm, 17 teeth, 4.25-in. pitch diameter, 4 diametral pitch.

Driven gear: $800/5 = 160$ rpm, $17 \times 5 = 85$ teeth, $4.25 \times 5 = 21.25$ pitch diameter, 4 diametral pitch. Center distance determined as above is found to be $12\frac{3}{4}$ in.

Q3-8. A 10-ton electric truck is to accelerate at the rate of 3 miles per hour per second (Mphps). Its rolling resistance is 25 lb per ton weight. If the truck employs a gear reduction between the motor and the wheels of 9 to 1, and the wheels are 36 in. in diameter,

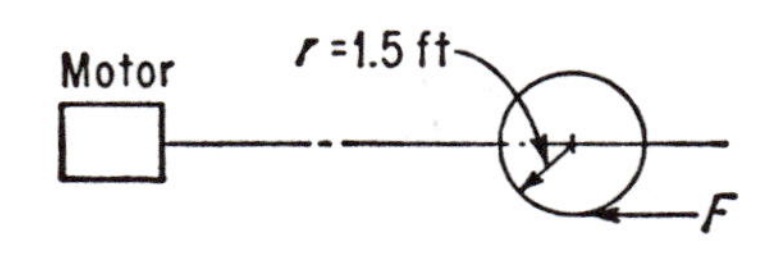

Fig. 3-1

what is the power and the torque required of the motor to drive the truck at 25 mph? Assume efficiency of the reduction gear at 90 per cent.

ANSWER. Refer to Fig. 3-1, and note that the total rolling resistance is as follows:

$$F' = 25 \times 10 = 250 \text{ lb}$$

Acceleration is found to be

$$3 \times 5{,}280/3{,}600 = 4.4 \text{ ft per sec}^2$$

On level road $F - F' = W/g \times a$, where F is the necessary propelling force. Therefore,

$$F = 2{,}000 \times 10 \times 4.4/32.2 + 250 = 3{,}000 \text{ lb in round numbers}$$

Required motor horsepower is simply

$$\text{hp} = F \times V/(550 \times \text{efficiency})$$

where V is velocity in fps, and here is equivalent to $25/60 \times 88 = 36.7$ fps.

$$\text{hp} = 3{,}000 \times 36.7/(550 \times 0.90) = \mathbf{222\ hp}$$

Wheel torque $= F \times$ wheel radius in ft $= 3{,}000 \times 1.5 = 4{,}500$ lb-ft
Required motor torque = wheel torque/(velocity ratio × efficiency)

$$4{,}500/(9 \times 0.90) = \mathbf{556\ lb\text{-}ft}$$

Q3-9. Two shafts are connected by spur gears, as shown in Fig. 3-2. The pitch radii of gears *A* and *B* are 4 in. and 20 in., respectively. If shaft *A* makes 800 rpm and is subjected to a resisting moment of 1,000 in.-lb, what is (*a*) rpm of *B*, (*b*) torque in

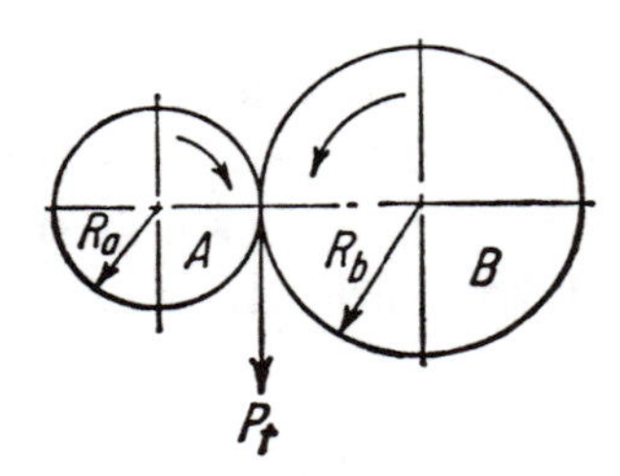

Fig. 3-2

shaft *B*, (*c*) speed reduction factor, (*d*) torque multiplication factor, and (*e*) tooth pressure of *A* and *B*?

ANSWER. Here R_a is 4 in., R_b is 20 in., and N_a is 800 rpm.

(*a*) Then

$$\frac{N_b}{N_a} = \frac{R_a}{R_b}$$

$$\frac{N_b}{800} = \frac{4}{20} = \frac{1}{5}$$

From which we obtain $N_b = 800/5 =$ **160 rpm.** This is the speed of *B*.

(*b*)

$$\frac{T_b}{T_a} = \frac{R_b}{R_a} = \frac{T_b}{1{,}000} = \frac{20}{4} = 5$$

Therefore $T_b =$ **5,000 in.-lb.** This is the torque in shaft *B*.

(*c*) The speed reduction factor is $N_a/N_b = 800/160 =$ **5.**

(*d*) The torque multiplication factor is $T_b/T_a = 5{,}000/1{,}000 =$ **5.**

(*e*) Tooth pressure $P_{ta} = T_a/R_a = 1{,}000/4 =$ **250 lb.**

Q3-10. (*a*) A tractor is to have a drawbar pull of 5,000 lb. The differential has a double-threaded worm acting on a wheel having 35 teeth. The rear wheels are 42 in. in diameter. The motor has a maximum torque of 200 lb-ft. The efficiency of the drive is assumed to be 90 per cent. What gear ratio is necessary in the transmission? (*b*) Calculate the stress in the rear axle if the diameter is 1.75 in.

ANSWER.

(*a*) For the differential (wheel teeth/worm teeth) $= 35/2 = 17.5$. Since the total wheel reaction is 5,000 lb, the total torque is

$$5{,}000 \times 21/12 = 8{,}750 \text{ lb-ft}$$

where the value 21 is half the rear-wheel diameter. Then from the relation

Motor torque × differential-gear ratio
× transmission-gear ratio × efficiency = drive-wheel torque

we can solve for the transmission-gear ratio:

$$\text{Transmission-gear ratio} = \frac{8{,}750}{200 \times 17.5 \times 0.90} = \mathbf{2.78}$$

(*b*) For each axle the torque is equal to $8{,}750 \times 12/2 = 52{,}500$ lb-in. Stress in axle is next determined in accordance with the following: $S = 16/(\pi \times d^3)(\text{torque in each axle}) = [16/(\pi \times 1.75^3)](52{,}500)$ where S is calculated as **49,500 psi.**

Q3-11. A worm-driven hoist raises a load of 5,000 lb at a speed of 100 fpm when the driving motor is exerting a torque of 130 lb-ft at a speed of 1,000 rpm. Assume the drum diameter and the type of worm, and calculate the number of teeth in the drum wheel. Determine the efficiency of the entire hoist mechanism.

ANSWER. Assume drum diameter to be 1.5 ft and number of worm teeth 2. Therefore, angular velocity of drum and drum wheel is

$$\frac{v}{\pi \times \text{drum diameter}} = \frac{100}{\pi \times 1.5} = 21.2$$

$$\text{Number of drum-wheel teeth} = \frac{\text{angular velocity of worm}}{\text{angular velocity of drum wheel}} \times n$$

$$\text{Number of drum-wheel teeth} = (1{,}000/21.2)(2) = 94.3, \text{ say 94 teeth}$$

$$\text{Efficiency} = \frac{\text{output}}{\text{input}} = \frac{Wv}{2\pi ST}$$

$$\text{Efficiency} = \frac{5{,}000 \times 100}{6.28 \times 1{,}000 \times 130} = 0.612, \text{ or } \mathbf{61.2 \text{ per cent}}$$

$$\text{hp} = \frac{2\pi ST}{33{,}000} = \frac{817{,}000}{33{,}000} = 24.7, \text{ say 25 hp}$$

In this solution n is number of worm teeth, N is number of drum wheel teeth, S is motor speed in rpm, W is load, v is load speed in fpm, D is drum diameter in feet, T is motor torque in lb-ft.

Q3-12. If one rear wheel of an automobile is jacked off the ground and the motor turns the drive shaft at 1,200 rpm, calculate speed of the floating rear wheel, of the ring gear, and of spider pinions. The numbers of teeth on the gears are drive-shaft pinion, 11; ring gear, 40; rear axle gear, 23; spider pinion, 12.

ANSWER. Refer to Fig. 3-3, and let

G = drive shaft
H = drive-shaft pinion, 11 teeth
K = ring gear, 40 teeth
B_1 = moving rear-axle gear, 23 teeth
CD = spider pinion, 12 teeth
B = stationary rear-axle gear, 23 teeth
N_G = drive-shaft speed in rpm
T_H = number of teeth on H; typical for other gear teeth

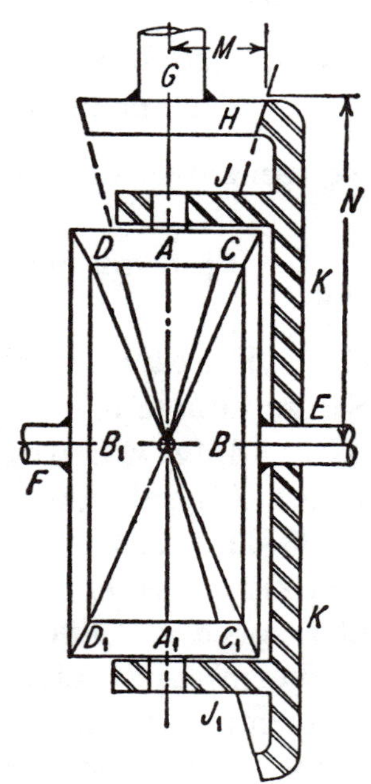

Fig. 3-3

With B stationary ($N_B = 0$) and B_1 rotating

$$N_K = N_G \frac{T_H}{T_K} = \frac{1{,}200}{40/11} = 330 \text{ rpm}$$

$$N_{CD} = N_K \left(\frac{T_B}{T_{CD}}\right) = (330)(23/12) = 632 \text{ rpm}$$

When vehicle is moving straight ahead, $N_B = N_K$, when there is equal traction. But under conditions of this problem the moving rear wheel will have twice N_{B_1} or $2 \times N_K$. So that it follows:

$$N_{B_1} = N_K + \left(N_{CD} \frac{T_{CD}}{T_{B_1}}\right) = 330 + [632(12/23)] = \mathbf{660 \text{ rpm}}$$

Chapter 4

MECHANISM

Questions and Answers

Q4-1. Make a sketch of a Whitworth slow-advance and quick-return motion, and calculate the length of the driving crank when the ratio of the time of advance to the time of return is 2. The distance between centers is 3 in. Also calculate the length of motion in the slotted crank.

ANSWER. Refer to Fig. 4-1. It is apparent that the sliding block will take twice as long in advancing than in returning when the driver crank OA, which rotates at a uniform speed, has turned through an arc ABC twice as long as arc CDA. This means that arc CDA equals 120°, or angle a is 60°. The length of the driving crank is

$$R = 3 \cos 60° = 3 \times \tfrac{1}{2} = 1.5 \text{ in.}$$

Length of motion of the sliding block on crank EF is diameter DB or **3 in.**

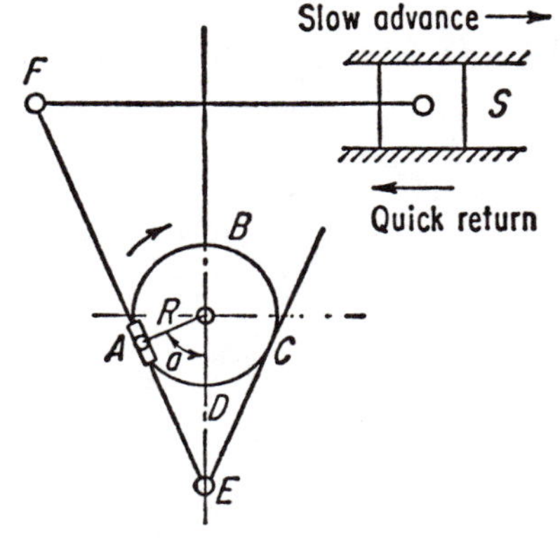

Fig. 4-1

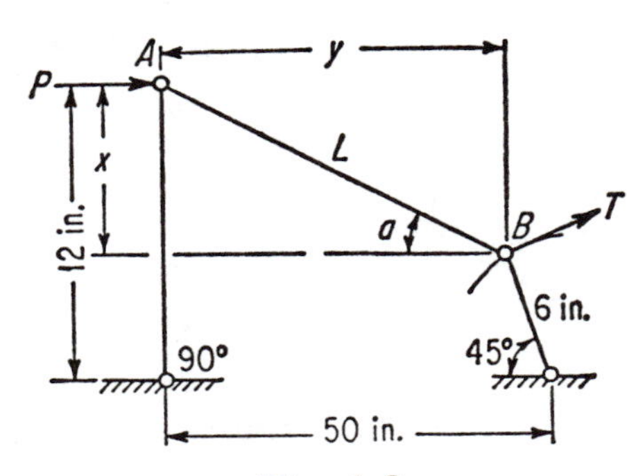

Fig. 4-2

Q4-2. Two rockers are connected at their free ends by a link. One rocker is 12 in. long and is perpendicular to the line of centers which is 50 in. long. The other rocker is 6 in. long and at an angle of 45° with the line of centers. A force of 100 lb is applied at the end of the 12-in. rocker and directed perpendicular to it. What torque is exerted on the shaft of the 6-in. rocker?

ANSWER. Refer to Fig. 4-2, and let

P = force on 12-in. rocker, lb.
R = reaction of connecting link, lb.
L, x, y = distances shown, in.
T = tangential force on 6-in. rocker, lb
a, b = angles shown on diagrams of forces, deg

From the diagram of forces at point A (Fig. 4-3) it may be seen that the horizontal component of R is

$$R_y = P = 100 \text{ lb}$$

Also, $y = (50 - 6 \cos 45°) = 45.76$ in.
$x = (12 - 6 \sin 45°) = 7.76$ in.
$L = (x^2 + y^2)^{1/2} = (7.76^2 + 45.76^2)^{1/2} = 46.4$ in.
$\cos a = 45.76/46.4 = 0.987$
$R = R_y/\cos a = 100/0.987 = 101.3$ lb
$a = \cos^{-1} 0.987 = 9°15'$ (see Fig. 4-4)
$b = 45° + a = 45° + 9°15' = 54°15'$
$T = R \cos b = 101.3 \times \cos 54°15' = 59.2$ lb
Torque on 6-in. rocker $= T \times 6/12 = 59.2 \times 0.5 =$ **29.6 lb-ft**

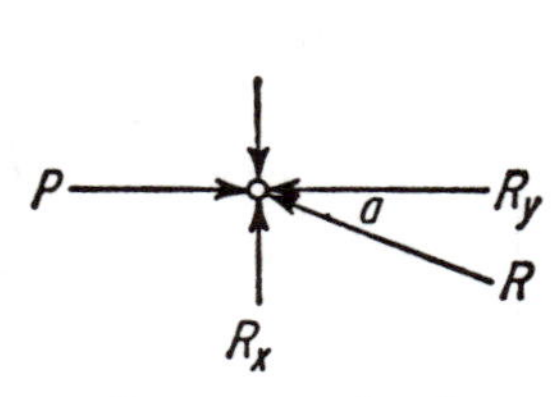

Forces at point A

Fig. 4-3

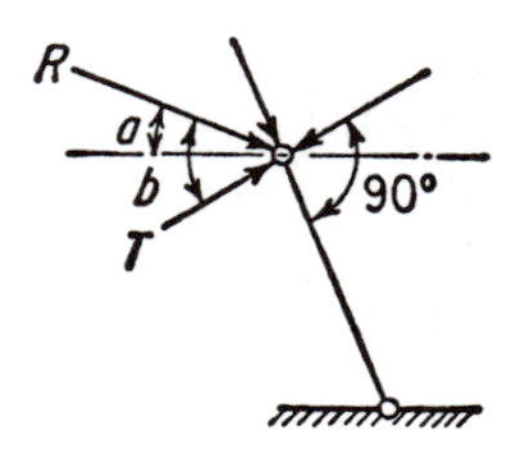

Forces at point B

Fig. 4-4

Q4-3. Two cranks, each 6 in. long, are connected by a drag link. Crank centers are 4 in. apart. The drag link connecting the free ends of the cranks is 7 in. long. Determine graphically the angular velocity ratio when one crank assumed as the driver has moved through 90° from the vertical position.

ANSWER. Let

ϕ_a, ϕ_b = angular velocity of links A and B, respectively, radians per sec

v_a, v_b = tangential velocity of links A and B, respectively, fps

r_a, r_b = length of links A and B, respectively, in.

Referring to the kinematic diagram (Fig. 4-5), let A be the driving link. Draw v_a, at point a, making it equal to unity on some chosen measuring scale. Then v_1 will be the component of v_a along link C.

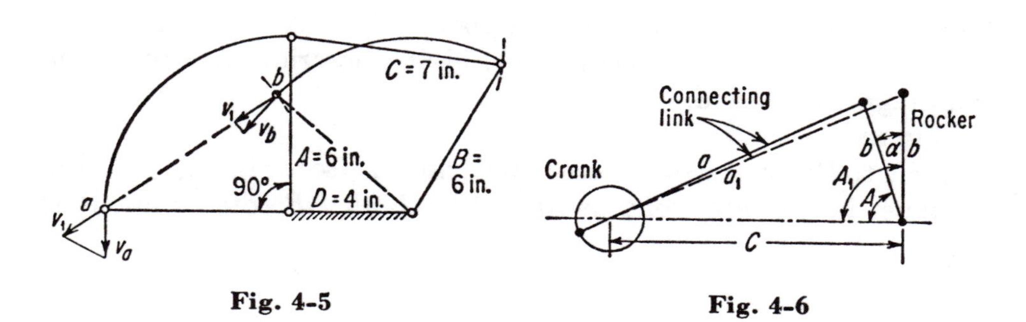

Fig. 4-5

Fig. 4-6

It will also be the rectangular component of v_b at point b. By means of dividers lay off v_1 at b and measure v_b with the chosen scale. It will be found that

$$v_b = 0.55v_a \qquad \text{or} \qquad \frac{v_b}{v_a} = 0.55$$

Now since $r_a = r_b$,

$$\frac{\phi_b}{\phi_a} = \frac{v_b}{r_b}\frac{r_a}{v_a} = \frac{v_b}{v_a} = \mathbf{0.55}$$

Q4-4. A crank 4 in. long oscillates a rocker 20 in. long through a connecting link 28 in. long. The distance between crank and rocker centers is 22 in. Calculate the angle through which the rocker oscillates.

ANSWER. Figure 4-6 shows the limiting positions for rocker travel. From the law of cosines we have

$$\cos A = \frac{b^2 + c^2 - a^2}{2bc} = \frac{20^2 + 22^2 - (28 - 4)^2}{2 \times 20 \times 22} = 0.35$$

$$A \cos^{-1} 0.35 = 69°31'$$

Similarly,

$$\cos A_1 = \frac{b^2 + c^2 - a^2}{2bc} = \frac{20^2 + 22^2 - (28 + 4)^2}{2 \times 20 \times 22} = 0.157$$

$$A_1 \cos^{-1} 0.157 = 99°9'$$

Thus, the angle moved through by rocker $A_1 - A =$ **29°38′.**

Q4-5. Name seven mechanisms, giving descriptive information and diagrams of each.

ANSWER.

1. Scott-Russell straight-line mechanism (see Fig. 4-7). Link AC must be twice as long as link DB. Link DB rotates about D

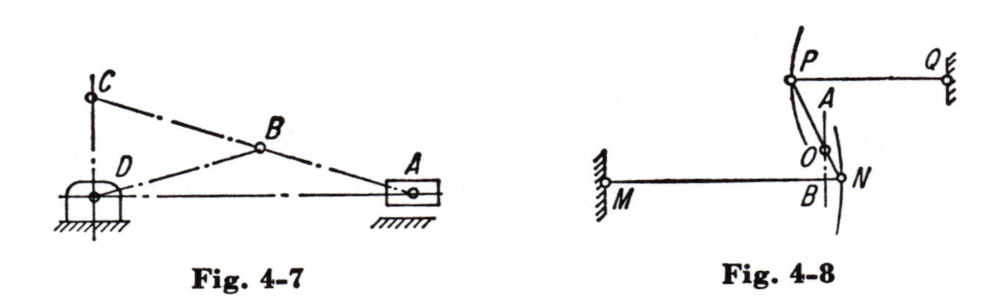

Fig. 4-7 **Fig. 4-8**

whereupon C moves up or down along CD and sliding block A moves along AD.

2. Watt straight-line mechanism (see Fig. 4-8). The tracer point O is located so as to divide NP inversely as the adjacent rotating link, or $ON:OP$ as $PQ:MN$. For small angles point O will give substantially a straight line along AB.

3. Pantograph with 1:3 ratio (see Fig. 4-9). The tracer point B must be placed so that BC is one-third of AC. For a given movement of A, B will now move one-third the distance that A moved and in the same direction.

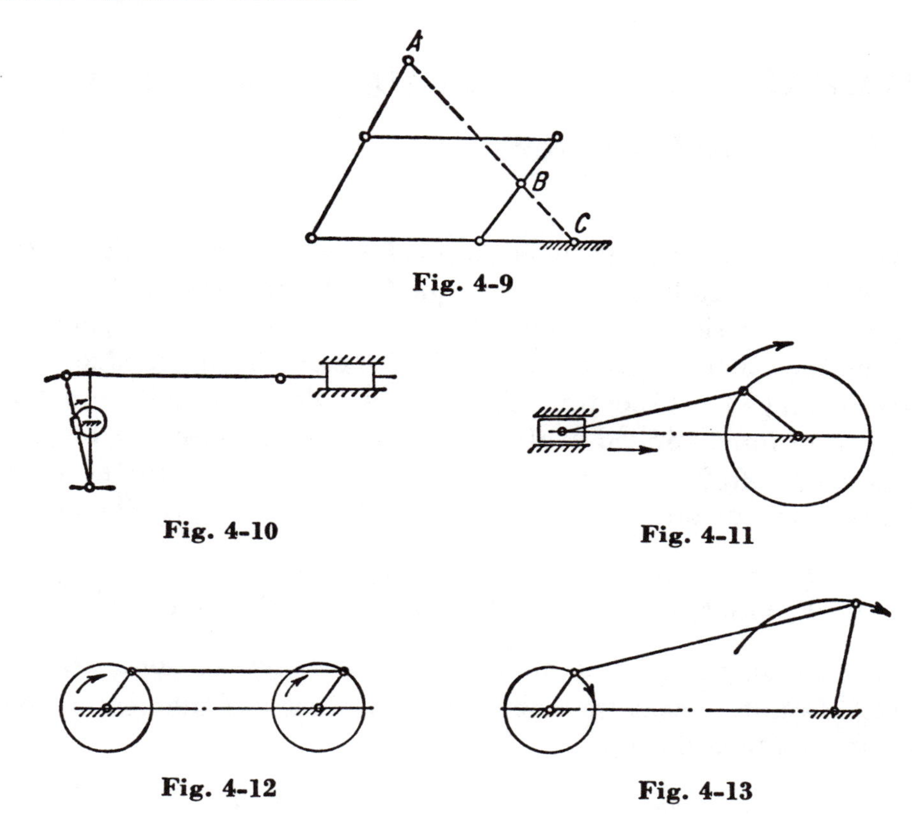

Fig. 4-9

Fig. 4-10

Fig. 4-11

Fig. 4-12

Fig. 4-13

4. Whitworth motion. Slow advance and quick return (see Fig. 4-10).

5. Rotary motion in straight-line harmonic (see Fig. 4-11).

6. Parallel straight-line mechanism (see Fig. 4-12).

7. Rotary rocking from continuous rotary (see Fig. 4-13).

Chapter 5

HYDRAULICS AND FLUID MECHANICS

Questions and Answers

Q5-1. A rectangular sluice gate is 6 ft wide and 9 ft high. It is immersed vertically in water with the 6-ft edges horizontal. Water stands on one side of the gate at a depth level with its upper edge and on the other side at a depth of 4.5 ft below the upper edge. The gate is hinged at the top edge and is held in equilibrium by a horizontal force applied at its lower edge. Calculate this force.

ANSWER. Refer to Fig. 5-1. The center of pressure acts two-thirds of the distance down from top (at water level). Then

$$F_1 = wh_1A_1 = 62.4 \times 4.5 \times (9 \times 6) = 15{,}200 \text{ lb}$$
$$F_2 = wh_2A_2 = 62.4 \times 2.25 \times (4.5 \times 6) = 3{,}800 \text{ lb}$$

Now take summation of moments about hinge at point equal to zero. Counterclockwise moments equal clockwise moments. Now set up this relationship.

$$F_1 \times 6 = (F_2 \times 7.5) + (F_0 \times 9)$$
$$15{,}200 \times 6 = (3{,}800 \times 7.5) + (F_0 \times 9)$$
$$F_0 = (91{,}200 - 28{,}500)/9 = \mathbf{6{,}980 \text{ lb}}$$

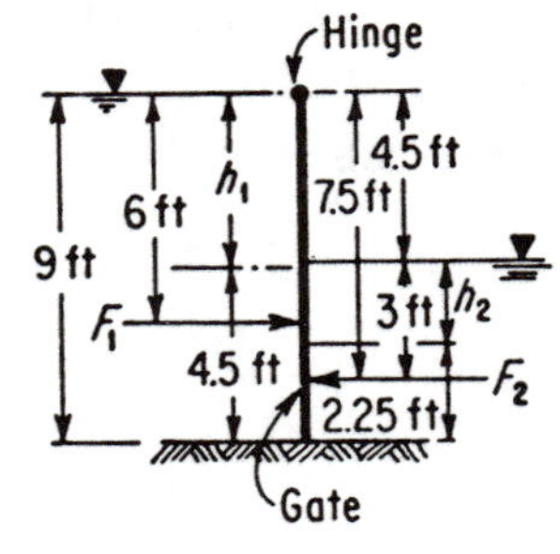

Fig. 5-1

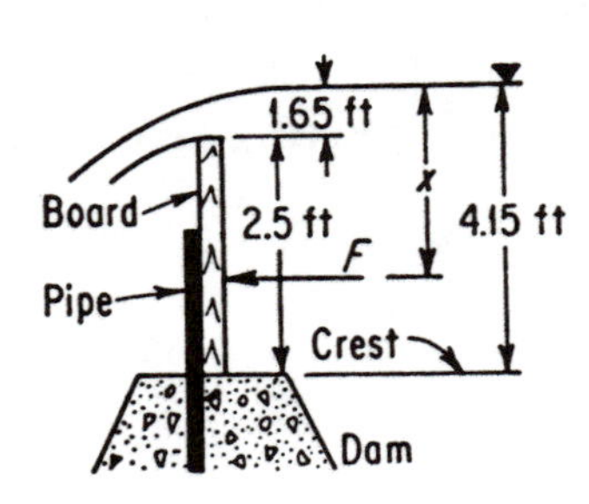

Fig. 5-2

Q5-2. A flashboard 2.5 ft high mounted above the crest of a dam is supported by a standard (Schedule 40) 2-in. pipe of steel placed on 8-ft centers. The pipes fail and the flashboard goes down at a head on the crest of the dam amounting to 4.15 ft of water. (*a*) What is the initial bending moment on each pipe in inch-pounds under static conditions? (*b*) What is the maximum pipe stress if the section modulus of the pipe is 0.616 in. cube?

ANSWER. Refer to Fig. 5-2.

$$(a)\ F = 62.4hA = 62.4(1.65 + 2.5/2)(2.5 \times 8) = 3{,}620 \text{ lb}$$

$$x = y + \frac{k^2}{y} = (1.65 + 2.5/2) + (2.5^2/12 \times 1/2.90) = 3.08 \text{ ft}$$

The bending moment on the pipe, M_0 is

$$M_0 = (4.15 - x) \times F \times 12 = (4.15 - 3.08) \times 3{,}630 \times 12 = \mathbf{46{,}600\ in.\text{-}lb}$$

(*b*) $M_0 = fS$, where f = maximum pipe stress in psi, and S = section modulus in inch cube. Rearranging and solving for f,

$$f = \frac{M_0}{S} = \frac{46{,}600}{0.616} = \mathbf{75{,}600\ psi}$$

Q5-3. A cylindrical log is 1 ft in diameter and 20 ft long. What weight of iron must be tied to one end of the log to keep it floating in an upright position in seawater with 18 ft of the log submerged. Specific gravity of log is 0.7, of iron 7.9, and of sea water 1.03.

ANSWER. The basic equation is given by

wt of log + wt of iron = buoyant force log + buoyant force of iron

$$(0.785 \times 1^2 \times 20 \times 62.4 \times 0.7) + \text{wt of iron}$$
$$= (0.785 \times 1 \times 62.4 \times 1.03) + (\text{vol of iron} \times 62.4 \times 1.03)$$
$$686 + \text{wt of iron} = 910 + (\text{vol of iron} \times 64.3)$$
$$\text{wt of iron} = (910 - 686) + (\text{vol of iron} \times 64.3)$$

This is so since weight of iron is equal to the volume of iron × 493. Then it follows that

$$\text{wt of iron} = 224 + \frac{64.3 \times \text{wt iron}}{493}$$
$$1 - 0.13 \times \text{wt iron} = 224$$

Finally

$$\text{wt iron} = 224/0.870 = \mathbf{258\ lb}$$

Q5-4. Assuming isothermal and steady flow, a pump takes suction from a large storage tank containing sulfuric acid (sp gr = 1.84) through a 3-in. line and discharges through a 2-in. line to a point 75 ft above the level in the storage tank. Tank is under slight positive pressure with a dry gas to prevent absorption of air moisture from the atmosphere. However, this pressure may be neglected. Friction losses may be taken as 26 ft of fluid flowing. What differential pressure must the pump develop and what horsepower motor would be necessary if the pump efficiency were 60 per cent and the velocity in the suction line were 3 fps?

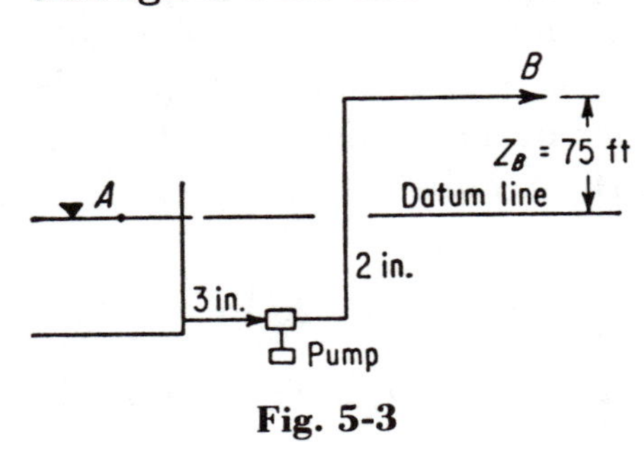

Fig. 5-3

ANSWER. Refer to Fig. 5-3. By observation Z_A is zero and v_A is zero because tank diameter is great compared to suction-pipe diameter. It may also be assumed that the water level is dropping so slowly that its velocity is negligible compared to the pipe velocity. Now, since both tanks are open to the atmosphere, $P_A = P_B$. Single fluid is flowing in the system and $w_A = w_B$, where w is density in pounds per cubic foot.

Bernoulli's equation breaks down to the following simple form:

$$-h_f + W_p = Z_B + \frac{v_B}{2g}$$

Now let us find the velocity in the 2-in. discharge line from that in the suction line

$$v_B = 3 \times (3/2)^2 = 6.75 \text{ fps}$$

Rearranging the previous form and solving for W_p, it is found that

$$W_p = 75 + (6.75^2/64.4) + 26 = 101.7 \text{ ft}$$

This is the total dynamic head required for the pump to operate properly. This is a differential head; it is neither the suction head nor the discharge head. It is the difference between the discharge and suction heads. We must convert this total dynamic head to pressure as follows:

$$(101.7/2.31) \times 1.84 = \mathbf{80.6\ psi}$$

For the second part of the problem we make use of the basic relation for horsepower: lb per min × total dynamic head/33,000. What the pump is really doing is, in fact, lifting so many pounds of sulfuric acid a minute through 101.7 ft.

$$\text{lb per min} = \text{gpm} \times 8.33 \times 1.84$$
$$(3 \times 3^2/0.408) \times 8.33 \times 1.84 = 1{,}010$$

where gpm is derived from the equation in simple form:

$$\text{gpm} \times 0.408/d^2 = \text{fps}$$

Here d is the nominal pipe size as given in the problem. There is no need to be concerned with actual inside diameter of pipe.

$$\text{Hydraulic horsepower} = 1{,}010 \times 101.7/33{,}000 = 3.1$$

We must adjust hydraulic horsepower to take into account pump efficiency.

$$\text{Hydraulic horsepower/pump efficiency} = \text{brake horsepower}$$
$$3.1/0.6 = 5.18 \text{ brake horsepower}$$

Since motor horsepower ratings are pump-shaft ratings, it would be safe to use a **5 hp motor.**

Q5-5. A pump takes 500 gpm from a sump and discharges the water from an 8-in. pipe at a point 20 ft higher. Assuming an over-all efficiency of 80 per cent, what horsepower is required?

ANSWER. Assume that discharge head of 20 ft includes pipe friction plus actual lift. Also h_f may be included in the efficiency. Then simply

$$\text{bhp} = 500 \times 8.33 \times 20/(33{,}000 \times 0.8) = \mathbf{3.15}$$

Q5-6. Compute the discharge through an orifice whose area is 1 sq in. Coefficient of discharge C may be taken as 0.61. The water reaches the orifice through a pipe whose area is 4 sq in. A pressure gauge reads 40 psi at a point 3 ft above the orifice. Neglect the velocity of approach.

ANSWER. Refer to Fig. 5-4 and the orifice equation

$$Q = CA\sqrt{2gh}$$

Determine h which is a composite of actual physical head *plus* the equivalent head due to the 40 psi pressure.

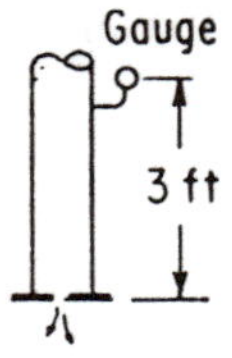

Gauge

3 ft

Fig. 5-4

$$h = 3 + (40 \times 2.31) = 95.4 \text{ ft}$$
$$Q = 0.61 \times \tfrac{1}{144}\sqrt{64.4 \times 95.4} = \mathbf{0.335\ cfs}$$

Q5-7. Discuss discharge from a tank with varying head.

ANSWER. A tank with constant cross section and water discharging through an orifice in the side or bottom of the tank into the air and neglecting velocity of approach the time in seconds to lower the levels between h_1 and h_2 is found by

$$\text{Volume} = tA_oC\sqrt{2g}\,\frac{h_1 - h_2}{2} \qquad \text{cu ft}$$

The lower limit h_2 should be chosen at some value greater than zero because the orifice flow equation ceases to rigidly hold at very low heads. This is due to vortex formation and surface tension. Nevertheless, questions in examinations often appear and the candidate is required to ignore this effect and apply the above equation and others to follow when the tank is being emptied.

Q5-8. Discuss the time required to lower water in a vessel. What formula may be used to determine this?

ANSWER. This application has many calls in industry. For instance, in the field during construction of chemical and refinery plants, large tanks, towers, and other vessels must be given a hydrostatic test before turning the plant on stream. The vessels are filled with water for this test and then must be emptied afterward. Because of close scheduling the time it takes to empty must be known.

The time to lower the water level in a vessel through an orifice or short pipe is given by

$$t = \frac{2A}{CA_0\sqrt{2g}}\left(\sqrt{h_1} - \sqrt{h_2}\right) \qquad \text{sec}$$

where A is cross-sectional area of vessel in square feet constant, h_1 is original head, and h_2 is final head. The equation neglects velocity of approach because of great differences between areas of orifice and cross section of tank. Refer to Fig. 5-5.

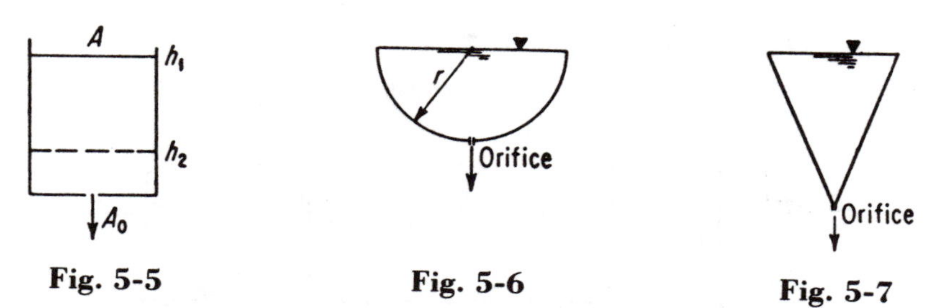

Fig. 5-5 **Fig. 5-6** **Fig. 5-7**

For a hemispherical tank at full level, the time to empty is

$$t = \frac{14}{15} \frac{\pi}{CA_0\sqrt{2g}} r^{5/2} \quad \text{sec} \qquad \text{(Fig. 5-6)}$$

where r is radius of hemisphere in feet and h_1 and h_2 apply as in the above.

Time to empty a v trough, Fig. 5-7:

$$t = \frac{4}{3} \frac{\text{initial volume}}{\text{initial flow rate}} \quad \text{sec}$$

The initial flow rate is determined by use of the equation with h_1 as the initial head.

Time to empty an inverted cone or inverted pyramid shape:

$$t = \frac{6}{5} \frac{\text{initial volume}}{\text{initial flow rate}} \quad \text{sec}$$

Time to empty a sphere at any level of liquid:

$$t = \frac{8}{5} \frac{\text{initial volume}}{\text{initial flow rate}} \quad \text{sec}$$

Time to empty a paraboloid of revolution:

$$t = \frac{4}{3} \frac{\text{initial volume}}{\text{initial flow rate}} \quad \text{sec}$$

Note that in all cases initial volume is cubic feet and initial flow rate is cubic feet per second.

Q5-9. A canal lock 300 ft long and 40 ft wide has a lift of 64 ft. Lock is emptied by twin culverts each 4 ft in diameter, feeding through a series of ports such that the coefficient of discharge may

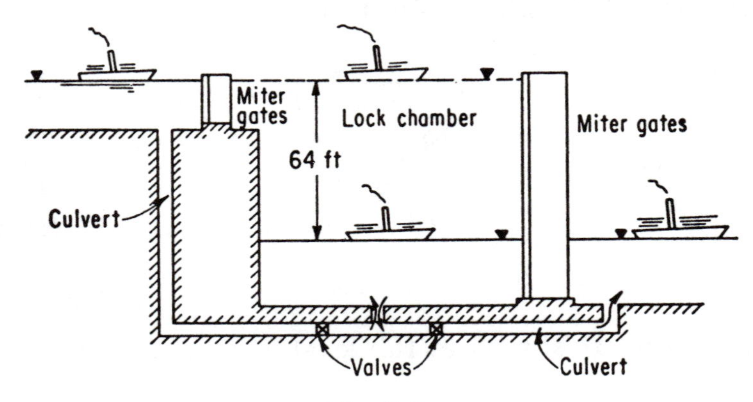

Fig. 5-8

be taken as 0.75. How long will it take to empty lock if full at start?

ANSWER. Refer to Q5-8 and apply the equation discussed (see Fig. 5-8).

$$t = \frac{2 \times 300 \times 40}{0.75 \times (0.785 \times 4^2) \times \sqrt{64.4}} (\sqrt{64} - \sqrt{0}) = 2{,}540 \text{ sec}$$

This is the time to empty for one culvert; for two culverts the time is halved to 1,270 sec, or **21.1 min.**

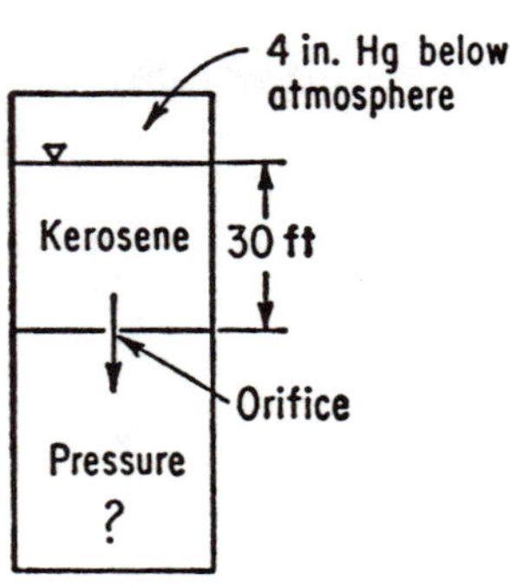

Fig. 5-9

Q5-10. Two closed tanks are connected by a 2-in. diameter circular orifice which has a coefficient of discharge of 0.63 for kerosene. One tank contains kerosene, specific gravity 0.78, to a depth of 30 ft above the center of the orifice and the other is empty. The tank with kerosene has a pressure of 4 in. of mercury below atmosphere in the space above the kerosene. What is the pressure in psi in the empty tank when the orifice is discharging 157.2 gpm?

ANSWER. Refer to Fig. 5-9. Convert 157.2 gpm to cubic feet per second.

$$157.2/60 \times 1/7.5 = 0.349 \text{ cfs}$$

The head causing the flow of 157.2 gpm may be found with the use of $Q = CA_0\sqrt{2gh}$. And by rearrangement and solving for h

$$h = \left[\frac{0.349}{0.63 \times 0.785 \times (2/12)^2 \times 8.03}\right]^2 = 10 \text{ ft of kerosene}$$

Convert the 30 ft of kerosene to its equivalent pressure in psi.

$$(30 \times 62.4 \times 0.78)/144 = 10.1 \text{ psi}$$

The 10-ft head of kerosene is equivalent to 10.1/3 = 3.36 psi. The 4 in. of mercury below atmosphere is equivalent to

$$4 \times 1/29.9 \times 14.7 = 1.97 \text{ psi}$$

The vacuum effect of the 4 in. of mercury reduces the head over the orifice. Thus, 10.1 − 1.97 = 8.13 psi, the pressure upstream of the orifice. Therefore, the pressure downstream of the orifice is the pressure upstream minus the differential pressure across the orifice.

$$8.13 - 3.36 = \mathbf{4.77\ psi}$$

Q5-11. A 3-in. thin plate or sharp-edged orifice is installed in a 16-in. pipe carrying a mineral oil having a specific gravity of 0.90. A vertical manometer with mercury and oil shows 4 in. difference in mercury levels. Calculate the flow of oil as barrels per hour and the power lost due to the orifice. Assume that 95 per cent of differential is lost. One barrel is equal to 42 gal.

ANSWER. For the lack of more complete information as to whether or not orifice flanges or pipe taps are used we can assume a coefficient of discharge equal to 0.61. Then

$$v_0 = C\sqrt{2g\,\Delta h} = 0.61\sqrt{64.4 \times 4/12 \times \frac{13.6 - 0.90}{0.90}} = 10.6 \text{ fps}$$

Specific gravity of mercury compared to that of water is 13.6. The latter portion of the equation corrects for the effect of the 4 in. of oil in the upstream leg of the manometer resting over the mercury in that leg. We also convert the 4 in. of mercury to feet of fluid flowing which in our case is the mineral oil. Then

$$10.6 \times 3{,}600 \times 3^2/144 \times 0.785 \times 1/42 \times 7.48 = \mathbf{334\ bbl\ per\ hr}$$

The power lost is found by the well-known formula lb per sec × ft/550.

$$\frac{334/3{,}600 \times 42 \times 8.33 \times 0.90 \;\frac{4}{12}\;\frac{13.6 - 0.90}{0.90}}{550} = \mathbf{0.248\ hp}$$

This assumes a total loss with no recovery. However, with 5 per cent recovery the more accurate answer would be

$$0.248 \times 0.95 = \mathbf{0.235\ hp}$$

Note. If air or any other light gas were flowing, the effect of head over the mercury (or water) in the upstream leg of the manometer would be negligible.

Q5-12. Water is flowing through an orifice meter in a pipeline 8 in. in diameter. The thin plate of the orifice has an opening 6 in. in diameter. Pipe taps just upstream and downstream from the orifice plate lead to a water–carbon tetrachloride (sp gr 1.6) gauge, and this shows a difference of 3 ft. How much water is flowing through the meter?

ANSWER. For lack of information on the orifice coefficient, 0.61 may be used. The effect of the water head over the carbon tetrachloride upstream leg of the manometer is to be corrected. Correct flow due to velocity of approach. Another useful formula to determine flow taking into account the velocity of approach is as follows:

$$Q = C \times 0.785 \times d^2 \sqrt{\frac{1}{1 - (d/D)^4}}\,\sqrt{2g\,\Delta h}$$

$$= 0.61 \times 0.785 \times (6/12)^2 \times \sqrt{\frac{1}{1 - (6/8)^4}} \times \sqrt{64.4 \times 3 \times \frac{1.6 - 1}{1}}$$

$$= \mathbf{1.56\ cfs}$$

Q5-13. Water flows through a venturi meter with throat diameter d equal to 3 in. and is installed in a 6-in. pipeline. Mercury in the manometer stands at 15 in. differential. The connecting tube is filled with water. Find the rate of discharge in gallons per minute. Do not correct for approach.

ANSWER. Coefficient of discharge may be taken as 0.98.

$$Q = CA_t \times \sqrt{2g\,\Delta h} = 0.98 \times 0.785 \times (3/12)^2 \sqrt{64.4 \times 15/12\, \frac{13.6 - 1}{1}}$$

$$= 1.58 \text{ cfs}$$

The flow in gpm is 1.58 × 7.48 × 60 = **710 gpm.**

Q5-14. At a water supply plant the raw water inflow is measured by means of a venturi meter located in a pipe 20 in. in diameter and the discharge coefficient is 0.95. If the difference in pressure between the upstream end and the throat is 5 psi, at what rate of flow in gpm will water flow through the meter, if throat diameter is 8 in.?

ANSWER. Note that in all the hydraulics problems where pipe diameters are mentioned the nominal size given in the wording of the problem may be taken as the inside diameter or working diameter. There is no need in most cases to concern oneself with the actual inside diameter, taking into account pipe-wall thickness. In the solution correct for effect of velocity of approach for accurate results.

$$Q = 0.95 \times 0.785 \times (8/12)^2 \sqrt{\frac{1}{1 - (8/20)^4}} \sqrt{64.4 \times 5 \times 2.31} = 9.19 \text{ cfs}$$

Flow in gpm is found in the usual manner.

$$\text{gpm} = 9.19 \times 7.48 \times 60 = \mathbf{4{,}130}$$

If the effect of velocity of approach were neglected, the flow would be 4,050 gpm, or an error of 1.93 per cent. This would be appreciable in practice under most conditions.

Q5-15. Discuss the selection of the differential producer for measuring flow in pipelines.

ANSWER. In the selection of suitable flowmeters price is all too frequently the sole determination, whereas a closer examination of the proposals may demonstrate that such a decision would be definitely improper from an economic standpoint. This does not mean that the lowest-priced equipment is never warranted, but simply that experience has repeatedly emphasized the importance of making an evaluation of design and specifications. These include the reputation of the meter, its anticipated life, materials, workmanship and finish, probable cost of maintenance, character of information furnished such as size of dials and charts, uniformity of graduations, whether direct reading in units of flow or indirect in nominal values of differentials, and most important, the friction loss induced.

For lines flowing by gravity under low heads, the friction loss of the proposed meter should be carefully studied to determine whether it is objectionable in its effect on the hydraulic gradient. In a pump-discharge line, friction loss alone may not be of significance, but when considered together with pipeline size and hours of operation, the combination of these three factors may be of great monetary importance in a 12-month period.

In general, flowmeters operate on the inferential principle. That is, the flow rate is "inferred" from the difference in pressure induced by a "restriction" placed in the pipeline and transmitted to the recording instrument (secondary instrument), through which no flow occurs. It is readily appreciated that elbows, tees, partly closed valves, or other fittings add to pipeline friction loss and must be reckoned with in the design of the system. So it is with meters. Whether water or any other liquid is first lifted to a reservoir or tank or pumped directly through the pipeline, the additional load is reflected back upon the pumps, requiring provision for more horsepower than would otherwise be required; and power always costs money.

The thin-plate orifice or primary instrument is always recommended wherever possible because of its high accuracy, low cost, and extreme flexibility. The following check list in Table 5-1 includes the commonly used primary devices and illustrates why various types are necessary. Special conditions dictate changes to final choices.

The pitot tube is another device used in industry but not to the extent that the others listed above are used. Its accuracy is good for one-point measurement. For measurement of total flow, a *velocity traverse* is necessary for good accuracy. It is not suitable for liquids carrying solids, but has excellent pressure recovery characteristics. It is low in cost for large sizes as well as small. It gives fair results for liquids carrying traces of gases or for gases carrying traces of liquids. It is impossible to change capacity.

TABLE 5-1. CHOICE OF DIFFERENTIAL PRODUCER*

Application consideration	Concentric orifice	Flow nozzle	Venturi tube
Accuracy	Excellent	Good	Good
Suitability for liquids containing solids in suspension	Poor	Poor	Excellent
Pressure recovery	Poor	Poor	Good
Suitability for viscous flows	Fair	Good	Good
Low cost in large sizes	Good	Fair	Poor
Low cost in small sizes	Excellent	Fair	Poor
Suitability for liquids containing traces of vapors	Excellent if flow upward	Excellent if flow upward	Excellent
Suitability for gases containing traces of condensate	Excellent if flow downward	Excellent if flow downward	Excellent
Ease of changing capacity	Excellent	Fair	Poor

* From The Foxboro Company.

Q5-16. Flow in a city water system is being tested by allowing the flow of water from a hydrant with an outlet 4 in. in diameter. The center of the outlet is 2 ft above the ground level and the issuing jet strikes the ground at a horizontal distance of 9 ft from the outlet. How many gpm are flowing from the outlet?

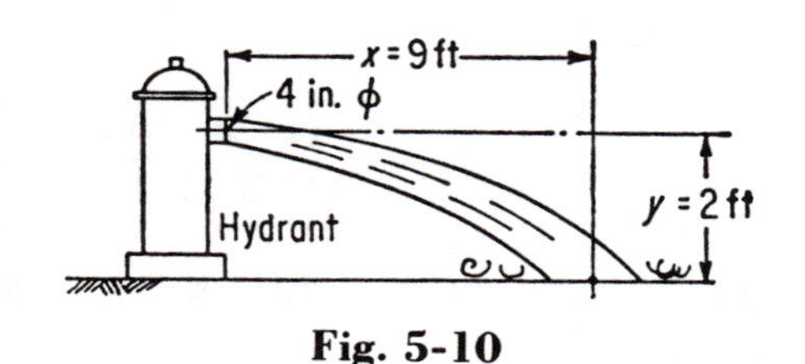

Fig. 5-10

ANSWER. Refer to Fig. 5-10. The issuing stream fills the entire cross section of the outlet.

$$v = \sqrt{\frac{x^2 g}{2y}} \quad \text{fps}$$

Then, by substituting the proper values of x and y from Fig. 5-10,

$$v = \sqrt{\frac{9^2 \times 32.2}{2 \times 2}} = 25.6 \text{ fps flow velocity}$$

From the law of continuity

$$Q = Av \quad \text{cfs}$$
$$= 0.785 \times 4^2/12^2 \times 25.6 = 2.24 \text{ cfs}$$

This is the volume of water flowing per second through the hydrant outlet. This may also be expressed in gpm as follows:

$$\text{gpm} = \text{cfs} \times 60 \times \text{gal per cu ft}$$
$$= 2.24 \times 60 \times 7.48 = \mathbf{1{,}004 \text{ gpm}}$$

Q5-17. In a filling operation a dredge is pumping sand, water, and mud through a steel pipe 12 in. in diameter. The discharge pipe is supported horizontally 4 ft above the ground. If the center of the jet from the pipe strikes the ground at a distance 4 ft away from the center end of the pipe horizontally, how many cubic yards of mud, water, and sand would be delivered by the pipe in 2 hr?

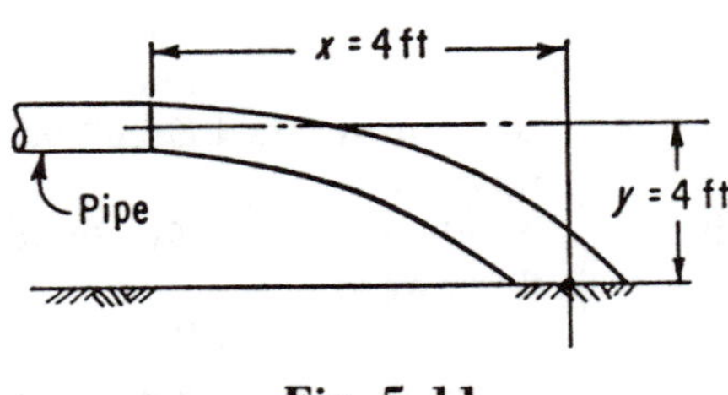

Fig. 5-11

ANSWER. Refer to Fig. 5-11. Here the mixture of sand, mud, and water may be considered as having the general characteristics of the water which is predominant. Now as in the previous problem

$$v = \sqrt{\frac{4^2 \times 32.2}{2 \times 4}} = 8.02 \text{ fps at the outlet}$$

Then $Q = Av = 0.785 \times {}^{12}\!/_{12} \times 8.02 = 6.3$ cfs. By dimensional analysis

$$\frac{\text{cu ft}}{\text{sec}}\,\frac{\text{sec}}{\text{hr}} \times \text{hr} \times \frac{\text{cu yd}}{\text{cu ft}} = \text{cu yd}$$

or $6.3 \times 3{,}600 \times 2 \times {}^1\!/_{27}$ = **1,680 cu yd.**

Problems of this type follow the reasoning outlined and should be readily recognizable in the examination. In the actual physical piping arrangement the jet must project horizontally and the exit pipe must be level. Another important consideration is to provide a straight run of pipe after any obstruction or elbow in the system upstream of the aperture.

Q5-18. A jet issuing under a head of 3 ft from a sharp-edged orifice of circular cross section 0.1 ft diameter in a vertical plane goes 5.9 ft horizontally to a point 3 ft below the center of the orifice. In a test 2,460 lb of water is collected from the jet in 10 min. Determine the diameter of the vena contracta of the jet.

ANSWER. Similarly as before

$$v = \sqrt{\frac{5.9^2 \times 32.2}{2 \times 3}} = 13.7 \text{ fps}$$

$$\begin{aligned} Q &= 2.460/10 \times \tfrac{1}{60} \times 1/62.4 = 0.0657 \text{ cfs} \\ &= A_0 v = 0.785 \times d^2 \times 13.7 \end{aligned}$$

From which d is found to be **0.0781 ft.** It has been assumed that v is velocity at the vena contracta. Actually velocity at the vena contracta is greater than at the orifice.

Q5-19. Are there any other convenient formulas for flow through pipes held at an angle? Discuss.

ANSWER. Previous problems were solved with pipe held horizontal. The following formulas* have been determined from actual

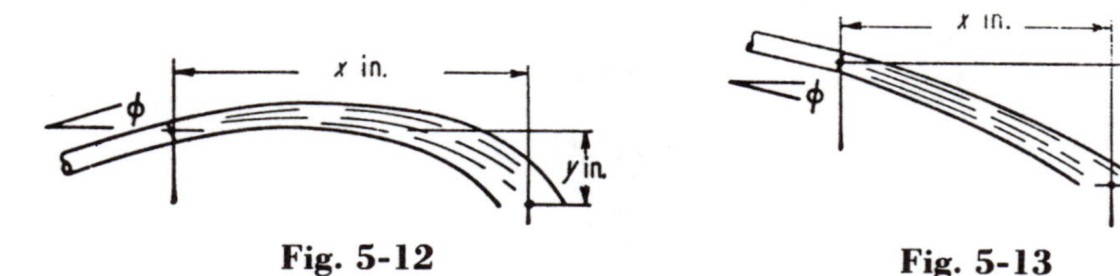

Fig. 5-12 Fig. 5-13

test and may be applied for flow through pipes held at angles other than horizontal.

For pipe running full and held upward at an angle (Fig. 5-12)

$$\text{gpm} = \frac{2.833d^2x/\cos\phi}{\sqrt{y + (x \tan\phi)}}$$

where d is diameter of pipe in inches, x is horizontal distance in inches, y is vertical distance in inches, ϕ is angle in degrees.

For pipe running full and held downward at an angle (Fig. 5-13)

$$\text{gpm} = \frac{2.833d^2x/\cos\phi}{\sqrt{y - (x \tan\phi)}}$$

* These are being published with the permission of L. D. Lamont, P.E.

For flow from a trough Lamont proposes the use of the following

$$\text{gpm} = \frac{3.607Ax/\cos\phi}{\sqrt{y - (x\tan\phi)}}$$

For sketch of system, see Fig. 5-14.

Q5-20. Discuss Reynolds number and its application to problems met in fluid flow.

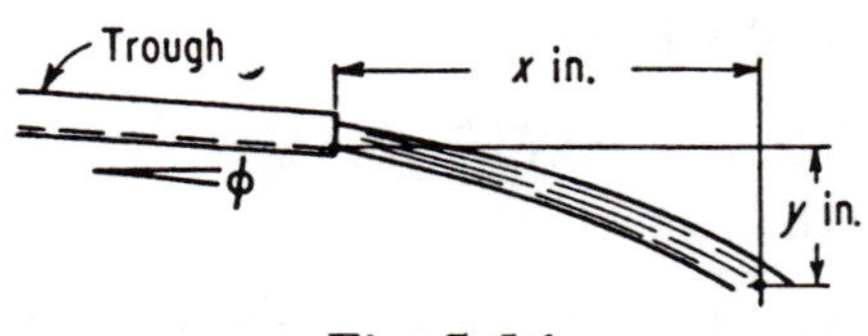

Fig. 5-14

ANSWER. Reynolds number is a nondimensional one which combines the physical quantities which describe the flow in either streamline or turbulent flow. The friction loss in a pipeline is also dependent upon this dimensionless factor. The Reynolds number may take on any one of the following forms:

$$\begin{aligned} \text{Re} &= \frac{Dv\rho}{\mu} \\ &= \frac{7{,}728dvs}{z} \\ &= \frac{124dv\rho}{z} \\ &= \frac{6.31W}{dz} \\ &= \frac{DG}{\mu} \end{aligned}$$

All the above equations are expressions of Reynolds number,
where D = pipe diameter, ft
v = average fluid velocity, fps
ρ = fluid density, lb per cu ft
μ = unit of absolute viscosity, lb per ft-sec
d = pipe diameter, in.
s = specific gravity at 60°/60°
z = viscosity, centipoise (also absolute viscosity)
W = flow rate, lb/hr

All values of the number determined by the above equations and below 2,100 may be considered to be in the viscous range; above 2,100, in the turbulent range.

For every pressure-drop calculation in pipeline flow of fluids the Reynolds number should be determined. From this the *friction factor* is found and then applied to the pressure-drop formula. Friction factor for streamline flow may be found directly by 16/Re (Fanning friction factor). Should Reynolds number appear in the area of transition, it would be safe to consider the flow turbulent and apply the proper curves for pipe roughness.

Q5-21. In addition to 37 ft of straight pipe 4 in. in diameter making up a run, three short radius elbows, two wide-open gate valves, and one wide-open globe valve are also included. What is the total run of pipe expressed in feet that will be used to calculate head loss in the Darcy (or Fanning) formula for friction drop?

ANSWER. Using Crane Chart (Fig. 5-15),

Pipe @ 37 ft..................	37
3 elbows @ 10 ft..............	30
2 gate valves @ 2.5 ft..........	5
1 globe valve @ 120 ft.........	120
Total (L).....................	**192 ft**

Note. When the straight pipe length is greater than 1,000 times nominal pipe diameter, the effect of restrictions such as valves and fittings may be neglected.

TABLE 5-2. FRICTION FACTORS FOR SMOOTH, CLEAN, COATED CAST-IRON PIPE

(Add 5 per cent for riveted steel and at least 50 per cent for old rusted pipe. Copper and brass and lead pipes have lower values.)

Pipe diameter, in.	Velocity in pipe, fps					
	1.0	2.0	3.0	4.0	5.0	10.0
1	0.035	0.032	0.030	0.029	0.027	0.024
2	0.033	0.030	0.028	0.027	0.026	0.024
4	0.031	0.028	0.026	0.025	0.025	0.023
6	0.029	0.026	0.025	0.024	0.024	0.022
12	0.025	0.023	0.022	0.022	0.021	0.020
16	0.023	0.022	0.021	0.021	0.020	0.019

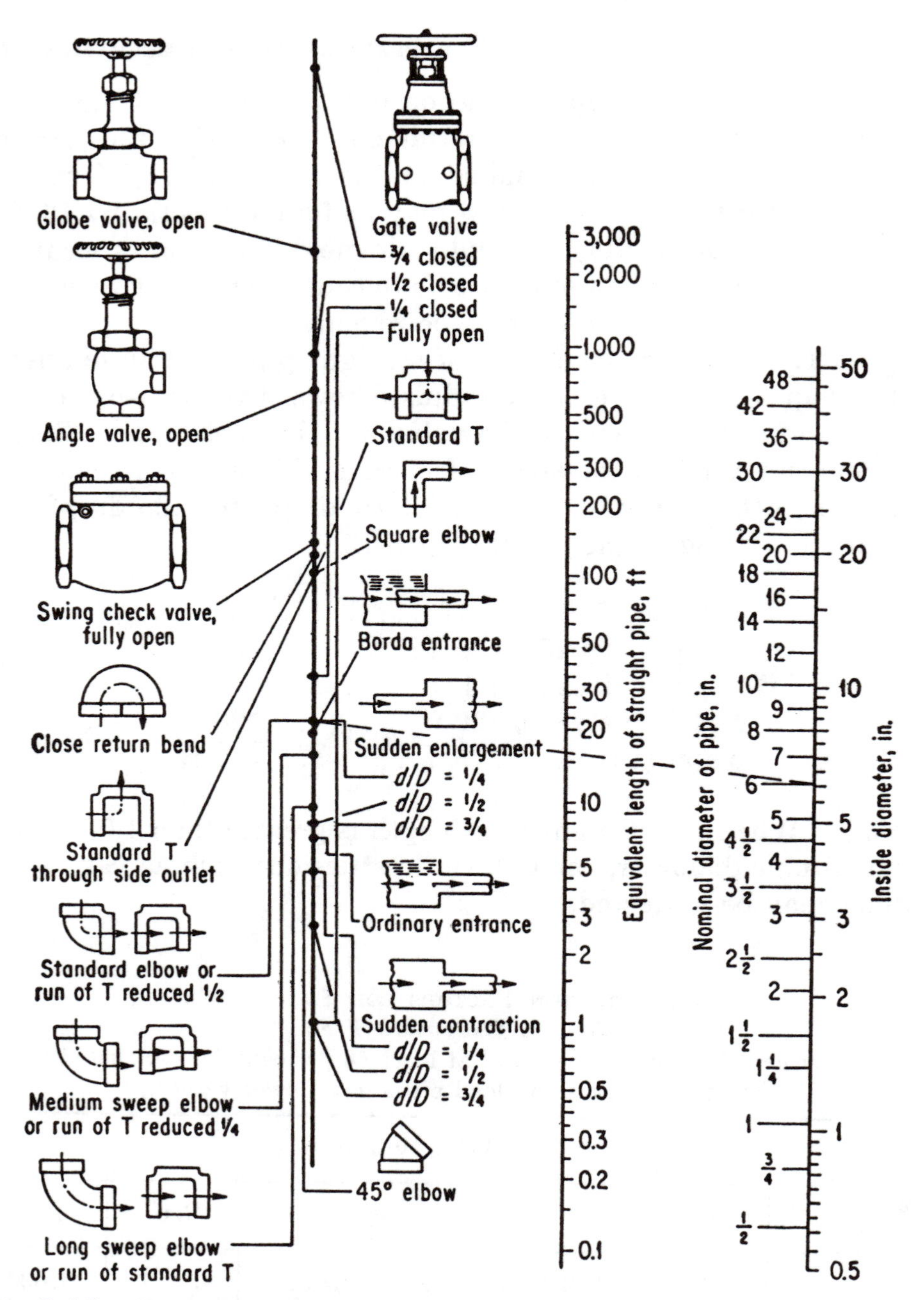

Fig. 5-15. Resistance of valves and fittings to flow fluids. A simple way to account for the resistance offered to flow by valves and fittings is to add to the length of pipe in the line a length which will give a pressure drop equal to that which occurs in the valves and fittings in the line. Example: The dotted line shows that the resistance of a 6-in standard elbow is equivalent to approximately 16 ft of 6-in standard steel pipe. Note: For sudden enlargements or sudden contractions, use the smaller diameter on the nominal pipe-size scale. Liquids and gases. ***(This chart is copyrighted by Crane Co., Chicago, and is reproduced here by permission of the copyright owner.)***

Q5-22. For the same size, length, and condition of pipe in the above question, what would be the head loss if the pipe size were increased to 8 in.? If it were reduced to 4 in.?

ANSWER. Of course the above question can be found by going through the same calculations as before, but normally in the design office it would be acceptable to take a short cut. It is well known that under same flow conditions the friction head loss (or pressure drop) increases as the size change to the 4.8 power. Thus, in our problem the original friction head may be multiplied by the factor

$$\left(\frac{\text{Smaller pipe size}}{\text{Larger pipe size}}\right)^{4.8} = \text{multiplier} = \left(\frac{6}{8}\right)^{4.8} = 0.25$$

Friction-head loss for 8-in. pipe $= 48 \times 0.25 =$ **12 ft**
Friction-head loss for 4-in. pipe $= 48 \times (6/4)^{4.8} = 48 \times 7 =$ **336 ft**

For the convenience of the engineer in the examination room the following chart (Table 5-3) has been developed by the author to show quickly multipliers for pipe sizes 1 to 20 in. in diameter. Nominal pipe sizes are used. Accuracy is within 5 per cent.

TABLE 5-3. FRICTION-HEAD CONVERSION

x (old), in.	Y (new), in.												
	1	1½	2	3	4	6	8	10	12	14	16	18	20
1	1.0	0.143	0.036	0.0051									
1½	7.0	1.0	0.25	0.036	0.0091	0.0013							
2	27.8	3.98	1.0	0.146	0.036	0.0051	0.0013						
3	195	27.8	7.0	1.0	0.25	0.036	0.0097						
4	770	112	27.8	3.94	1.0	0.146	0.036						
6	....	770	195	27.8	7.0	1.0	0.25	0.086	0.036				
8	....	3100	770	112	27.8	3.94	1.0	0.343	0.146	0.067			
10	....	.....	.....	320	82	11.8	2.92	1.0	0.416	0.20	0.105		
12	....	.....	.....	......	......	27.8	7.0	2.4	1.0	0.483	0.25	0.146	
14	....	.....	.....	......	......	58	14.7	5.02	2.08	1.0	0.526	0.30	0.181
16	....	.....	.....	......	......	112	27.8	9.6	3.94	1.88	1.0	0.57	0.341
18	....	.....	.....	......	......	195	492	16.8	7.0	3.34	1.74	1.0	0.601
20	....	.....	.....	......	......	......	82	27.8	11.8	5.6	2.9	1.65	1.0

Q5-23. What is the Fanning equation for head loss? How does it tie in with Darcy equation? Do both give the same results?

ANSWER. The Fanning equation gives the same results as the Darcy, but must be used in conjunction with the Reynolds number correlation.

$$h_f = \frac{2 \times f_2 \times L \times v^2}{g \times d}$$

All terms except f_2 are the same as for the Darcy equation. By examination and comparison between the Darcy and Fanning equations it may be seen that

$$f_1 = 4f_2$$

For streamline flow we saw that $f_2 = 16/\text{Re}$, and from the above we see that f_1 (Darcy) $= 64/\text{Re}$. Here a word of caution would not be amiss. *Be sure that the proper friction factor is used with its friction loss formula.*

Head loss may be converted to pressure drop by use of the relation

$$\Delta P = \frac{2f_2L\rho v^2}{144gd} \qquad \text{psi}$$

This conversion is so because $\Delta P/\rho \times 144 = h_f$, where ρ is density.

Q5-24. Determine in feet of oil the loss in static head when oil is pumped through a wrought-iron pipe 1 in. in diameter inside and 1,000 ft long. Average velocity of oil is 1 fps at 60°F. Absolute viscosity is 4 centipoise and specific gravity of oil is 0.85 at the flowing temperature.

ANSWER. First let us determine the Reynolds number. For friction-loss calculations other than water it is best to use Fanning's formula and friction factor.

$$\text{Re} = \frac{\frac{1}{12} \times 1 \times 62.4 \times 0.85}{0.000672 \times 4} = 1{,}640 \qquad \text{(streamline flow)}$$

The Fanning friction factor f_2 may be calculated or determined directly. By calculation we obtain $16/1{,}640 = 0.0098$. Then

$$h_f = \frac{2 \times 0.0098 \times 1{,}000 \times 1^2}{32.2 \times \frac{1}{12}} = \mathbf{7.25\ ft\ of\ oil}$$

Q5-25. Crude oil is flowing through a pipeline 12 in. in diameter and 20 miles long and at an average velocity of 2 fps. The viscosity of the oil may be taken as 2×10^{-3} lb-sec/sq ft and specific gravity of 0.92 with reference to water at 60°F. How much pressure would be required at a pumping station at one end of the line if the other end were 100 ft higher?

ANSWER. Since length of pipeline is greater than 1,000 times pipe diameter, the effect of fittings, etc., may be ignored. Pump-discharge pressure would be made up of friction head plus the static discharge heads. Now first determine Reynolds number.

$$\text{Re} = \frac{{}^{12}\!/_{12} \times 2 \times 62.4 \times 0.92}{0.002 \times 3.2174 \times 10} = 1{,}790 \qquad \text{(streamline flow)}$$

However, we can simply calculate f_2 as $16/1{,}790 = 0.00895$. Then

$$h_f = \frac{2 \times 0.00895 \times 20 \times 5{,}280 \times 2}{{}^{12}\!/_{12} \times 32.2} = 117 \text{ ft of oil}$$

The total dynamic head (W_p in Bernoulli's equation) is

$$117 + 100 = 217 \text{ ft}$$

The pressure required to support a head of 0.92 specific gravity oil 335 ft high is

$$\frac{217}{2.31} \times 0.92 = \mathbf{93.7\ psi}$$

It is assumed that the pump is operating "flooded" with no suction pressure.

Q5-26. What are the four rules for approximating head loss in pipe?

ANSWER. The following rules may be used to advantage in the design office as well as in the examination room:

Rule 1. At constant head, capacity is proportional to $d.^{2.5}$

Example. A 4-in. pipe discharges 100 gpm. How much would a 2-in. pipe discharge under same conditions?

$$4^{2.5} = 32 \qquad 2^{2.5} = 5.66$$

Hence, capacity = (100 × 5.66)/(32) = **17.6 gpm.**

Rule 2. At constant capacity, head is proportional to $1/d^5$.

Example. Capacity is 500 gpm and pipe size is 3 in. in diameter. What is the friction head?

Let us assume a 4-in. pipe flowing 240 gpm. The friction will be greater for 3-in. pipe. Hence

$$h = 240(4^5/3^5) = \mathbf{1{,}010\ ft\ per\ 1{,}000\ ft\ of\ pipe}$$

Rule 3. At constant diameter, head is proportional to gpm.[2]

Example. One hundred gpm produces 50-ft friction in a pipe. How much will 200 gpm produce?

$$h = 50(200^2/100^2) = \mathbf{200\ ft}$$

Rule 4. At constant diameter, capacity is proportional to $\sqrt{h}$.

Example. Diameter is 12 in. Friction is 200 ft per 1,000 ft. What is the capacity?

From friction charts the nearest figure is 84 ft, giving 5,000 gpm. Capacity will be greater. Hence

$$\text{gpm} = 5{,}000\,\frac{\sqrt{200}}{\sqrt{84}} = \mathbf{7{,}730\ gpm}$$

All the above rules assume turbulent flow conditions with a Reynolds number above 2,100.

Q5-27. Discuss pressure drop in the flow of compressible fluids. Focus attention on the subject by providing a sample calculation.

ANSWER. Changes in fluid density and kinetic energy are wholly negligible in pressure-drop calculations for liquids flowing in conduits of uniform cross section. In the flow of compressible fluids, on the other hand, these two factors may affect the pressure drop appreciably, particularly for high-velocity flow at low pressures. Complete calculation of pressure drop in this latter case involves integration and is rather complicated. For this reason a rapid means of making the calculation accurately should be used.

Lobo, Friend, and Skaperdas* showed that the pressure drop in the isothermal flow of a compressible fluid may be readily obtained by means of the equation

$$\frac{p_E}{p_1} = 1 - \frac{1}{2}\left[1 - \left(\frac{p_2}{p_1}\right)^2\right] + \frac{G^2 v_1}{g p_1} \ln\left(\frac{p_1}{p_2}\right) \qquad (1)$$

* W. E. Lobo, Leo Friend, and G. T. Skaperdas, "Pressure Drop in the Flow of Compressible Fluids," Industrial Engineering Chemistry, vol. 34, p. 821, 1942.

where

$$1 - \frac{p_E}{p_1} = 2f \frac{L}{D} \frac{G^2 v_1}{g p_1} \tag{2}$$

A plot of p_E/p_1 vs. p_2/p_1 for various values of G^2v_1/gp_1 was used to solve Eq. (1) for p_2/p_1. To avoid the difficulty of interpolation between the curves on their plot, an alignment chart is presented which gives a direct solution for p_2/p_1 (see Fig. 5-16). Calculations indicate clearly that serious error can result if kinetic energy changes

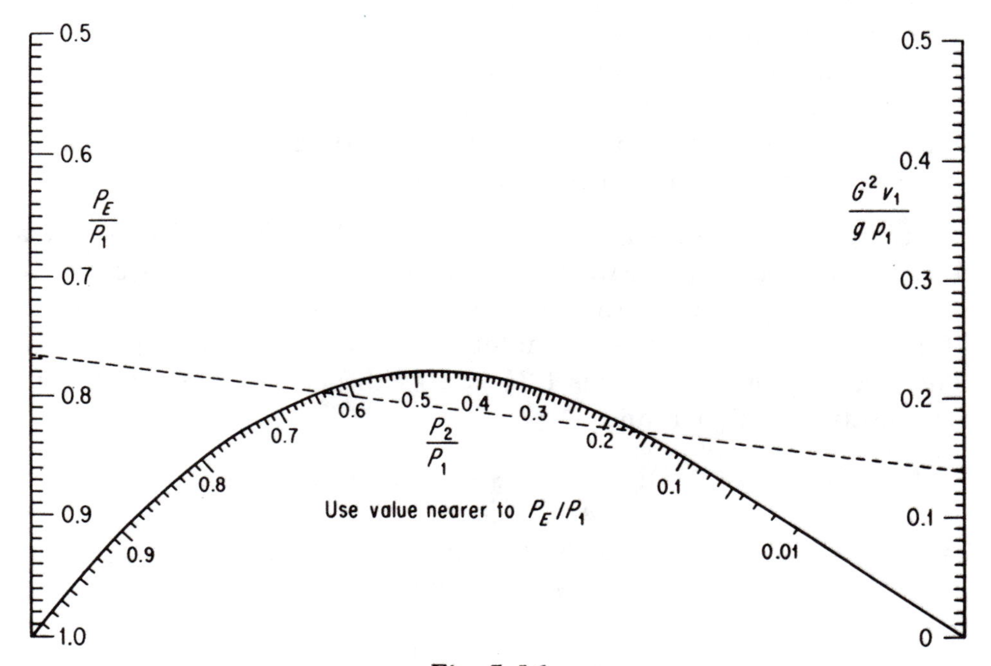

Fig. 5-16

are always neglected. However, as the term G^2v_1/gp_1 decreases, or p_2/p_1 or p_E/p_1 increases, the importance of kinetic energy changes decreases. Thus, for low-velocity flow at high pressures, good results can be obtained with the standard formulas of flow, but for high-velocity flow at low pressures, the approximate formulas are inadequate. The value of G^2v_1/gp_1 can serve as a useful index of the need for considering kinetic energy changes.

In the case of high velocity and/or low pressure, where $G^2v_1/144gp_1$ is appreciably greater than zero in Fig. 5-16, say greater than 0.1, multiply ΔP obtained in the usual manner, neglecting the effect of compressibility, by the factor $1/[1 - (G^2v_1/144gp_1)]$, where p_1 is inlet pressure, psia.

The following nomenclature applies to this study; any consistent set of units may be used.

D = diameter of pipe, ft
f = Fanning equation friction factor
g = acceleration due to gravity, 32.2 ft per sec²
G = mass velocity, lb per sec-ft
L = length of pipe, ft
p_1 = inlet pressure, lb per sq ft
p_2 = outlet pressure, lb per sq ft
p_E = pseudoterminal pressure, lb per sq ft, defined by Eq. (2)
v_1 = specific volume of fluid, ft^{-3}-lb^{-1}

Example. A hydrocarbon vapor at 990°F is flowing through a 4.0-in.-ID pipe at the rate of 15,900 lb per hr. It is required to calculate the pressure drop in 75 ft of pipe if the inlet pressure is 35 psia and the density at the inlet to the pipe is 0.115 lb per cu ft. The viscosity of the vapor is 1.31×10^{-5} lb-per ft-sec at the flowing temperature of 990°F and 35 psia.

$$G = (15{,}900/3{,}600)(4 \times 144/\pi 4^2) = 50.6 \text{ lb per sec per sq ft}$$

$$\frac{DG}{\mu} = \frac{4 \times 50.6}{12 \times 1.31 \times 10^{-5}} = 1{,}290{,}000$$

$$f = 0.0038$$

$$v_1 = \frac{1}{\rho_1} = 8.70 \text{ cu ft per lb}$$

$$p_1 - p_E = \frac{2fG^2Lv_1}{gD} = \frac{2 \times 0.0038 \times 50.6^2 \times 75 \times 8.70 \times 12}{32.2 \times 4}$$

$$= 1180 \text{ lb per ft}^2 = 8.2 \text{ psi}$$

Since $p_1 = 35$ psia, $p_E = 35 - 8.2 = 26.8$ psia, and $P_E/p_1 = 26.8/35 = 0.765$.

$$G^2v_1/gp_1 = 50.6^2 \times 8.70/(32.2 \times 35 \times 144) = 0.137$$

From Fig. 5-17,

$$p_2/p_1 = 0.638 \qquad p_2 = 0.638 \times 35 = 22.3 \text{ psia}$$

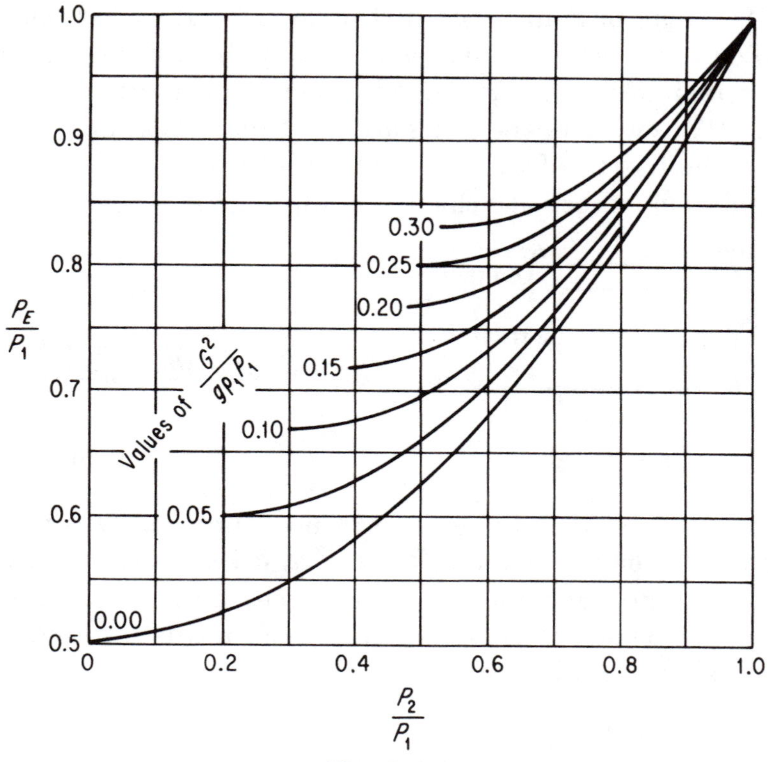

Fig. 5-17

Pressure drop is

$$\Delta P = 35 - 22.3 = \mathbf{12.7\ psi}$$

Let's check our calculation using the nomograph, Fig. 5-16. Since $P_E/p_1 = 0.765$ and $G^2v_1/gp_1 = 0.137$, use these two points, connect them on the nomograph by a line, and find p_2/p_1 to be equal to 0.638, in agreement with the solution given in the example calculation.

ANSWER. A quick visual check of the question obviously indicates no need to correct for velocity of approach. This is a rectangular weir with no end contractions (suppressed weir). Then, using the simplified Francis formula,

$$Q = 3.33bH^{3/2} = 3.33 \times 10 \times 1.44^{3/2} = \mathbf{57.6\ cfs}$$

Q5-28. The flow in a channel is 35 cfs. It is desired to dis-charge 5 cu ft per sec (cfs) over a 90° triangular weir in one channel and the remainder over a standard (no end contractions) rectangular weir. The crests of the weirs are to be set at the same elevation. Calculate the length of the rectangular weir and the head on both weirs, neglecting effect of velocity of approach.

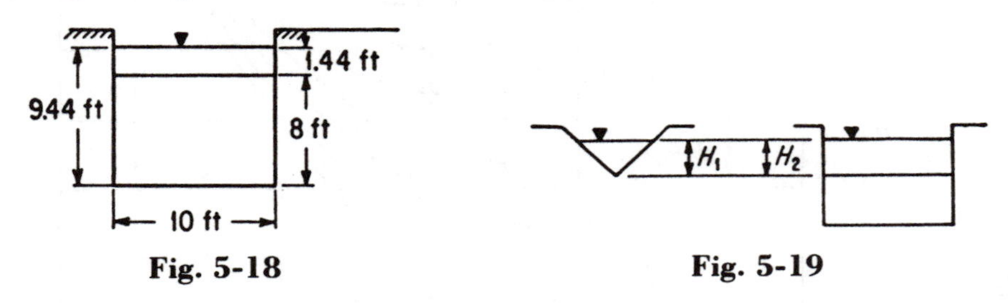

Fig. 5-18 Fig. 5-19

ANSWER. Note V-notch weirs are not appreciably affected by **velocity of approach. Now refer to Fig. 5-19. Let H_1 and H_2 be** the heads on the triangular weir (V-notch) and rectangular weir, respectively. Then $H_1 = H_2$. Since the heads are equal, we calculate value of H_1 by use of

$$H_1 = \left(\frac{Q_1}{2.53}\right)^{1/2.5} = \left(\frac{5}{2.53}\right)^{1/2.5} = \mathbf{1.319\ ft} = H_2$$

Since $H_1 = H_2$,

$$b = \frac{Q_2}{3.33 \times H^{3/2}} = \frac{30}{3.33 \times 1.319^{3/2}} = \mathbf{5.94\ ft}$$

Q5-29. In a certain processing plant it is desired to determine the flow rate of a gas flowing through an irregularly shaped duct in which no measuring devices are installed and in which they cannot very well be used. Gas analysis shows that the flowing gas contains 0.24 per cent CO_2 by volume. It is decided to determine the flow rate by bleeding CO_2 into the gas stream from a small weighed cylinder or bottle. A constant rate is obtained by means of a flowmeter, and after thorough mixing of the gases by passing them through donut sections and bends, the average analysis of the mixture shows that it contains 1.41 per cent CO_2. The loss in weight of the cylinder is 7.94 lb in 5 min, measured by a stopwatch. If the temperature of the gas stream and the mix is 120°F, what is the volume flowing in cubic feet per minute?

ANSWER. The ratio of CO_2 to CO_2-free gas in the original is 0.24/99.76 = 0.0024, and the same ratio after mixing is 1.41/98.59 = 0.0143. The increase in this ratio, 0.0143 − 0.0024 = 0.0119, represents moles CO_2 bled into the stream per mole CO_2-free gas. Using this figure and the data given in the problem, the flow rate is

$$7.95/5 \times 1/44 \times 1/0.0119 \times 100/99.76 \times 379 \times (120 + 460)/(60 + 460)$$

This calculates out to **1,380 cfm.**

Chapter 6

THERMODYNAMICS, HEAT AND POWER

Questions and Answers

Q6-1. If the specific heat for a certain process is given by the equation $c = 0.2 + 0.00005T$ Btu per lb-deg how much heat should be transferred to raise the temperature of 1 lb from 500°R to 2,000°R?

ANSWER.

$$Q = w \int_{500}^{2,000} c\,dT = \int_{500}^{2,000} (0.2 + 0.00005T)\,dT$$

$$= 0.2T + \frac{0.00005T^2}{2}\Bigg]_{500}^{2,000} = \mathbf{394\ Btu\ per\ lb}$$

Q6-2. A water-jacketed air compressor receives 300 cfm of air at 14 psia and 70°F and discharges the air at 70 psia and 280°F. Cooling water enters the jacket at 60°F and leaves at 70°F, or a 10°F rise for a flow rate of 24 lb per min. Determine the horsepower required for the driving motor for this reciprocating compressor. In your calculations assume reasonable efficiencies.

ANSWER. Conditions given in problem:

$$p_1 = 14 \text{ psia}$$
$$T_1 = 530°\text{R}$$
$$p_2 = 70 \text{ psia}$$
$$T_2 = 740°\text{R}$$

$$\text{Density of entering air} = (29/379) \times 14.7/14.0 \times 520/530 = 0.079 \text{ lb per cu ft}$$

$$\text{Air flow rate in lb per min} = 0.079 \times 300 = 23.7$$

Compression heat to be removed = 23.7 × 0.24 × (740 − 530)
= 1,195 Btu per min
Heat absorbed by cooling water = 24 × 1 × (70 − 60)
= 240 Btu per min

Total power = 1,195 + 240 = 1,435 Btu per min, and horsepower required is given by

1,435/(2,545/60) = **33.83 hp**

Note. It appears that not all the friction losses in the system are absorbed by the cooling water. Technical literature indicates that bearing losses and windage can account for as much as 5 per cent of the total power, so that actual horsepower requirements are greater than those calculated above.

Q6-3. Estimate the steam flow rate required to drive a boiler feedpump which requires 100 bhp at 5,000 rpm. The turbine is a velocity-compounded impulse turbine with two rows of moving blades. Steam is available at 500 psi and a total steam temperature of 560°F. The turbine back pressure is 60 psi. State required efficiencies assumed.

ANSWER. Make the following assumptions:

Turbine mechanical efficiency = 95 per cent
Stage efficiency = 80 per cent
Throttle conditions:
Enthalpy = 1,273.4 Btu per lb
Ideal back pressure conditions = 1,095 Btu per lb. isentropic expansion
Ideal heat drop = 1,273.4 − 1,095 = 178 Btu per lb
Actual heat drop = 0.80 × 178 = 142.4 Btu per lb

This is the energy delivered to the turbine shaft.

Power required at the turbine shaft
= 100 × 2,545/0.95 = 267,894 Btuh

Finally,

Steam required = 267,894/142.4 = **1,881 lb per hr**

Q6-4. If 4,670 cfm of air saturated at 95°F enters a four-stage adiabatic compressor, having a compression ratio of 2.33 to 1, at atmospheric pressure, how much heat must be removed in the first-stage intercooler?

ANSWER. Inlet air = 4,670 cfm and, using steam tables,

Saturation partial pressure of water vapor at 95°F = 0.8153 psia
Saturation specific volume of water vapor at 95°F
= 404.3 cu ft per lb

The air and the water both occupy the same volume at their respective partial pressures. The pounds of water entering with the air per hour =

$$4{,}670 \times 60/404.3 = 692 \text{ lb per hr}$$

First Stage. After the 2.33 to 1 compression ratio, $P_2 = 14.7 \times 2.33 = 34.2$ psi. Then, using $\gamma = 1.4$,

$$\left(\frac{T_1}{T_2}\right)_{abs} = \left(\frac{p_2}{p_1}\right)^{(\gamma - 1/\gamma)}$$

$(T_{2abs})/(460 + 95) = 2.33^{(1.4-1/1.4)}$ and

$$T_{2abs} = 705°\text{R or } 245°\text{R}$$

Intercooler.

Final gas volume = 4,670 × 60 × 14.7/34.2 = 120,000 cu ft per hr
Water remaining in air = 120,000/404.3 = 297 lb per hr
Condensation in intercooler = 692 − 297 = 395 lb per hr
Specific volume of atmospheric air
= (359/29)(555/492)(14.7/14.7 − 0.8153)
= 14.8 cu ft per lb
Air in inlet gas = 4,670 × 60/14.8 = 18,900 lb per hr

Heat load. (245°F to 95°F)
Sensible heat:

$$Q_{air} = 18{,}900 \times 0.25(245 - 95) = 708{,}000 \text{ Btuh}$$
$$Q_{water} = 692 \times 0.45(245 - 95) = 46{,}700 \text{ Btuh}$$

Latent heat:

$$Q_{water} = \quad 395 \times 1040.1 \quad = \quad 411,000 \text{ Btuh}$$

Total **1,165,700 Btuh**

If condensation had not been accounted for, an error of 33 per cent would have resulted. Note that over half of the water condenses in the first-stage intercooler.

Q6-5. A cylinder of nitrogen has been returned to the filling plant for recharging. It had originally contained the gas in the filled condition at 80°F and a pressure of 2,000 psig. When weighed for water content, it was found to hold 96 lb of water at 70°F. How many cubic feet of "free" gas will the cylinder discharge as its pressure returns to atmospheric. Compressibility factor for nitrogen at 2,000 psig and 80°F may be taken as 1.125.

ANSWER. First find number of moles contained in the nitrogen bottle. But in order to do this the volume of the bottle must be calculated. Density of water at 70°F is 62.22 lb per cu ft. Then the volume of the bottle is simply found by $V = 96/62.22 = 1.54$ cu ft. Proceeding,

$$N = \frac{pV}{ZKT} = \frac{(2{,}000 + 14.7) \times 1.54}{1.125 \times 10.71 \times (460 + 80)}$$

$$= 0.476 \text{ lb mole of nitrogen}$$

This quantity of gas will occupy the following volume at 70°F and 14.7 psia (standard for the compressed gas industry)

$$\left(0.476 \times 379 \frac{460 + 70}{460 + 60}\right) - 1.54 = 183.83 - 1.54 = \mathbf{182.29 \text{ cu ft}}$$

Note that we have subtracted the volume of the bottle because this is the amount remaining within the bottle itself after internal pressure has dropped from 2,000 psig to atmospheric 14.7 psia. There is no longer a differential of pressure to drive the remainder out.

Q6-6. Calculate the volume occupied by 30 lb of chlorine at a pressure of 743 mm of mercury (Hg) and a temperature of 70°F. Molecular weight of chlorine may be taken as 71.

ANSWER. One lb mole of chlorine will occupy 379 cu ft volume at standard conditions (60°F and 14.7 psia). This pressure of 14.7

psia is also equivalent to 760 mm Hg. We can say, in effect, that the molal volume (379 cu ft) will vary directly as the ratio of the absolute temperatures and indirectly as the ratio of the absolute pressures. Thus,

$$379 \times \frac{30}{71} \times \frac{760}{743} \times \frac{460 + 70}{460 + 60} = \mathbf{166.2\ cu\ ft}$$

Whenever there is a decrease in pressure compared to standard conditions, the pressure ratio is greater than unity; for an increase, the ratio becomes less than unity. As for change in temperature, its effect on volume is to increase it with increase in temperature or decrease volume at lowered temperatures.

Q6-7. A 1,000-gal tank filled with acetylene at 15 psia is supercharged with 4.3 lb of the gas. What would be the final pressure? Assume no change in temperature (60°F) takes place during the charging period.

ANSWER. The 1,000-gal tank has the equivalent volume of 1,000/7.48 = 133.6 cu ft. The weight of gas originally in the tank is given by

$$(133.6/379)\ 26 = 9.5 \text{ lb}$$

Thus, the total weight of the gas under the new pressured conditions is 9.5 + 4.3 = 13.8 lb. We then proceed to say, "What happens to the original 15 psia when we add the additional weight of gas?"

$$15 \times \frac{379}{133.6} \times \frac{13.8}{26} = 15 \times 2.83 \times 0.53 = \mathbf{22.5\ psia}$$

The above equation is to say, if one mole is involved and compressed into a smaller volume what would be the final pressure?

Q6-8. A mixture of gases has the following composition by volume: oxygen, 6.3 per cent; sulfur dioxide, 14.6 per cent; nitrogen, 79.1 per cent. Calculate the composition by weight of this mixture. Assume gases will not react chemically.

ANSWER. Basis: 1 lb mole of the mixture. Now, since volume per cent equals mole per cent or mole fraction, and mole fraction multiplied by molecular weight gives weight in pounds, we can set up the following table.

Gas	Mole fraction ×	Molecular weight M =	Pounds
Oxygen	0.063	32	2.02
Sulfur dioxide	0.146	64	9.36
Nitrogen	0.791	28	22.18
Total weight	.....	...	33.56

Since this is the weight of 1 lb mole, it is the average molecular weight. Composition by weight is found simply.

For oxygen 2.02/33.56, or **6 per cent by weight**
For sulfur dioxide 9.36/33.56, or **27.9 per cent by weight**
For nitrogen 22.18/33.56, or **66.1 per cent by weight**

Q6-9. A mixture of gases has the following composition by weight: oxygen, 10.7 per cent; carbon monoxide, 0.9 per cent; nitrogen, 88.4 per cent. Calculate the composition by volume (volumetric analysis) of the mixture.

ANSWER. Use as a basis 1 lb weight of the mixture, setting up the following table:

Gas	Wt. fraction	÷ M =	Lb mole
Oxygen	0.107	32	0.00335
Carbon monoxide	0.009	28	0.000321
Nitrogen	0.884	28	0.0315
Total			0.03517

Now, since mole per cent is equal to volume per cent, all we need to do is to calculate mole per cent:

For oxygen: 0.00335/0.03517, or **9.5 per cent by volume**
For carbon dioxide: 0.000321/0.03517, or **0.9 per cent by volume**
For nitrogen: 0.0315/0.03517, or **89.6 per cent by volume**

This same treatment may be given to any composition of gases so long as no chemical reaction takes place.

Q6-10. Air having a total volume of 490.3 cu ft at 212°F and 1 atm pressure contains:

0.20 mole of oxygen
0.78 mole of nitrogen
0.02 mole of water vapor

(*a*) What is the average molecular weight of the gas?
(*b*) What is the weight per cent of the constituents?
(*c*) What are the partial pressures of the constituents?
(*d*) If the constituents were separated, what volume would they occupy at the same conditions of temperature and pressure?

ANSWER.

(*a*) Assume one lb mole of the mixture. Then set up table as in Q6-8.

Gas	Mole fraction	× M	= Pounds	
Oxygen	0.20	32	6.4	
Nitrogen	0.78	28	21.84	
Water vapor	0.02	18	0.36	
Total			**28.60**	**Average molecular weight**

(*b*) As in Q6-8, the weight per cent is:

For oxygen 6.4/28.6, or **22.8 per cent by weight**
For nitrogen 21.84/28.60, or **76.4 per cent by weight**
For water vapor 0.36/28.60, or **0.8 per cent by weight**

(*c*) The partial pressures are:

For oxygen 14.7 × 0.20 = **2.94 psia**
For nitrogen 14.7 × 0.78 = **11.47 psia**
For water vapor 14.7 × 0.02 = **0.294 psia**

This total equals 14.704 psia. Merely drop the last digit.

(*d*) For oxygen: $379 \times 0.20 \times \frac{212 + 460}{60 + 460} =$ **98 cu ft**

For nitrogen: $379 \times 0.78 \times \dfrac{212 + 460}{60 + 460} =$ **382 cu ft**

For water vapor: $379 \times 0.02 \times \dfrac{212 + 460}{60 + 460} =$ **9.8 cu ft**

Q6-11. A boiler flue-gas analysis shows, after converting to per cent composition by weight, CO_2, 0.1 per cent; O_2, 6.2 per cent; CO, 0.1 per cent; N_2, 74.2 per cent. Find the instantaneous specific heat at constant pressure at 500°F if c_p (specific heat at constant pressure) for CO_2 at 500°F is 0.235, for CO is 0.251, for O_2 is 0.22, for N_2 is 0.256.

ANSWER. Basis: 100 lb of flue gas.

$$\begin{aligned} CO_2 &= (19.5/100)0.235 = 0.0458 \\ CO &= (0.1/100)0.251 = 0.000251 \\ O_2 &= (6.2/100)0.22 = 0.01364 \\ N_2 &= (74.2/100)0.256 = 0.1864 \\ \text{Total} &= \mathbf{0.2461} \end{aligned}$$

Q6-12. Discuss the energy equation, giving the formula and its terms.

ANSWER. In mechanical engineering the forms of energy are many and varied. However, in this work only internal energy, flow, kinetic, and potential energy are of any real significance. The energy equation is set up in accordance with the law of the conservation of energy (energy balance). If, in a device or system, energy is not stored therein, the energy entering it must be equal to the energy leaving. The general energy equation may be shown to take the form

$$\frac{v_1^2}{2gJ} + \frac{y_1}{J} + H_1 + Q = \frac{W}{J} + \frac{v_2^2}{2gJ} + \frac{y_2}{J} + H_2$$

where v_1 = velocity entering system, fps
J = 778 ft-lb per Btu
y_1 = potential distance above a certain datum, ft
H_1 = enthalpy of working substance entering system, Btu
Q = heat transferred into or out of system, Btu
W = flow work, ft-lb
v_2 = velocity leaving system, fps

y_2 = potential distance above a certain datum, ft

H_2 = enthalpy of working substance leaving system, Btu

This is the energy balance for 1 lb of the working substance. Each term has the same unit of energy; in this form, Btu. In most heat engines, the change in potential energy from y_1 to y_2 is so small as to be negligible. Practically, that is to say that there is very little difference in the elevations of inlet and exit openings. In hydraulic turbines, however, the difference in potential energy from the reservoir to the tail race represents a significant figure; so that this potential difference must be retained in the general equation.

Q6-13. A steam turbine receives 3,600 lb of steam per hour at 110 fps velocity and 1525 Btu per lb enthalpy. The steam leaves at 810 fps and 1300 Btu per lb enthalpy. What is the horsepower output?

ANSWER. Let basis be 1 lb steam flow per second.

$$\frac{v_1^2}{2gJ} + \frac{y_1}{J} + H_1 + Q = \frac{W}{J} + \frac{v_2^2}{2gJ} + \frac{y_2}{J} + H_2$$

Assume centerline of nozzle passes through datum horizontally so that $y_1 = y_2$. Also $Q = 0$, because adiabatic process assumed. Then

$$\frac{v_2^2 - v_1^2}{2gJ} + (H_1 - H_2) = \frac{-W}{J}$$

By rearranging for W, we obtain

$$-W = \frac{v_2^2 - v_1^2}{2g} - J(\Delta H) = \frac{810^2 - 110^2}{64.4} + 778(1{,}300 - 1{,}525)$$

$$= -165{,}000 \text{ ft-lb} \qquad \text{or} \qquad W = 165{,}000 \text{ ft-lb/lb}$$

$$\text{hp} = (165{,}000)(1 \text{ lb steam per sec})/550 = \mathbf{300\ hp}$$

Q6-14. A vessel with a capacity of 5 cu ft is filled with air at a pressure of 125 psia when at a temperature of 600°F. It is desirable to lower the pressure to 60 psia. (*a*) What amount of heat will have to be extracted and what will be the final temperature of the gas, assuming that the vessel does not change in size with change in temperature and that the air may be treated as an ideal gas? (*b*) Find the gain in entropy during this process.

ANSWER. As the problem indicates, this is a constant-volume process involving a gas. Refer to Figs. 6-1 and 6-2. However, the

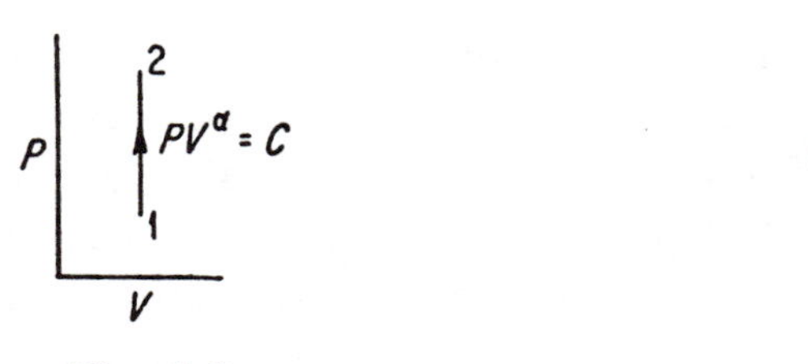

Fig. 6-1 **Fig. 6-2**

numbers must be reversed for pressure is being lowered and the system will suffer a drop in temperature also. Now, since no work is done,

$$Q = \Delta U = w \int_{T_1}^{T_2} c_v \, dT \qquad \text{Btu}$$

From the reading of the problem, we know that $p_1 = 125$ psia, $p_2 = 60$ psia, $V_1 = V_2 = 5$ cu ft. $M = 0.1705$, $B = 0$,

$$C = 4.138 \times 10^{-9}$$

These constants will be used to calculate the instantaneous specific heat of air to be applied here. Since T_1 is given, we must find T_2 by use of the perfect-gas equation modified as we shall see.

$$\frac{125 \times 144}{600 + 460} = \frac{60 \times 144}{T_2}$$

from which $T_2 = (60 \times 144)/(125 \times 144) \times 1{,}060 = 508°$ R. For a variable specific heat,

$$\begin{aligned} Q &= w\left[M(T_2 - T_1) + \frac{B}{2}(T_2^2 - T_1^2) + \frac{C}{3}(T_2^3 - T_1^3)\right] \\ &= w\,[0.1705(508 - 1{,}060) + 0 + 4.138 \times 10^{-9}/3(508^3 - 1{,}060^3)] \\ &= w(-95.302) \end{aligned}$$

But we must find the value of w to complete the calculation. We remember

$$w = \frac{P_1V_1}{RT_1} = \frac{(125 \times 144)5}{53.3 \times 1{,}060} = 1.61 \text{ lb}$$

Finally,

$$Q = 1.61 \times (-95.302) = \mathbf{-153\ Btu}$$

indicating heat has been removed from the substance.

The equation $S = wc_v \ln(T_2/T_1)$ applies, and by substitution of the proper values in it we find $\Delta S = $ **$-$0.193 Btu per lb $-$ °R.**

Q6-15. A mole of a gas at a pressure of 500 psia undergoes a constant pressure nonflow process with the temperature changing from 1000 to 1100°R. What is the maximum amount of work that can be obtained from this expansion process?

ANSWER. The maximum work is the reversible work:

$$W = P(V_2 - V_1) \qquad \text{ft-lb per lb}$$

At any state

$$pV = RT$$

And upon substitution in the above

$$\begin{aligned} W &= R(T_2 - T_1) \\ &= (1.986 \text{ Btu/mole °R})(1100 - 1000)\text{°R} \\ 198.6 \text{ Btu/mole} &= \mathbf{154{,}500 \text{ ft-lb per mole}} \end{aligned}$$

Note that gas constant was in molal dimensions. Another form would be as previously shown to be R = 1,544 molecular weight, approximately.

Q6-16. A volume of gas having an initial entropy of 2800 Btu per °F is heated at constant temperature of 1000°F until the entropy is 4300. How much heat is added and how much work is done during the process?

ANSWER. *Heat added* is the area beneath the *TS* curve for the process.

$$Q_{1-2} = T(S_2 - S_1) = (460 + 1{,}000)(4{,}300 - 2{,}800)$$
$$= \mathbf{2.19 \times 10^6 \text{ Btu}}$$

Work done. Since work of an isothermal process is exactly compensated for by transferred heat, then

$$W_{1-2} = Q_{1-2} = 2.19 \times 10^6 \text{ Btu} \qquad \text{or} \qquad 2.19 \times 778 \times 10^6$$
$$= \mathbf{1.705 \times 10^9 \text{ ft-lb}}$$

Q6-17. How much work will be required to compress 2 lb of an ideal gas from an initial volume of 25 cu ft and an initial pressure of 13.5 psia to a final pressure of 75 psia, according to the equation $PV^{1.35}$ equals a constant? What will be the final temperature if the initial temperature is 55°F?

ANSWER.

$$(a)\quad W = 144\,\frac{(75 \times V_2) - (13.5 \times 25)}{1 - 1.35} = \mathbf{-78{,}000\ ft\text{-}lb}$$

The above equation was solved with V_2 determined first from the treatment below and then inserted.

$$\frac{V_2}{V_1} = \left(\frac{P_1}{P_2}\right)^{1/1.35} = \frac{V_2}{25} = \left(\frac{13.5}{75}\right)^{1/1.35} = 0.2811$$

from which

$$V_2 = 0.2811 \times 25 = 7.03 \text{ cu ft}$$

$$(b)\quad \frac{T_1}{T_2} = \left(\frac{P_1}{P_2}\right)^{(1.35-1)/1.35} = \frac{55 + 460}{T_2} = \left(\frac{13.5}{75}\right)^{0.35/1.35} = 0.6414$$

from which

$$T_2 = 515/0.6414 = 803°\text{R} \qquad \text{or} \qquad 803 - 460 = \mathbf{343°F}$$

Q6-18. Air drawn into a compressor is at 61°F and 14.7 psia. Flash point of the lubricating oil used is 350°F. (*a*) If compression is a reversible adiabatic (isentropic), what pressure could be attained in the compressor if the maximum allowable temperature is 50°F below the flash point of the oil? (*b*) If compression follows the law PV^n with n equal to 1.3, what is the maximum allowable pressure?

ANSWER.

(*a*) After first finding T_2 and letting $k = 1.4$ for air,

$$T_2 = (350 - 50) + 460 = 760°\text{R}$$
$$P_1 = P_2(T_1 - T_2)^{k/(k-1)} = 14.7 = P_2(521/760)^{3.5}$$

from which

$$P_2 = 14.7/0.267 = \mathbf{55\ psia}$$

(*b*) For air during a reversible polytropic process of compression (polytropic)

$$\frac{P_1}{P_2} = \left(\frac{T_1}{T_2}\right)^{n/(n-1)} = \left(\frac{521}{760}\right)^{1.3/0.3} = 0.195$$
$$P_2 = 14.7/0.195 = \mathbf{75.4\ psia}$$

Q6-19. It is desired to have air delivered from a nozzle at a velocity of 1,800 fps, a pressure of 15 psia and 40°F. The nozzle coefficient is 0.98. Neglecting the velocity of approach, assume air to be an ideal gas and find (a) the initial temperature of the air and (b) the initial pressure. Assume constant c_p.

ANSWER.

$$(a)\quad v_2 = 223.8\sqrt{0.240(T_1 - 460 + 40)} \times 0.98 = 1{,}800$$

Solving for T_1, this is found to be equal to 770°R, or **310°F.**

(b) Assuming adiabatic process where k equals 1.4 for air and $k/(k-1)$ is equal to 3.5

$$\left(\frac{T_1}{T_2}\right)^{3.5} = \frac{P_1}{P_2} = \left(\frac{770}{500}\right)^{3.5} = 1.54^{3.5} = 4.532 = \frac{P_1}{15}$$

Thus $$P_1 = 15 \times 4.532 = \mathbf{68.5\ psia}$$

Q6-20. Steam at a pressure of 170 psia arrives at a steam engine with a quality of 97 per cent. For 1 lb, what are (a) the enthalpy, (b) the volume, (c) the entropy, (d) the internal energy, and (e) the temperature of this steam?

ANSWER. Refer to Fig. 6-3. With the use of steam tables at 170 psia

(a) $H_1 = H_f + 0.97(H_{fg})$
$= 341.11 + 0.97(855.2)$
$= \mathbf{1171.11\ Btu\ per\ lb}$

(b) $\bar{v}_1 = \bar{v}_f + 0.97\bar{v}_{fg}$
$= 0.01821 + 0.97(2.656)$
$= \mathbf{2.5945\ cu\ ft\ per\ lb}$

(c) $S_1 = S_f + 0.97(S_{fg})$
$= 0.5266 + 0.97(1.0327)$
$= \mathbf{1.5283\ units}$

(d) $U_1 = H_1 - AP_1\bar{v}_1 = 1171.11 - (1/778 \times 170 \times 144 \times 2.5945) = \mathbf{1089.11\ Btu/lb}$

(e) From the steam tables, the temperature is **368.42°F.**

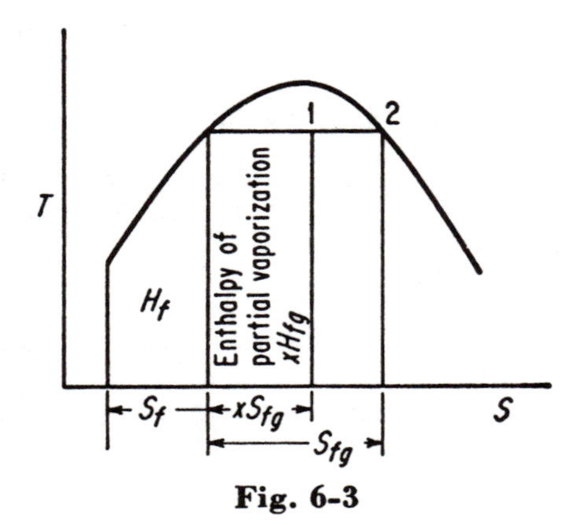

Fig. 6-3

Q6-21. Two boilers discharge equal amounts of steam into the same main. The steam from one is at 200 psia and 420°F, and from the other, at 200 psia and 95 per cent quality. (a) What is the equilibrium condition after mixing? (*b*) What is the loss of entropy by the higher-temperature steam? Assume no pressure drop in the pipeline.

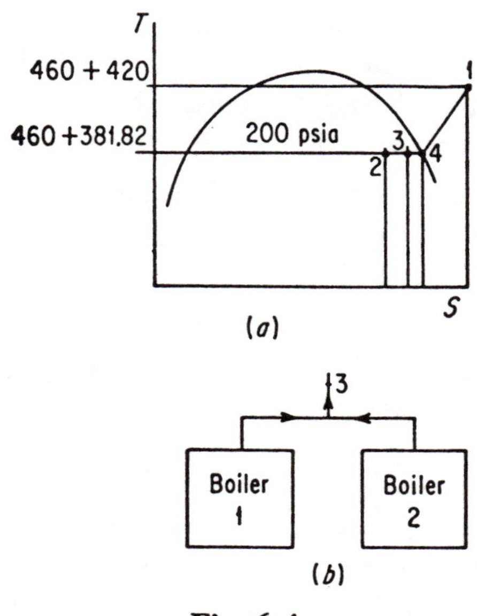

Fig. 6-4

ANSWER. Refer to Fig. 6-4 and use the Mollier diagram.

$$(a)\ H_3 = \frac{H_1 + H_2}{2} = \frac{1{,}225 + 1{,}164}{2} = \mathbf{1194.5\ Btu\ per\ lb}$$

At 200 psia, steam tables give data to determine the quality of the mixture.

$$1{,}194.5 = H_f + x_3 H_{fg} = 355.4 + x_3(843.3)$$

Solving for x_3,

$$x_3 = 0.995, \text{ or } \mathbf{99.5\ per\ cent\ quality}$$

$$(b)\ S_1 - S_3 = 1.575 - 1.541 = \mathbf{0.034\ entropy\ units\ loss}$$

Q6-22. Water at 70°F and atmospheric pressure is pumped into a boiler, evaporated at 300 psia pressure, and superheated to a total steam temperature of 600°F at constant pressure. How much heat must be supplied per pound of water?

ANSWER. Except for the step of pumping the water into the boiler, the entire process occurs at constant pressure. The effect of pressure on the enthalpy of the liquid is very small (and will be calculated), but ordinarily we may assume that the enthalpy of the liquid water at 70°F and atmospheric saturation pressure is the same as the heat at 70°F and 300 psia. Now, from the steam tables and the Mollier diagram

$$H_1 = 38.05 \text{ Btu per lb or } (70 - 32 = 38) \qquad \text{(enthalpy at 70°F)}$$

$$H_2 = 1314.7 \text{ Btu per lb} \qquad \text{(enthalpy of superheated vapor)}$$

$$H_2 - H_1 = 1{,}314.7 - 38.05$$

$$= \mathbf{1276.65\ Btu\ per\ lb\ water\ vaporized}$$

The effect of pressure on the liquid to get it into the boiler may be determined from the following expression:

$$H_1 = H_a + A(P_1 - P_a)\bar{v}_a$$

where the subscript a refers to the saturated liquid at the temperature t_1 at which the water is pumped into the boiler, i.e., 70°F in this case. The saturation pressure corresponding to 70°F from the steam tables is 0.3628 psia. From the steam tables giving the properties of the saturated liquid, $H_a = 38.05$ Btu per lb and $\bar{v}_a = 0.01605$ cu ft per lb; hence

$$H_1 = 38.05 + 144/778(300 - 0.3628)(0.01605)$$

$$= 38.05 + 0.88 = 38.93 \text{ Btu per lb}$$

From the tables for superheated steam, with steam pressure at 300 psia and temperature at 600°F, the enthalpy is 1314.7 Btu per lb. Then the heat added to each pound of water becomes

$$1314.7 - 38.93 = 1275.77 \text{ Btu}$$

By comparison the difference between both methods is very small. Note that enthalpies were obtained from the Mollier diagram and not directly from the steam tables for greater precision. However, the use of the Mollier diagram is acceptable for the written examination.

Q6-23. Saturated steam at 200 psia expands continuously through a throttle to atmosphere (14.7 psia). What are the state of expanded steam and the change in entropy in the process if no heat is lost (constant enthalpy) to the surroundings, and if all kinetic energy due to any high-velocity jets is dissipated?

ANSWER. A throttling process is one occurring at constant enthalpy. The Mollier diagram is the handiest tool to use for the solution of this problem. Locate the point on the Mollier diagram where the 200-psia pressure line crosses the saturation line. This is the initial state, and from the coordinates of the diagram we obtain

$$H_1 = 1198.4 \text{ Btu per lb}$$
$$S_1 = 1.545 \text{ Btu/(lb)(°F)}$$

Now proceed horizontally to the right or left as the case may be at constant enthalpy to the intersection with the 14.7 psia line. This is the final state. Read

$$H_2 = \mathbf{1189.4 \text{ Btu per lb}}$$
$$S_2 = \mathbf{1.823 \text{ Btu/(lb)(°F)}}$$
$$t_2 = \mathbf{311.5°F}$$

Change in entropy $S_2 - S_1 = 1.823 - 1.545 =$ **0.278**. Degrees superheat are **99°F**.

Q6-24. Five pounds of steam expand isentropically (nonflow) from p_1 equal to 300 psia and t_1 equal to 700°F to t_2 equal to 200°F. Find x_2 and the work.

ANSWER. From the superheat steam tables, we find $S_1 = 1.6751$, $H_1 = 1{,}368.3$, and $\bar{v}_1 = 2.227$. From the saturated steam tables, we find $S_{f2} = 0.2938$, $H_{f2} = 167.99$, $S_{fg2} = 1.4824$, $H_{fg2} = 977.9$, $v_{g2} = 33.64$, $p_2 = 11.526$. With $S_1 = S_2$ we see that

$$1.6751 = 0.2938 + 1.4824x_2$$

from which $x_2 \equiv$ **93.2 per cent**. Then we can proceed to find

$$H_2 = 167.99 + 0.932(977.9) = 1079 \text{ Btu per lb}$$
$$\bar{v}_2 = 0.932 \times 33.64 = 31.3 \text{ cu ft}$$
$$U_2 = H_2 - \frac{P_2V_2}{J} = 1{,}079 - \frac{(11.526)(144)(31.3)}{778}$$
$$= 1012.2 \text{ Btu per lb}$$
$$U_1 = 1{,}368.3 - \frac{(300)(144)(2.227)}{778} = 1244.6 \text{ Btu per lb}$$
$$W = \Delta U = 1{,}244.6 - 1{,}012.2 = 232.4 \text{ Btu per lb}$$
$$W = 5 \times 232.4 = \mathbf{1162 \text{ Btu}} \text{ for 5 lb of steam}$$

Q6-25. A vapor with a quality of 100 per cent and a temperature of 100°F is supplied to suitable heat-transfer coils for special heating, or as the case may be, cooling purposes. The condensate leaves the coils as a saturated liquid at the same pressure. The heat load is 1 million Btu per hr. Find the weight of vapor to be supplied to the coils in pounds per hour if the vapor used is steam, ammonia, or sulfur dioxide. The enthalpy of vaporization or condensation for steam (from steam tables) 1036.4 Btu per lb; for ammonia (from ammonia tables) 477.79 Btu per lb; for sulfur dioxide (from sulfur dioxide tables) 140.8 Btu per lb.

ANSWER.

For steam: $\frac{1 \times 10^6}{1{,}036.4} = \mathbf{965\ lb\ per\ hr}$

For ammonia: $\frac{1 \times 10^6}{477.79} = \mathbf{2{,}093\ lb\ per\ hr}$

For sulfur dioxide: $\frac{1 \times 10^6}{140.8} = \mathbf{7{,}100\ lb\ per\ hr}$

Q6-26. In determining the quality of steam in a main with a throttling calorimeter, the following readings were taken: pressure in main 110 psig, barometer 30.6 in. Hg, manometer reading 2.04 in. Hg, and thermometer after stem correction 220°F. Find the quality of the steam in the main.

ANSWER. Use the Mollier chart. Convert barometer to psia reading:

$$30.6 \times 0.4912 = 15.04 \text{ psia}$$

Main pressure is

$$110 + 15.04 = 125.04 \text{ psia} = P_1$$

Manometer pressure is

$$(2.04 \times 0.4912) + 15.04 = 16.04 \text{ psia} = P_2$$

From the Mollier diagram steam at 16.04 psia and a total steam temperature of 220°F the enthalpy is equal to 1163.4 Btu per lb. This is also the enthalpy existing in the main (constant enthalpy process). Now on the Mollier diagram move horizontally left or right (depending on the diagram used) along the enthalpy line of

1,163.4 and where this line intersects the pressure line (interpolation) of 125.04 psia read constant moisture content of 4.2 per cent. Then quality is 100 − 4.2 = **95.8 per cent.**

Q6-27. Water from a boiler operating at 150 psig is blown down to a flash tank held at a pressure of 25 psig. Neglecting pressure drop in the blowdown line and assuming insulation of line and flash tank, how many pounds of water are flashed into vapor per pound of water fed to the tank? Use 15 psia as atmospheric pressure.

ANSWER. This is a throttling process at constant enthalpy taking place on the left side of the *TS* diagram. One method of solution is to use a *TS* diagram for steam, beginning at a point on the saturated liquid line and to expand the liquid into the area under the dome at constant enthalpy (see Fig. 6-5). Where the expansion meets the horizontal 25 psig (40 psia) line, read quality from the x lines. However, we shall calculate the results, and the candidate should check using the *TS* diagram. On the basis of an energy balance

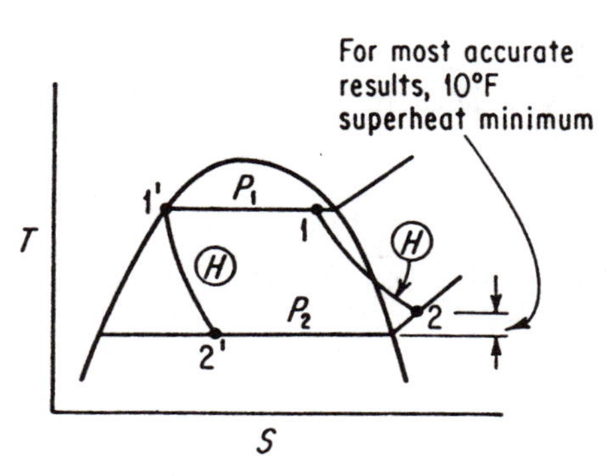

Fig. 6-5

Enthalpy of saturated water = enthalpy of flashed vapor
+ enthalpy of saturated liquid at final state

From the steam tables:

Enthalpy of saturated water at 150 psig......	338.53 Btu per lb
Enthalpy of saturated water at 25 psig.......	236.02 Btu per lb
Enthalpy of saturated vapor at 25 psig.......	1,169.7 Btu per lb

Let x equal the weight of water flashed into saturated vapor at the flash tank pressure of 25 psig; then the enthalpy balance is

$$338.53 \times 1 = 1{,}169.7(x) + [236.02(1 - x)]$$

Solving for x, we find it to be equal to **0.107 lb water flashed** into saturated vapor. Then, 1 − 0.107 = 0.893 lb of water remains as saturated liquid at 25 psig. Finally,

$$\text{lb water/lb vapor} = 0.893/0.107 = 8.34$$

Note that this treatment may be applied to all other pure fluids: ammonia, carbon dioxide, methyl chloride, Freon-12, etc.

Q6-28. If the upstream pressure in each case is 300 psig and 15 psia downstream, how much of each of the following fluids will flow through a circular hole in a thin plate 1 in. in diameter (*a*) water, (*b*) air, and (*c*) steam? Assume the water and air temperatures at 70°F and the steam at 700°F total steam temperature.

ANSWER.

(*a*) Assume coefficient of discharge for the orifice to be 0.61 for all cases.

$$G = CA_0\sqrt{2g\,\Delta h} \times 62.4$$
$$= 0.61 \times 0.785 \times 1/144\sqrt{64.4 \times 300 \times 2.31} \times 62.4$$

For water:

$G =$ **43.2 lb per sec**

(*b*) For air $P_c = 0.53P_1 = 0.53(300 + 15) = 167$ psia. Now since $P_2 = 15$ psia, P_2 is less than P_c and represents unretarded flow.

$$G = C\frac{0.532}{\sqrt{T_1}}A_0P_1 = 0.61\frac{0.532}{\sqrt{530}}0.785\frac{1}{144}315 \times 144$$
$$= \mathbf{3.48\ lb\ per\ sec}$$

Note that pressure was expressed in pounds per square foot absolute since orifice area was expressed in square feet. Results are the same as if the pressure were psia and area in square inches.

(*c*) For steam:

$$P_c = 0.55P_1 = 0.55 \times 315 = 174 \text{ psia}$$

$$G = A_0P_1C\frac{1}{70}\frac{1}{1 + 0.00065\,\Delta t}$$
$$= 0.785 \times \frac{1}{144} \times 315 \times 144 \times 0.61 \times \frac{1}{70}\frac{1}{1 + (0.00065 \times 278)}$$
$$= \mathbf{1.83\ lb\ per\ sec}$$

Note that degrees superheat are $700 - 422 = 278$°F.

Q6-29. What is acoustic velocity? How is a usable expression derived? Give an example.

ANSWER. Acoustic or critical velocity as it is called is the maximum that any gas can attain flowing through pipes or apertures. This is equal to the velocity of sound in the gas.

The velocity of sound in a gas varies:

(a) Directly as the $\sqrt{\text{ratio of specific heats}}$
(b) Directly as the $\sqrt{\text{absolute temperature}}$
(c) Inversely as the $\sqrt{\text{specific gravity}}$

The velocity of sound in air at 32°F is approximately 1,087 fps, and at 60°F it is equal to

$$1{,}087\sqrt{\frac{520}{492}} = 1{,}117 \text{ fps}$$

The ratio of specific heats for air may be taken as 1.4. The velocity of sound in any gas would therefore be equal to

$$\frac{1{,}117}{1.4}\sqrt{\frac{k}{\text{sp gr}}} = 943\sqrt{\frac{k}{\text{sp gr}}}$$

Let us assume a natural gas with a k value of 1.3, specific gravity of 0.6, at 60°F. The acoustic velocity or velocity of sound in the gas is

$$V_s = 943\sqrt{1.3/0.6} = 1{,}386 \text{ fps}$$

This is the maximum velocity that the gas can attain. The acoustic velocity may also be found with the following equation:

$$V_s = 223.8\sqrt{\frac{kT}{M}}$$

where M is molecular weight of the gas.

Q6-30. A Terry steam turbine with six nozzles develops 30 hp. This is 25 per cent of the power delivered to the steam jets. Steam is supplied at 150 psia and 50°F superheat. The pressure in the casing is 20 psia. Calculate the diameter of a nozzle throat.

ANSWER. The power developed by each of the six nozzles is

$$(30/0.25)/6 = 20 \text{ hp}$$

Refer also to Fig. 6-6 for the process involved. The pressure at the throat is the critical pressure or

$$P_t = P_c = 0.55P_1 = 0.55 \times 150 = 83 \text{ psia}$$

The throat velocity may now be determined, remembering that the

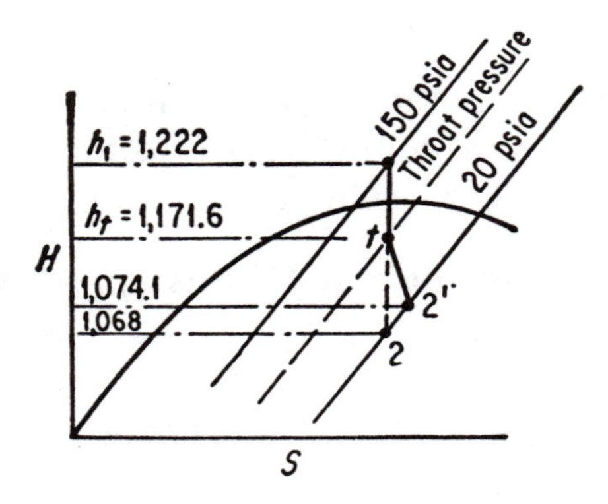

Fig. 6-6

process of nozzle expansion is isentropic for this portion.

$$v_t = 223.8 \sqrt{1{,}222 - 1{,}171.6} = 1{,}580 \text{ fps}$$

Note that the nozzle efficiency N_n is unity for the isentropic expansion. In order to find the throat area, we must find among other things the flow of steam in pounds per second.

$$G = \frac{20 \times 550}{778(1{,}222 - 1{,}068)0.98^2} = 0.0955 \text{ lb per sec}$$

A velocity coefficient of 0.98 has been assumed. Now, in order to apply the continuity equation for the throat area the specific volume of the steam at the throat must be found. From the Mollier diagram and steam tables (subscript t denotes throat section)

$$\bar{v}_t = (\bar{v}_f + xv_{fg})_t = 0.01756 + (0.989 \times 5.264) = 5.234 \text{ cu ft per lb}$$

Applying the law of continuity,

$$A_t = \frac{G\bar{v}_t}{v_t} = \frac{0.0955 \times 5.234}{1{,}580} \times 144 = 0.0455 \text{ in}^2.$$

$$0.785\, d_t^2 = A_t = 0.0455$$

$$d_t = \sqrt{\frac{0.0455}{0.785}} = \textbf{0.242 in. diameter}$$

Determine the discharge outlet diameter by finding the enthalpy at 2′ by

$$H_{2'} = H_1 - [(H_1 - H_2)N_n]$$

Figure 6-6 shows this to be 1074.1 Btu per lb. With the flow rate the same and a change in specific volume d_2 could be determined for the actual nozzle. The exit velocity would be found with the use of

$$v_{2'} = 223.8\sqrt{(H_1 - H_2)N_n}$$

Note that in determining v_t, N_n was equal to unity because it may be assumed that for the purposes of this work the entire loss due to friction occurs between the throat and the exit section. Although it was not required for the answer it may be worth noting that

$$\text{Rankine efficiency} = \frac{H_1 - H_{2'}}{H_1 - H_f} \text{ (with } H_f \text{ at 20 psia)}$$
$$= \frac{1{,}222 - 1{,}074.1}{1{,}222 - 196.09}$$
$$= 0.146, \text{ or } 14.6\%$$
$$\text{Rankine steam rate} = \frac{H_1}{H_1 - H_2} = 17.2 \text{ lb per hp-hr}$$

Q6-31. Compute the work per pound of steam and the steam consumption in pounds per indicated horsepower-hour when steam is used in a Carnot cycle between zero and a quality of unity and 150 psia and 3 psia.

ANSWER. Refer to Fig. 6-7 and the Mollier diagram.

$$\text{Net work (area of rectangle } abcd) = (H_b - H_a)JE_t$$
$$\text{Steam rate} = \frac{2{,}545}{(H_b - H_a)(1/E_t)} \text{ lb per hp-hr}$$

Then $P_a = 150$ psia
$t_a = 358.43°F$
$H_a = 330.53$ Btu per lb
$P_b = 150$ psia
$t_b = 358.43°F$
$H_b = 1{,}194.4$ Btu per lb
$P_d = 3$ psia
$t_d = 141.5°F$

From this we can obtain the thermal efficiency of the cycle as follows

$$E_t = (358.43 - 141.5)/(358.43 + 460) = 0.265$$

$$\text{Net work} = (1{,}194.4 - 330.53)778 \times 0.265 = \mathbf{178{,}000\ ft\text{-}lb\ per\ lb}$$

$$\text{Water rate} = \frac{2{,}545}{(1{,}194.4 - 330.53)(1/0.265)} = \mathbf{11.1\ lb\ per\ hp\text{-}hr}$$

Q6-32. An actual engine supplied with steam at 115 psia and containing 1 per cent moisture was found to use 21 lb of steam per

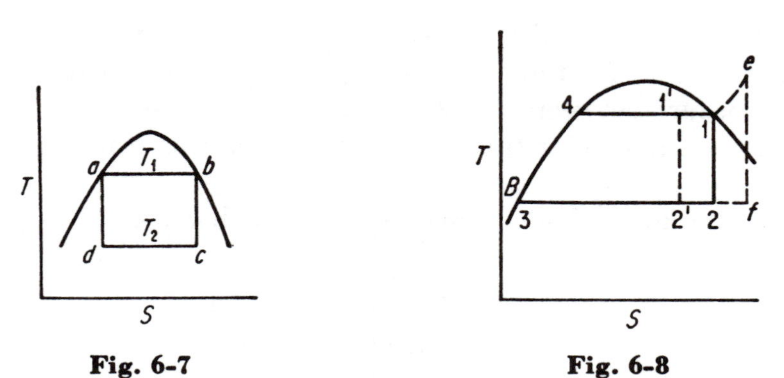

Fig. 6-7 **Fig. 6-8**

indicated horsepower-hour. This occurred when the steam in the condenser was 140°F. Compute (*a*) engine efficiency, (*b*) heat rate, and (*c*) actual thermal efficiency of the engine.

ANSWER. Refer to Fig. 6-8 and use the Mollier diagram and steam tables.

(*a*) Engine efficiency:

$$E_e = \frac{2{,}545}{21 \times (1{,}180 - 936)} \times 100 = 0.495 \times 100 = \mathbf{49.5\ per\ cent}$$

(*b*) Heat rate:

$$21 \times (1{,}180 - 107.9) = \mathbf{22{,}500\ Btu\ per\ hp\text{-}hr}$$

(*c*) Actual thermal efficiency:

$$E_{ea} = \frac{2{,}545}{(1{,}180 - 107.9) \times 21} \times 100 = 0.113 \times 100$$

$$= \mathbf{11.3\ per\ cent}$$

Q6-33. Process heat exchangers are supplied with 290 psig saturated steam, which is trapped and flows to a flash tank maintained at 20 psig. The flash tank is close by the trapping system. Flash steam is carried off as flash steam to a 20-psig steam system, and the remaining condensate is pumped back to the power plant. The steam load for heating is 29,000 lb per hr. Assume no subcooling of condensate after trapping. Determine in pounds per hour how much 20-psig steam is produced and how much condensate is returned to the power plant as condensate return.

ANSWER. Refer to Fig. 6-9. Assume a steady flow rate and no flashing in the pipeline to the flash tank. On the basis of an energy balance,

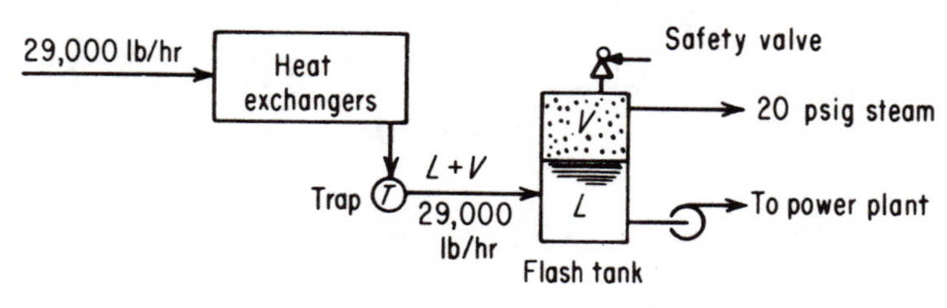

Fig. 6-9

Enthalpy of saturated water as condensate leaving exchangers
= enthalpy of flashed saturated vapor leaving trap
+ enthalpy of saturated liquid at final state

From steam tables:

Enthalpy of saturated water at 290 psig = 395.49 Btu per lb
Enthalpy of saturated water at 20 psig = 227.82 Btu per lb
Enthalpy of saturated vapor at 20 psig = 1166.7 Btu per lb

Let x equal the weight of water flashed into saturated vapor at the flash-tank pressure of 20 psig. Then the enthalpy balance is

$$395.49 \times 1 = 1166.7(x) + [227.82(1 - x)]$$

Solving for x, we find it to be equal to 0.18 lb water flashed into saturated vapor. Then, $1 - 0.18 = 0.82$ lb water remains as saturated liquid at 20 psig. Finally,

20-psig steam to process = $0.18 \times 29{,}000$ = **5,220 lb per hr**
Condensate returned to boiler house
= $0.82 \times 29{,}000$ = **23,780 lb per hr**

Q6-34. State whether or not the following process violates either of the laws of thermodynamics and show the basis for your answer. Air at 100 psig and 70°F enters an apparatus which is thermally and mechanically isolated from the surroundings; one-half of the air issues from the apparatus at 180°F and atmospheric pressure, and the other half issues at minus (−) 40°F and one atmosphere. There is no change in the composition of the working substance.

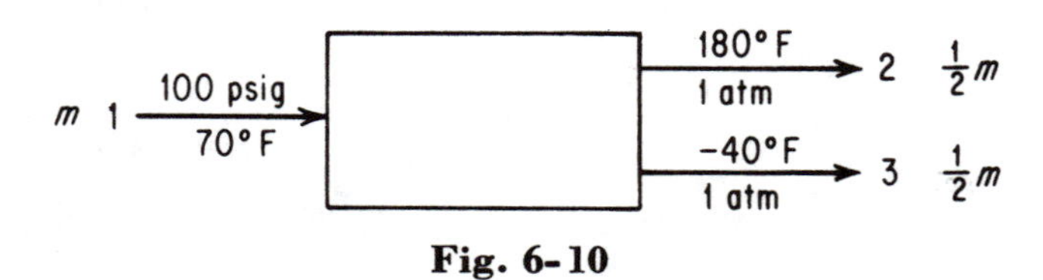

Fig. 6-10

ANSWER. Please refer to Fig. 6-10.

$H_1 = 127$ Btu per lb
$H_2 = 0.5 \times 153 = 76.5$ Btu per lb
$H_3 = 0.5 \times 101 = 50.5$ Btu per lb
$H_1 - (H_2 + H_3) = 0$
$Q = 0;\ W = 0$
First law satisfied.

$S_1 = 0.765$ Btu/(lb)(°F)
$S_2 = 0.5 \times 0.97 = 0.485$ Btu/(lb)(°F)
$S_3 = 0.5 \times 0.86 = 0.43$ Btu/(lb)(°F)
$S_2 + S_3 - S_1 = 0.915 - 0.765 = 0.15$
$S = 0$
Second law satisfied.

$Q - W = 0 \qquad \Delta U = 0 \qquad \Delta TS = \Delta PV$

$$\left.\begin{array}{l} T_1S_1 = 530 \times 0.765 = 405 \\ \left.\begin{array}{l} T_2S_2 = 640 \times 0.485 = 310 \\ T_3S_3 = 420 \times 0.43 \ = 180 \end{array}\right\} 490 \end{array}\right\} \Delta TS = 85 \text{ Btu per lb}$$

$P_1V_1 = {}^{144}\!/_{778} \times 115 \times {}^{379}\!/_{29} \times {}^{530}\!/_{520} = 285$

$P_2V_2 + P_3V_3$

$= {}^{1}\!/_{2} \times {}^{144}\!/_{778} \times 15 \times {}^{379}\!/_{29}({}^{640}\!/_{520} + {}^{420}\!/_{520}) = 370$

$\Delta TS = \Delta PV = 85 = 85$. Thus, **both laws are satisfied.**

Q6-35. **A real incompressible fluid enters a machine at areas A_1** and A_2 and leaves at area A_3. The temperature is constant. The mass density of the fluid is 2.5 slugs per cu ft. Consider all openings to be at the same elevation. Calculate the shaft horsepower and indicate whether the horsepower is in or out.

$V_1 = 30$ fps	$p_1 = 115$ psia	$A_1 = 0.2$ sq ft
$V_2 = 40$ fps	$p_2 = 75$ psia	$A_2 = 0.1$ sq ft
$V_3 = 65$ fps	$p_3 = 40$ psia	

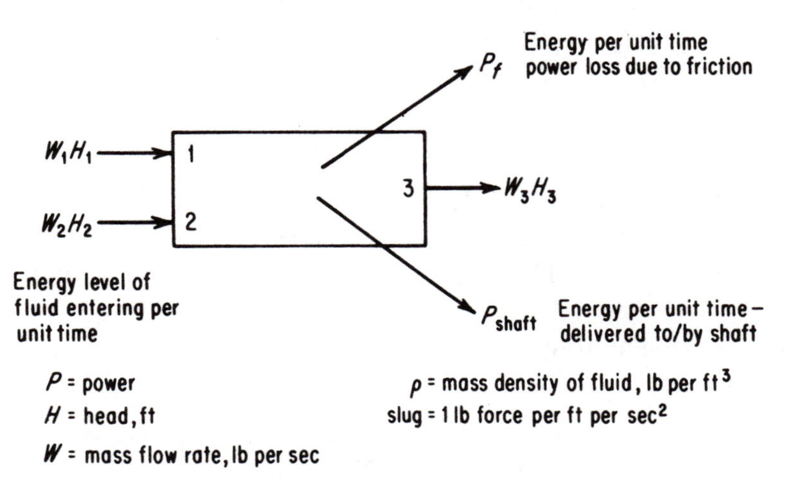

Fig. 6-11

ANSWER. Refer to Fig. 6-11.

$$H = \frac{P}{\rho} + \frac{V^2}{2g} + Z_{\text{elevation}} = \text{total head, ft}$$

$$W = \rho A V$$

where ρ = fluid density, lb per cu ft
A = area, sq ft
V = velocity, fps

$$H_1 = \frac{p_1}{\rho_1} + \frac{V_1^2}{2g}$$

$$p_1 = 115 \times 144 = 16{,}560 \text{ lb per sq ft}$$

$$\rho = 2.5 \text{ slugs per cu ft}$$

$$32.2 \text{ lb force} = 1 \text{ slug} \times 32.2$$

$$2.5 \text{ slugs per cu ft} = 80.5 \text{ lb per cu ft}$$

$$H_1 = (16{,}560/80.5) + (30^2/64.4) + 0 = 219 \text{ ft}$$
$$H_2 = (75 \times 144)/(80.5) + (40^2/64.4) + 0 = 159 \text{ ft}$$
$$H_3 = (40 \times 144)/(80.5) + (65^2/64.4) + 0 = 137 \text{ ft}$$

$$W_1 = \rho_1 A_1 V_1 = 80.5 \times 0.2 \times 30 = 483 \text{ lb per sec}$$
$$W_2 = \rho_2 A_2 V_2 = 80.5 \times 0.1 \times 40 = 322 \text{ lb per sec}$$
$$W_1 + W_2 + W_3 = 805 \text{ lb per sec}$$

Then

$$W_1H_1 = 483 \times 219 = 105{,}800 \text{ lb-ft per sec}$$
$$W_2H_2 = 322 \times 159 = 51{,}200 \text{ lb-ft per sec}$$
$$W_3H_3 = 805 \times 137 = 110{,}300 \text{ lb-ft per sec}$$

Thus

$$\text{Fluid power in} = W_1H_1 + W_2H_2 = 157{,}000 \text{ lb-ft per sec}$$
$$\text{Fluid power out} = W_3H_3 = 110{,}000 \text{ lb-ft per sec}$$
$$\text{Net power in} = 157{,}000 - 110{,}000 = 46{,}700 \text{ lb-ft per sec}$$

Assume a reasonable efficiency (mechanical and hydraulic) = 70 per cent.

$$\text{Efficiency} = \frac{\text{output}}{\text{input}}$$

Therefore,

$$\text{Output} = \text{shaft power} = 0.70 \times 46{,}700 = 32{,}690 \text{ lb-ft per sec}$$
$$\text{Shaft horsepower} = 32{,}690/550 = 59 \text{ shaft horsepower}$$

This is a **turbine** since power is **out. Answer.**

Q6-36. Superheated steam at 200 psia and a total steam temperature of 500°F enters an ideally insulated nozzle with negligible velocity and expands to a back pressure of 20 psia with a quality of 98 per cent.

(*a*) Calculate the velocity leaving the nozzle in feet per second.

(*b*) Calculate the nozzle efficiency per cent and the nozzle velocity coefficient.

ANSWER. Refer to Fig. 6-12 and an h-S diagram for steam.

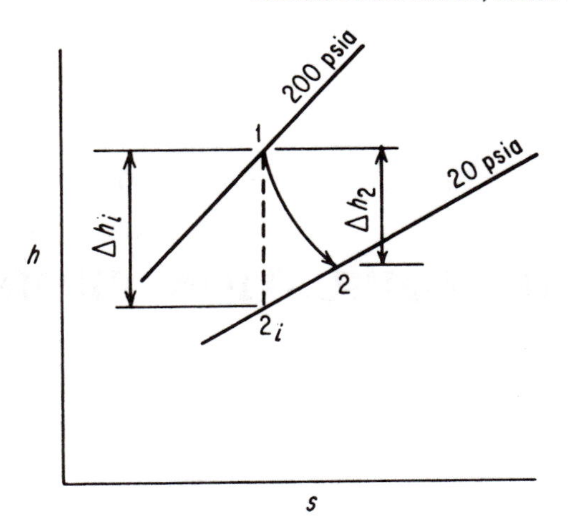

Fig. 6-12

$$p_1 = 200 \text{ psia}$$
$$t_1 = 500°\text{F}$$
$$x_2 = 0.98$$
$$h_1 = 1{,}268.9 \text{ Btu per lb}$$
$$S_1 = 1.624$$
$$S_1 = S_{2i} = 1.624$$
$$S_{f2} + x_2 \times S_{fg2} = S_2$$
$$S_2 = 0.3356 + 0.98 \times 1.3962 = 1.7038$$

There is an attendant entropy increase.

$$h_2 = h_{f2} + x_2 \times h_{fg2} = 196.16 + 0.98 \times 960.1 = 1{,}137.1$$

Actual heat drop $= \Delta h_a = h_1 - h_2 = 1{,}268.9 - 1{,}137.1$
$= 131.8$ Btu per lb

$$S_1 = S_{2i} = 1.624 = S_{f2i} + x_{2i} \times S_{fg2i} = 0.3356 + x_{2i} \times 1.3962,$$

from which

$$x_{2i} = 0.923$$

Then $h_{2i} = 196.16 + 0.923 \times 960.1 = 1{,}082.33$ Btu per lb

Ideal heat drop $= \Delta h_i = 1{,}268.9 - 1{,}082.33 = 186.57$ Btu per lb

(*a*) V_2 exit nozzle velocity $= 223.84(h_1 - h_2)^{1/2} =$ **2,570 fps**

(*b*) Nozzle efficiency = actual heat drop/ideal heat drop =
131.8/186.57 — 0.7066 or **70.66%**

Nozzle velocity coefficient $= 0.7066^{1/2} =$ **0.84**

Chapter 7

FUELS AND COMBUSTION PRODUCTS

7-1. Fuels. The source of heat which is used to produce steam in a boiler is the fuel. The cost of fuel is by far the greatest single item of expense in the production of power. Fuels may be *solid, liquid,* or *gaseous.* The principal solid fuels are wood, peat, lignite, and coal. For liquid fuels we have fuel oil (Bunker C for industrial use and No. 2 oil for commercial use), tar, and other unrefined petroleum oils. The gaseous fuels consist of natural gas, refinery and betterment gases, coke-oven gas, and blast-furnace gas. Sawdust, bagasse (sugar-cane pulp), and garbage and sewer disposal wastes may be included among others. The principal ingredients of all these fuels are carbon and hydrogen.

In atomic power plants the one natural fuel is U^{235}, making up 0.7 per cent of natural uranium with the balance U^{238}. U^{235} is the only naturally occurring, readily fissionable nuclear fuel. Enriched uranium fuels may also be used.

Q7-1. A coal has the following composition by weight: carbon, 74.79 per cent; hydrogen, 4.98 per cent; oxygen, 6.42 per cent; nitrogen, 1.20 per cent; sulfur, 3.42 per cent; water, 1.55 per cent; ash, 7.82 per cent. Calculate its heating value, using the Dulong formula.

ANSWER. The formula does not apply when the fuel contains carbon monoxide.

$$Q = 14{,}500 \times 0.7479 + 62{,}000\left(0.0498 - \frac{0.0642}{8}\right) + 4{,}000 \times 0.0324$$

$$= \textbf{13,650 Btu per lb} \text{ very nearly}$$

A calorimeter test showed this fuel to have 13,480 Btu per lb, showing a remarkable degree of accuracy for this formula.

7-2. Theoretical Air for Combustion. In order to select certain equipment in a boiler installation the amount of theoretical air required for combustion is of great significance. Let us take, for example, the combustion of carbon to carbon dioxide with the theoretical amount of air.

$$C + O_2 \rightarrow CO_2$$

One mole of each of the reactants (fuel and air) and the product of combustion are involved. This means that 12 lb carbon will react with 32 lb of oxygen to form 44 lb of carbon dioxide. Thus, for each pound of carbon for theoretical combusiton:

$$^{32}/_{12} = 2.666 \text{ lb oxygen required}$$
$$^{44}/_{12} = 3.666 \text{ lb carbon dioxide formed}$$

Now, since air by weight has 23.13 per cent oxygen,

$$2.666/0.2313 = 11.541 \text{ lb air needed}$$

11.541 × 0.7687 = 8.87 lb nitrogen are in the products of combustion. If hydrogen and hydrocarbons are involved in the combustion reaction, similar treatment may be accorded them with their products of combustion, carbon dioxide and water. In order to facilitate calculations see Table 7-1.

7-3. Excess Air. In actual practice an amount of air in excess of the theoretical is introduced with the fuel to influence complete combustion. While the theoretical air is readily calculated when the fuel analysis is given, the amount actually required for a given fuel and installation depends on experience and economical

Table 7-1. Consolidated Combustion Data: Air Required and Combustion Products

Table gives required combustion air and products for common combustibles burned with theoretical air requirements. Air and products are given in moles, cu ft and lb (see right-hand column) for 1 mole, 1 cu ft and 1 lb of fuel.

Fuel	For 1 mole of fuel					For 1 cu ft of fuel					For 1 lb of fuel					
	Air		Other products (than N_2)			Air		Other products (than N_2)			Air		Other products (than N_2)			
	O_2	N_2	CO_2	H_2O	SO_2	O_2	N_2	CO_2	HO_2	SO_2	O_2	N_2	CO_2	H_2O	SO_2	
C	1.0	3.76	1.0								0.0833	0.313	0.0833			Moles
	379	1425	379								31.6	118.8	31.6			Cu ft
	32.0	105	44.0								2.67	8.78	3.67			Pounds
H_2†	0.5	1.88		1.0		0.00132	0.00496		0.00264		0.250	0.940		0.5		Moles
	189.5	712		379*		0.5	1.88		1.0*		94.8	356		189.5*		Cu ft
	16.0	52.6		18		0.0422	0.139		0.0475		8.0	26.3		9.0		Pounds
S	1.0	3.76			1.0						0.0312	0.1176			0.0312	Moles
	379	1425			379						11.84	44.6			11.84	Cu ft
	32.0	105			64						1.0	3.29			2.0	Pounds
CO	0.5	1.88	1.0			0.00132	0.00496	0.00264			0.179	0.0672	0.0357			Moles
	189.5	712	379			0.5	1.88	1.0			6.77	25.4	13.53			Cu ft
	16.0	52.6	44.0			0.0422	0.139	0.116			0.571	1.88	1.57			Pounds
CH_4	2.0	7.52	1.0	2.0		0.00528	0.0198	0.00264	0.00528		0.125	0.470	0.0625	0.125		Moles
	758	2850	379	758*		2.0	7.52	1.0	2.0*		47.4	178	23.7	47.4*		Cu ft
	64.0	210	44.0	36.0		0.169	0.556	0.116	0.0950		4.0	13.17	2.75	2.25		Pounds
C_2H_2	2.5	9.40	2.0	1.0		0.0066	0.0248	0.00528	0.00264		0.0962	0.362	0.0769	0.0385		Moles
	947	3560	758*	379*		2.5	9.40	2.0	1.0*		36.4	137	29.15	14.58*		Cu ft
	80.0	263	88.0	18.0		0.211	0.694	0.232	0.0475		3.08	10.13	3.38	0.692		Pounds
C_2H_4	3.0	11.29	2.0	2.0		0.00792	0.0298	0.00528	0.00528		0.1071	0.403	0.0714	0.0713		Moles
	1137	4280	758	758*		3.0	11.29	2.0	2.0*		40.6	153	27.1	27.1*		Cu ft
	96.0	316	88.0	36.0		0.253	0.834	0.232	0.0950		3.43	11.29	3.14	1.286		Pounds
C_2H_6	3.5	13.17	2.0	3.0		0.00923	0.0347	0.00528	0.0079		0.1167	0.439	0.0667	0.10		Moles
	1326	4990	758	1137*		3.5	13.17	2.0	3.0*		44.2	166.3	25.3	37.9*		Cu ft
	112.0	369	88.0	54.0		0.296	0.972	0.232	0.1425		3.73	12.29	2.93	1.8		Pounds

* The volumes shown for H_2O apply only where the combustion products are at such high temperature that all the H_2O is a gas.

† Varying assumptions for molecular weight introduce a slight inconsistency in the values of air and combustion products from the burning of hydrogen. True molecular weight of hydrogen is 2.02 but the approximate value of 2 is used in figuring the air and combustion products.

Courtesy of *Power Magazine*.

limitations. Table 7-2 below gives practical excess air in per cent by weight for various fuels.

TABLE 7-2

Fuel	*Excess per cent by weight*
Bunker C oil	20
Natural gas	15
Refinery gas	15
Coke-oven gas	20
Blast-furnace gas	20
Bagasse	50
Wood	50
Tar	30
Coal (pulverized)	25
Coal (stoker)	40

The amount of air required to burn a fuel completely to CO_2 and water vapor is very small, but in order to obtain maximum steam output, ensure complete combustion, avoid smoke, avoid overheating, minimize slagging (coal burning), and ensure adequate mixing of the combustion gases and air, it is necessary to use from 10 to 75 per cent excess air. If this is not done, some of the oxygen will escape without combining with its allotted share of the combustible matter so that some of the hydrogen will escape unburned or there will take place only partial combustion of the carbon into CO. Under these undesirable conditions the Orsat analysis will show CO, O_2, and nitrogen.

Theoretically, maximum furnace efficiency is highest when a minimum of excess air is used. Practical difficulties in securing close regulation of air may, however, dictate a larger amount of air to ensure some excess air at all times, particularly when the air flow is subject to varying conditions of draft.

The degree of efficiency of fuel combustion is determined by calculations involving the flue-gas analysis. This gas analysis is usually obtained by the fundamental Orsat principle which involves absorbing the oxygen, carbon dioxide, and carbon monoxide individually from a gas sample, thus obtaining their per cent of the original sample. Other methods of analysis are based on the specific gravity or the electrical conductivity of the flue gas.

When the rate of fuel firing is held constant, the effect of decreasing excess air is as follows:

(*a*) Amount of flue gas will decrease, thus saving heat which would be swept up the stack.

(*b*) Stack temperature may either increase or decrease, depending upon the number of tubes in the convection section of the boiler.

(*c*) Firebox temperature will increase because there is less cooling by the excess air.

(*d*) The capacity of the furnace will increase. The increased firebox temperature allows a higher charge rate. The decreased flue-gas volume allows a higher fuel-fired rate. This does not hold if firebox temperature is already at its maximum.

(*e*) Furnace efficiency increases. The ideal condition is approached where there is a ratio of fuel and air perfectly mixed and burned so that there will be neither an excess of fuel nor an excess of air.

Q7-2. Calculate the pounds of air and the products of combustion required and formed, respectively, when 1 lb of a gasoline composed of 85 per cent carbon and 15 per cent hydrogen is burned in the theoretical amount of air. What percentage of CO_2 exists in the products of combustion by volume?

ANSWER. Basis: 1 lb fuel and standard gas conditions. Make up table below and refer to Table 7-1. Assume combustion is complete.

Fuel	Per cent weight	Weight fraction lb	Theo. air, lb per lb fuel	Theo. air, lb	CO_2, lb per lb fuel	CO_2, lb	N_2, lb per lb fuel	N_2, lb	Water, lb per lb fuel	Water, lb
C.........	85	0.85	11.45	9.73	3.67	3.12	8.78	7.45		
H_2.........	15	0.15	34.3	5.15	....	....	26.3	3.94	9.0	1.35
Total......	100	1.0	.....	14.88	....	3.12	.....	11.39	...	1.35

Orsat analysis will show no water vapor (dry basis) because it will condense out at room temperature. The 3.12 lb of CO_2 will occupy at standard conditions

$$3.12/44 \times 379 = 27 \text{ cu ft}$$

For the nitrogen

$$11.39/28 \times 379 = 154 \text{ cu ft}$$

Volume per cent of CO_2 is $27/(27 + 154) \times 100 =$ **14.9 per cent.**

Q7-3. An analysis of the flue gases of a boiler shows 14 per cent CO_2, 2 per cent CO, 5 per cent O_2, and 79 per cent nitrogen. Calculate the percentage of heat lost in the unburned CO up the stack.

ANSWER. The solution to this problem may take the following form.

$$\frac{\text{Heat generated if all C burned to } CO_2 - \text{actual heat generated}}{\text{Heat generated if all C burned to } CO_2} \times 100$$

Combustions reactions for carbon in problem:

$$C + O_2 \rightarrow CO_2 + 14{,}544 \text{ Btu per lb C}$$
$$2C + O_2 \rightarrow 2CO + 4{,}350 \text{ Btu per lb C}$$

From the wording of the problem, for every 100 moles of flue gas there are included 14 moles of CO_2 and 2 moles of CO. There are 14 lb atoms C in 14 moles of CO_2 and 2 lb atoms C in 2 moles of CO. Then, substituting in the heat-generation ratio

$$\frac{[(14 + 2) \times 12 \times 14{,}544] - [(2 \times 12 \times 4{,}350) + (14 \times 12 \times 14{,}544)]}{(14 + 2) \times 12 \times 14{,}544} \times 100$$

or **9.6 per cent heat lost up stack.**

Q7-4. Given complete analysis of coal and flue gas, determine per pound of coal (*a*) the theoretical air, (*b*) actual air, (*c*) dry flue gas, (*d*) moisture in flue gas, (*e*) per cent excess air, and (*f*) per cent loss from incomplete combustion.

ANSWER. The ultimate analysis of the coal as fired, per cent by weight,

Ash	10.49
Sulfur	1.20
Hydrogen	6.47
Carbon	71.98
Nitrogen	1.16
Oxygen	8.70
Total	100.00

Heating value as higher heating value of the coal is 13,800 Btu per lb coal. The flue gas analysis, per cent by volume is:

CO_2	10.2
CO	0.6
O_2	9.2
N_2	80.0
Total	100.0

Moles of carbon and nitrogen in flue gas:

Flue gas	%	Moles gas	Moles C	Moles N_2
CO_2............	10.2	10.2	10.2	
CO.............	0.6	0.6	0.6	
O_2.............	9.2	9.2		
N_2.............	80.0	80.0	....	80.0
Total moles.....	....	100.0	10.8	80.0

Weight of carbon burned = 10.8 × 12 = 129.6 lb
Weight of coal burned = 129.6/0.7198 = 180.0 lb
Moles air supplied = 80.0/0.79 = 101.3 moles
Weight of air supplied = 101.3 × 29 = 2,938 lb
Weight of air supplied per lb coal = 2,938/180 = **16.32 lb**
Moisture produced per lb coal = 0.0647 × 9 = **0.58 lb**
Dry flue gas per lb coal = coal + air − ash
− moisture 1 + 16.32 − 0.1049 − 0.58 = **16.64 lb**
Fraction carbon burning to CO
= 0.6/(10.2 + 0.6) = 0.0555
Weight of carbon burning to CO
= 0.0555 × 0.7198 = 0.040 lb
Heat loss from CO = 0.040 × 10,160 = 406 Btu
Per cent heat loss from CO = (406/13,800) × 100 = **2.94 per cent**

Theoretical air per lb coal:

Combustible	Lb of combustible	Lb O_2 required per lb combustible (from Table 7-1)	Lb O_2 required
Carbon....................	0.7198	2.67	1.922
Hydrogen..................	0.0647	8.00	0.518
Sulfur....................	0.012	1.00	0.012
Total.....................	......	....	2.452
Less oxygen in the coal......	......	....	−0.087
Net oxygen needed..........	......	....	2.365

Air required: 2.365 × 4.32 = **10.22 lb**
Excess air: 16.32 − 10.22 = 6.10 lb
Per cent excess air: 6.10/10.22, or **59.7 per cent**

Q7-5. Given the volumetric analysis of a natural gas, determine the volumetric analysis of the flue gas if the fuel is burned with 40 per cent excess air by volume.

ANSWER.

Gas analysis by volume per cent:

CO	0.60
H_2	1.62
CH_4	94.30
C_2H_4	0.15
H_2S	0.25
O_2	0.40
CO_2	0.85
N_2	1.83
Total	100.00

Oxygen, combustion products of 100 moles of fuel:

Gas	Moles	Moles O_2 needed	Moles produced		
			CO_2	H_2O	SO_2
CO	0.60	0.30	0.60		
H_2	1.62	0.81	.....	1.62	
CH_4	94.30	188.60	94.30	188.60	
C_2H_4	0.15	0.45	0.30	0.30	
H_2S	0.25	0.37	.....	0.25	0.25
CO_2	0.85	......	0.85		
Subtotal	.....	190.53	96.05	190.77	0.25
Minus O_2 in fuel	.....	−0.40			
Total	.....	190.13	96.05	190.77	0.25

Nitrogen in flue gas:

N_2 required for 190.13 moles O_2 = 190.13 × 3.76 = 714.9 moles

40% excess N_2 (by volume) = 0.4 × 714.9 = 286.0 moles

N_2 in fuel = 1.8 moles

N_2 in flue gas by addition = 1,002.7 moles

Unburned O_2 in flue gas: 0.40 × 190.13 = 76.0 moles

Volumetric gas analysis (H_2O disappears):

Gas	Moles	Mole %	Vol %
CO_2............	96.1	8.2	8.2
O_2..............	76.0	6.5	6.5
SO_2.............	0.2	0.0	0.0
N_2..............	1,002.7	85.3	85.3
Total...........	1,175.0	....	100.0

Q7-6. An Orsat analysis of a flue gas shows that it contains 12.5 per cent CO_2, 4.1 per cent O_2, 83.4 per cent N_2. From this information determine (*a*) gravimetric analysis, (*b*) per cent excess air, (*c*) volume of air per pound of fuel fired, (*d*) volume of total flue gases per pound of fuel fired, and (*e*) percentage of gross heating value lost through noncondensation of water vapor in flue gases. Air leakage into furnace is negligible; fuel has no nitrogen content. Gross heating value of fuel may be taken as 18,300 Btu per lb. Flue-gas temperature may be taken as 490°F and 29.5 in. Hg. Barometer is 30.0 in. Hg.

ANSWER. By inspection of the heating value the fuel appears to be a fuel oil. Now let us take as a basis 100 moles of the flue gas. Then we see that there are 83.4 moles of N_2 leaving and the same entering because there is no leakage. Also N_2 undergoes no chemical reaction in the combustion process. We also know that mole per cent equals volume per cent. The moles of air coming in

$$83.4/0.79 = 105.4 \text{ moles air needed to give 100 moles of flue gas}$$

Moles of oxygen in flue gas consist of 12.5 moles in CO_2 and 4.1 moles in oxygen, or a total of 16.6 moles. This was determined on the basis of the theoretical combustion equations. Moles of O_2 entering furnace are $105.4 \times 0.21 = 22$ moles. But we can account for only 16.6 moles from the flue-gas analysis. Therefore, there are $22 - 16.6 = 5.4$ moles of O_2 coming in with the air which react chemically with another substance within the furnace. This must be hydrogen because as water vapor it would not show in the Orsat analysis.

Referring to the theoretical combustion of H_2 and O_2, 2 moles of H_2 plus 1 mole O_2 gives 2 moles of water vapor, the 5.4 moles of O_2

must give 10.8 moles of water vapor. Thus, the total fuel composition is made up of carbon and hydrogen, and nothing else.

Fuel analysis:

Weight of carbon in 12.5 moles flue gas

$$= 12.5 \times 12 = 150 \text{ lb}$$

Weight of hydrogen in 10.8 moles water vapor

$$= 10.8 \times 2 = 21.6 \text{ lb}$$

$$\text{Total weight of fuel} = 171.6 \text{ lb}$$

$$\text{Per cent } carbon = 150/171.6 \times 100 = \mathbf{87.5 \text{ per cent}}$$

$$\text{Per cent } hydrogen \text{ by difference} = 100 - 87.5 = \mathbf{12.5 \text{ per cent}}$$

Per cent excess air: This may be found from the following relation:

$$\frac{\text{Unnecessary } O_2 \text{ from flue-gas analysis}}{\text{Necessary } O_2 \text{ for combustion}}$$

$$\frac{4.1}{12.5 + 5.4} \times 100 = \mathbf{22.9 \text{ per cent}}$$

Volume of air required per pound of fuel fired taken at 78°F and 30 in. Hg for boiler-room air:

$$105.4 \times 1/171.6 \times 379 \times \frac{78 + 460}{520} \times \frac{30}{30} = \mathbf{240 \text{ cu ft}}$$

Volume of total flue gases per pound of fuel fired:

$$\frac{\text{Moles total flue gases}}{\text{lb fuel}} = \frac{100 + 10.8}{171.6} = 0.654$$

$$0.654 \times 379 \times \frac{490 + 460}{520} \times \frac{30}{29.5} = \mathbf{455 \text{ cu ft}}$$

Percentage gross heating value of fuel lost through noncondensation of water vapor formed during combustion:

$$\text{Mole water vapor per lb fuel} = 10.8/171.6 = 0.063$$

$$\text{Weight of hydrogen so involved} = 0.063 \times 18 = 1.13 \text{ lb}$$

Thus $1.13 \times 1{,}058.2/18{,}300 \times 100 = \mathbf{6.55 \text{ per cent}}$

Note that heat of condensation is taken as 1058.2 Btu per lb vapor.

Q7-7. A steam boiler is required to deliver 50,000 lb of steam per hour (evaporation rate) at 400 psia and a total steam temperature of 600°F. Assuming an efficiency of 80 per cent, calculate how many tons of coal containing 5 per cent moisture and having a dry calorific value of 14,800 Btu per lb must be supplied to the furnace per hour.

ANSWER. Some high-volatile coals are wetted down (tempered) to improve their burning characteristics. Now, the required heat output is the evaporation rate × the enthalpy of the steam at the superheater outlet.

$$50{,}000 \times (1{,}306.9 - 68) = 61.9 \times 10^6 \text{ Btu per hr}$$

The value of 68 is the enthalpy of the feed water assumed to enter the boiler at 100°F. Thus, it is perfectly acceptable to say 100 − 32 = 68 Btu per lb. The Btu fired per hour under actual fuel conditions is found in accordance with

$$14{,}800 \times 0.80 \times (1 - 0.05) \times 2{,}000 = 22.5 \times 10^6 \text{ per ton coal}$$

Finally, the firing rate is

$$61.9/22.5 = \mathbf{2.76\ tons\ tempered\ coal}$$

Up to 5 per cent moisture is added to reduce the rate of combustion of volatile matter in a coal so as to reduce smoking. Western coals usually show this characteristic to "smoke."

Q7-8. The air entering a boiler has a relative humidity of 70 per cent. Atmospheric pressure is 14.42 psia. Boiler-room temperature is 90°F. The gas used as fuel is 75 per cent methane and 25 per cent ethane by volume. The gas enters the burners with a relative humidity of 80 per cent and at a pressure of 5 psig and 76°F. The amount of air actually available for combustion is 15 per cent in excess of that theoretically required. Calculate the volumetric percentage of water vapor in the flue gas leaving through the boiler breeching.

ANSWER. It is assumed that the boiler setting is tight and that all air used in the combustion calculations enters with the fuel. Using the concept of moles or pound moles, we shall take as a basis 1 mole of the fuel. Then the amount of "bone-dry" air used per mole of fuel may be calculated.

$$\frac{(0.75 \times 2) + (0.25 \times 3.5)}{0.21} \times 1.15 = 12.95 \text{ moles "bone-dry" air}$$

This follows since 2 moles of oxygen and 3.5 moles of oxygen are needed theoretically to combine with a mole of methane and ethane, respectively. This gives the amount of theoretical total oxygen, and dividing by the percentage by volume of oxygen in dry air gives the theoretical dry air needs. Increasing this by 15 per cent gives actual dry air amount as 12.95 moles.

The flue gas from the boiler contains nitrogen, oxygen, carbon dioxide, and water vapor. The nitrogen comes through the combustion reaction chemically unchanged with the incoming air stream. The oxygen appears in the flue gas due to the fact that air in excess of theoretical requirements was used to promote combustion. If no excess air were used and intimate mixing of air and fuel took place, then no oxygen would appear in the flue gas. Carbon dioxide appears because of the complete combustion of the carbon portion of the ethane molecule and methane molecule with the oxygen entering with the air. Water vapor is the result of the combustion of the hydrogen portions of the fuel. The absence of carbon monoxide indicates complete combustion of the carbon portion of the fuel.

We shall now calculate the relative amounts of each of the constituents in the flue gas.

$$\begin{aligned} N_2 &= 0.79 \times 12.95 = 10.23 \text{ moles per mole fuel} \\ O_2 &= (0.21 \times 12.95) - [(0.75 \times 2) + (0.25 \times 3.5)] \\ &= 0.345 \text{ mole per mole fuel} \\ CO_2 &= 0.75 + (2 \times 0.25) = 1.25 \text{ moles per mole fuel} \end{aligned}$$

This last factor is true since an analysis of the chemical equation after balancing shows that for each mole of methane burned 1 mole of CO_2 is formed.

$$CH_4 + 2O_2 \rightarrow CO_2 + 2H_2O$$

And that 2 moles of CO_2 are formed when 1 mole of ethane is burned.

$$2C_2H_6 + 7O_2 \rightarrow 4CO_2 + 6H_2O$$

The water content in the flue gas is derived from three sources, namely: moisture in the fuel, moisture in the entering air, and water formed by combustion. Let us first calculate the water entering with the fuel. From steam tables the saturation vapor pressure of water at 76°F is 0.444 psia. Then the water-vapor content in the fuel is by humidity ratio

$$\frac{0.444}{(5 + 14.42) - 0.444} \times 0.80 = 0.019 \text{ mole vapor per mole dry fuel}$$

The water-vapor content of the incoming combustion air is

$$\frac{0.698}{14.42 - 0.698} \times 0.70 \times 12.95 = 0.461 \text{ mole vapor per mole dry fuel}$$

where vapor pressure of water at 90°F is 0.698 psia from steam tables.

The water formed through combustion is

$$(0.75 \times 2) + (0.25 \times 3) = 2.25 \text{ moles vapor per mole dry fuel}$$

This may be seen by referring back to the combustions formulas above.

Then the total water vapor in the flue gas is

$$0.019 + 0.461 + 2.25 = 2.73 \text{ moles per mole dry fuel}$$

The volumes of the various constituents in the flue gas are:

N_2	10.23	moles per mole dry fuel
O_2	0.345	mole per mole dry fuel
CO_2	1.25	moles per mole dry fuel
H_2O	2.73	moles per mole dry fuel

The volumetric percentage of water vapor may be now determined as follows

$$(2.73)/(10.23 + 0.345 + 1.25 + 2.73) \times 100 = \mathbf{18.8 \text{ per cent}}$$

An examination of the above problem and solution would provide the necessary information to calculate the dew point of the flue gas leaving the boiler. Assuming that the breeching pressure is 13 psia due to stack effect or because of an induced draft fan, then the partial pressure of the water vapor in the mixture is $0.18 \times 13 = 2.34$ psia.

Now refer to the steam tables; this corresponds to a saturation temperature of approximately 132°F. This is the dew point of the mixture below which water would condense out. In practice an air heater or economizer placed in the exit flue-gas stream would be designed not to cool the exit gases below 132°F. In actual installations, the flue gas leaving such a heat reclaimer would hardly ever drop below 300°F for good operating and maintenance reasons. This lower temperature limit of 300°F is far enough above the safe limit to prevent corrosion of the cooling surfaces.

Q7-9. A producer gas made from coke has the following composition by weight: $CO = 27.3$ per cent, $CO_2 = 5.4$ per cent, $O_2 = 0.6$ per cent, $N_2 = 66.7$ per cent. This gas is burned with 20 per cent excess air. If combustion is 98 per cent complete, calculate the weight and composition of the gaseous products formed per 100 lb of producer gas burned.

ANSWER. The combustion equation is $CO + \frac{1}{2}O_2 = CO_2$. Take the basis of calculation to be 100 lb of the original producer gas.

CO present $= 27.3/28 = 0.975$ lb mol
O_2 required for combustion $= 0.975/2 = 0.487$ lb mol
O_2 supplied with 20 per cent excess air $= 1.2 \times 0.487 = 0.585$ lb mol
O_2 already present in gas $= 0.6/32 = 0.019$ lb mol
O_2 to be supplied from air $= 0.566$ lb mol or 18.2 lb
Weight of air used $= 18.2/0.232 = 78.2$ lb
Weight of N_2 introduced $= 78.2 - 18.2 = 60.0$ lb
CO_2 formed in combustion $= 0.98 \times 0.975 = 0.995$ lb mol or 42.1 lb
O_2 assumed in combustion $= 0.487 \times 0.98 = 0.477$ lb mol or 15.3 lb
Total N_2 present in gases $= 60 + 66.7 = 126.7$ lb
Total CO_2 present in gases $= 42.1 + 5.4 = 47.5$ lb
Total O_2 present in gases $= 18/1 + 0.6 - 15.3 = 3.4$ lb
CO present in gases $= 0.02 \times 27.3 = 0.5$ lb
Total weight of flue gases $= 126.7 + 47.5 + 3.4 + 0.5 =$ **178.1 lb**
Total weight of gas plus air used $= 178.1$ lb
Composition of flue gases by weight:

CO_2	47.5/178.1 or **26.6 per cent**
O_2	3.4/178.1 or **1.9 per cent**
CO	0.5/178.1 or **0.3 per cent**
N_2	126.7/178.1 or **71.2 per cent**

Chapter 8

THE STEAM POWER PLANT

8-1. Testing of Steam-generating Units. The Power Test Codes of the American Society of Mechanical Engineers (ASME) go into quite a detailed discussion for the determination of the efficiency of steam-generating units. There are simpler test forms that may be used, but only for routine plant operating purposes. The ASME method should be used for contract testing and where more precise data are needed.

In order to determine the efficiency of a steam-generating unit, the *direct method* involves merely the measurement of the energy input and output. This is given in equation form as follows:

$$\text{Efficiency} = \frac{\text{Wt of steam} \times (\text{enthalpy of steam} - \text{heat of feedwater})}{\text{wt of fuel} \times \text{gross heating value of fuel}} \times 100 \qquad (8\text{-}1)$$

This determination is usually checked by means of a *heat balance.*

Because of the inaccuracy of instrumentation (weighing scales and flowmeters) it has become accepted practice to determine efficiency by the *indirect method.* Thus,

$$\text{Efficiency} = \frac{\text{heating value of the fuel} - \text{losses}}{\text{heating value of the fuel}} \times 100 \qquad (8\text{-}2)$$

In this method steam is flowmetered or the fuel is weighed only to establish the capacity at which the test is made.

In the performance test report the following should be covered: (*a*) the apparatus under test should be described; (*b*) purpose of test should be stated; (*c*) conditions of the test should be indicated, for example, capacity, type of fuel, feed-water temperature, steam temperature, excess air; (*d*) duration of test should be decided upon—a 24-hr test for direct method and especially if boiler is stoker fired; otherwise a 12-hr test would be acceptable for other fuels and methods of firing. For the indirect method, the duration may be from 4 to 6 hr.

For the *heat balance*, the following points must be determined:

(*a*) Heat absorbed by boiler unit
(*b*) Heat lost in dry flue gas
(*c*) Heat lost due to moisture in fuel
(*d*) Heat lost due to moisture from H_2 in fuel
(*e*) Heat lost due to moisture in air
(*f*) Heat lost due to incomplete combustion of carbon
(*g*) Heat lost due to unburned carbon
(*h*) Heat lost due to radiation

For formulas and methods for determining the above, the license candidate or student is referred to the Power Test Code or the standards of the industry.

Steam quality may be determined with throttling calorimeter or the separating calorimeter.

Boiler output may be obtained from the following relationship:

$$\frac{\text{Evaporation lb per hr } (H_2 - H_1)}{1{,}000} = \text{kilo Btu per hr} \quad (8\text{-}3)$$

Also

$$\text{Boiler hp} = \frac{\text{evaporation lb per hr } (H_2 - H_1)}{33{,}480} \quad (8\text{-}4)$$

$$\text{Factor of evaporation} = \frac{H_2 - H_1}{970.3} \quad (8\text{-}5)$$

where 970.3 is heat of vaporization of steam at atmospheric pressure, H_1 is enthalpy of feedwater, and H_2 is enthalpy of steam at outlet.

Q8-1. A steam generator evaporated 40,000 lb of water from a feed-water temperature of 220°F to steam at 180 psia and a quality of 97 per cent. The weight of coal fired per hour is 4,500 lb, gross heating value of coal as fired is 11,800 Btu per lb. Determine the rate of heat absorption in Btu per hr and the boiler horsepower developed.

ANSWER. From steam tables H_1 is 188.06 Btu per lb and H_2 is 1,171 Btu per lb. Then the rate of heat absorption is found from

$$40{,}000(1{,}171 - 188.06) = \mathbf{39.32 \times 10^6 \text{ Btu per hr}}$$
$$\text{Boiler hp} = 39.32 \times 10^6/33{,}480 = \mathbf{1{,}174 \text{ boiler hp}}$$

8-2. Performance of Boilers. The over-all efficiency of a boiler at any operating condition is the percentage of the heating value of the fuel which is transferred after combustion to the working substance (steam and water). Efficiency is output over input based on the gross heating value of the fuel. If we let W_f be equal to pounds of fuel fired per hour and H_f be the gross heating value of the fuel as fired in Btu per lb, then

$$\text{Over-all boiler efficiency} = \frac{\text{evaporation lb per hr } (H_2 - H_1)}{W_f H_f} \times 100 \quad (8\text{-}6)$$

Equation (8-6) includes the effect not only of the furnace, boiler, and grate, but also of heat-transfer accessories, such as superheater, water walls, economizer, and air preheater. Heat release, pounds of steam generated per pound of fuel fired, rate of heat transfer per square foot of heating surface, temperature of stack gases, and per cent CO_2 in stack gases all must be noted.

After installation and drying out of the boiler setting an efficiency test is made on the boiler. The amount of CO_2 in the flue gases at the point of maximum efficiency is noted. This is then used as the control point and is an indication of excess air. Curves plotting all these variables against rate of evaporation or boiler horsepower may be made for future reference.

Q8-2. Calculate the efficiency of a steam-generating unit consisting of an economizer, boiler, and superheater. One hundred thousand pounds of feed water enter the economizer per hour at a pressure of 200 psia and a temperature of 190°F. The steam leaves the superheater at 200 psia and a total steam temperature of 500°F. The coal as fired has a gross heating value of 12,500 Btu per lb and 5.8 short tons are fired per hour.

ANSWER. With the use of the steam tables and Mollier diagram and letting $H_1 = (190 - 32) = 158$ Btu per lb, $H_2 = 1269$ Btu per lb, $W_f = 5.8 \times 2{,}000 = 11{,}600$ lb of coal fired per hr,

$$H_f = 12{,}500 \text{ Btu per lb}$$

the over-all efficiency is

$$\frac{100{,}000 \times (1{,}269 - 158)}{11{,}600 \times 12{,}500} \times 100 = \mathbf{76.5\ per\ cent}$$

Q8-3. A gas-fired steam boiler delivers 250 lb of steam per hour of 98 per cent quality. Calculate the boiler efficiency at the following conditions: feed-water temperature 60°F, steam pressure 25 psig, barometer 30.2 in. Hg, fuel-gas pressure 4 in. water gauge (WG) at burner, gas temperature 80°F, gas consumption 800 cu ft per hr, gross heating value of fuel gas 540 Btu per cu ft at 60°F and 30 in. Hg.

ANSWER. The approach to this problem is quite similar to the previous one except for the determination of the proper heating value to use in the efficiency formula. The output of the boiler may be found in accordance with

$$250[236 + (0.98 \times 934) - 28] = 281{,}000 \text{ Btu per hr}$$

The input may be found by first correcting the gas volume from the "as-fired" condition to standard conditions, for the heating value given in the problem is for standard conditions. Another way to do this would be to correct the heating value to the gas at the actual as-fired condition. Our method of solution is to follow the first approach, i.e., correct the gas volume to standard conditions. Now, the actual gas pressure in absolute measurements is

$$30.2 + 4/13.6 = 30.5 \text{ in. Hg}$$

Then the gas volume as corrected is

$$800 \times \frac{30.5}{30.0} \times \frac{460 + 60}{460 + 80} = 782 \text{ cu ft per hr}$$

Therefore, the input will be $782 \times 540 = 422{,}000$ Btu per hr and the efficiency

$$\frac{281{,}000}{422{,}000} \times 100 = \mathbf{66.7\ per\ cent}$$

Q8-4. Exhaust steam from an engine goes into an open-type feed-water heater at 1 psig and a moisture content of 12 per cent. How many pounds of this steam are needed to raise the temperature of 1,000 lb of feedwater from 60 to 200°F?

ANSWER. If this were a closed feed-water heater the solution to the problem would be rather a simple heat balance, i.e., heat absorbed by the feedwater being equal to the heat released by the steam. However, since this is an open feed-water heater the steam mixes with the feed water so that both weights of steam condensed and feed water are combined in the solution. The 1,000 lb of heated feed water are composed of actual initial feed water plus the weight of steam condensed. Let w equal weight of steam condensed, W equal weight of actual feed water, enthalpy of feed water equal to $200 - 32 = 168$ Btu per lb at final conditions of 200°F, enthalpy of initial feed water at 60°F equal to $60 - 32 = 28$ Btu per lb. Then the heat balance is

$$(W + w)168 = (1{,}000 - w)28 + w(185 + 0.88 \times 967)$$

Solving for w, this is found to be equal to **140 lb of steam.**

Q8-5. A single-effect horizontal tube evaporator is designed to produce 30,000 lb per hr of distillate (water) at a pressure of 30 psig using steam at a pressure of 45 psig. Assume feed water at 100°F containing 150 ppm of solids and continuous blowdown limiting the concentration to 2,000 ppm. How much steam will be required?

ANSWER. Refer to Fig. 8-1. The following is the heat balance across the system.

Heat in = heat in steam + heat in feed water

Heat out = heat in distillate + heat in continuous blowdown + heat in steam heating the condensate

Thus

Heat in = heat out, neglecting radiation and other losses

If we let W be the amount of steam used, then we strike the balance

$$(W \times 1{,}177) + (32{,}250 \times 68) = (30{,}000 \times 1{,}171.5) + (2{,}250 \times 243) + (W \times 262)$$

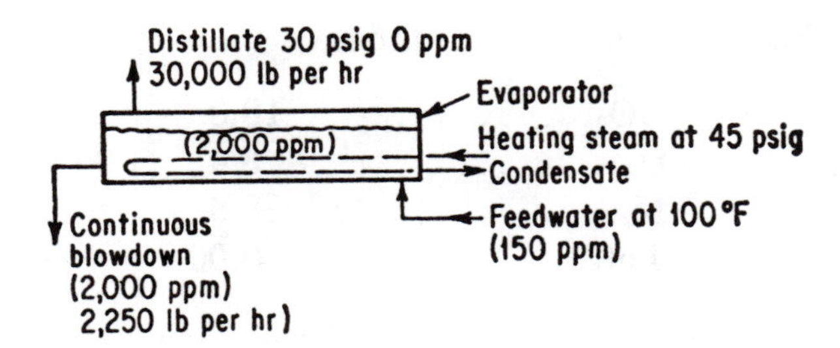

Fig. 8-1

Solving for W, this is found to be **36,600 lb** of steam. However, since the feed-water rate is given by

$$30{,}000 + [(150/2{,}000)(30{,}000)] = 30{,}000 + 2{,}250 = \mathbf{32{,}250\ lb\ per\ hr}$$

the continuous blowdown must be the difference

$$32{,}250 - 30{,}000 = 2{,}250 \text{ lb per hr}$$

Obtain enthalpies from the steam tables and Mollier diagram.

Q8-6. The following data were obtained during a test of a boiler using natural gas as fuel:

Volume of fuel used, scfh	2,895
Water rate of boiler, lb per hr	1,690
Steam pressure, psia	170
Steam temperature, °F	485
Flue-gas temperature, °F	505
Feed-water temperature, °F	145
Temperature of fuel entering burners, °F	90
Gross heating value of fuel, Btu per scf	1,024
Standard conditions of gas measurement, psia at 60°F	14.7

Fuel analysis, per cent by volume:

CH_4	93.7	per cent
C_2H_6	5.3	
CO_2	1.0	
Total	100.00	

Flue-gas analysis, per cent by volume:

CO_2	10.0
O_2	3.4
N_2	86.6
Total	100.00

Prepare a heat balance based on the above data.

ANSWER. Assume as a time basis a 1-hr period. A temperature of 60°F will be selected as a datum above which to calculate heat input and output.

Heat input.

(*a*) Heat in the fuel: volume of fuel in standard cubic feet (scf) multiplied by Btu per scf

$$2{,}895 \times 1{,}024 = 2{,}964{,}480 \text{ Btu}$$

(*b*) Heat in feed water: pounds of water used per hour multiplied by the difference between the actual feed-water temperature and 60°F.

$$1{,}690 \times (145 - 60) = 143{,}650 \text{ Btu}$$

$$\textbf{Total heat input} = 2{,}964{,}480 + 143{,}650 = \mathbf{3{,}108{,}130\ Btu}$$

Heat output.

(*a*) Heat in steam: from the steam tables, it is found that saturated steam at an absolute pressure of 170 has a temperature of 368.4°F. Then the degree superheat is the difference

$$485 - 368.4 = 116.6°\text{F}$$

The enthalpy of saturated steam (above 32°F on which the steam tables are based) at 170 psia is 1196.3 Btu per lb. In obtaining steam-table values or with the use of the Mollier diagram there may be slight differences from values in this presentation, but this is a natural occurrence and depends on the tables or diagram used. In so far as the examination is concerned, this does not make the slightest bit of difference in the credit given the written solution.

The heat of superheat may be either obtained from the steam tables or from the Mollier diagram. It may also be calculated, as we shall see here. The heat of superheat is calculated by multiplying the degrees of superheat by the mean specific heat of steam which may be obtained from tables. Then the heat of superheat is

$$116.6 \times 0.582 = 67.9 \text{ Btu per lb}$$

The total heat of the steam above the 32°F datum is

$$1196.3 + 67.9 = 1264.2 \text{ Btu per lb}$$

Now, since the steam tables are based on a datum of 32°F and these calculations are based on a datum of 60°F, a correction must be made corresponding to the heat of the liquid at 60°F, namely, 28.1 Btu per lb. Thus, the heat in each pound of steam above the 60°F datum is 1264.2 − 28.1 = 1236.1 Btu per lb.

The total heat in the steam on the hourly basis is 1,236.1 × 1,690, or 2,089,000 Btu.

(*b*) Sensible heat loss in dry flue gases: We remember that one mole of gas at 14.7 psia and 60°F will occupy 379 cu ft. Then, on this test, 2,895/379 = 7.638 moles of gas are burned per hour. It may be found that the ratio of flue gas to fuel is 10.53. So that 7.638 × 10.53 = 80.43 moles of flue gas leave the boiler each hour. By breakdown:

$$\begin{aligned} CO_2 &= 80.43 \times 0.1 = 8.043 \text{ moles} \\ O_2 &= 80.43 \times 0.034 = 2.735 \text{ moles} \\ N_2 &= 80.43 \times 0.866 = 69.652 \text{ moles} \end{aligned}$$

Then the sensible heat loss in the dry flue gas may be obtained by converting moles to pounds and multiplying by the specific heat at constant pressure and by the temperature difference.

$$\begin{aligned} CO_2 &= 8.043 \times 44 \times 0.202 \times (505 - 60) = 31{,}811 \text{ Btu} \\ O_2 &= 2.735 \times 32 \times 0.219 \times (505 - 60) = 8{,}529 \text{ Btu} \\ N_2 &= 69.652 \times 28 \times 0.248 \times (505 - 60) = 215{,}230 \text{ Btu} \end{aligned}$$

$$\text{Total sensible heat loss} = 31{,}811 + 8{,}529 + 215{,}230 = 255{,}570 \text{ Btu}$$

(*c*) Heat loss to water vapor formed during combustion: It may be found that for each mole of flue gas 0.193 mole of water vapor was formed. Then the total water formed per hour is

$$0.193 \times 80.43 = 15.52 \text{ moles water}$$

The sensible heat in the water formed by combustion may be determined as

$$15.52 \times 18 \times 0.30 \times (505 - 60) = 37{,}295 \text{ Btu}$$

Since the gross heating value was used in calculating the heat input, the heat of vaporization of the water must be included as a part of the heat output. From the steam tables the heat of vaporization of water at 60°F is 1059 Btu per lb, then this item of heat loss is $15.52 \times 18 \times 1059 = 295{,}842$ Btu.

The total heat loss to water formed during combustion, then, is $37{,}295 + 295{,}842 = 333{,}137$ Btu.

(*d*) Heat loss due to incomplete combustion: None, since the flue gas contains no CO.

(*e*) Radiation and other losses: The difference between the heat input and other items of heat output is the radiation and other losses which in this problem are

$$3{,}108{,}130 - (2{,}089{,}000 + 257{,}200 + 333{,}137) = 428{,}793 \text{ Btu}$$

The completed heat balance is shown in Table 8-1.

TABLE 8-1

	Btu per hr	Per cent
Input:		
Heat in fuel	2,964,480	95.38
Heat in feed water	143,650	4.62
Total	3,108,130	100.00
Output:		
Heat in steam	2,089,000	67.21
Heat loss in flue gas		
Sensible heat in dry gas 257,200; Heat in water vapor 333,137	590,337	19.05
Due to incomplete combustion (none)		
Radiation and unaccounted for losses	428,793	13.74
Total	3,108,130	100.00

Q8-7. A boiler plant can burn coal costing \$5 per ton, or oil costing 4 cents per gallon, or natural gas costing 15 cents per 1,000 cu ft. Assume Btu values and combustion efficiency ratings for the fuels listed and choose the most economical fuel for producing 1,000 lb of steam at 250 psig and 150°F superheat, using feed water at a temperature of 180°F.

ANSWER. The solution to this problem is the same even though the unit fuel costs may vary. These stated costs are low, but the problem may be worked out using the same approach as will be shown.

Steam at 250 psig and 150°F superheat.........	1292 Btu per lb
Feed water at 180°F (180 − 32)...............	−148 Btu per lb
	1144 Btu per lb

Total Btu to be absorbed = 1,144 × 1,000 = 1,144,000

Coal. Assume 13,500 Btu per lb and 85 per cent efficiency:

$$\frac{1{,}144{,}000 \times 5.00}{13{,}500 \times 0.85 \times 2{,}000} = \$0.248$$

Oil. Assume 18,500 Btu per lb and 85 per cent efficiency:

$$\frac{1{,}144{,}000 \times 0.04}{18{,}500 \times 0.85 \times 8.33 \times 0.80} = \$0.437$$

Gas. Assume 1000 Btu per cu ft and 75 per cent efficiency:

$$\frac{1{,}144{,}000 \times 0.15}{1{,}000 \times 0.75 \times 1{,}000} = \$0.228$$

Therefore, *gas* is the most economical fuel under these conditions.

Q8-8. A stoker-fired boiler plant having a Dutch-oven type of furnace was operated for a year with an average flue-gas analysis of 13 per cent CO_2, 0 per cent CO, and 6.25 per cent O_2. Combustible matter to ashpit was 10 per cent. The second year an effort was made to raise the efficiency and an average flue-gas analysis of 15 per cent CO_2, 0.1 per cent CO, and 3.9 per cent O_2 was maintained with 16 per cent combustible matter in the ashpit. A high grade of Pennsylvania bituminous slack coal was burned both years, having 7 per cent ash and 14,200 Btu per lb bone dry. At the end of the second year it was found that the efficiency was the same as for the first year of operation, and yet the cost per 1,000 lb of steam had increased 2 per cent. Explain.

ANSWER. The steam load is assumed to be the same for both years. The analysis of flue gas indicates that a lower percentage of excess air was used for the second year. This would result in a higher furnace temperature, which would increase the maintenance cost of the furnace and stoker. There would also be somewhat more ash to remove on account of the higher combustible content in the ashpit. These factors would explain the increased cost of steam generation for the second year.

Q8-9. Why and when are high pressures and temperatures justified from the standpoint of fuel economy in steam generating practice? Assume your own figures and illustrate your answer.

ANSWER. High pressures and temperatures are justified from the standpoint of fuel economy in steam-generating practice, primarily from the resulting higher heat cycle efficiency. High pressures and temperatures increase the amount of heat available for transformation to work with comparatively slight increase or even less fuel consumption as compared with low pressures and temperatures. This is illustrated below.

(*a*) *Pressure study.* Compare steam at 215 psia with 1,200 psia. Steam is dry and saturated in both cases and with 2 in. Hg absolute exhaust pressure after adiabatic expansion.

Initial pressure, psia	215	1,200
Btu per lb initially	1,199	1,181
Btu per lb finally	859	761
Btu available for work (difference)	340	420

$$\text{Thermodynamic gain} = \frac{420 - 340}{340} \times 100 = 23.5 \text{ per cent}$$

If the same feed-water temperature, namely, 212°F, is assumed for both pressures, the heat added per pound of steam can be found.

$$\begin{aligned} \text{Btu per lb of steam} &= 1{,}199 = 1{,}181 \\ \text{Btu per lb feed water } (212 - 32) &= \phantom{1{,}}180 = \phantom{1{,}}180 \\ \text{Btu added} &= 1{,}019 = 1{,}001 \end{aligned}$$

Less amount of heat supplied

$$= \frac{1{,}019 - 1{,}001}{1{,}019} \times 100 = 1.75 \text{ per cent}$$

(*b*) *Temperature study.* Compare saturated steam with superheated steam (200°F superheat) both at 215 psia and with 2 in. Hg absolute exhaust pressure after adiabatic expansion.

$$\begin{aligned} \text{Superheat, °F} &= \quad 0 = \quad 200 \\ \text{Btu per lb initially} &= 1{,}199 = 1{,}313 \\ \text{Btu per lb finally} &= \underline{\ \ 859} = \underline{\ \ 927} \\ \text{Btu available for work} &= \ \ 340 = \ \ 386 \end{aligned}$$

$$\text{Thermodynamic gain} = \frac{386 - 340}{340} \times 100 = 13.5 \text{ per cent}$$

If the same feedwater temperature, namely 212°F is assumed for both cases, the heat added per lb of steam can be found.

$$\begin{aligned} \text{Btu per lb steam} &= 1{,}199 = 1{,}313 \\ \text{Btu per lb feed water} &= \underline{\ \ 180} = \underline{\ \ 180} \\ \text{Btu added} &= 1{,}019 = 1{,}133 \end{aligned}$$

Greater amount of heat supplied

$$= \frac{1{,}133 - 1{,}019}{1{,}019} \times 100 = 11.2 \text{ per cent}$$

It will be noted from the above that more heat is available for work with less fuel consumption for the higher pressure and more heat available for work with slightly greater fuel consumption for the higher superheat. While this is not always true, it is in general so. Of course, there are other considerations in going to higher pressures and temperatures, such as increasing steam velocity to turbine blades, prevention of turbine-blade erosion, smaller pipes due to smaller specific volume of steam, etc.

Chapter 9

STEAM ENGINES

9-1. Description. A steam engine is a reciprocating machine having displacement in which work is done by a piston acted upon by pressure and moving it. Steam engines are generally classified as to construction, operation, or type of valve gear. The construction may be horizontal, vertical, or angular. They may be single-acting, double-acting, reciprocating, or rotary. Their operation may be condensing, noncondensing, bleeder, or extraction, simple or multiple stage.

According to the type of valves employed, the classification may be divided into simple, compound, uniflow, D-shaped slide valve, rotary valves, poppet, and piston valves.

Q9-1. A simple double-acting steam engine of 15-in. bore, 16-in. stroke, and 2.5-in.-diameter piston rod is operating at 225 rpm. Find the indicated horsepower if the scale or range of the spring used is 80 psi and the mean effective pressure determined by the usual method is 25.3 at the head end and 25.9 at the crank end.

ANSWER. Note carefully that the area of the rod must be subtracted from the piston area when determining the crank-end horsepower.

Piston area at head end = $15^2 \times 0.785$ = 176.5 sq in.
Piston-rod area = $2.5^2 \times 0.785$ = 4.9
Piston area at crank end by difference = 171.6 sq in.
Stroke = $^{16}/_{12}$ = 1.333 ft
Head end ihp = $25.3 \times 1.333 \times 176.5 \times 225/33{,}000$ = 40.6 ihp
Crank end ihp = $25.9 \times 1.333 \times 171.6 \times 225/33{,}000$ = 40.4 ihp
Total ihp = 40.6 + 40.4 = **81.0 ihp**

If friction hp were 10, then the brake (shaft) hp would be by difference

$$\text{bhp} = 81 - 10 = 71 \text{ bhp}$$

Q9-2. Indicator diagrams taken from a single-cylinder, double-acting steam engine showed the following measurements:

	Head end	Crank end
Length of diagram, in.	2.95	2.95
Height of steam line, in. (YY' in Fig. 9-3)	1.47	1.41
Height of point of admission, in. (DD')	0.50	0.46
Length of steam line, in. (AB)	1.86	1.86
Indicator spring scale	120	120
Other data include:		
Clearance ratio on both ends, per cent	16	16
Diameter of piston rod, in.	2.5	2.5
Cylinder bore, in.	....	14
Piston stroke, in.	....	14

Calculate the steam consumption of the engine under the given conditions, pounds per hour, when operating at a constant speed of 200 rpm.

ANSWER. Consider the head end. The piston displacement on the head end is calculated as follows:

$$D_h = \frac{\pi \times 14^2 \times 14}{4 \times 1{,}728} = 1.247 \text{ cu ft}$$

The cutoff C is $\frac{1.86}{2.95} = 0.63$. The pressure of the steam, as indicated along the stream line AB, is $1.47 \times 120 = 176$ psi.

The pressure of the steam in the clearance space at the point of admission is $0.5 \times 120 = 60$ psi.

From steam tables, the specific volume of dry saturated steam at 176 psig or 190 psia is 2.404 cu ft per lb. Then the density d_1 is $\frac{1}{2.404}$, or 0.416 lb per cu ft.

The specific volume of dry saturated steam at 60 psig is 5.916 cu ft per lb or density of $1/5.916 = 0.169$ lb per cu ft.

Substituting values in Eq. (9-2),

$$W = 60 \times 200 \times 1.247[0.416(0.16 + 0.63) - (0.169 \times 0.16)]$$
$$W = 4{,}513 \text{ lb per hr}$$

Similarly, values for the crank end are calculated and substituted in the equation

$$W = 60 \times 200 \times 1.207[0.400(0.16 + 0.63) - (0.158 \times 0.16)]$$
$$W = 4{,}215 \text{ lb per hr}$$

Then the total steam consumption of the engine is

$$4{,}513 + 4{,}215 = \mathbf{8{,}728\ lb\ per\ hr}$$

Q9-3. A test of a steam engine shows that it uses 1,560 lb of steam per hr when developing 40 bhp. At the same time, the engine is developing an indicated horsepower of 50. The inlet steam is 95 per cent quality at 250 psia. Exhaust or back pressure is 14.7 psia. Calculate (*a*) brake thermal efficiency and (*b*) indicated thermal efficiency.

ANSWER.

(*a*) Steam consumption of engine is W_s = 1,560/40, or 39 lb per bhp-hr. The enthalpy in 1 lb of steam at throttle conditions is given by

$$H_1 = 376.04 + 0.95(825.4) = 1{,}160.17 \text{ Btu per lb}$$

The enthalpy of the liquid for 1 lb of exhaust steam at 14.7 psia and operating condensing is 180.07 Btu.
Then, *brake thermal efficiency* is

$$E_t = \frac{2{,}545}{39(1{,}160.17 - 180.07)} \times 100 = \mathbf{6.66\ per\ cent}$$

(*b*) The steam consumption based on the indicated hp is 1,560/50, or 31.2 lb per bhp-hr. Then *indicated thermal efficiency* is

$$E_t = \frac{2{,}545}{31.2(1{,}160.17 - 180.07)} \times 100 = \mathbf{8.32\ per\ cent}$$

Q9-4. What is the horsepower constant of an 18 by 48-in. engine running at 74 rpm, and what would be the mean effective pressure for the development of 375 indicated horsepower?

ANSWER.

$$\text{Horsepower constant} = \frac{1 \times 48/12 \times (18 \times 18 \times 0.7854) \times 74 \times 2}{33{,}000}$$
$$= 4.565$$

Development of 375 ihp would require $375/4.565 =$ **82.15 mep.**

Q9-5. An engine uses 25 lb of steam per indicated horsepower-hour and the evaporative economy of the boiler, under the operating conditions, is 8 lb of steam per pound of coal. Assume that the heating value of the coal is 10,000 Btu per lb, what percentage of energy contained in the coal is realized by the engine?

ANSWER. The fuel consumption is $25/8$, or 3.125 lb coal per hp-hr. Now, since 1 hp-hr is equal to 33,000 × 60, or 1,980,000 ft-lb, there is 1,980,000/3.125, or 633,600 ft-lb realized per lb of coal. One Btu is equivalent to 778 ft-lb, and since each lb of coal has 10,000 Btu, the energy in a pound of coal would be 7,778,000 ft-lb. Finally, the energy realized by the engine per pound of coal used would be

$$\frac{633{,}600 \times 100}{7{,}778{,}000} = \mathbf{8.15\ per\ cent}$$

Q9-6. A simple steam engine direct-connected to a 200-kw generator uses 47 lb of steam per kilowatt-hour under full load. Steam pressure is 115 psia. Initial quality is 98 per cent with back pressure of 17 psia. Calculate the thermal efficiency of the engine.

ANSWER.

$$\frac{3{,}415}{47 \times (1{,}171 - 187)} \times 100 = \mathbf{7.3\ per\ cent}$$

Note that we use the enthalpy of the saturated liquid (187 Btu per lb) since the cycle is considered as a whole, including the condenser. If the machine operated noncondensing, the enthalpy of the steam at the back-pressure condition is used.

Chapter **10**

STEAM TURBINES AND CYCLES

10-1. General Considerations. Steam turbines have replaced steam engines in many applications in power plant services, for driving centrifugal pumps, for driving generators in electric power stations, and in many industrial plant applications. This is due primarily to the turbine's compactness, higher speed, and fewer moving parts. The steam turbine has about the same economy as the steam engine when operating with vacuums from 24 to 26 in. Hg. However, the turbine can operate successfully with 29 in. Hg vacuum. These latter conditions do not materially improve performance of the engine because of the increased specific steam volumes. The economy of noncondensing steam turbines is considerably below that of high-grade reciprocating steam engines. The steam turbine gives better speed regulation and its exhaust is not contaminated by cylinder lubricating oils. The small turbine is found useful where naturally its speed can be utilized advantageously as in driving rotary pumps and blowers. Turbine exhaust steam may be used for process and heating purposes such as is common in oil refinery practice. Over-all heat economy is helped considerably in this way. Very often steam turbines are used for standby service for electric motors and electric generators for emergency purposes.

TABLE 10-1. STEAM CONSUMPTION OF PRIME MOVERS

	Lb/hp-hr
Simple noncondensing engines	29–45
Simple noncondensing corliss engines	26–35
Compound noncondensing engines	19–28
Compound condensing engines	12–22
Simplex-duplex steam pumps	120–200
Turbines noncondensing	21–45
Turbines condensing	9–32

Q10-1. Determine the efficiency of a condensing turbine operating on a Rankine cycle as follows:

Throttle pressure	200 psia
Steam temperature	600°F
Exhaust (back) pressure	2 in. Hg
Exhaust temperature	101°F
Steam rate	10.5 lb/hp-hr

ANSWER. From the Mollier diagram:

Enthalpy at inlet (H_1)	1322 Btu/lb
Enthalpy at outlet (H_2)	936 Btu/lb
Available energy	386 Btu/lb

The theoretical steam rate is 2,545/386 = 6.6 lb/hp-hr. Thus, the *Rankine efficiency* of the turbine is 6.6/10.5, or **63 per cent.**

The *theoretical thermal efficiency* of the cycle is the ratio between the available (1322 − 936 = 386 Btu per lb) and the energy supplied to the condensate above 32°F, which is

$$1322 - 101 - 32 = 1253 \text{ Btu per lb}$$

Then the *ideal Rankine cycle* is determined as 386/1,253, or **30.8 per cent.**

The Rankine cycle may be improved by increasing the initial pressure and temperature. Then, in order to avoid excessive condensation in the lower pressure stages, this pressure increase must go hand-in-hand with an increase in temperature.

Q10-2. In a power plant, the steam supply is at 400 psia and a temperature of 700°F. After expansion in the turbine until the pressure is reduced to 80 psia, the steam is reheated to 700°F. Upon completing the expansion in the last stages of the turbine, the steam is exhausted at 1 in. Hg absolute. (*a*) For the corresponding ideal vapor cycle, find the cycle efficiency. (*b*) If the steam were not reheated, what would be the corresponding Rankine vapor cycle efficiency? (*c*) What is the chief advantage due to reheating?

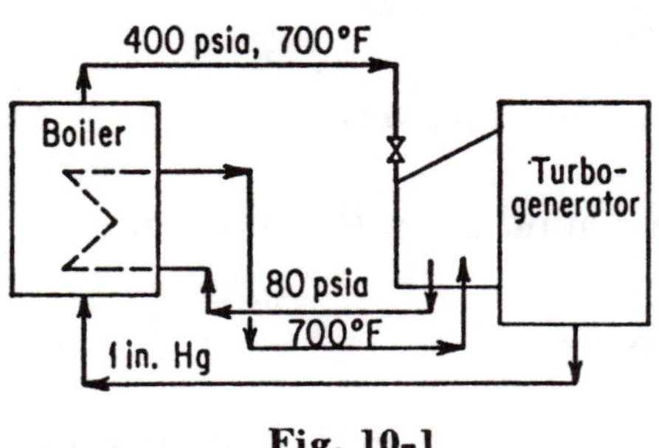

Fig. 10-1

ANSWER. Refer to Figs. 10-1 to 10-3. Make free use of Mollier diagram and steam tables. Now note the following points:

$$H_a = 1362.3 \text{ Btu/lb} \qquad S_d = S_e = 1.8274$$
$$H_b = 1199.38 \text{ Btu/lb} \qquad S_a = S_b = 1.6396$$
$$H_d = 1379.2 \text{ Btu/lb}$$

Quality at point e may be obtained in the usual manner for mixture of liquid and vapor. This is found to be 89.2 per cent. In like

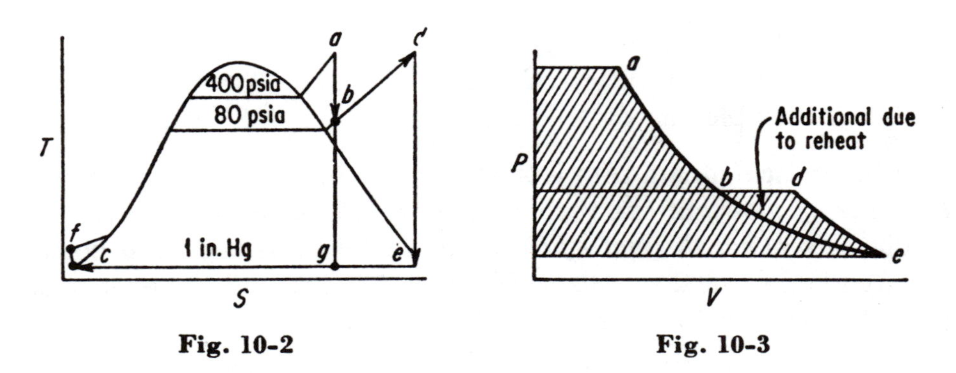

Fig. 10-2

Fig. 10-3

manner, H_e is found to be equal to 981.06 Btu/lb. Also H_c is equal to 47.06 Btu/lb.

$$F = A(P_j - P_c)(144)\, \bar{V}_c$$
$$= (1/778)(400 - 0.4912)(144)(0.01607) = 1.19 \text{ Btu/lb}$$

(*a*) The *ideal cycle efficiency* is found in accordance with the following formula:

$$\frac{H_a - H_b + H_d - H_e - F}{H_a - H_c + H_d - H_b - F} \times 100$$

And by substitution in this formula the efficiency is found to be **37.4 per cent**

(*b*) The Rankine cycle efficiency is found from this relation:

$$\frac{H_a - H_g}{H_a - H_c} \times 100$$

And by substitution in this formula the *Rankine efficiency* is found to be equal to **36.5 per cent,** with the help of the following relations:

$$S_a = S_g = 1.6396 = S_f + x_g \times 1.9452$$

From which x_g is calculated out as 0.7959. Note that H_g is equal to 881.06 Btu/lb.

(*c*) The chief advantage due to reheating is in the increased quality at the lower pressures, improving the engine efficiency markedly. But an optimum must be reached where too high a quality must not be given to the steam. In this manner too much energy is not given to the condenser (increasing unavailability) and blade erosion is reduced in the turbine if the quality is too low.

Q10-3. An actual test on a turbogenerator gave the following data: 29,760 kw delivered with a throttle flow of 307,590 lb steam per hour under the following conditions: throttle pressure 245 psia, superheat at throttle 252°F, exhaust pressure 0.964 in. Hg abs., absolute pressure at the one bleeding point 28.73 in. Hg, temperature of the feed water leaving the bleeder heater 163°F. For the corre-

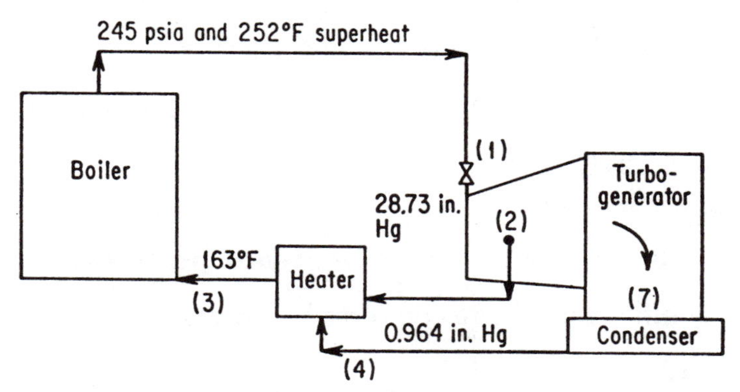

Fig. 10-4

sponding ideal unit find (*a*) per cent throttle steam bled at the one stage, (*b*) net work for each pound of throttle steam, (*c*) ideal steam rate, (*d*) cycle efficiency. For the actual unit find (*e*) the combined

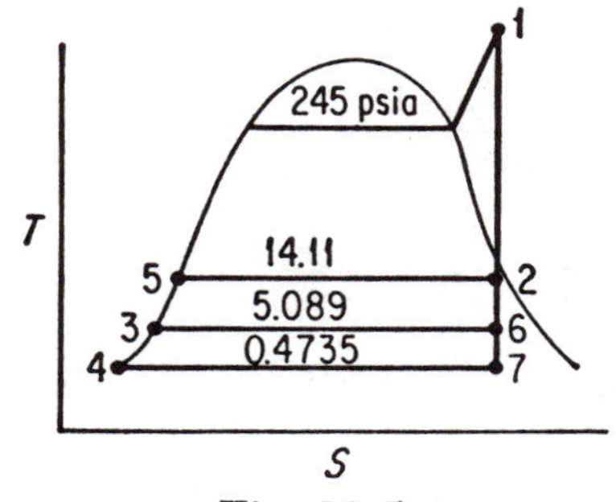

Fig. 10-5

steam rate, (*f*) combined thermal efficiency, and (*g*) combined engine efficiency.

ANSWER. Refer to the steam flow sheet Fig. 10-4 and the *TS* diagram Fig. 10-5 above. Now with the use of the steam tables and Mollier diagram determine the following:

$S_1 = 1.676$
$H_1 = 1366$ Btu per lb
$H_2 = 1106$ Btu per lb
$P_2 = 14.11$ psia
$H_3 = 130.85$ Btu per lb
$P_3 = 5.089$ psia
$H_4 = 46.92$ Btu per lb
$P_4 = 0.4735$ psia
$H_5 = 177.90$ Btu per lb

(*a*) Per cent throttle steam bled:

$$\frac{H_5 - H_4}{H_2 - H_4} \times 100 = \mathbf{12.41\ per\ cent}$$

(*b*) Heat converted to work:

$$H_1 - H_2 + (1 - m_2)(H_2 - H_7) = \mathbf{419.1\ Btu/lb}$$

where $m_2 = 0.1241$.

(*c*) Ideal steam rate:

$$(3{,}413/419.1) = \mathbf{8.14\ lb\ steam\ per\ kwhr}$$

(*d*) Cycle efficiency (heat converted into work/heat supplied):

$$\frac{419.1}{H_1 - H_3} = \frac{419.1}{1{,}366 - 130.85} \times 100 = \mathbf{36\ per\ cent}$$

(*e*) Combined steam rate: lb steam consumed/kwhr generated

$$307{,}590/29{,}760 = \mathbf{10.34\ lb\ per\ kwhr}$$

(*f*) Combined thermal efficiency:

$$\frac{3{,}413}{\text{heat supplied}} = \frac{3{,}413}{10.34 \times (H_1 - H_3)}$$

$$\frac{3{,}413}{10.34 \times 1{,}235.15} \times 100 = \mathbf{27.3\ per\ cent}$$

(*g*) Combined engine efficiency = (27.3/36) × 100 = **75.7 per cent**

Q10-4. A turbogenerator is operated on the reheating-regenerative cycle with one reheat and one regenerative feed-water heater.

Throttle steam at 400 psia and 700°F. total steam temperature is used. Exhaust is at 2 in. Hg abs. Steam is taken from the turbine at a pressure of 63 psia for both reheating and feedwater heating. Reheat is to 700°F. For the ideal turbine working under these conditions find: (*a*) percentage of throttle steam bled for feed-water heating, (*b*) heat converted to work per pound of throttle steam, (*c*) heat supplied per pound of throttle steam, (*d*) ideal thermal efficiency, (*e*) draw *TS* diagram and layout showing boiler, turbine, condenser, feed-water heater and piping and using the same letters to designate corresponding points on the two diagrams.

ANSWER. Refer to Figs. 10-6 and 10-7. Use the steam tables and Mollier diagram. Values obtained by the student may differ

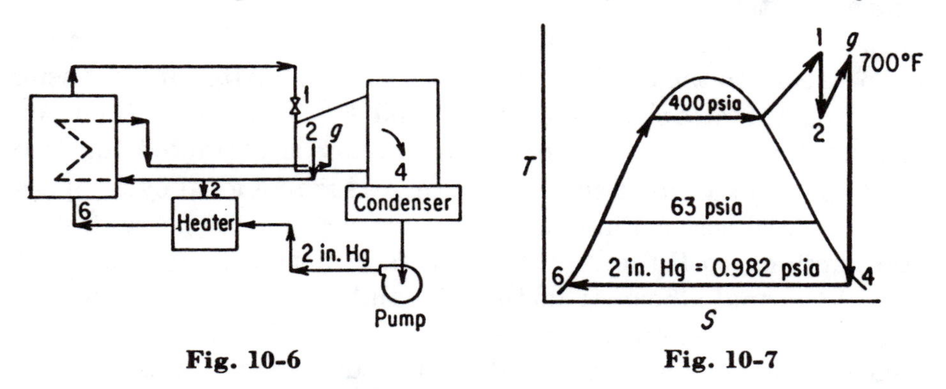

Fig. 10-6   Fig. 10-7

slightly because of the steam tables used and interpolations within the Mollier diagram but this is not to be disturbing in any way.

$P_1 = 400$ psia $H_2 = 1178$ Btu per lb
$t_1 = 700$°F $H_g = 1380.1$ (from steam tables)
$H_1 = 1362.3$ Btu per lb $S_g = 1.8543$ (from steam tables)
$S_1 = 1.6396$

(*a*) Per cent throttle steam bled:

$$\frac{H_6 - H_5}{H_2 - H_5} = \frac{196.17}{1107.9} = 0.1771, \text{ or } \mathbf{17.71 \text{ per cent}}$$

(*b*) Heat converted to work per pound throttle steam:

$$(H_1 - H_2) + (0.8229)(H_g - H_4) = \mathbf{467.3 \text{ Btu/lb}}$$

(*c*) Heat supplied per pound throttle steam:

$$(H_1 - H_6) + (H_g - H_2) = \mathbf{1299.1 \text{ Btu/lb}}$$

(*d*) Ideal thermal efficiency:

$$(467.3)/(1{,}299.13) \times 100 = \mathbf{36.1\ per\ cent}$$

In Q10-3, p. 151, the factor m_2 is the decimal part of the original bleed at point 2. If we let w_1 be the original weight of steam and sufficient steam is bled so that it is just condensed, the heat removed from the bleed steam is given by

$$m_2 w_1 = (H_2 - H_f)$$

where H_f is the enthalpy of liquid at the corresponding pressure. For more penetrating coverage see standard handbooks and texts on the subject.

Q10-5. Steam is supplied at 600 psia and 740°F to a steam turbine operating on the reheat-regenerative cycle. After an expansion to 100 psia the steam is reheated to 700°F. Expansion then continues to 1 psia but at 30 psia some steam is extracted for feed-water heating in a direct contact heater. Assume ideal cycle operation with no losses. Find (*a*) steam extracted as a percentage of steam supplied to throttle, (*b*) steam rate in pounds per kilowatt-hour, (*c*) thermal efficiency of turbine, and (*d*) quality or superheat of exhaust if in the actual turbine combined efficiency is 72 per cent, generator efficiency is 94 per cent, and actual extraction is the same as the ideal.

ANSWER. From the steam tables and Mollier diagram the following hold.

$P_1 = 600$ psia	$P_e = 1$ psia	$S_3 = 1.802$
$t_1 = 740$°F	$H_1 = 1372$ Btu per lb	$H_x = 1245$ Btu per lb
$P_2 = 100$ psia	$S_1 = 1.605$	$H_e = 1007$ Btu per lb
$t_3 = 700$°F	$H_2 = 1188$ Btu per lb	$H_f = 70$ Btu per lb
$P_x = 30$ psia	$H_3 = 1377$ Btu per lb	$H_{fx} = 219$ Btu per lb

Also refer to Fig. 10-8*a* of skeleton Mollier diagram and flow sheet Fig. 10-8*b*.

(*a*) Steam extracted for feed-water heater:

$$\frac{H_{fx} - H_f}{H_x - H_f} \times 100 = \mathbf{12.68\ per\ cent,}\ \text{or } x = 0.1268$$

(*b*) For the Rankine-cycle steam rate:

$$w_s = \frac{3413}{H_1 - H_c}$$

For this cycle:

$$w_s = \frac{3413}{(H_1 - H_2) + x(H_3 - H_x) + (1 - x)(H_3 - H_c)}$$

$$= \mathbf{6.52\ lb\ per\ kwhr}$$

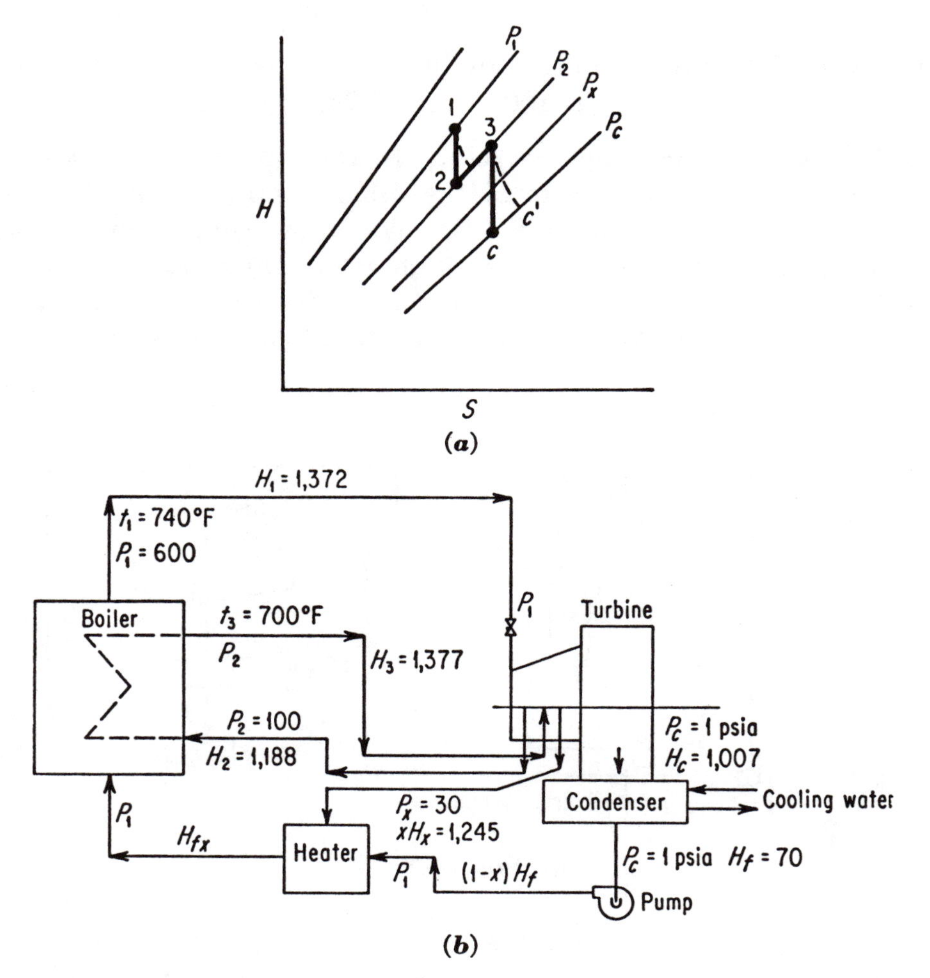

Fig. 10-8

(*c*) Thermal efficiency:

$$E_t = \frac{(H_1 - H_2) + x(H_3 - H_x) + (1 - x)(H_3 - H_c)}{(H_3 - H_2) + (H_1 - H_{fx})} \times 100$$

$$= \mathbf{39\ per\ cent}$$

It is of interest to note that in an ideal cycle the thermal efficiency of the turbine is the same as that of the cycle.

(*d*) Condition of exhaust. Now the engine efficiency of the turbine alone is 0.72/0.94 = 0.765. Then

$$H_3 - H_{c'} = 0.765(H_3 - H_c) = 283$$
$$H_{c'} = H_3 - 283 = 1{,}094$$

From the Mollier diagram the condition at $H_{c'}$ is 1.1 per cent moisture, from which quality is 100 − 1.1 = **98.9 per cent.**

10-2. The Mercury-vapor Cycle. In the mercury-vapor-steam cycle the basis is the effects of the difference in thermodynamic properties of the two pure fluids. We know that steam works under relatively high pressures with an attendant relative low temperature. Mercury, on the other hand, has its vapor characteristic as operating under low pressures with attendant high temperature.

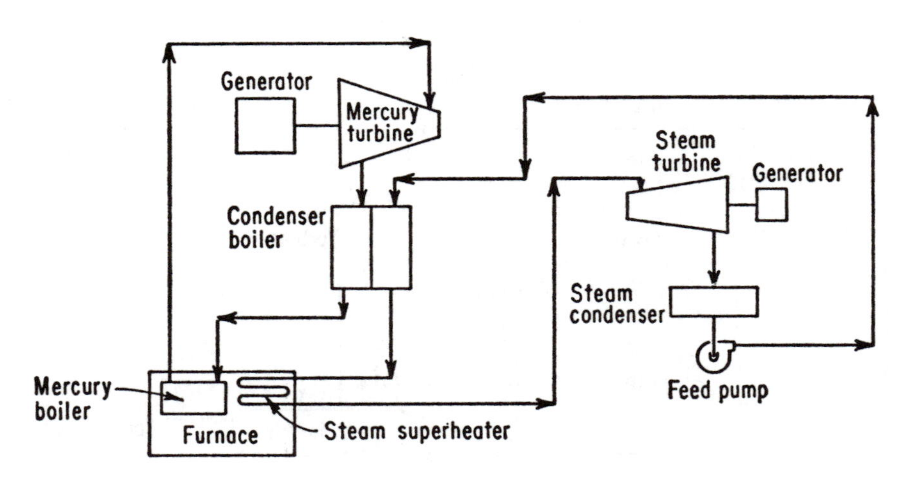

Fig. 10-9

In the cycle the pressures are so selected that the mercury vapor condenses at a temperature higher than that at which steam evaporates. The processes of mercury vapor condensation and steam evaporation take place in a common vessel called the *condenser-boiler*, which is the heart of the cycle. In the steam portion of the cycle, condenser water carries away the heat of steam condensation; in the mercury portion of the cycle, it is the steam which picks up the heat of condensation of the mercury vapor. Thus, there is a great saving in heat and the economies effected reflect consequent improvement in cycle efficiency.

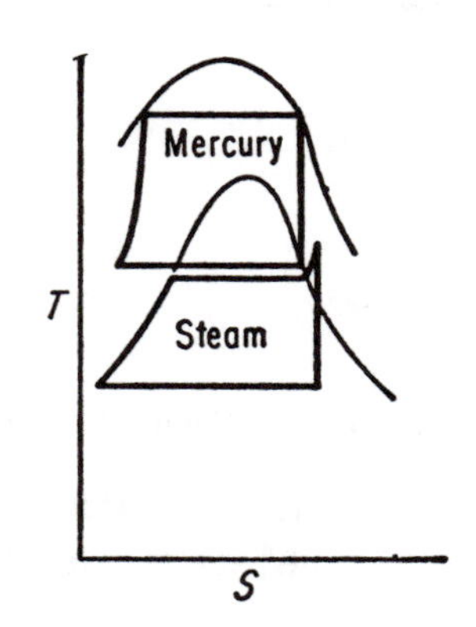

Fig. 10-10

Refer to Fig. 10-9 for the flow sheet hookup and to Fig. 10-10 for the process shown on the *TS* diagram.

The same furnace serves the mercury boiler and the steam superheater. Mercury vapor is only condensed, not superheated. Now, if the condenser-boiler is high enough above the mercury boiler, the head of mercury is great enough to return the liquid mercury to the boiler by gravity, making the use of a mercury feed pump unnecessary.

Q10-6. A binary cycle steam and mercury plant has a maximum temperature of 1000°F for both. The mercury is condensed in the steam boiler at 10 psia and the steam pressure is 1,200 psia. Condenser pressure is 1 psia. Expansions in both turbines are assumed to be constant entropy. Steam cycle has superheat, but no reheat or regeneration. Find the efficiency of the cycle as described. Find its efficiency without the mercury.

ANSWER. Refer to Fig. 10-11 for flow sheet. Now set up two columns.

Mercury cycle

$H_{m1} = 151.1$ Btu per lb
$S_{m1} = 0.1194$
At 10 psia, $S_{me} = 0.1194$

Quality x is found as follows:

$$0.1194 = 0.0299 + x(0.1121)$$
$$x_m = 0.798$$
$$H_{me} = 22.6 + 0.798(123) = 120.7$$

Steam cycle

$H_{s1} = 1499.2$ Btu per lb
$S_{s1} = 1.6293$
At 1 psia, $S_{se} = 1.6293$

Quality x is found as follows:

$$1.6293 = 0.1326 + x(1.8456)$$
$$x_s = 0.81$$
$$H_{se} = 69.7 + 0.81(1{,}036.3) = 69.7 + 839 = 908.7$$

Mercury cycle

$H_{mf} = 22.6$

Steam cycle

$H_{sf} = 69.7$

Assume 98 per cent quality steam leaving the mercury condenser. Then

$$H_{sw} = 571.7 + 0.98(611.7) = 1{,}171.7$$

Balance around the mercury condenser:

$$\text{Steam heat gain} = 1171.7 - 69.7 = 1102 \text{ Btu per lb}$$
$$\text{Mercury heat loss} = 120.7 - 22.6 = 98.1 \text{ Btu per lb}$$

Therefore, weight of mercury per lb steam = 1,102/98.1 = 11.23

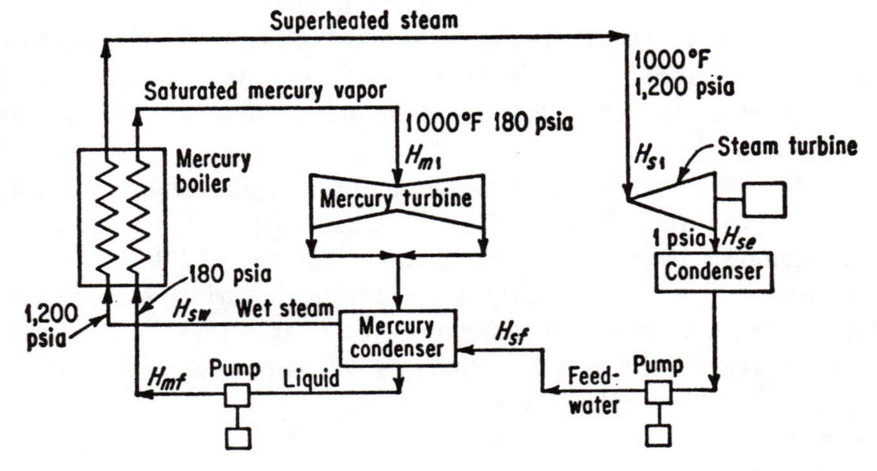

Fig. 10-11

Heat input per pound of steam:

$$\text{For mercury} = 11.23(151.1 - 22.6) = 1442$$
$$\text{For steam} = 1{,}499.2 - 908.7 = \underline{327.5}$$
$$\text{Total} = 1769.5 \text{ Btu}$$

Work done per pound of steam:

$$\text{For mercury} = 11.23(151.1 - 120.7) = 342$$
$$\text{For steam} = 1{,}499.2 - 908.7 = \underline{590.5}$$
$$\text{Total} = 932.5 \text{ Btu}$$

Binary cycle efficiency

$$(932.5/1{,}769.5)100 = \mathbf{52.7\ per\ cent}$$

Steam cycle efficiency without mercury topping turbine

$$(590.5)/(1{,}499.2 - 69.7) \times 100 = \mathbf{41.3\ per\ cent}$$

Symbols used:

Mercury cycle:

H_{m1} = enthalpy at turbine inlet
S_{m1} = entropy at turbine inlet
S_{me} = entropy of exhaust
H_{me} = enthalpy of exhaust
H_{mf} = enthalpy of condensed mercury

Steam cycle:

H_{s1} = enthalpy at turbine inlet
S_{s1} = entropy at turbine inlet
S_{se} = entropy of exhaust
H_{se} = enthalpy of exhaust
H_{sf} = enthalpy of condensed steam
H_{sw} = enthalpy of wet steam leaving mercury condenser

Q10-7. (*a*) Draw a diagram showing the principal paths of fluid flow for a mercury-steam power plant. Indicate on the diagram the state of the fluid, whether liquid or saturated vapor or superheated vapor. (*b*) What over-all station efficiency would you expect for a 1000°F mercury steam cycle? (*c*) Approximately how many pounds of mercury per pound of steam per hour are circulated? Why?

ANSWER.

(*a*) See Fig. 10-11.

(*b*) Refer to Q10-6, and note that for the steam cycle alone without the mercury topping turbine the efficiency is about **40 per cent,** for the binary cycle, between **50** and **55 per cent.**

(*c*) About 10 lb of mercury are circulated for each pound of steam vaporized. Heat of condensation of mercury is about 100 Btu per lb while heat of vaporization of water is about 1000 Btu per lb. Thus to vaporize 1 lb of steam, 1,000/100 = 10 lb of mercury must be circulated.

Chapter 11

GAS TURBINES AND CYCLES

11-1. Fundamental Considerations. The gas turbine is a prime mover embodying a compressor, a combustion chamber, and a turbine as illustrated in Fig. 11-1. Any combination of these elements working together is considered a gas turbine; however, the use of these components in conjunction with an Otto-cycle engine, a Diesel engine, a steam power plant, or other common forms of prime movers is not considered a gas-turbine power plant. Other components besides the compressor, combustion chamber, and turbine may be added to these and the resulting power plant is still considered a gas turbine. Such components are regenerators, compressor interstage coolers, turbine interstage combustion chambers, and split compressors or split turbines. The addition of these other components merely improves the performance or mechanical operation of the gas turbine and does not change its fundamental principles. It is well to note here that whereas in the internal-combustion-engine cycles, all processes occur in the same cylinder, in the gas-turbine cycle compression occurs in one machine (the compressor) and expansion occurs in another machine (the turbine). Thus, the *flow work* of getting the air into and out of the compressor and the burned gases into and out of the turbine must be considered.

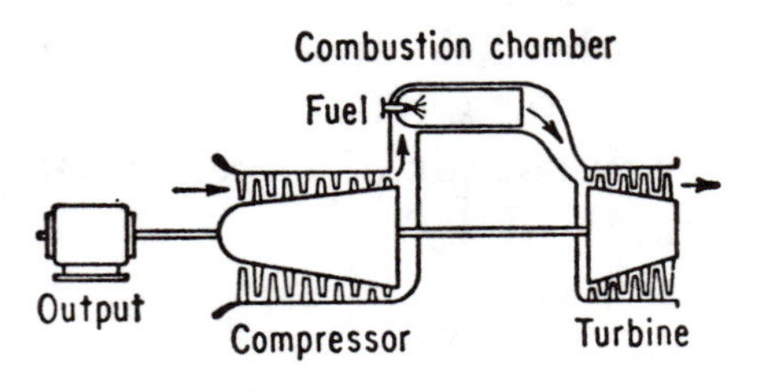

Fig. 11-1. Schematic diagram of simple-cycle gas turbine.

Q11-1. Describe the operation of the gas turbine. What are the principal factors affecting its power and efficiency?

ANSWER. Refer to Fig. 11-2 and the enthalpy—entropy diagram in Fig. 11-3. Air is taken into the compressor at point 1 and delivered at point 2. At C part of the compressed air mixes with the fuel

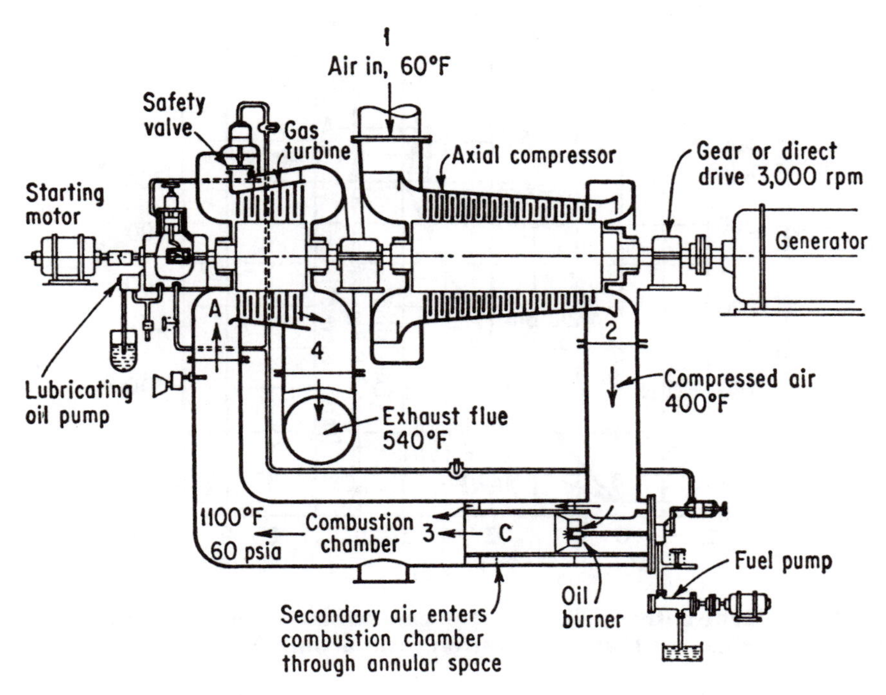

Fig. 11-2. The continuous-combustion gas-turbine unit includes a high-efficiency air compressor, a combustion chamber, and a multistage turbine.

which burns in the combustion chamber, the remainder passing around the chamber. The mixture of hot gases at point 3 passes on and enters the turbine at A. After expanding to point 4, the gases pass out of the exhaust at E. The principal factors affecting the gas turbine's efficiency are air temperatures, combustion process, and heat exchange.

The effects of altitude on performance is beyond the scope of this presentation and the student is referred to the technical literature for details.

Q11-2. The components of an "open" type of gas-turbine cycle are a turbine-generator, a combustor, a compressor, and a regenerator.

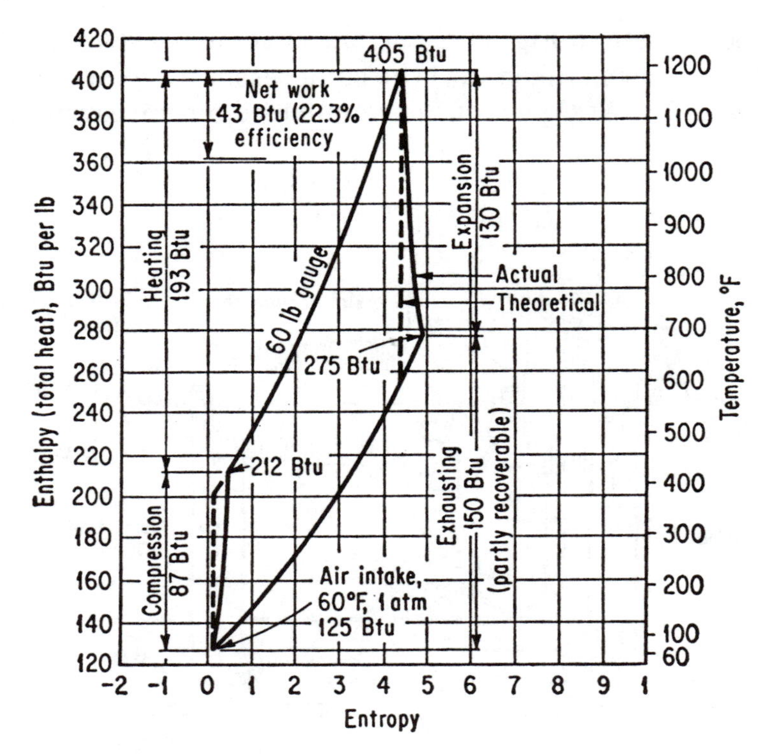

Fig. 11-3. Heat-entropy diagram for gas-turbine cycle resembles that for a diesel cylinder except that compression ratio is only 5:1.

(*a*) Make a schematic diagram of the apparatus. Label the equipment, and indicate the kind of working fluid and its direction of flow.

(*b*) Sketch pressure-volume and temperature-entropy diagrams of the equivalent ideal cycle. Number the corresponding thermodynamic states on the diagrams shown and the diagram of part (*a*). Draw the temperature-entropy diagram to conform to a regenerative effectiveness of 100 per cent.

(*c*) Consider the working fluid of the ideal cycle of part (*b*) to have the properties of air at all time with c_p equal to 0.24 Btu/(lb)(F°) and k equal to 1.40. The pressure

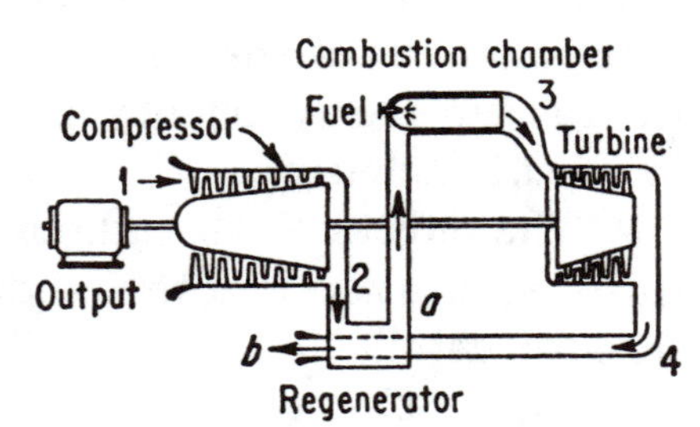

Fig. 11-4. Schematic diagram of a regenerative-cycle gas turbine.

ratio is 4, atmospheric conditions of temperature is 60°F, combustor outlet temperature is 1200°F, and regenerator effectiveness is 100 per cent. Compute the cycle thermal efficiency.

ANSWER.

(*a*) Refer to Fig. 11-4. This is self-explanatory.

(*b*) Refer to Fig. 11-5 for the *PV* diagram and to Fig. 11-6 for the *TS* diagram. Dotted lines represent 100 per cent regenerative effectiveness.

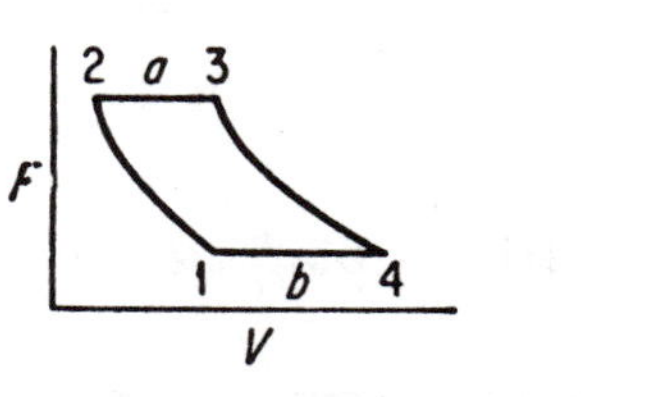

Fig. 11-5

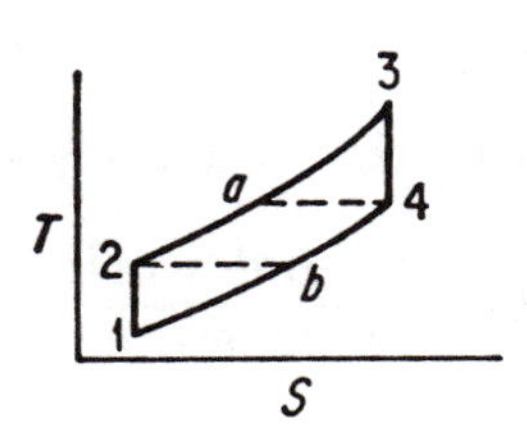

Fig. 11-6

(*c*) Net work is equal to turbine work minus compressor work. This is the same as for the simple cycle.

$$\text{Heat supplied} = wc_p(T_3 - T_a) = wc_p(T_3 - T_4)$$

$$\text{Efficiency} = \frac{(T_3 - T_4) - (T_2 - T_1)}{T_3 - T_4} = 1 - \frac{T_2 - T_1}{T_3 - T_4}$$

$$= 1 - \frac{T_1}{T_4} = 1 - \frac{T_2}{T_3}$$

Assume process of compression is adiabatic. Then

$$T_1 = 60 + 460 = 520°\text{R}$$

$T_3 = 1{,}200 + 460 = 1660°\text{R}$. With the use of the relationship for adiabatic compression

$$\frac{T_2}{T_1} = \left(\frac{P_2}{P_1}\right)^{(k-1)/k}$$

Solve for T_2.

$$T_2 = 520(4)^{(1.4-1)/1.4} = 520 \times 1.486 = 772°\text{R}. \quad \text{Likewise}$$

$$T_4 = \frac{1{,}660}{1.486} = 1120°\text{R}$$

Finally, the efficiency is found to be

$$E_t = 1 - \frac{520}{1{,}120} = 1 - \frac{772}{1{,}660} = 0.535, \text{ or } \mathbf{53.5 \text{ per cent}}$$

The net work is determined to be for 1 lb of working substance

$$1 \times 0.24[(1{,}660 - 1{,}120) - (772 - 520)]$$
$$= 129.8 - 60.5 = 69.3 \text{ Btu}$$

The heat supplied is next found in accordance with

$$wc_p(T_3 - T_4) = 1 \times 0.24(1{,}660 - 1{,}120) = 129.8 \text{ Btu}$$

And finally the efficiency is determined:

$$E_t = \frac{\text{net work}}{\text{heat supplied}} = \frac{69.3}{129.8} \times 100 = \mathbf{53.5 \text{ per cent}}$$

Q11-3. A gas turbine consists of a compressor, a combustor, and an expander. Air enters the compressor at 60°F and 14.0 psia and is compressed to 56 psia; the isentropic efficiency of the compressor is 82 per cent. Sufficient fuel is injected to give the mixture of fuel vapor and air a heating value of 200 Btu per lb. Combustion may be assumed complete, and the weight of fuel may be neglected. The expander reduces the pressure to 14.9 psia, with an engine efficiency of 85 per cent. Assume that the combustion products have the same thermodynamic properties as air, with c_p equal to 0.24 and constant. The isentropic exponent may be taken as 1.4.

(*a*) Sketch complete cycle on both *PV* and *TS* diagrams.

(*b*) Find the temperature after compression, after combustion, and at exhaust.

(*c*) Determine Btu per lb of air supplied, the work required to drive the compressor, the work delivered by the expander, the net work produced by the gas turbine, and its thermal efficiency.

ANSWER.

(*a*) Refer to Figs. 11-7 and 11-8. The *ideal cycle* is given by 1-2-3-4-1. *Actual compression* takes place along 1-2′. *Actual heat added* lies along 2′-3′. The *ideal expansion* process path is 3′-4′. For the *actual expansion* see path taken along 3′-4″. *Ideal work* is equal to c_p × ideal temperature difference. *Actual work* is equal to c_p × actual temperature difference. See also Fig. 11-2.

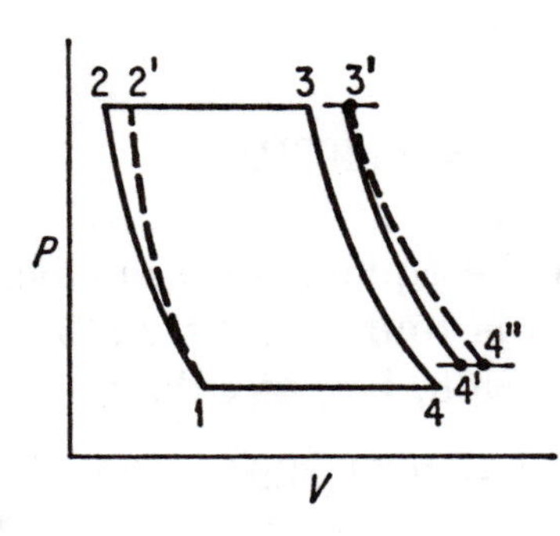

Fig. 11-7

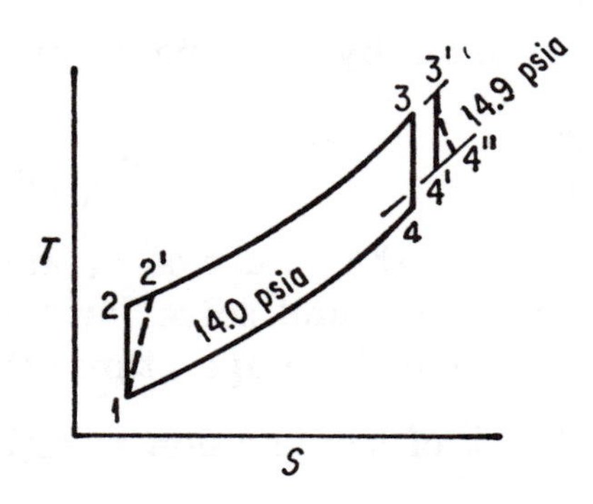

Fig. 11-8

(*b*)

$$\text{Efficiency (isentropic)} = \frac{\text{ideal work of compression}}{\text{actual work of compression}}$$

$$\text{Efficiency (isentropic)} = \frac{c_p(T_2 - T_1)}{c_p(T_{2'} - T_1)} = 0.82$$

Entering air temperature is 60°F or 520°R. Then on the same basis as the previous problem determine T_2.

$$T_2 = 520\left(\frac{56}{14}\right)^{(1.4-1)/1.4} = 773°\text{R}$$

Then insert the values of T_1 and T_2 in the equation for efficiency above and solve for $T_{2'}$. This is found to be 829°R, or **369°F**. This is temperature after compression.

For temperature after combustion use the following approach. $Q = c_p(T_{3'} - T_{2'}) = 200 = 0.24(T_{3'} - 829)$. From which $T_{3'}$ is found to be 1663°R. This temperature is equivalent to **1203°F**.

$$\text{Engine efficiency} = \frac{\text{ideal work of expansion}}{\text{actual work of expansion}}$$

$$\text{Engine efficiency} = \frac{c_p(T_{3'} - T_{4''})}{c_p(T_{3'} - T_{4'})} = 0.85$$

Using a similar approach as above, determine T_4 by use of the relation

$$\frac{T_{4'}}{T_{3'}} = \left(\frac{P_{4'}}{P_{3'}}\right)^{(k-1)/k}$$

From which by rearrangement

$$T_{4'} = 1663\left(\frac{14.9}{56}\right)^{(1.4-1)/1.4} = 1140°\text{R}$$

Now insert the values of $T_{4'}$ and $T_{3'}$ in the equation for engine efficiency above, and solve for $T_{4''}$. This is found to be 1218°R, or **758°F**. This is temperature after expansion, i.e., at exhaust.

(*c*) Work of compression $= c_p(T_{2'} - T_1) = 0.24(829 - 520)$ $= \mathbf{74.2\ Btu}$

Work delivered by expander $= c_p(T_{3'} - T_{4''})$ $= 0.24(1663 - 1218) = \mathbf{106.9\ Btu}$

Net work $= 106.9 - 74.2 =$ **32.7 Btu.** And the thermal efficiency is

$$E_t = \frac{\text{net work}}{\text{heat supplied}} = \frac{32.7}{200} \times 100 = \mathbf{16.4\ per\ cent}$$

Chapter 12

INTERNAL-COMBUSTION ENGINES AND CYCLES

12-1. Actual Indicator Cards—Otto Cycle. Although they approximate the polytropic curve $PV^n = C$, the compression and expansion curves of the actual PV or indicator cards are not truly adiabatic. This is largely due to the fact there is a heat transfer between the gases and the cooled cylinder walls, but it is further complicated by a variation in the specific heat as the temperature of the gases changes. For an engine using gasoline as fuel, the polytropic n has a value of 1.3 for compression and expansion curves.

The compression pressures for engines operating on this cycle vary from 50 to 250 psi, depending on type of fuel for which the engine is designed. For automobile engines pressures range from 80 to 120 psi. Lower pressures are required for kerosene; blast furnace gas, 150 psi; alcohol, 250 psi. Figure 12-1 shows an actual indicator card for the four-cycle process.

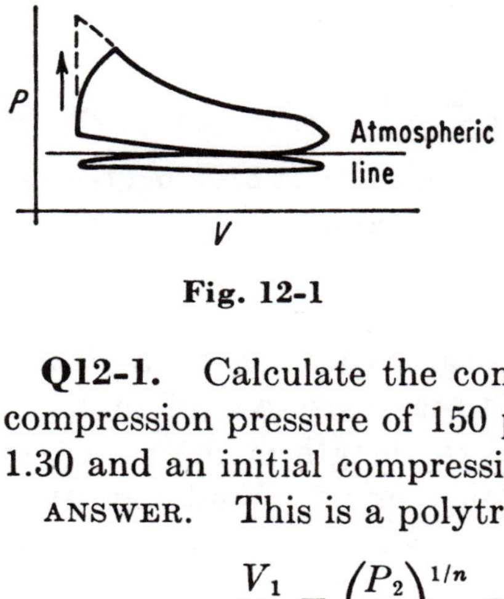

Fig. 12-1

The thermal efficiency of the Otto cycle is determined by

$$E_t = 1 - \left(\frac{V_2}{V_1}\right)^{k-1} \qquad (12\text{-}1)$$

Q12-1. Calculate the compression ratio necessary to produce a compression pressure of 150 psia with a compression curve index of 1.30 and an initial compression of 13.7 psia.

ANSWER. This is a polytropic process and the following holds:

$$\frac{V_1}{V_2} = \left(\frac{P_2}{P_1}\right)^{1/n} = r_c = \left(\frac{150}{13.7}\right)^{1/1.3} = \mathbf{6.29}$$

Q12-2. Calculate the area of the intake port of one of the cylinders of an automobile motor so that the velocity through it is 100 fps. The engine bore is 3.06 in. and the stroke 3.75 in., engine speed is 3,800 rpm. The intake port opens at top dead center and closes at 40° past bottom center, or 220° of crank travel. Volumetric efficiency is 60 per cent.

ANSWER. Volumetric efficiency is obtained from the following:

$$E_v = \frac{\text{actual intake charge (cu ft of mixture)}}{\text{piston displacement (cu ft)}}$$

$$\text{Piston displacement} = 0.785D^2L = 0.785 \times 3.06^2 \times 3.75$$
$$= 27.5 \text{ cu in.} = V$$

$$\text{Actual intake charge} = \text{vol efficiency} \times PD$$
$$= 0.60 \times (27.5/1{,}728) = 0.0096 \text{ cu ft} = Q$$

$$Q = \text{velocity through port} \times \text{port area} \times \text{time intake is open}$$

$$\text{Time intake valve open} = \text{time interval for receiving charge} = t$$

$$t = \text{time for one revolution} \times 220/360$$

$$t = \frac{60}{3{,}800} \times \frac{220}{360} = \frac{\text{sec}}{\text{rev}} \times \text{rev} = 0.0096 \text{ sec}$$

During the next (140 + 360) degrees, the intake port is inactive, so that the acceptance of the charge has nothing to do with this period. Therefore

$$Q = 0.0096 = 100 \text{ fps} \times \text{port area} \times 0.0096 \text{ sec}$$

from which port area $A = \dfrac{0.0096}{100 \times 0.0096} = 0.01$ sq ft, or **1.44 sq in.**

Note that A is the average port area. The full port area is greater. The average port area is only correct if we assume instantaneous full port opening and closing.

Q12-3. A 10- by 18-in. single-acting gas engine runs 200 rpm and makes 96 explosions per minute. The gross weight on a Prony brake arm was 140 lb, tare of 20 lb, and brake arm was 60 in. long. Indicator card area was 1.18 sq in. and the length of the card was 3 in. Scale on the spring was 200 psia per in. Find indicated horsepower, brake horsepower, friction horsepower, and mechanical efficiency.

ANSWER.

$$(a)\ \text{ihp} = \frac{PLAN}{33{,}000} = \frac{78.7 \times 18/12 \times 78.5 \times 96}{33{,}000} = \mathbf{27\ ihp}$$

The value of 78.7 in the above equation is the mean effective pressure MEP determined so that $1.18/3 \times 200 = 78.7$ psi. The value of A is $0.785 \times 10^2 = 78.5$ sq in.

$$(b)\ \text{bhp} = \frac{2\pi NT}{33{,}000} = \frac{2 \times 3.1416 \times 200 \times (60/12 \times 140 - 20)}{33{,}000} = \mathbf{22.85}$$

(c) Friction hp = ihp − bhp = 27 − 22.85 = **4.15**

(d) Mechanical efficiency = 22.85/27 × 100 = **84.6 per cent**

12-2. Gasoline Engine—Compression Ratio and Fuel Economy. The thermal efficiency increases with compression ratio. Since the cycle efficiency is an index of the work done on the piston by the air-fuel mixture for each Btu of fuel burned, it represents the fuel economy of the engine. Thus, an increase in efficiency due to a higher compression ratio improves the economy by lowering the fuel consumption per unit of power output.

12-3. Gasoline Engine—Compression Ratio and Exhaust Temperature. From the technical literature it may be seen that the cold-air standard of the Otto cycle is indicated as in terms of thermal efficiency to be

$$E_t = 1 - \left(\frac{1}{r_c}\right)^{0.4} \tag{12-2}$$

It is possible to derive an expression from Eq. (12-2) to take the form of

$$T_e = \frac{k}{r_c^{0.4}} + T_i \tag{12-3}$$

where T_e is exhaust temperature in degrees Rankine, k is a factor mainly dependent on the heat supplied to the air during combustion, and T_i is intake temperature in degrees Rankine. Thus we see that with k constant and T_i constant, *an increase in the compression ratio tends to lower the exhaust temperature* T_e.

The effect of opening the throttle is to speed up the engine. Therefore, the time required to burn the fuel in each cylinder is less, and partial combustion may be occurring at point of exhaust, with consequent raising of the exhaust temperature. Furthermore, the loss of heat through the cylinder walls is a small fraction of the heat input, because the gases are in contact with the cylinder for a shorter time. This fact further tends to increase the temperature of the exhaust.

12-4. Work for Otto Cycle. Below is given the work formula for the Otto cycle. This is derived from the equations for work from the isentropic and isobaric processes for gases. Heat is supplied in the action from *B* to *C* in the cycle of operation.

$$\text{Work} = \frac{(P_3V_3 - P_4V_4) - (P_2V_2 - P_1V_1)}{k - 1} \quad \text{ft-lb} \qquad (12\text{-}4)$$

In terms of temperatures

$$\text{Work} = \frac{WR}{k - 1}[(T_3 - T_4) - (T_2 - T_1)] \quad \text{ft-lb} \qquad (12\text{-}5)$$

Heat supplied in the Otto cycle is given by the expression already familiar

$$Q = Wc_v(T_3 - T_2) \quad \text{Btu} \qquad (12\text{-}6)$$

Q12-4. Calculate the volumetric efficiency of a six-cylinder automobile motor of $3\frac{5}{16}$-in. bore and $3\frac{3}{4}$-in. stroke when running at 2,000 rpm, with 60 cfm of entering air.

ANSWER.

$$PD = \text{piston displacement} = \frac{0.785D^2L}{1{,}728} = \frac{0.785(3.3125)^2 \times 3.75}{1{,}728}$$

$$= 0.0187 \text{ cu ft}$$

No. of suction strokes per min $= N = 2{,}000/2$

Volume displaced per min $= V \times N \times$ number of cylinders

$= 0.0187 \times 1{,}000 \times 6 = 112.2$ cu ft

Volumetric efficiency $= 60/112.2 \times 100 =$ **53.5 per cent**

12-5. The Diesel Cycle. After proving mathematically that such an engine was practical, Diesel built the first engine of its kind in 1892. When he tried to start it, the first explosion wrecked the engine. Unfortunately, while in the midst of a successful career Dr. Rudolph Diesel mysteriously disappeared from a cross-channel steamer while on a trip from Antwerp to London in 1913.

It may be seen from an inspection of Eq. (12-2) that as r_c increases, the bracketed factor increases and the efficiency decreases. Therefore, a low cutoff fuel ratio r_0 is desirable for best thermal efficiency. The point of cutoff seldom occurs later than 10 per cent of the stroke or r_0 equal to 2.4, usually earlier. We may also observe that a working substance with a high value of k is advantageously helpful since efficiency increases with k. However, k for a real gas actually decreases with an increase in temperature.

As in the Otto cycle, the value of k in the cold-air standard is 1.4. Lower values, about 1.35, would be used in the hot-air standard.

Q12-5. Calculate the compression ratio required in a diesel engine to obtain a compression pressure of 450 psia, assuming the air at the beginning of compression to exist at 13 psia and 150°F. Assume a compression curve exponent of 1.35; calculate the temperature at the end of compression.

ANSWER. Assume 1 lb of air and apply the perfect-gas law. But before proceeding let us set up nomenclature as follows:

P_2 = initial pressure, psia
P_3 = final pressure, psia
V_2 = initial volume, cu ft
V_3 = final volume, cu ft
R = gas constant for air, 53.3
n = compression-curve exponent
T_2 = initial temperature, °R
T_3 = final temperature, °R
w = weight of air, lb

$$V_2 = \frac{wRT_2}{P_2 \times 144} = \frac{1 \times 53.3 \times (150 + 460)}{13 \times 144} = 17.4 \text{ cu ft}$$

Refer to the polytropic process of gases. Then

$$V_3 = V_2\left(\frac{P_2}{P_3}\right)^{1/n} = 17.4\left(\frac{13}{450}\right)^{1/1.35} = 1.255 \text{ cu ft}$$

Compression ratio $r_c = V_2/V_3 = 17.4/1.255 =$ **13.9.** Proceeding to the next part of the question, the determination of the final temperature, set up the equation

$$T_2 = T_3\left(\frac{P_2}{P_3}\right)^{(1.35-1)/n} = T_3\left(\frac{13}{450}\right)^{0.35/1.35} = (150 + 460)$$

$$= T_3 \times 0.398$$

from which $T_3 = 610/0.398 = 1530°R$, or the equivalent of **1070°F.**

Q12-6. A four-cycle six-cylinder diesel engine of 4¼-in. bore and 6-in. stroke running at 1,200 rpm, has 9 per cent CO_2 present in the exhaust gases. The fuel consumption is 28 lb per hr. Assuming that 13.7 per cent CO_2 indicates an air-fuel ratio of 15 lb of air to 1 lb of fuel, calculate the volumetric efficiency of the engine. The intake air temperature is 60°F and the barometric pressure is 29.80 in. Hg.

ANSWER. Atmospheric air contains 76.9 per cent nitrogen by weight. If an analysis of the fuel oil shows zero nitrogen before combustion, all of the nitrogen in the exhaust gases must come from the air. Therefore, with 13.7 per cent CO_2 by volume in the dry exhaust the nitrogen content is

$$N_2 = 76.9/100 \times 15 = 11.53 \text{ lb } N_2 \text{ per lb of fuel oil}$$

or

$$11.53/28 = 0.412 \text{ mole } N_2 \text{ per lb fuel oil}$$

$$\text{Percentage of } CO_2 \text{ in exhaust gases} = \frac{CO_2}{N_2 + CO_2} \text{ moles}$$

Then we can say

$$\frac{13.7}{100} = \frac{CO_2}{CO_2 + 0.412}$$

Solving for CO_2 in the above relation, we obtain $CO_2 = 0.0654$ mole. Now since mole per cent is equal to volume per cent, for 9 per cent CO_2 in the exhaust gases

$$0.9 = \frac{CO_2}{CO_2 + N_2} = \frac{0.0654}{0.0654 + N_2}$$

Solve for N_2, which is found to be equal to 0.661 mole. The weight of N_2 is determined in the usual manner to be $0.661 \times 28 = 18.5$ lb. The air for combustion is

$$\frac{N_2}{0.769} = \frac{18.5}{0.769} = 24.1 \text{ lb air per lb fuel oil}$$

The specific volume of air at 60°F and 29.8 in. Hg is 13.02 cu ft per lb. Thus, the actual charge drawn into the cylinder is found in accord with

$$24.1 \times \frac{13.02}{3{,}600} \times 28 = 2.45 \text{ cu ft per sec}$$

We remember that the volumetric efficiency is the ratio of actual charge drawn into the cylinder divided by the piston displacement. Now the piston displacement for a single cylinder is

$$0.785(4.25)^2 \times 6 \times 1/1{,}728 = 0.0492 \text{ cu ft}$$

For the six cylinders the volume displaced is

$$\frac{6N}{60} \times PD = 6 \times \frac{1{,}200}{2} \times \frac{1}{60} \times 0.0492 = 2.95 \text{ cu ft per sec}$$

Finally, volumetric efficiency is $2.45/2.95 \times 100 =$ **83 per cent**

Q12-7. Sketch the following indicator diagrams: (1) a four-cycle Otto engine with a wide-open throttle and with a nearly closed throttle; (2) a two-cycle Otto engine under the same conditions; (3) a four-cycle diesel engine under full-load and under light load; and (4) a two-cycle diesel engine under the same conditions.

ANSWER. Refer to Figs. 12-2 to 12-5. These are self-explanatory.

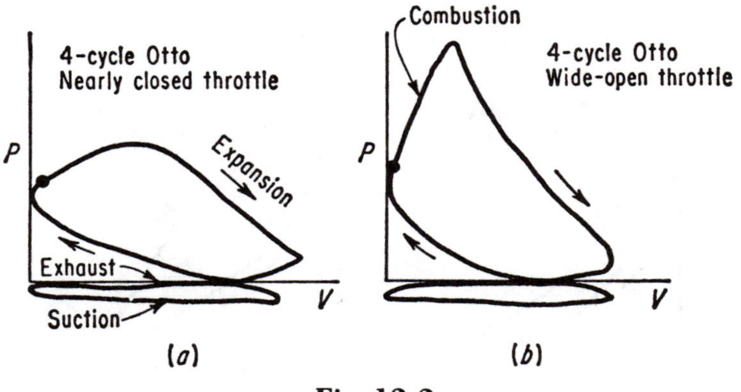

Fig. 12-2

Q12-8. A diesel engine of the air-cell type has a cylinder bore of 4.25 in. and a stroke of 6 in. Assuming flat surfaces for both cylinder head and piston face, calculate the distance in inches between these two surfaces under the following conditions: pressure at the end of compression at 500 psia, pressure at the beginning of com-

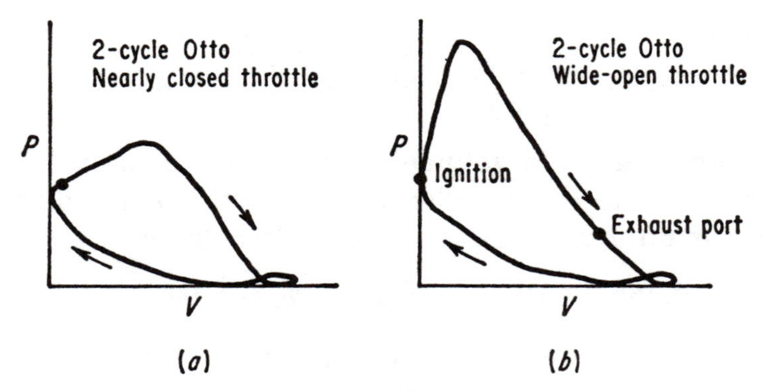

Fig. 12-3

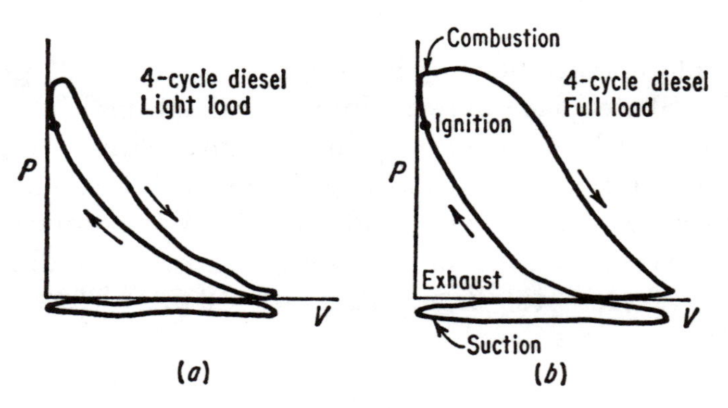

Fig. 12-4

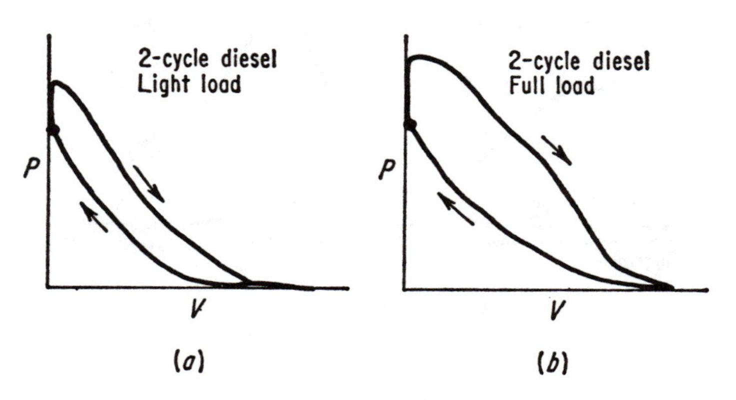

Fig. 12-5

pression at 13.7 psia, volume between piston and head is 30 per cent of the total clearance volume with 70 per cent being in the air cell. Refer to Fig. 12-6.

ANSWER. Displacement $V_2 - V_3 = 0.785 \times 4.25^2 \times 6 = 85$ cu in.

Refer to Fig. 12-3 and assume compression curve exponent equal to 1.35.

$$r_c = \frac{V_2}{V_3} = \left(\frac{P_3}{P_2}\right)^{1/n} = \left(\frac{500}{13.7}\right)^{1/1.35} = 14.32$$

Now, since $V_2 - V_3 = 85$ cu in. and $V_2 = 14.32 \times V_3$, then by substitution

$$14.32 \times V_3 - V_3 = 85$$

From which $V_3 = 85/(14.32 - 1) = 6.4$ cu in. From the problem,

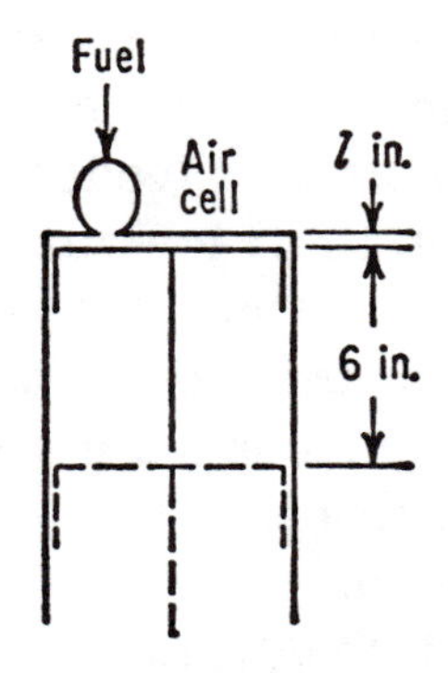

Fig. 12-6

the volume at end of compression is 30 per cent of total clearance volume. Then

$$6.4 \times 0.3 = 0.785 \times (4.25)^2 \times l$$

The distance l is calculated from the above as

$$l = \frac{6.4 \times 0.3}{0.785(4.25)^2} = \mathbf{0.135\ in.}$$

Per cent clearance may be determined although not required here. With c designated as clearance fraction and relation between c and r_c being $r_c = (1 + c)/c$,

$$r_c = 14.32 = \frac{1 + c}{c}$$

from which $c = 1/13.32 = 0.0752$, or 7.52 per cent.

Chapter 13

PUMPS AND PUMPING

13-1. Classification of Pumps. Classified by use, pumps are called: low service, high service, deep well, booster, sewage, sludge, boiler feed, chemical, proportional feeders, air blowers, etc.

Low-service pumps operate at low discharge heads to lift water from sources of supply to water-treatment works. High-service pumps operate at high discharge heads to deliver water to distribution systems. Proportional feeders are used for dosage of solutions of chemicals or liquid chemicals.

The standard classification of pumps may be divided into four general classes with a section devoted to each class. This is shown in Fig. 13-1.

Q13-1. Name the type of pump best suited for the various services listed below:

(*a*) Storm drainage—large capacity—low head
(*b*) Sanitary sewage—small capacity—low head
(*c*) Water supply—small capacity—high head—suction lift
(*d*) Sewage sludge—at 150 gpm and 75 ft head
(*e*) Deep well pump—500 gpm and 200 ft head

Describe each type and preferred method of installation.

ANSWER.

(*a*) Mixed-flow or axial-flow pump recommended. These are screw pumps with propellor revolving in a straight cylindrical section. This pump should be installed so that head is maintained as stable as possible because of tendency to increase discharge with decrease in head.

(*b*) Automatically operated plunger or diaphragm-type pumps are best suited for this service because of small space requirements. Such pumps are to be kept "flooded" at all times.

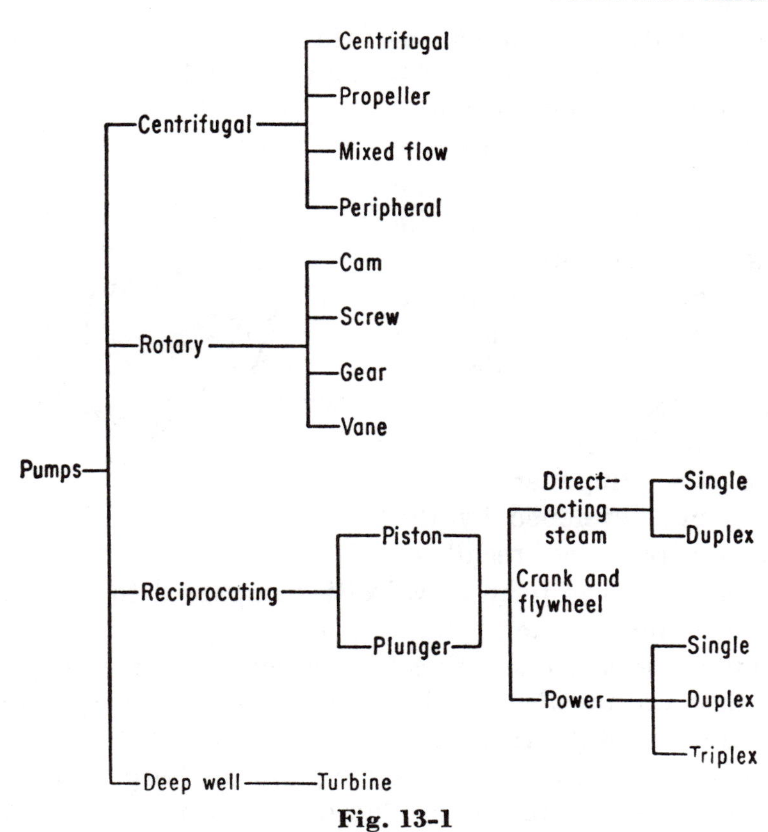

Fig. 13-1

(*c*) A volute type of centrifugal pump or a piston or gear pump is recommended for this service. Care should be taken so that NPSH is properly compensated for in suction lift considerations.

(*d*) Use a pump similar to a Fairbanks-Morse trash pump. Be sure pump does not run overloaded.

(*e*) Use a deep-well turbine-type pump of about six stages. Install fully suspended and impeller casings well below the low water line in the casing.

13.2. The Centrifugal Pump. In this pump the centrifugal pumping principle consists, essentially, of an impeller arranged to rotate within a case so that the liquid will enter at the center and be thrown out by centrifugal force to the outer periphery of the impeller and discharged into the outer case. Figure 13-2 shows the volute-type centrifugal pump. The volute converts velocity energy of the liquid into static pressure at the discharge connection.

In the centrifugal impeller the intake at the center is smaller than at its outer diameter. The liquid flows in at the center by suction, or from a low pressure, and is whirled by the impeller, gathering pressure from the kinetic energy by virtue of the centrifugal force, and is discharged almost tangentially in the direction of rotation with high velocities, which velocities are decelerated and converted into pressure in the casing surrounding the impeller. The pressure head developed by the pump is entirely the result of kinetic energy in the form of velocities imparted to the water or liquid by the impeller and is not due to impact or displacement.

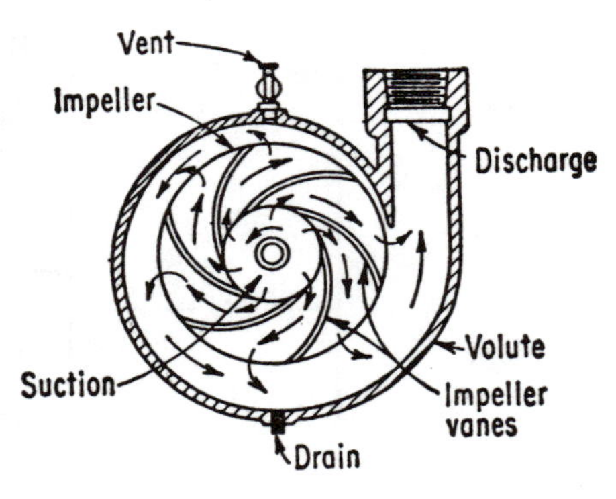

Fig. 13-2

This pumping principle differs from others (reciprocating and the like) in that its impeller can be whirled freely in the pump even though the discharge valve has been closed. When the shutoff head has been reached (pump running and discharge valve closed) no higher pressure can be produced within the pump without increasing the speed.

In a diffuser-type pump stationary guide vanes surround the impeller. These gradually expanding passages change the direction of liquid flow and convert velocity head into pressure head.

In a turbine pump the liquid is picked up by the vanes of the impeller and whirled at high velocity for nearly a complete revolution in an annular channel in which the impeller turns. Energy is added to the liquid in a number of impulses and it enters the discharge at high velocity.

Mixed-flow pumps develop their head partly by centrifugal force and partly by the lift of the vanes on the liquid being pumped.

Propeller pumps develop most of their head by the propelling or lifting action of the vanes on the liquid.

13-3. Pump Head. The pressure head against which the pump will operate depends upon the diameter of the impeller and the speed at which it is rotated. A centrifugal pump may actually raise the liquid or force it into a pressure vessel. It may merely give it enough head to overcome pipe friction. No matter what the service of a centrifugal pump all forms of energy imparted to the liquid in performing this service must be accounted for in establishing the work performed. In order that all these forms of energy may be added algebraically, it is customary to express them all in terms of head in feet of liquid flowing.

Total dynamic head (W_p in Bernoulli's theorem) is made up of the total static head plus the friction head plus the pressure required at the discharge point of the system. The pressure then must be converted to feet of head in the usual manner.

A pump operating at "flooded" suction may be considered as having neither static suction lift nor static suction head. However, a slight suction head must be carried to ensure a full supply of liquid at the eye of the impeller. In most pumping applications velocity head may be neglected, except when large pumping volume rates are being handled.

13-4. Pump Performance Curve. As we noted before the pressure head against which a centrifugal pump will operate depends upon the diameter of the impeller and the speed at which it is rotated. In order to determine what operating conditions a centrifugal pump is good for, a curve known as a performance curve is used. A typical performance (characteristic) curve is shown in Fig. 13-3. This is a plot for an impeller of a fixed diameter rotating at a single constant speed.

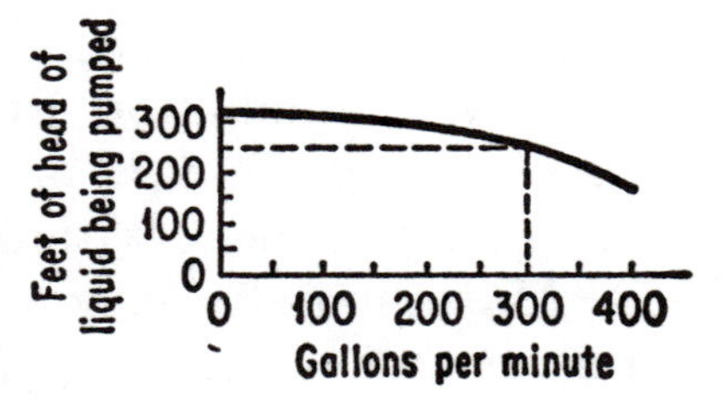

Fig. 13-3

The capacity of the pump, usually in gpm, is plotted on the horizontal axis, while the pressure rise through the pump (stated in feet of head of the liquid pumped) is plotted on the vertical axis. A performance curve is determined by running the pump under actual test conditions. By throttling a valve on the discharge, various flows from zero at full shutoff to full flow at wide open can be obtained. At each flow condition, the gpm is measured with a meter and the corresponding discharge pressure is noted.

From these test results, a number of points are located and a smooth curve is drawn connecting the points. From this curve, the corresponding capacity and head for any set of operating conditions can be read. As shown in Fig. 13-3, for 300-gpm flow, the pressure rise would be equal to 250 ft of the liquid being pumped. Performance curves are plotted on all centrifugal pumps.

Different curves may be obtained with the same pump by using different diameter impellers. An impeller with a larger diameter than that for which Fig. 13-3 has been drawn would have a curve above that shown and parallel to it. An impeller having a smaller diameter would have a curve parallel but below. It may be noted, also, that a flat or horizontal head-capacity curve is usually most desirable from the standpoint of giving a wide range of capacities without much change in discharge pressure. This is especially so if the piping system to which the pump is connected is made up primarily of friction head and none or very little of vertical lift. It must be remembered that a certain amount of slope in the curve is necessary to give proper control of flow and pressure. When a piping system is made up primarily of vertical lift and very little pipe friction, then a steep characteristic is recommended.

There is a good reason for the slope of the curve. Friction within the casing caused by the liquid (called disc friction) increases with increased flow and is reflected as actual loss in pressure or head.

A complete performance curve for a centrifugal pump is shown in Fig. 13-4. In addition to the head-capacity curve we see an efficiency curve and a brake-horsepower curve. The efficiency curve rises to a peak within certain capacity limits and then falls off. Maximum efficiency lies within the design range.

Figure 13-3, we said, applied to a specified impeller diameter at constant speed. One of the great advantages of a centrifugal pump over the positive displacement pump is its flexibility of operation. This flexibility lies in the fact that with a single casing size the capacity rating of a single pump at constant rpm may be varied by simply trimming the impeller diameter to the required size. Therefore, with one set of patterns and castings for one size pump the manufacturer is able to meet any one combination head-capacity requirements within the operating range by merely machining the impeller.

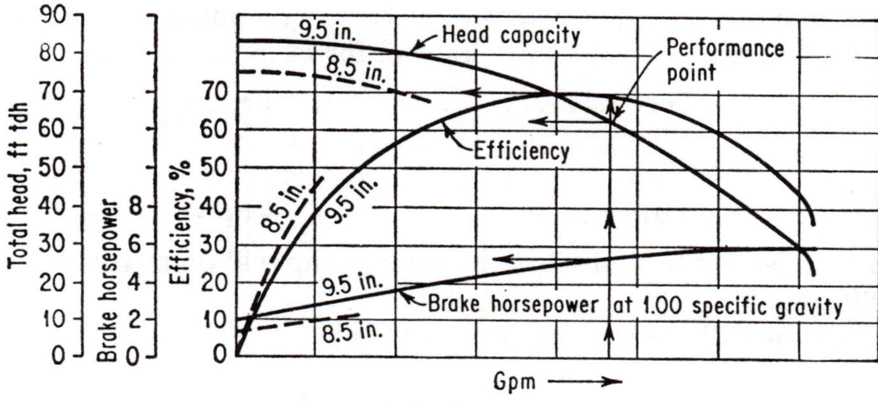

Fig. 13-4

If a pump is handling water and is discharging a certain flow at a certain total dynamic head requiring a definite brake horsepower, this same pump handling a liquid lighter than water such as gasoline (sp gr 0.75) or a liquid heavier than water such as a brine (sp gr 1.2) would discharge the same quantity of liquid at the same total dynamic head as it does for water. From the characteristic curve submitted by the manufacturer (always clear cold water) the brake horsepower is obtained for water. The horsepower requirements for gasoline would be 75 per cent of the curve reading for water; for the brine, 120 per cent. This treatment holds for cases when viscosities are in the order of water. Where viscosities are much greater, special treatment of the problem is necessary since capacity and horsepower are greatly affected. See M-7, p. 444.

13-5. Change of Performance. Again we repeat, the head developed by a centrifugal pump depends upon the impeller diameter and its rotative speed. According to the fundamental laws of physics, the capacity flow of the centrifugal pump will vary directly as the speed; the pressure head will vary as the square of the speed; and the power required will vary as the cube of the speed. These are known as the *laws of affinity* and apply to all velocity machines, i.e., centrifugal pumps, centrifugal fans, blowers, etc.

Thus, if the speed is doubled, the flow will be doubled, the pressure head will be multiplied four times, and the power will be increased eight times. Likewise, in reducing the speed one-half, the capacity will be cut in half, the pressure head will be only one-fourth, and the horsepower one-eighth.

To compute the performance of a centrifugal pump at some other speed when present values are known, for example, assume:

Speed................	1,160 rpm
Flow.................	300 gpm
Head.................	40 ft
Power input..........	4 hp

and it is desired to find what the same pump will do at the speed of 1,760 rpm.

The flow will increase proportionately:

$$\frac{1{,}760}{1{,}160} \times 300 = 455 \text{ gpm}$$

The head will increase as the square:

$$\left(\frac{1{,}760}{1{,}160}\right)^2 \times 40 = 92 \text{ ft}$$

The power will increase as the cube:

$$\left(\frac{1{,}760}{1{,}160}\right)^3 \times 4 = 13.98 \text{ hp}$$

and vice versa. To compute from 1,760 rpm to 1,160 rpm, reverse the order of the divisor, place the 1,160 above the line and 1,760 below. Any given difference in the performance for any variation in speed can be found in the same way. In a field test, if the total head in feet, capacity in gpm, and horsepower input of the pump can be measured, its efficiency can be computed.

$$\frac{\text{gpm} \times \text{total dynamic head in ft}}{3{,}960} = \text{water hp} \qquad (13\text{-}1)$$

$$\frac{\text{Water hp}}{\text{bhp}} \times 100 = \text{pump efficiency} \qquad (13\text{-}2)$$

Q13-2. An acceptance test was conducted on a centrifugal pump having a suction pipe 10 in. in diameter and a discharge pipe 5 in. in diameter. Flow was 818 gpm of clear cold water. Pressure at suction pipe was 4.5 in. mercury vacuum and discharge pressure was 15.5 psig at a point 3 ft above that point where the suction pressure was measured. Input to pump was 15 bhp. Find pump efficiency.

ANSWER. We shall apply Bernoulli's theorem. Friction head h_f is considered zero because data were taken close to pump flanges.

If datum line is taken through the point of suction measurement, then Z_A is zero and Z_B is 3. Velocity in suction pipe is found to be 3.32 fps; velocity in discharge pipe is 13.2 fps, v_A and v_B, respectively. Static suction head is now determined as

$$4.5/12 \times 13.6/1 = - \text{ (minus) } 5.1 \text{ ft of water}$$

This is the term $P_A w_A$ in Bernoulli's theorem. Static discharge head is

$$15.5 \times 2.31 = 35.8 \text{ ft. of water} = \frac{P_B}{w_B}$$

The total dynamic head W_p is found to be 46.6 ft. The hydraulic hp can be calculated

$$\frac{818 \times 8.33 \times 46.6}{33{,}000} = 9.65$$

Pump efficiency is equal to hydraulic horsepower divided by brake horsepower.

$$9.65/15 \times 100 = \mathbf{64.2 \text{ per cent}}$$

Q13-3. If the pump in Q13-2 ran at 1,750 rpm, what new gpm, head, and brake horsepower would be developed and required if the pump speed were increased to 3,500 rpm? Assume constant efficiency.

ANSWER. In all problems involving the change in performance and without the use of pump curves, pump efficiency is assumed to be constant for the pumping ranges involved. Now let us apply the laws of affinity.

gpm change:

$$\frac{817}{x} = \frac{1{,}750}{3{,}500} = 0.5 \qquad x = \mathbf{1{,}634 \text{ gpm}}$$

Head change:

$$\frac{46.4}{y} = \left(\frac{1{,}750}{3{,}500}\right)^2 = 0.25 \qquad y = \mathbf{186.4 \text{ ft}}$$

bhp change:

$$\frac{15}{z} = 0.5^3 = 0.125 \qquad z = \mathbf{124 \text{ bhp}}$$

The laws of affinity are theoretical. Not only do they apply to a change in rotative speed (constant impeller diameter), but they also apply to a change in impeller diameter for a particular pump. These laws give approximate results only but their results are practical for every-day use in design and application. At constant rotative speed:

(*a*) Capacity varies directly as impeller diameter.
(*b*) Head varies directly as (impeller diameter)^2.
(*c*) Horsepower varies directly as (impeller diameter)^3.

13-6. Pump Selection and System-head Curve. In pumping problems it is often convenient to show graphically the relation between flow and friction head in the piping system. Such a curve is shown in Fig. 13-5 and is known as a system friction-head curve. It is obtained by plotting the calculated friction head in the piping, valves, and fittings of the suction and discharge lines, against the flow on which the calculation was based. In the figure, at 200 gpm the friction head is 9 ft, point A, and at 400 gpm the friction head is 36 ft, point B. Friction head varies roughly as the flow squared. This approach is acceptable whenever the flow in the pipeline is in turbulence (Reynolds number 2,100 or over).

The system-head curve for a particular piping system is obtained by combining friction-head curve with static head and any difference of pressures in a pumping system. In Fig. 13-6 the friction-head

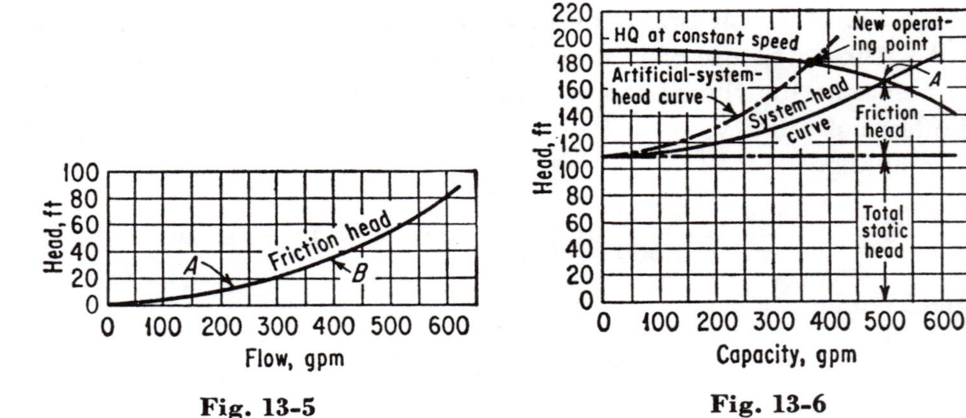

Fig. 13-5

Fig. 13-6

curve (Fig. 13-5) is combined with a total static head of 110 ft to obtain the system-head curve. Here the static head is assumed to be constant, as it is where the suction source and discharge levels are constant. Friction head increases with the flow. Superimposing the pump's head-capacity curve on the system-head curve, as in Fig. 13-6, shows the head and capacity at which the pump will operate, such as point A where the two curves cross. In this system the pump discharges 500 gpm at a 165-ft head.

Where static or pressure varies, system-head curves may be plotted for minimum and maximum static heads or pressures. Superimposing a pump's head-capacity curve on such a system-head curves permits predicting the capacity the pump delivers at different static heads.

Reduced pump capacity can be obtained with constant-speed operation by throttling its discharge to increase the friction head in the system. The capacity at which a centrifugal pump operates is determined by the intersection of its head-capacity characteristic curve with the system-head curve. Throttling the discharge of the pump increases the friction head in the system. Consequently, a new system-head curve results, such as that indicated by the artificial system-head curve in Fig. 13-6. Point of intersection of the pump's head-capacity curve with the new system-head curve determines the pump's new operating capacity, as indicated.

Throttling a centrifugal's pump discharge will not build up excessive pressure. It cannot do so. For example, in Fig. 13-6 if a valve is slowly closed, the head developed by the pump increases until at shutoff it reaches 190 ft, or 16.5 per cent over that at design capacity. Head at shutoff varies with the pump type, or specific speed, but under no conditions will a centrifugal pump develop an excessive head at shutoff. However, a centrifugal pump should not be operated at shutoff for any great length of time, for the pump will overheat.

As we have already indicated, a system-head curve is a graphical representation of the relationship between flow and hydraulic losses in a given piping system. Since hydraulic losses are functions of rate of flow, size and length of pipe, and size, number, and type of fittings, each system has its own characteristic curve and specific values.

In virtually all pump applications at least one point on the system curve is given to the pump manufacturer in order to help him select the pump properly. It is well to repeat here that the manufacturer will guarantee but the one point given to him by the customer. In many cases, however, it is highly desirable graphically to superimpose the entire system curve over the head-capacity curve of the candidate pump as in Fig. 13-6.

13-7. Centrifugal Pumps in Parallel or Series Operation. Frequently where there is a wide range in demand two or more pumps may be operated in parallel or series to satisfy the high demand, with just one of the pumps used for low demands. For proper specification of the pumps and evaluation of their performance under various conditions, the system curve should be used in conjunction with the composite pump performance curves.

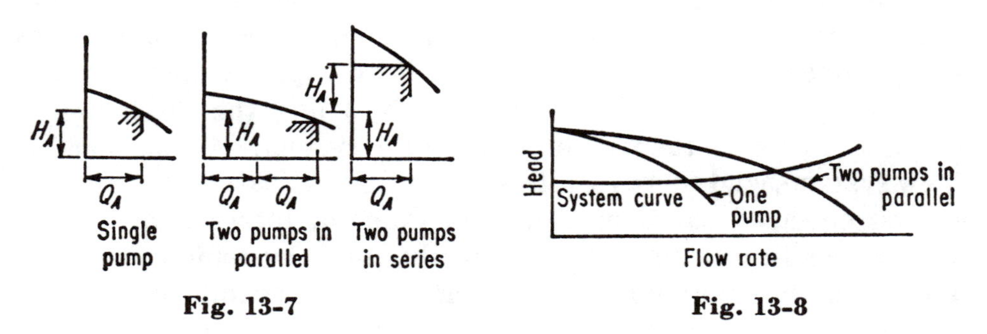

Fig. 13-7

Fig. 13-8

For *pumps in parallel*, performance is obtained by adding the capacities *at the same head.* For *pumps in series*, performance is obtained by adding the heads *at the same capacity.* Figure 13-7 shows single pump in operation, two pumps in parallel, and two pumps in series. Figure 13-8 shows pump performance in parallel. Here superimposing the system curve on the pump performance curves clearly indicates what flow rates can be expected and at what heads each of the pumps will be operating.

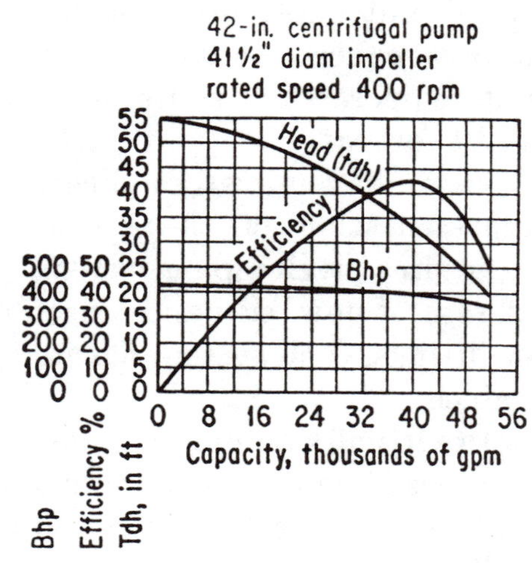

Fig. 13-9

Q13-4. Each of two variable speed centrifugal pumps to be used in a water-pumping station has the characteristic curve shown in Fig. 13-9 when operating at a speed of 400 rpm. The pumps are to be arranged in series or parallel operation. Compute the following:

(*a*) The brake horsepower input for each pump when the pump is opearting at the point of maximum efficiency at 50 per cent of it rated speed.

(*b*) The total discharge when both pumps are operated in parallel at rated speed and when the total dynamic head is 40 ft. (*c*) The total discharge when both pumps are operated in series at rated speed, and when the total dynamic head is 50 ft.

ANSWER.

(*a*) Brake horsepower is proportional to speed cubed. Apply laws of affinity and assume efficiency constant. Brake horsepower at full rated speed (from curve) is 400. Then

$$\frac{400}{z} = \left(\frac{400}{200}\right)^3 = 8$$

$$z = 400/8 = \mathbf{50\ bhp}$$

(*b*) Pumps in parallel: each delivers half total flow rate at constant total dynamic head. From curve: 32,000 gpm and 40 ft head. Then for two pumps the total flow is 32,000 × 2 = **64,000 gpm.**

(*c*) Pumps in series: each delivers same flow rate but at one-half of total dynamic head. From curve: at 50/2 head, flow rate is **48,000 gpm.**

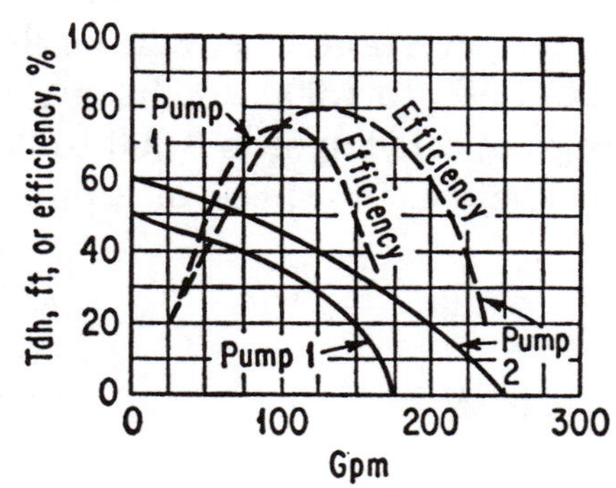

Fig. 13-10

Q13-5. Two centrifugal pumps have head-capacity curves and efficiencies as shown in Fig. 13-10. Pump 1 has a speed of 950 rpm and pump 2, a speed of 1,150 rpm.

(*a*) What would be the combined discharge when both pumps work in parallel against a total dynamic head of (1) 40 ft and (2) 20 ft?

(*b*) Against what total dynamic head could the pumps deliver 75 gpm when working in series?

ANSWER.

(*a*) Placing several pumps in parallel on the same line decreases the capacity of each in a way that varies with the pump characteristics. There is, of course, increased friction in the discharge lines when more water is put through; and, as this increases the head on the pump, there will be a corresponding decrease in discharge. To find data for parallel operation, add the discharge of each at the same pressure (head). The answers are read directly from the curve: **200 gpm** for the 40-ft head, **350 gpm** for the 20-ft head.

(*b*) For pumps in series, keep gpm constant and build up vertically on head. From curve read **90 ft head.**

13-8. Specific Speed. Specific speed is an indication of centrifugal-pump type. For units of capacity and head (feet) used in the United States, specific speed of a centrifugal pump is the speed at which an exact model of the pump would have to run if it were designed to deliver 1 gpm against 1-ft head per stage. Thus all pump sizes can be indexed by the rotative speed of their unit capacity-head model. Specific-speed index of a pump is a guide in determining maximum head against which it can operate, modified by suction conditions. The lower the specific speed is, the higher the head per stage that can be developed. Low specific speeds range from 500 to 1,000, medium specific speeds from 1,500 to 4,000, and high specific speeds 5,000 to 20,000.

The specific speed of a pump can be calculated from the formula

$$N_s = \frac{N\sqrt{Q}}{H^{3/4}} \tag{13-3}$$

where N = rpm
Q = gpm
H = ft head per stage

For the correct typing of a pump design use Q and H values that give maximum efficiency.

The specific speed for efficient centrifugal pumps should never be below 650 or greater than 5,000 at its rated point. For specific speeds below 1,000, the impeller diameter is large and narrow having excessively high disc friction and excessive hydraulic losses. If the specific speed for a set of given conditions becomes less than 650, the head should be divided between several stages. For values of specific speed above 2,000, a mixed-flow impeller (Francis type) is generally used. Best efficiencies, in general, are obtained from pumps having specific speeds from 1,500 to 3,000. Pumps should be selected to fall within this range.

Figure 13-11 shows that specific speed is approximately related to impeller shape and efficiency. There is no real sharp dividing line between various impeller designs. Ranges shown are approximate. Suction limitations of different pumps bear a relation to the specific speed. The Hydraulic Institute publishes charts giving recommended specific speed limits for various conditions.

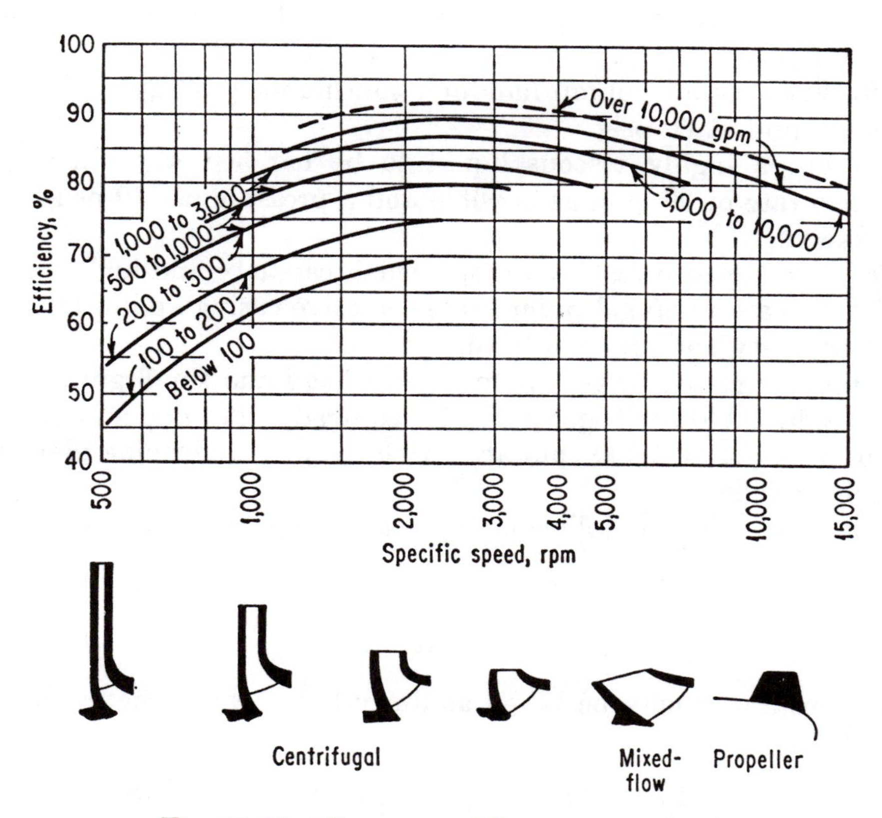

Fig. 13-11. (*Courtesy of Power magazine.*)

Q13-6. Required to pump 1,000 gpm water at a total head of 500 ft. Pump will be directly connected to a 60-cycle induction motor. Assuming a speed of 1,150 rpm and a single-stage pump, select the pump.

ANSWER. Calculate N_s. This is found to be 517, which is too low. Use two stages and calculate new specific speed. This works out to 878 and is acceptable. Use a two-stage centrifugal pump.

Q13-7. Required to pump 5,500 gpm water at a total dynamic head of 15 ft. Rpm is to be determined with a N_s of 5,000.

ANSWER. Rearrange Eq. (13-3) and solve for N. This is found to be equal to 515 rpm. This is the highest allowable speed for a centrifugal pump for efficient operation.

Q13-8. What types of pumps would you select for the following purposes? State engineering reasons for your choice.

(*a*) For a recirculating cooling-water system handling 10 gpm at 100 ft.

(*b*) For a supply pump handling continuously 50 gpm pf a corrosive liquid at 60 psig.

(*c*) For a highly viscous liquid to be handled at as nearly a uniform rate of 30 gpm as possible and a pressure of 150 psig.

ANSWER.

(*a*) Use a gear or rotary pump—small capacity and high head.

(*b*) Use a centrifugal pump—noncorrosive casing and internals—high capacity and medium head.

(*c*) Use a piston or gear pump—high head and small gpm.

Q13-9. A centrifugal pump is required to develop a discharge head of 30 psi at 2,000 rpm at shutoff head. Determine diameter of this impeller.

ANSWER. Use the following equation: Impeller diameter (in.) is equal to

$$d = C \frac{1,840}{N} H^{1/2}$$

The constant C may be taken as unity but it varies between 0.95 and 1.09.

$$d = (1,840/2,000) \times (30 \times 2.31)^{1/2} = \mathbf{7.65\ in.}$$

Q13-10. What steps should be taken in selecting a pump?

ANSWER. (*a*) Sketch layout, (*b*) determine capacity, (*c*) figure total dynamic head, (*d*) study liquid conditions, and (*e*) choose class and type of pump.

13-9. Cavitation. Cavitation describes a cycle of phenomena that occurs in flowing liquid because the pressure falls below the vapor pressure of the liquid. When this occurs, liquid vapors are released in the low-pressure area and a bubble or bubbles form. If this happens at the inlet to a centrifugal pump, the bubbles are carried into the impeller to a region of high pressure where they suddenly collapse. Perhaps a good descriptive term for this is "implosion," the opposite of explosion.

Formation of these bubbles in a low-pressure area, and later their sudden collapse, is called *cavitation*. Erroneously the word is frequently used to designate the effects of cavitation rather than the phenomenon itself.

How does cavitation manifest itself in a centrifugal pump? Usual symptoms are noise and vibration in the pump, a drop in head and capacity with a decrease in efficiency, accompanied by pitting and corrosion of the impeller vanes. The pitting is a physical effect, which is produced by the tremendously localized compression stresses caused by the collapse of the bubbles. Corrosion follows the liberation of oxygen and other gases originally in solution in the liquid. When centrifugal pumps are being considered for handling liquids that are extremely volatile and close to their bubble points, the characteristic curve should also show the *NPSH* requirements.

There is a tendency at times to specify a small, less expensive pump at excessive speeds to compete with the higher initial cost of a larger pump operating at a lower speed. However, most manufacturers rate their pumps at safe operating speeds even in the face of low price competition.

Practical rotative speeds are inversely relative to the diameter of the impeller, therefore, a smaller pump will operate at a greater rpm than a larger pump, but will also deliver less water at peak performance than a larger one. When an impeller is lifting water, there is a much greater pressure on the upper or working side of the blades than on the under side, and at safe speeds this pumping action is normal and will give indefinite service. However, if the pump is operated at excessive speeds, this differential pressure becomes too great and causes a powerful pulsating vacuum on the under side of the blade tips and as each particle of water is pulled away from the blade, it takes with it a small particle of the metal and produces a peculiar pitting or grooved effect. Thus, cavitation takes its toll of the metal, and its repeated erosive action finally results in complete honey-combing and total destruction of the blade, with resultant loss in pump performance. Figure 13-12 shows the effect of cavitation on efficiency and discharge of a centrifugal pump.

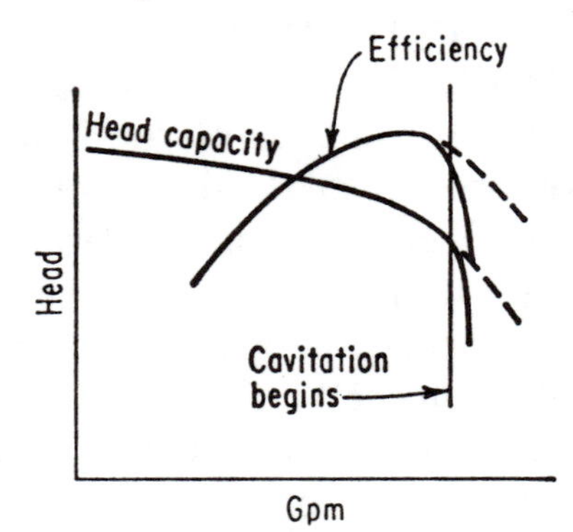

Fig. 13-12

13-10. NPSH and the Centrifugal Pump. A centrifugal pump is designed to handle liquids which for all practical purposes are incompressible. It will not function as a normal pump when handling compressible fluids such as gases or vapors. If such gases are present in the pump, they interfere with its normal operation and have a marked effect on its characteristics as we have previously seen. To prevent vaporization at the impeller of a centrifugal pump, it is necessary to keep the pressure at this point above the vapor pressure of the liquid at all times.

What is NPSH? The pump manufacturer knows from experience, calculations, and experiments the values of the pressure drops from the suction flange to the impeller vanes. This *internal* pressure drop can be called "suction loss in the pump," since it is a loss in available pressure to prevent vaporization at the point of low pressure. The important point to keep in mind is to maintain the pressure at the entrance to the impeller vanes above the vapor pressure of the liquid at all times. The energy available at the pump suction flange to do this and to overcome pump suction losses is called *net positive suction head over the vapor pressure.* This term is commonly abbreviated NPSH, and was originally used in conjunction with pumps handling boiling liquids where all the available energy came from the static elevation of the liquid above the pump. It is also used to describe conditions where a volatile liquid is not at its vapor pressure at the pumping temperature, as we will show in an example following.

The suction limitations for a particular pump is often shown by the manufacturer plotted in the form of a curve giving minimum NPSH requirements for all capacities in the operating range. As long as the available energy equals or exceeds these figures there will be no undue vaporization causing limited capacity, cavitation, and accompanying troubles. This means that for all practical purposes all values stated by the pump manufacturer for minimum NPSH requirements are based on:

(*a*) The pressure drop from the suction flange to the impeller vane. This pressure drop includes the velocity head at the pump suction flange.

(*b*) All values of NPSH are referred to the pump center line. This is assumed to be 3 ft above the pump house floor.

Most pump applications resolve themselves into one of a few basic types of suction conditions. Details vary with every installation but the general method of calculation is perhaps best illustrated by a few simplified examples. In these installations the setting is at sea level and all friction losses are at the maximum flow. To simplify formulas, the following symbols are used:

S = vertical distance from liquid surface to pump house floor, ft
B = distance from pump house floor to shaft center line, ft (assumed to be 3 ft in all examples, a good average figure)
p_a = atmospheric pressure, 14.7 psia
p_{vp} = vapor pressure of liquid at pumping temperature, psia
p = pressure on surface of liquid, psig
h_f = all losses in suction line up to pump suction flange, not including drop in pressure transformed into velocity head at suction flange

The basic formula for calculating NPSH available is

$$\text{NPSH (ft)} = \frac{(p + p_a - p_{vp})2.31}{\text{sp gr}} \pm S - B - h_f \qquad (13\text{-}4)$$

For boiling liquids the first term of this equation automatically becomes zero. The plus (+) sign is used when there is a static suction head and the minus (−) sign when there is a static suction lift.

Q13-11. A liquid is boiling in an open tank at atmospheric pressure. Refer to Fig. 13-13*a*. Liquid temperature is at 212°F boiling water whose surface is 14 ft above the floor. Suction line losses are 5 ft. Calculate the NPSH available.

ANSWER.

$$S = 14 \text{ ft} \qquad p_a = p_{vp} \qquad p = 0 \qquad h_f = 5 \text{ ft}$$

Then

$$\text{NPSH} = S - B - h_f = 14 - 3 - 5 = \mathbf{6\ ft\ available}$$

As long as a liquid is boiling at the surface the same NPSH is available regardless of whether the working pressure is 14.7 psia, or 200 psia. It is evident that if the liquid is boiling, the NPSH available is derived solely from the static head. The working pressure has no bearing on the static head, hence it cannot affect the NPSH.

Q13-12. Assume the tank in Fig. 13-12*a* is a closed one and contains cold liquid propane of 0.58 sp gr with the tank at 200 psia which is the equilibrium vapor pressure of the propane at the

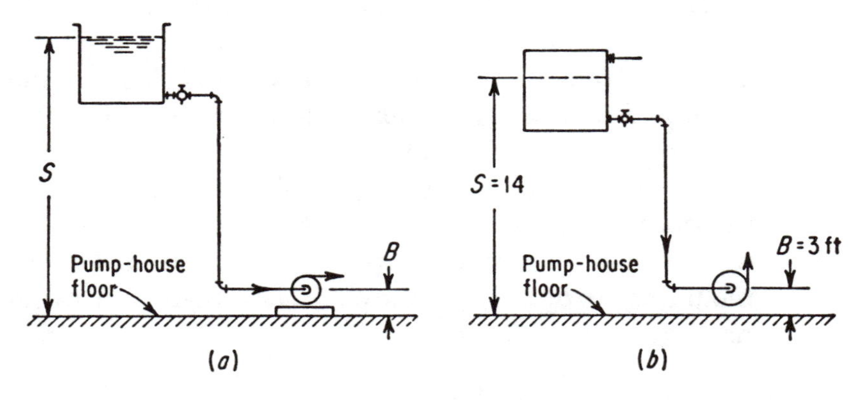

Fig. 13-13

pumping temperature. The suctionline losses are 5 ft and the static head is 14 ft. Calculate the available NPSH.

ANSWER.

$$S = 14 \text{ ft} \qquad h_f = 5 \text{ ft} \qquad p + p_a = p_{vp} = 200 \text{ psia}$$

Then

$$\text{NPSH} = S - B - h_f = 14 - 3 - 5 = \mathbf{6\ ft\ available}$$

Q13-13. When the liquid in the suction tank is not at its boiling point, it is possible to take advantage of the energy in the difference between the working pressure and the vapor pressure. Let us assume that the tank in Fig. 13-13*b* is a closed accumulator at a working pressure of 30 psig and contains a liquid of 0.65 sp gr whose vapor pressure is 40 psia. Again, if the static head is 14 ft and the losses in the suction line are 2 ft, calculate the NPSH available.

ANSWER.

$$S = 14 \text{ ft} \qquad h_f = 2 \text{ ft} \qquad p_{vp} = 40 \text{ psia} \qquad p = 30 \text{ psig}$$

Then

$$\text{NPSH} = \frac{(30 + 14.7 - 40)2.31}{0.65} + 14 - 3 - 2 = 16.8 + 9$$

$$= \mathbf{25.8\ ft\ available}$$

Q13-14. Tank farms, marine terminals, and bulk stations service present limiting suction conditions due to long lengths of piping. Appearances are deceiving and sometimes when there is apparently ample suction head, the piping conditions are such that produce a suction lift condition at the pump flange. Refer to Fig. 13-14. If it is full, the liquid surface is 28 ft above the pumphouse

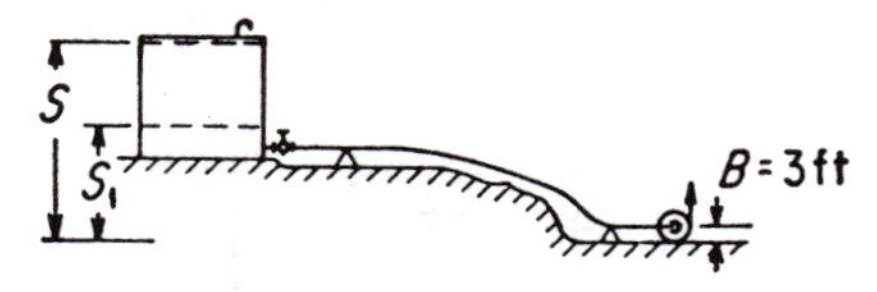

Fig. 13-14

floor, and when almost empty it is 18 ft. The tank contains gasoline at 0.74 sp gr and whose vapor pressure is 7.3 psia at the pumping temperature. Tank is open to the atmosphere and the line losses are 29 ft. Calculate the NPSH available.

ANSWER.

$$S = 28 \text{ ft and } 18 \text{ ft} \qquad h_f = 29 \text{ ft} \qquad p_{vp} = 7.3 \text{ psia}$$

Then

$$\text{NPSH} = \frac{(p_a - p_{vp})2.31}{\text{sp gr}} + S - B - h_f$$

$$= \frac{(14.7 - 7.3)2.31}{0.74} + 28 - 3 - 29 = \textbf{19.1 ft available}$$

with a full tank

$$= 19.1 - 10 = \textbf{9.1 ft available with an empty tank}$$

The worst condition is for the empty tank and the pump selected should be good for these conditions. Care must be taken to prevent any pockets from developing in the suction line where vapor might collect and vapor-bind the line.

The situation often arises when pumping takes place with a centrifugal pump and there is a suction lift as from an underground tank. Such an arrangement should be avoided wherever possible. There is no static head; hence, all the energy necessary to get the liquid into the impeller eye must come from the difference between the working pressure at the surface of the liquid and its vapor pressure.

Q13-15. The storage tank in Fig. 13-15 is vented to the atmosphere and contains gasoline as in the tank in Q13-14. Vapor pressure is 6 psia at the pumping temperature. The surface of the

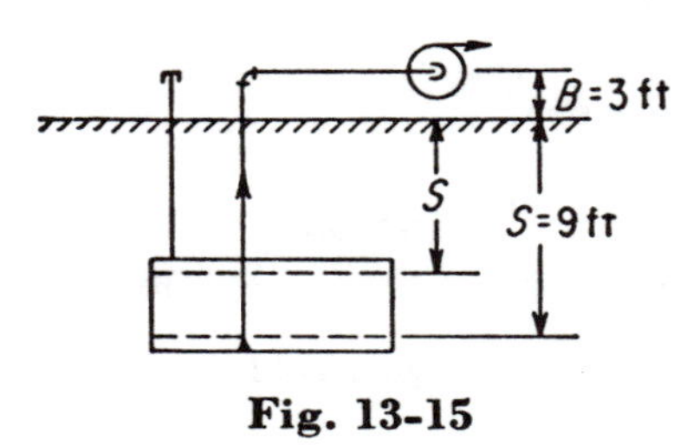

Fig. 13-15

liquid is 6 ft below the floor level when the tank is half full and 9 ft when at the entrance to the suction pipe. Suction line losses are 2 ft. Calculate NPSH available.

ANSWER.

$$S = 9\text{ ft} \qquad h_f = 2\text{ ft} \qquad p_{vp} = 6\text{ psia}$$

Then

$$\text{NPSH} = \frac{(p_a - p_{vp})2.31}{\text{sp gr}} - S - B - h_f$$

$$\text{NPSH} = \frac{(14.7 - 6)2.31}{0.74} - 9 - 3 - 2 = \textbf{13 ft available}$$

A pump with suitable NPSH characteristics will operate under these conditions but care must be taken to make sure that the capacity does not exceed that allowed by the NPSH available. Vapor binding is sure to develop in such an installation and it is recommended that a positive displacement type of pump be installed.

13-11. Maximum Viscosities for Centrifugal Pumps. As the size of a centrifugal pump increases, so does the maximum permissible viscosity it can handle. An approximate indication of this effect is given below from actual practice.

TABLE 13-1

Kinematic viscosity (stokes)	*Minimum pump size, in. (discharge nozzle)*
0.01–0.1	No limit
0.2 –0.3	¾–1 in.
0.4 –0.7	1–1½
0.8 –0.9	1½–2
1.0	2–3
2.0	3–4
3.0	5–6
4.0	6–8
5.0	8–10
7.0 –9.0	12–14
10.0 –20.0	14–16

13-12. Estimating Efficiency of Centrifugal Pumps. The following table has practical significance:

TABLE 13-2

Gpm	*Per cent*	*Gpm*	*Per cent*
Less than 100	30	500 to 1,000	60
100 to 200	40	1,000 to 2,000	70
200 to 500	50	2,000 and up	80

Q13-16. A motor-driven pump is required to deliver 400 gal of water per minute (gpm) against a head of 200 ft for 2,000 hr each year. Bids are offered by two concerns for pumps having an expected life of 15 years with a salvage value equal to the cost of removal. The cost of interest, taxes, and insurance may be taken as 8 per cent of the purchase price, and the cost of power is 2.5¢ per kwhr. Bid A guarantees an overall efficiency of 79 per cent at full load, whereas bid B guarantees an overall efficiency of 75 per cent. How much more is the client justified in paying for the pump and motor with the higher efficiency?

ANSWER.

$$\text{Hydraulic hp required} = (400 \times 8.33 \times 200)/33{,}000 = 20.2 \text{ hhp}$$
$$\text{Electrical kilowatt equivalent} = 20.2 \times 746 = 15.1 \text{ kw}$$
$$\text{At 79 per cent efficiency, kw for bid } A = 15.1 \times 1/0.79 = 19.1$$
$$\text{At 75 per cent efficiency, kw for bid } B = 15.1 \times 1/0.75 = 20.1$$

Annual cost saving in electrical energy by Bid A

$$= (20.1 - 19.1) \times 2000 \times 0.025 = \$50 \text{ per year saved by Bid } A$$

If we let $x =$ the additional capital investment made by Bid A, then $0.08x =$ additional annual fixed charges.

$$x/15 = \text{annual amount for recovery of } x$$

Finally, the balance or "breakeven" point:

$$\text{Additional fixed annual charges} + \text{recovery} = 50$$
$$0.08x + x/15 = 50$$

Solving for x,

$$x = \mathbf{\$341}$$

Chapter 14

FANS, BLOWERS, COMPRESSORS

14-1. General. With the exception of the fact that gases are compressible the theory of design for equipment to move gases and vapors is similar to that of pumps for liquids. Heads developed in terms of feet of fluid being pumped are almost identical for equivalent circumferential speeds as for centrifugal pump designs. Axial-flow impellers impart similar heads to both liquids and gases and valve actions in positive displacement compressors are quite similar to those of piston water pumps.

When considering the compression of gases, the machines are usually referred to as fans (centrifugal and propeller), blowers (centrifugal and lobular), and compressors (multistage centrifugal, lobular, and piston).

FANS AND BLOWERS

Q14-1. Find the motor size needed to provide the forced-draft service to a boiler that burns coal at the rate of 10 tons per hr. The air requirements are 59,000 cfm, air is being provided under 6 in. water gauge (WG) by the fan which has a mechanical efficiency of 60 per cent. Assume fan to deliver at a total pressure of 6 in. WG.

ANSWER. The horsepower is determined by the basic formula given as

$$\text{hp} = \frac{\text{cfm} \times \text{pressure, psf}}{33{,}000 \times \text{efficiency (decimal)}} \tag{14-1}$$

We must convert the 6 in. of water to represent a pressure in pounds per square foot.

$$6/12 \times 62.4 = 31.2 \text{ psf}$$

Finally,

$$\text{hp} = \frac{59{,}000 \times 31.2}{33{,}000 \times 0.60} = 93 \text{ required; use } \mathbf{100\text{-hp motor}}$$

Q14-2. A blower with the inlet open to the atmosphere delivers 3,000 cfm of air at a pressure of 2 in. WG through a duct 11 in. in diameter, the manometer being attached to the discharge duct at the blower. Air temperature is 70°F, and the barometer pressure is 30.2 in. Hg. Calculate the air horsepower.

ANSWER. Air horsepower may be determined in another way by the use of

$$\text{hp} = \frac{\text{lb air per min} \times \text{total head}}{33{,}000} \tag{14-2}$$

First find air density at the flowing condition by correcting the standard density.

$$0.075 \times \frac{30.2}{29.92} = 0.0758 \text{ lb per cu ft}$$

Next the total head, remembering that total head is equal to sum of static plus velocity heads.

$$\text{Static head} = (2/12)(62.4/0.0758) = 137 \text{ ft of air}$$

$$\text{Velocity head} = \{[(3{,}000/60) \times 183]/11^2\}^2/(2 \times 32.2) = 89.2 \text{ ft}$$

$$\text{Air hp} = \frac{3{,}000}{60} \times 0.0758 \times \frac{(137 + 89.2)}{550} = \mathbf{1.56}$$

If fan efficiency were assumed to be 70 per cent, the bhp required would be 1.56/0.7, or 2.22. Use a 3-hp motor.

14-2. Parallel Operation. What about parallel operation? It is a well-known fact that the dip in the pressure curve is a serious drawback to parallel operation when fans are operating at peak efficiency. When fans are individually motored, it is possible to have one fan carrying more than its share of the load and the other fan much less. In unitary equipment, however, where two fans are running on the same shaft, this problem is not as serious. When the air from the two fans in the unitary equipment is carried *in individual ducts for a distance before joining*, the imposed series resistance will help to stabilize operation at the design conditions.

14-3. Fan Laws. In order to determine the effect of changes in the conditions of fan operation, certain fan laws are used and apply to all types of fans. Their application is necessarily restricted not only to fans of the same shape but also to the same point of rating on the performance curve. In the majority of cases system resistance is so small that there is no need to correct horsepower for difference in pressure. However, if there is a great difference or change in temperature through a system, the fan exhausting hot gases may require more power than a blower furnishing the same weight of air to the system. It would be helpful to review the laws of affinity on pumps.

The following constitute the several fan laws:

(*a*) Air or gas capacity varies directly as the fan speed.

(*b*) Pressure (static, velocity, and total) varies as the square of the fan speed.

(*c*) Power demand varies as the cube of the fan speed.

The above apply to a fan having a constant wheel diameter. When air or gas density vary, the following apply:

(*d*) At constant speed and capacity the pressure and power vary directly as the air or gas density.

(*e*) At constant pressure the speed, capacity, and power vary inversely as the square root of density.

(*f*) For a constant weight of air or gas: the speed, capacity, and pressure vary inversely as the density. Also, the horsepower varies inversely as the square of the density.

Q14-3. A certain fan delivers 12,000 cfm at a static pressure of 1 in. WG when operating at a speed of 400 rpm and requires an input of 4 hp. If in the same installation 15,000 cfm are desired, what will be the new speed, new static pressure, and new power needs?

ANSWER.

$$\text{New speed} = 400 \times \frac{15{,}000}{12{,}000} = \mathbf{500\ rpm}$$

$$\text{New static pressure} = 1 \times \left(\frac{500}{400}\right)^2 = \mathbf{1.56\ in.}$$

$$\text{New Power} = 4 \times \left(\frac{500}{400}\right)^3 = \mathbf{7.81\ hp}$$

Q14-4. A certain fan delivers 12,000 cfm at 70°F and normal barometric pressure at a static pressure of 1 in. WG when operating at 400 rpm, and requires 4 hp. If the air temperature is increased to 200°F (density 0.06018 lb per cu ft) and the speed of the fan remains the same, what will be the new static pressure and power?

ANSWER.

$$\text{New static pressure} = 1 \times \frac{0.06081}{0.075} = \mathbf{0.80\ in.}$$

$$\text{New power} = 4 \times \frac{0.06081}{0.075} = \mathbf{3.2\ hp}$$

Q14-5. If the speed of the fan in Q14-4 above is increased so as to produce a static pressure of 1 in. WG at 200°F, what will be the new speed, new capacity, and new power needs?

ANSWER.

$$\text{New speed} = 400 \times \sqrt{\frac{0.075}{0.06018}} = \mathbf{446\ rpm}$$

$$\text{New capacity} = 12{,}000 \times \sqrt{\frac{0.075}{0.06018}} = \mathbf{13{,}392\ cfm\ (at\ 200°F)}$$

$$\text{New power} = 4 \times \sqrt{\frac{0.075}{0.06018}} = \mathbf{4.46\ hp}$$

Q14-6. If the speed of the fan of the previous examples is increased so as to deliver the same weight of air at 200°F as at 70°F, what will be the new speed, new capacity, new static pressure, and new power?

ANSWER.

$$\text{New speed} = 400 \times \frac{0.075}{0.06018} = \mathbf{498\ rpm}$$

$$\text{New capacity} = 12{,}000 \times \frac{0.075}{0.06018} = \mathbf{14{,}945\ cfm}$$

$$\text{New static pressure} = 1 \times \frac{0.075}{0.06018} = \mathbf{1.25\ in.}$$

$$\text{New power} = 4 \times \left(\frac{0.075}{0.06018}\right)^2 = \mathbf{6.20\ hp}$$

Q14-7. At what point should a fan be selected for operation, and why?

ANSWER. At the point of maximum efficiency. Here cost of operation and the noise produced will be the least.

14-4. Duct-fan Characteristics. As in pumping, all system resistance curves pass through the origin. The curves will intersect all fan performance curves at some point. However, where

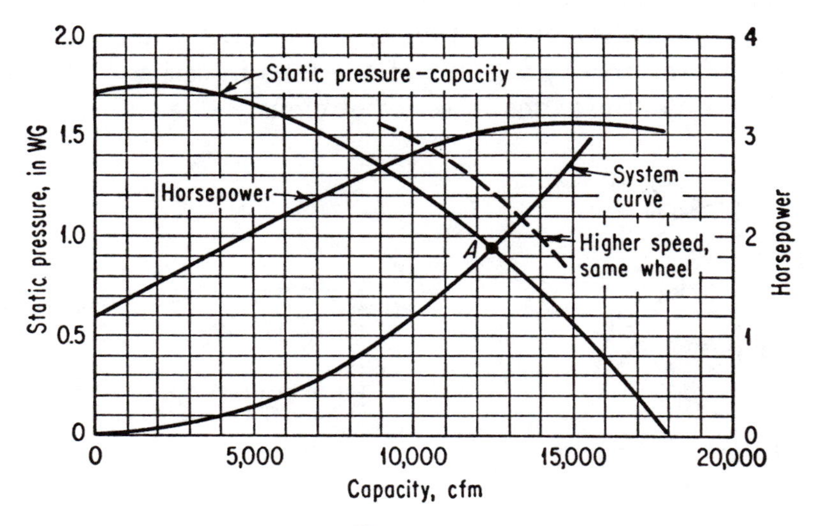

Fig. 14-1

intersect takes place, operation must be stable and the efficiency must be high.

A static pressure curve can easily be drawn through a given point based on the fact that the pressure required to overcome system resistance to flow varies for all practical purposes as the square of the flow rate. When a certain fan is connected to a given system, the system characteristic may be used to learn what will happen. Suppose we want to know what such an arrangement will look like when handling 12,500 cfm against a static pressure of 0.95 in. WG. Figure 14-1 shows this. We see that but one condition satisfies both fan and system. This is point A. A higher-speed fan having a definite wheel diameter when attached to the same system would show higher cfm capacity, higher static pressure, and greater horsepower requirements. In order to realize the needed capacity at the static pressure resulting from this flow, the duct system must be dampered or the fan speed reduced. Normally, in practice the fan is selected to produce the correct flow and develop the desired static pressure at the selection point where both curves cross. Since dampering is a waste of power when carried on in the duct system, this mode of operation should be avoided as much as possible. For best total results, radial dampering at fan inlet is suggested.

14-5. Fans in Series. When low-pressure fans are in series, the weight of flow is substantially that of one fan but the total pressure is the sum of the total pressures of the fans in series. The installation of two identical fans in series obviously does not double the

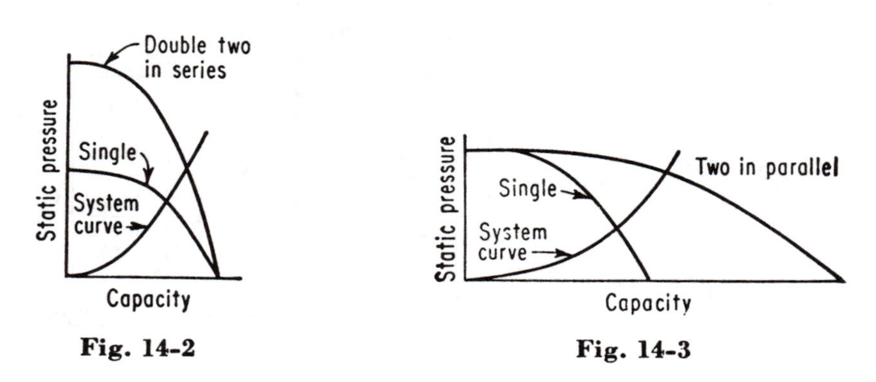

Fig. 14-2 **Fig. 14-3**

quantity of flow through a given system. Double the air quantity would require some four times the pressure and eight times the horsepower. Two identical fans in series a little more than doubles the horsepower and increases the capacity and static pressure as indicated in Fig. 14-2. This placing of two identical low-pressure fans in series is often termed staging.

14-6. Fans in Parallel. Previously we passed over paralleling of fans lightly. We shall now show how two identical fans can be hooked up characteristically. The characteristic of the combined set is such that the capacity is the sum of the separate capacities for a given static or total pressure (see Fig. 14-3). It is well to note that the placing of fans in parallel has been a more common practice than placing them in series.

14-7. Estimating Fan Horsepower. The design engineer is often called upon to estimate horsepower requirements for moving a given volume of normal atmospheric air through ductwork with resistance measured in inches of water. With the flow known in cubic feet per minute and the system resistance known, the chart in Fig. 14-4 may be used. Move across the chart from the vertical scale right to the system-resistance line and then vertically down to read horsepower, i.e., air horsepower. Since a certain degree of inefficiency is involved in a fan, the brake or shaft horsepower from which the fan motor size is selected may be estimated by assuming a fan efficiency of 50 to 75 per cent, depending on fan design and capacity. Thus, for 20,000 cfm and 1 in. static pressure, the chart reads 3.2 air horsepower, or use a 7½-hp fan motor.

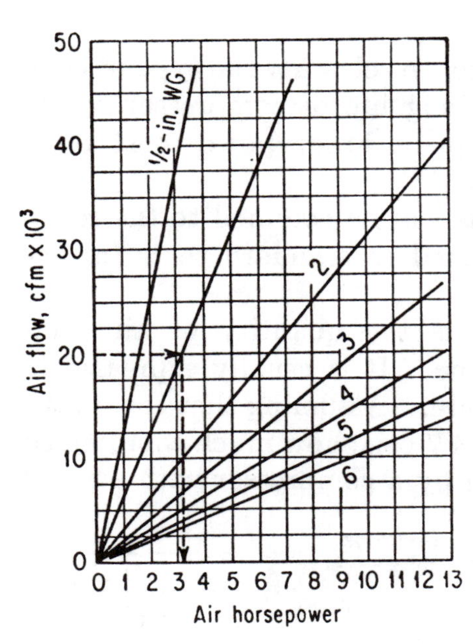

Fig. 14-4. Air horsepower for fans. Basic formula: 0.16 air hp per WG per 1,000 cfm. *(Courtesy of Power Engineering.)*

AIR AND GAS COMPRESSORS

14-8. General Considerations. Standard types of stationary reciprocating compressors include: (*a*) vertical and V-type single-acting units in sizes up to 100-hp single-stage, and two-stage, either air- or water-cooled; (*b*) vertical or V-type semi-radial and right-angle type double-acting machines, available in sizes of 60 hp and above, single-stage and multistage, and water-cooled only; (*c*) single-frame horizontal or vertical double-acting compressors, available in sizes up to 125 hp, single-stage and multistage, and water-cooled only; (*d*) duplex, horizontal or vertical double-acting machines in sizes of 75 hp and over, single-stage and multistage and water-cooled only.

Choice between single- and two-stage compression depends upon many varying factors such as size of cylinders, speed, ratio of compression, discharge temperature limitation, cost of power, continuity of service, method of cooling, permanence of installation, etc. In general, the dividing line between single- and two-stage air compression for double-acting compressors may be drawn as follows:

For pressures below 60 psig, single-stage
For pressures above 100 psig, two-stage
For pressures between 60 and 100 psig, single-stage
For capacities below 300 cfm and two-stage for greater requirements

Cost of power is an important factor in selecting the type of compressor. Because of the relatively long life of compressor equipment, higher efficiency with consequent lower power cost often justifies a higher initial investment.

Treatment of compressor problems will be confined to the reciprocating type. The student is referred to standard texts for the many other types.

Q14-8. Air-compressor capacity is the quantity of air compressed and delivered per unit of time. It is usually expressed in cfm at intake pressure and temperature. Assuming that the intake temperature is at 68°F (528°R) and 14.7 psia, how much compressed air at 250 psia and 80°F is delivered by a compressor with a rated capacity of 700 cfm?

ANSWER. Assuming that the perfect-gas law applies, we can proceed as follows:

$$\frac{PV}{T} = \frac{14.7 \times 700}{528} = \frac{250 \times V}{540}$$

Solving for V, we find it to be equal to **42 cfm.**

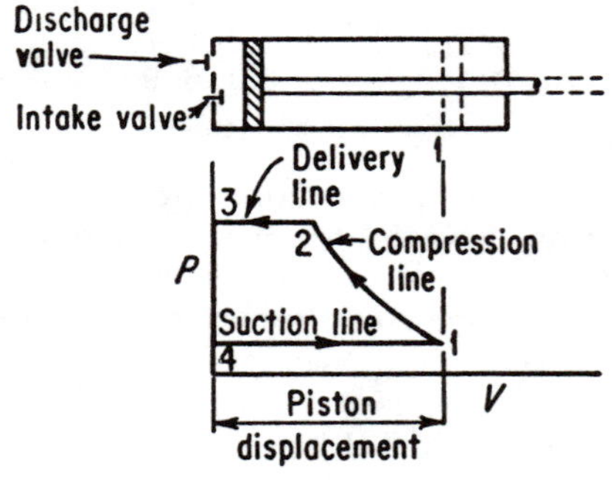

Fig. 14-5

14-9. Work of Compressor without Clearance. Refer to Fig. 14-5. The work done on the air (or gas) is the area enclosed within the diagram and is expressed by the following equation:

$$W = \frac{nwR(T_2 - T_1)}{1 - n} \quad \text{ft-lb} \qquad (14\text{-}3)$$

Or a more convenient form

$$W = \frac{nP_1V_1}{1 - n}\left[\left(\frac{P_2}{P_1}\right)^{(n-1)/n} - 1\right] \quad \text{ft-lb} \qquad (14\text{-}4)$$

where V_1 is volume drawn into cylinder measured in cubic feet and w is the weight of this charge of gas which passes through the compressor. For isentropic compression

$$W = \frac{kP_1V_1}{1-k}\left[\left(\frac{P_2}{P_1}\right)^{(k-1)/k} - 1\right] \quad \text{ft-lb} \qquad (14\text{-}5)$$

For isothermal compression

$$W = wRT_1 \ln \frac{P_1}{P_2} \quad \text{ft-lb} \qquad (14\text{-}6)$$

Q14-9. Calculate the power required to compress 5,000 cu ft of "free" air per hour initially at 14.5 psia and 70°F to 100 psia, assuming nonclearance compressor (Fig. 14-5) and single-stage compression, (*a*) isothermal compression and (*b*) adiabatic compression.

ANSWER.

(*a*) Use Eq. (14-6), but first find the weight of air w. Assume the validity of the perfect-gas law.

$$w = 5{,}000 \times \frac{28.8}{379} \times \frac{(460+60)}{(460+70)} \times \frac{14.5}{14.7} = 370 \text{ lb}$$

For isothermal compression the work is

$$W = wRT_1 \ln \frac{P_1}{P_2} = 370 \times 53.3 \times (460+70) \times \ln \frac{14.5}{100}$$
$$= -20.2 \times 10^6 \quad \text{ft-lb}$$

$\text{hp} = 20.2 \times 10^6/(3{,}600 \times 550) =$ **10.2**

(*b*) To find adiabatic hp, use Eq. (14-5).

$$W = \frac{1.4 \times 14.5 \times 5{,}000}{1-1.4}\left[\left(\frac{100}{14.5}\right)^{0.4/1.4} - 1\right] \times 144$$
$$= -26.8 \times 10^6 \quad \text{ft-lb}$$

$\text{hp} = 26.8 \times 10^6/(3{,}600 \times 550) =$ **13.5**

It takes more work to compress and deliver a gas isentropically (adiabatically) than isothermally. As the exponent n decreases, the work of compression decreases. Somewhere in between $n = 1$ (isothermal compression) and $n = k$ (adiabatic compression) there is a value of n that represents polytropic compression. Intimate intercooling of the gases during compression gives a value of n close to unity; poor intercooling (as in multistage compression) because of scaling, warm and little use of cooling water results in values of n approaching the adiabatic. For centrifugal compressors and no intimate intercooling values of n greater than the adiabatic (c_p/c_v) have been recorded. However, good intercooling will bring down the value of n approaching 1.2 for air. Figure 14-6 gives a quick

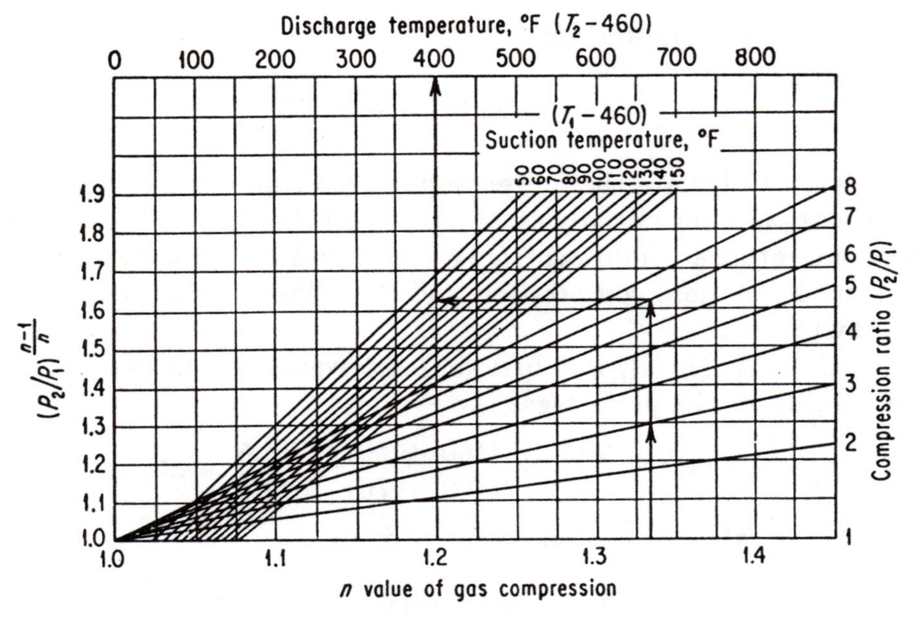

Fig. 14-6. Gas-compression temperatures. Basic formula is $T_2 = T_1[P_2/P_1]^{n-1/n}$, where T_1 = suction temperature, °R; T_2 = discharge temperature, °R; P_1 = suction pressure, psfa; P_2 = discharge pressure, psfa; psfa = pounds per square feet absolute.

representation of the effect of compression of a gas given suction temperature, compression ratio, and value of the compression exponent n.

14-10. Actual Compression Curves. As mentioned above, polytropic compression and values of $n = k$ are developed by circulating cooling water or air around the compression cylinder. In small cheap compressors, cooling is inadequate and n would be 1.35 (for air) or greater. Under favorable conditions a value of n equal to 1.3 or less may be realized. Values of n equal to 1.25 to 1.3 represent the best results for water-jacketed compressors.

In isothermal compression, all the heat equivalent of the work done on a gas is carried away, and the air leaves the compressor with the same internal energy as it had upon entering the compressor. In an isentropic compression, no heat is carried away, so that the gas leaves with an increase in internal energy equal to the work done on it. In a polytropic compression, some heat is carried away but some is also retained to give some increase in internal energy. Although more energy remains in the gas after isentropic compression, the gas is not conveniently utilized until its temperature has been reduced, and the result is defeated anyway.

14-11. Isothermal Compression Horsepower. Although isothermal compression is never attained in practice, the work and horsepower required by the process are quite often used as ideal conditions with which the actual performance of a compressor may be compared.

As we have indicated previously, the net work represented by an ideal P-V diagram is equal to the area enclosed in the diagram. By integration, it can be shown that the area is equal to the work expressed in foot-pounds.

$$W = 2.3P_1V_1 \log R \tag{14-7}$$

where W = work required for isothermal compression, ft-lb
P_1 = intake pressure, lb per sq ft abs
V_1 = volume of gas compressed measured at intake conditions, cu ft
R = ratio of discharge pressure to intake pressure, P_2/P_1

As we know, gas volumes are ordinarily measured in standard cubic feet so that Eq. (14-7) can be made into a more useful form by substituting pressure and volume at standard conditions of 14.7 psia and cubic feet measured at 60°F. Then Eq. (14-7) becomes

$$W = 144 \times 14.7 \times (V_s) \times 2.3 \times \log R \tag{14-8}$$

Combining all constants,

$$W = 4{,}869\ V_s \log R \tag{14-9}$$

This equation can be used to calculate the work required to compress any volume of gas isothermally or the horsepower according to

$$\text{hp} = \frac{4{,}869\ V_s \log R}{33{,}000} \tag{14-10}$$

$$= 0.1475\ V_s \log R \tag{14-11}$$

Q14-10. What is the isothermal horsepower required for compressing 1,000 scf of air per minute from an intake pressure of 14.0 psia to a discharge pressure of 60 psia?

ANSWER. $R = 60/14.0 = 4.285$. The log of this ratio is found by slide rule or logarithms as equal to 0.63195. Then

$$\text{hp} = 0.1475 \times 1{,}000 \times 0.63195 = \mathbf{93.21\ hp}$$

Q14-11. A compressor will handle 1 million intake cu ft of gas per day of natural gas. Based on isothermal conditions, what horsepower would be required when compressing the gas from 50 to 250 psig? Atmospheric pressure is 14.0 psia.

ANSWER. Using the perfect-gas law, the standard volume of gas in 1 million cu ft per day (Mcfd) is found to be

$$V_s = \frac{1{,}000{,}000}{1{,}000} \times \frac{50 + 14}{14.7} = 4{,}350 \text{ Mcfd}$$

and

$$R = \frac{250 + 14}{50 + 14} = 4.1$$

The horsepower equation based on the Mcfd term is given by the following:

$$\text{hp} = 0.1024\ V_s \log R \tag{14-12}$$

so that

$$\text{hp} = 0.1024 \log R \times V_s = 0.06275 \times 4{,}350 = \mathbf{275\ hp}$$

14-12. Polytropic and Adiabatic Compression Horsepower. With a given value of n the actual compressor cylinder horsepower required for the compression of a given quantity of gas in a given length of time can be calculated. This horsepower in the cylinder of the compressor should not be confused with the brake hp of compression. The brake horsepower of compression is somewhat larger than the compressor cylinder horsepower and represents the brake horsepower that an engine must deliver to a compressor not only to compress the gas but also to overcome the friction in the moving parts of the compressor.

Because of the similarity of the calculations involved, the determination of both polytropic and adiabatic hp will be discussed jointly.

The relation for work for both situations is given by

$$W = \frac{n}{n-1} P_1 V_1 (R^{(n-1)/n} - 1) \tag{14-13}$$

If V_1 is the volume handled per minute the above equation will give foot-pounds of work done per minute.

Since a common standard pressure for gas measurement is 14.7 psia, Eq. (14-13) can be changed to a more convenient form by substituting this standard pressure in the equation as follows:

$$W = \frac{n}{n-1} \times 144 \times 14.7 V_s (R^{(n-1)/n} - 1) \tag{14-14}$$

or

$$W = 2{,}117 V_s \frac{n}{n-1} (R^{(n-1)/n} - 1) \tag{14-15}$$

where V_s is the volume of gas handled measured at 14.7 psia. When V_s is in terms of cubic feet per minute, the horsepower required for compression is

$$\text{hp} = \frac{2{,}117}{33{,}000} V_s \frac{n}{n-1} (R^{(n-1)/n} - 1) \tag{14-16}$$

In Eqs. (14-13), (14-14), (14-15), and (14-16), if the value of the exponent n is replaced by the ratio of specific heats $k = c_p/c_v$, the equations also hold true for adiabatic compression. However, since no actual compressor ever operates under adiabatic conditions, such a substitution is seldom required in working out practical problems. Nevertheless, in licensing examinations, this is often a requirement.

Values of $R^{(n-1)/n} - 1$ corresponding to various values of R and n are given in Table 14-1 below.

TABLE 14-1. VALUES OF THE FACTOR ($R^{(n-1)/n} - 1$) FOR USE IN CALCULATING WORK AND HORSEPOWER OF COMPRESSION

Pressure ratio, R	Value of exponent, n									
	1.05	1.10	1.15	1.20	1.25	1.30	1.35	1.40	1.45	1.50
1.3	0.0125	0.0241	0.0348	0.0447	0.0539	0.0625	0.0704	0.0778	0.0849	0.0913
1.4	0.0163	0.0310	0.0449	0.0578	0.0697	0.0810	0.0914	0.1010	0.1100	0.1185
1.5	0.0195	0.0375	0.0544	0.0700	0.0845	0.0982	0.1110	0.1228	0.1343	0.1447
1.6	0.0226	0.0436	0.0631	0.0814	0.0985	0.1145	0.1297	0.1435	0.1572	0.1694
1.7	0.0256	0.0494	0.0716	0.0926	0.1140	0.1304	0.1474	0.1635	0.1790	0.1933
1.8	0.0284	0.0549	0.0797	0.1030	0.1246	0.1453	0.1645	0.1824	0.2000	0.2162
1.9	0.0310	0.0601	0.0873	0.1128	0.1370	0.1598	0.1810	0.2010	0.2205	0.2380
2.0	0.0336	0.0650	0.0947	0.1224	0.1487	0.1738	0.1970	0.2185	0.2405	0.2600
2.1	0.0360	0.0698	0.1014	0.1318	0.1600	0.1870	0.2120	0.2360	0.2595	0.2805
2.2	0.0382	0.0743	0.1083	0.1404	0.1710	0.1997	0.2266	0.2520	0.2775	0.3000
2.3	0.0404	0.0786	0.1147	0.1490	0.1824	0.2120	0.2410	0.2680	0.2950	0.3200
2.4	0.0426	0.0828	0.1220	0.1570	0.1915	0.2240	0.2545	0.2840	0.3125	0.3390
2.5	0.0446	0.0868	0.1270	0.1650	0.2010	0.2365	0.2680	0.2985	0.3290	0.3650
2.6	0.0466	0.0908	0.1328	0.1725	0.2106	0.2470	0.2815	0.3135	0.3455	0.3755
2.7	0.0484	0.0945	0.1384	0.1800	0.2200	0.2580	0.2935	0.3280	0.3615	0.3920
2.8	0.0503	0.0981	0.1438	0.1870	0.2285	0.2685	0.3060	0.3415	0.3775	0.4100
2.9	0.0520	0.1017	0.1490	0.1945	0.2375	0.2790	0.3182	0.3555	0.3920	0.4260
3.0	0.0537	0.1050	0.1540	0.2005	0.2460	0.2890	0.3300	0.3685	0.4070	0.4420
3.1	0.0554	0.1083	0.1591	0.2075	0.2543	0.2990	0.3410	0.3810	0.4210	0.4580
3.2	0.0570	0.1113	0.1640	0.2140	0.2620	0.3085	0.3520	0.3940	0.4355	0.4730
3.3	0.0585	0.1145	0.1687	0.2203	0.2695	0.3175	0.3625	0.4060	0.4490	0.4885
3.4	0.0600	0.1175	0.1730	0.2260	0.2775	0.3270	0.3735	0.4180	0.4630	0.5040
3.5	0.0615	0.1205	0.1777	0.2320	0.2850	0.3360	0.3840	0.4300	0.4760	0.5180
3.6	0.0629	0.1233	0.1820	0.2380	0.2920	0.3440	0.3940	0.4410	0.4890	0.5320
3.7	0.0642	0.1263	0.1860	0.2435	0.2990	0.3530	0.4040	0.4520	0.5010	0.5450
3.8	0.0656	0.1290	0.1902	0.2495	0.3060	0.3620	0.4140	0.4640	0.5140	0.5600
3.9	0.0670	0.1317	0.1942	0.2545	0.3125	0.3695	0.4240	0.4750	0.5260	0.5740
4.0	0.0683	0.1344	0.1980	0.2600	0.3190	0.3775	0.4320	0.4860	0.5380	0.5860
4.1	0.0696	0.1368	0.2020	0.2650	0.3265	0.3855	0.4410	0.4960	0.5500	0.6000
4.2	0.0707	0.1393	0.2058	0.2705	0.3325	0.3930	0.4510	0.5060	0.5620	0.6130
4.3	0.0719	0.1417	0.2095	0.2750	0.3385	0.4000	0.4600	0.5160	0.5730	0.6250
4.4	0.0731	0.1440	0.2135	0.2800	0.3450	0.4080	0.4680	0.5260	0.5840	0.6380
4.5	0.0743	0.1464	0.2165	0.2850	0.3510	0.4160	0.4760	0.5360	0.5950	0.6500
4.6	0.0753	0.1490	0.2200	0.2890	0.3565	0.4220	0.4850	0.5460	0.6060	0.6620
4.7	0.0765	0.1509	0.2240	0.2945	0.3630	0.4300	0.4940	0.5560	0.6170	0.6740
4.8	0.0776	0.1532	0.2275	0.2990	0.3690	0.4370	0.5020	0.5650	0.6280	0.6860
4.9	0.0786	0.1554	0.2305	0.3035	0.3745	0.4430	0.5100	0.5740	0.6380	0.6980
5.0	0.0796	0.1577	0.2335	0.3080	0.3800	0.4500	0.5180	0.5840	0.6480	0.7090
5.5	0.0845	0.1676	0.2490	0.3285	0.4060	0.4820	0.5560	0.6260	0.6990	0.7650
6.0	0.0890	0.1770	0.2635	0.3480	0.4310	0.5130	0.5910	0.6680	0.7440	0.8170
6.5	0.0932	0.1855	0.2765	0.3660	0.4540	0.5410	0.6240	0.7070	0.7880	0.8650
7.0	0.0971	0.1933	0.2890	0.3830	0.4760	0.5670	0.6560	0.7420	0.8300	0.9120
7.5	0.1007	0.2010	0.3005	0.3995	0.4960	0.5930	0.6860	0.7770	0.8690	0.9570
8.0	0.1041	0.2080	0.3115	0.4140	0.5160	0.6160	0.7140	0.8110	0.9070	1.0000
8.5	0.1072	0.2145	0.3220	0.4290	0.5340	0.6390	0.7410	0.8420	0.9420	1.0400
9.0	0.1102	0.2210	0.3320	0.4420	0.5520	0.6610	0.7670	0.8720	0.9780	1.0790
9.5	0.1131	0.2270	0.3415	0.4560	0.5690	0.6820	0.7930	0.9020	1.0120	1.1170
10.0	0.1158	0.2327	0.3500	0.4680	0.5840	0.7020	0.8160	0.9300	1.0430	1.1530

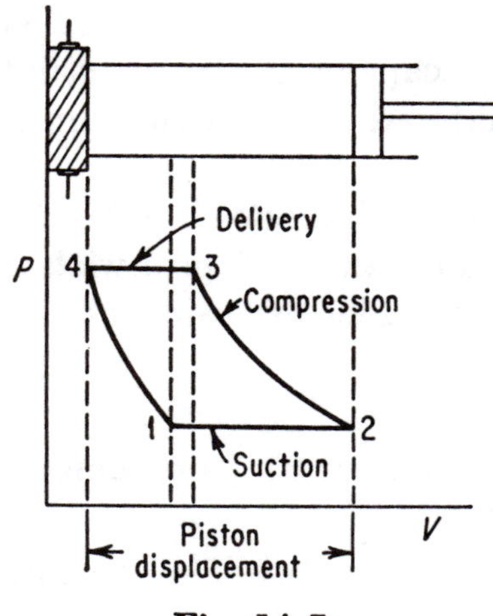

Fig. 14-7

14-13. Gas Compressor with Clearance. Refer to Fig. 14-7. The events shown there are the same for those in the case of no clearance, except that since the piston does not force all the gas from the cylinder at pressure P_2, the remaining gas must reexpand to the intake pressure, process 3-4, before intake starts again.

Since the value of n on the expansion curve has little effect on the results, it is taken as being the *same for both compression and expansion,* although actually the values are different. Without clearance, the volume of air drawn into the cylinder is the same as the piston displacement. The work of compression required is given by

$$W = \frac{n}{n-1} w'RT_1 \left[\left(\frac{P_2}{P_1}\right)^{(n-1)/n} - 1\right] \quad \text{ft-lb} \qquad (14\text{-}17)$$

where w' is the weight of actual volume drawn into cylinder along $4 - 1$. In an actual compressor with clearance, the piston displacement must be greater than volume drawn in, for a given capacity; this means a larger machine than without clearance, costing more and having greater mechanical friction.

14-14. Volumetric Efficiency. Volumetric efficiency ranges from 65 to 85 per cent and is determined from the following:

$$E_v = \frac{\text{actual volume drawn into compressor, cu ft}}{\text{piston displacement, cu ft}} \qquad (14\text{-}18)$$

or

$$E_v = 1 + c - \left[c\left(\frac{P_2}{P_1}\right)^{1/n}\right] \qquad (14\text{-}19)$$

where c is clearance per cent as a decimal.

14-15. Capacity of Gas Compressor. Let us assume the gas is air. The capacity is actual volume of "free" air delivered. Altitude and ambient air temperature affect capacity. The higher above sea level and the greater the ambient temperature, the lower the delivered capacity. In these cases, a booster compressor is required to bring the main compressor up to capacity. In an actual compressor installation the air must be filtered and drawn from a cool environment to realize best performance.

Q14-12. An air compressor with 6 per cent clearance handles 50 lb per min of air between the pressures of 14.7 and 64.7 psia with n equal to 1.33. What is the weight in the cylinder per minute?

ANSWER. The volumetric efficiency is

$$E_v = 1 + 0.06 - \left[0.06\left(\frac{64.7}{14.7}\right)^{1/1.33}\right] = 0.877, \text{ or } \mathbf{87.7 \text{ per cent}}$$

Piston displacement weight = 50/0.877 = 57.1 lb. The weight corresponding to total volume including clearance volume is 57.1 × 1.06 = **60.5 lb.**

14-16. Effect of Clearance on Compressor Performance. In a single-stage compressor, clearance reduces volumetric efficiency. The per cent of capacity reduction is greater than the per cent of cylinder clearance because the piston must travel back part of its return stroke before clearance-space air has expanded to atmospheric pressure and permitting free air to flow into the cylinder. Clearance may be so great that no air is discharged from the compressor. This characteristic is sometimes used to control the output of the compressor by increasing the clearance when a reduced output is desired. Observe from Eq. (14-19) that volumetric efficiency goes down as the pressure ratio goes up.

The volume occupied by expanded clearance air is in proportion to its discharge pressure; and the loss in compressor capacity due to clearance is less for two-stage than for single-stage compression, volume and terminal pressure being equal.

We can see that it is desirable to have the clearance as small as practicable. However, it has been shown that since there is no significant variation in the actual horsepower required for small variations of the clearance, there is no need to increase the cost of manufacturing just to reduce the clearance by a small amount. Clearances vary from about 1 per cent in some very large compressors to 8 per cent or more in other compressors, with clearances of 4 to 8 per cent being common. Neither clearance nor volumetric efficiency is a reliable indicator of the quality of a compressor. The user is most concerned about the power consumed for a given capacity. An increase in clearance requires a larger compressor to deliver the same amount of gas, hence requires more power.

Q14-13. An acetylene gas compressor of three-stage design is indicated as a 10 by 7¼ by 4 by 10-in. stroke and has a rating of 50 scfm. For particular operating reasons it is suggested that its clearance be increased in the low-pressure cylinders, i.e., the ones with the 7¼ and 10-in. bores, as will be indicated in the answer below. What will have to be the increased speed to realize the same capacity?

ANSWER. Let us first determine the original volumetric efficiency by assuming a clearance and a value of n equal to 1.15 for acetylene. Let c equal 0.07. The original volumetric efficiency is before change with R of 2.6 actual

$$E_{v1} = c(1 - R^{1/n}) + 1 = 0.07(1 - 2.6^{0.87}) + 1$$
$$= 0.91, \text{ or } 91 \text{ per cent}$$

Increase the clearance space in accordance with suggestions.

Low-pressure cylinder		*Crank end*		*Head end*
10 in.		78.5		78.5
7¼ in.		−41.2		−12.6 (4 in.)
	Total	37.3	+	65.9 = +103.2 cu in.

Thus, the total displacement originally is 103.2 cu in. per inch stroke or with a 10-in. stroke machine the total is 103.2 × 10, or 1032 cu in. On the original assumption basis of a 7 per cent clearance, the clearance volume is 0.07 × 1,032, or 72.24 cu in. Displacement determined from actual measurement.

Now to determine the clearance increase:

	Crank end		*Head end*
	78.5		78.5
	−12.6		− 3.14
Total	65.9	+	75.4 = 141.3 cu in./in.

New displacement is 141.3 × 10, or 1,413 cu in. The increase in clearance is 1,413 − 1,032, or 381 cu in. Total clearance is 381 +

72.24 = 453 cu in. Per cent new clearance is (453)/(1,413 + 72), or 31 per cent. This is equal to an increase in clearance of 31 − 7, or 24 percent.

The volumetric efficiency under the suggested changes is now found.

$$E_{v2} = 0.31(1 - 2.6^{0.87}) + 1 = 0.60, \text{ or } 60 \text{ per cent}$$

This reflects an appreciable drop in volumetric efficiency. Next step is to find the new speed to operate the compressor. The capacity at 100 rpm under the original condition when the volumetric efficiency is 91 per cent is

$$1{,}032/1{,}728 \times 100 \times 0.91 = 54.3 \text{ scfm}$$

With the reduced volumetric efficiency conditions

$$1{,}032/1{,}728 \times 100 \times 0.60 = 35.8 \text{ scfm}$$

This a reduction of (54.3 − 35.8)/(54.3) × 100 = 34 per cent. Finally, to realize same capacity as before (52.5 scfm), the compressor must be speeded up to

$$54.3/35.8 \times 100 = \mathbf{152\ rpm}$$

For acetylene work this speed was considered dangerous and change was not made.

14-17. Actual Indicator Card for Compressor. Because of fluid friction and inertia of the valves and frictional resistance of valves

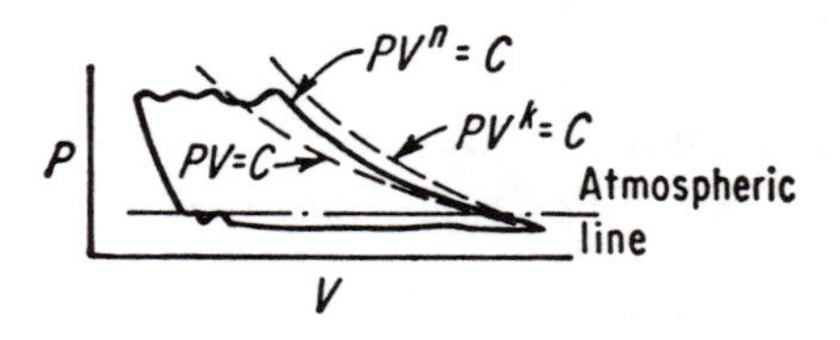

Fig. 14-8

to motion, the actual indicator card differs from the ideal (*PV* diagram). Figure 14-8 shows the effect of the fluttering of valve operation so that the area of the diagram is increased; and thus the work of the cycle.

14-18. Efficiency of Compressor. Compression efficiency is found from the air horsepowers we have been calculating and given by previous equations.

$$\text{Compression efficiency} = \frac{\text{air hp}}{\text{ihp}} \tag{14-20}$$

For mechanical efficiency:

$$E_m = \frac{\text{ihp}}{\text{bhp}} \tag{14-21}$$

For compressor efficiency:

$$E_c = \text{compression eff.} \times \text{mech eff.} = \frac{\text{air hp}}{\text{bhp}} \tag{14-22}$$

Air horsepower may be required for isothermal, isentropic, or polytropic compression. There are times when the efficiency given in an examination problem may be called adiabatic efficiency. If such a problem involves a reciprocating compressor and mechanical efficiency is mentioned, the adiabatic efficiency is the compression efficiency referred to isentropic compression. If the problem involves a centrifugal compressor, or other rotary compressor, as in a gas-turbine application and where mechanical friction is small and mechanical efficiency is high and indicated horsepower is not mentioned, then adiabatic efficiency is compressor efficiency referred to isentropic compression.

The indicated horsepower for a compressor of the reciprocating type may be determined from the relation very familiar to all

$$\text{ihp} = \frac{PLAN}{33{,}000} \tag{14-23}$$

where P = mean effective pressure, psi
L = length of stroke, ft
A = piston area, sq in.
N = number of working strokes per minute (equal to two times the rpm for a double-acting single-cylinder compressor)

14-19. Effect of Supercompressibility Upon Compressor Performance. Up to the present discussion we have at all times assumed the validity of the perfect-gas law in all gas-compressor calculations. Let us now consider the effect when the gas being compressed deviates from the perfect-gas law. For our example we shall take natural gas.

When pressures and temperatures are such that deviations of the ideal-gas law exists, the effect is known as supercompressibility and the operation of the compressor cylinder must be thoroughly understood so that we can predict the volume throughout and horsepower involved.

The effect of supercompressibility is very pronounced at high pressures. Wet gas mixtures and those which are not principally methane are greatly affected at the high pressures. When we find a cylinder handling, in some instances, a throughput of 150 to 200 per cent of the calculated throughput, neglecting supercompressibility, with a horsepower requirement in the same order of magnitude, we face a condition which we cannot afford to overlook. It is an important consideration to evaluate supercompressibility; it cannot be ignored.

When a gas is compressed, it is the space between the molecules that is reduced and not the molecules themselves. Up to a pressure of 100 psig the ideal-gas law is valid and the volumes so calculated are correct for all practical purposes. However, when pressures of over 100 psig are encountered, this reduction in volume with pressure does not follow the ideal- or perfect-gas law. The *compressibility factor* is a multiplier used in finding the volume occupied by a given weight of "deviated" gas from the volume occupied by the same weight of gas under the ideal-gas law.

To elucidate, let us solve a problem involving the capacity of a gas-compressor cylinder at high pressure.

Q14-14. A gas of 0.65 gravity (air = 1) and 80°F is to be compressed from 1,500 psig to 4,500 psia in a cylinder of 15 per cent clearance whose displacement is 45,000 cu ft per day. It is required to compare the capacity of the cylinder when deviation from the perfect-gas law is neglected and when it is considered in the calculation.

ANSWER. Neglecting deviation and using another recognized form of the volumetric efficiency for this application to be,

$$E_{v_1} = K - [c(R^{1/n} - 1)] = 0.97 - [0.15(3^{1/1.2} - 1)]$$
$$= 0.745, \text{ or } 74.5 \text{ per cent}$$

where we have assumed the reexpansion to have n of 1.2, K of 0.97, and R equal to 4,500/1,500, or 3. The capacity in standard cubic feet (sffd) per day is calculated from

$$\text{Displacement, cu ft per day} \times E_{v_1} \times \frac{\text{suction pressure}}{14.7} \times \frac{520}{\text{suction °R}}$$

or $$45{,}000 \times 0.745 \times \frac{1{,}500}{14.7} \times \frac{520}{540} = 3{,}290{,}000 \text{ cu ft}$$

For the computations when deviation is recognized, we have the following relation:

$$\text{Actual capacity} = \text{capacity, no deviation} \times \frac{E_{v_2}}{E_v \times Z_s}$$

where E_{v_2} = volumetric efficiency compressibility factor
Z_s = compressibility factor at suction conditions
Z_d = compressibility factor at discharge

Now the deviated volumetric efficiency is calculated, using

$$E_{v_2} = 1 + c - \left[(1 + c - E_{v_1})\frac{Z_s}{Z_d}\right]$$

or $$E_{v_2} = 1 + 0.15 - [(1 + 0.15 - 0.745)(0.78/0.98)]$$
$$= 0.828, \text{ or } 82.8 \text{ per cent}$$

$$\text{Actual capacity} = 3{,}290{,}000 \times \frac{0.828}{0.745 \times 0.78} = 4{,}690{,}000 \text{ scfd}$$

Thus, because of supercompressibility, the capacity of the cylinder under these conditions has been shown to be *greater by*

$$\frac{4.69 - 3.29}{3.29} \times 100 = \mathbf{42.5 \text{ per cent}}$$

There is also an increase in brake horsepower of about 25 per cent.

Chapter 15

HEAT TRANSMISSION

15-1. Basic Principles of Heat Transmission. Problems of greatest importance and most frequently occurring in the entire field of engineering are unquestionably those that involve the strength of materials. Second in importance to these only are those problems that involve the flow of heat. Most engineers readily recognize the significance of the first group, and are rather well equipped to work out methods for their solution.

On the other hand, when confronted with a problem involving the flow of heat, the average engineer will look up the formulas in his elementary physics text, and a table of thermal conductivities in a handbook, and considers that he has the complete data for solving the problem. Just to clarify the point, in the published proceedings of a technical society some years ago it was stated that since copper had seven times the thermal conductivity of iron, substituting copper tubes for iron tubes in a piece of heat-transfer equipment should increase the capacity to seven times the original figure using iron. This, of course, is not true. It is doubtful if the substitution would actually increase the capacity 10 per cent.

There are many such similar contradictions that only add to the confusion. Much has been published on the subject of heat transmission,* but we shall attempt to help clarify the situation somewhat in this chapter. Our treatment, because of the lack of space, will be systematic and in compact form.

15-2. Methods of Heat Transfer. Heat may flow by three methods:

*J. P. Holman, "Heat Transfer," 5th ed., McGraw-Hill Book Company, Inc., New York, 1981.

(*a*) *Conduction.* Energy may flow through a conducting body by transfer from one molecule to the next, without visible movement of the body as, for example, through an iron bar which is hotter at one end than at the other.

(*b*) *Convection.* Energy may be carried by movement of heated fluids away from a hot body, as in the heating of water by a hot surface. In this case each particle of water moves away with some heat and another takes its place.

(*c*) *Radiation.* Energy may be transmitted from a hot to a cold body by "electromagnetic" waves, this energy being converted to heat when absorbed by the other body.

15-3. Conduction. This is the simplest case. It is the confusion between the complicated cases occurring in practice and the simple cases of pure conduction that causes many errors.

Assuming a steady state and a homogeneous material, the following formula may be expressed:

$$Q = \frac{kA(t_2 - t_1)}{L} \tag{15-1}$$

where Q = heat transferred, Btu per hr
k = thermal conductivity, English units, Btu/(sq ft)(°F)(ft) thickness
A = area of path of heat flow, sq ft
t_1 = lower temperature, °F
t_2 = higher temperature, °F
L = length of path through which heat flows, ft

The coefficient of conductivity k gives the number of Btu that will flow in 1 hr through a conductor 1 ft long and 1 sq ft in cross section, when there is a temperature difference of 1 °F from one end of the conductor to the other. This coefficient is a characteristic of the material of which the conductor is made and varies with temperature directly. Some texts give thermal conductivity in terms per inch of thickness. Then the length of path is in inches.

Table 15-1 gives the values of thermal conductivity for various common materials. More extensive tables may be found in the literature and in engineering handbooks.

TABLE 15-1. THERMAL CONDUCTIVITIES

Material	Experimental temperature range, °F	k per ft thickness
Aluminum	32–212	118
Brass, yellow	32–212	63
Copper, pure	At 32	226
Wrought iron	32–527	35
Cast iron, 3.5% carbon	At 212	28
Lead	At 59	20
Nickel	32–212	34
Platinum	64–212	41
Silver	At 64	243
Steel, mild	32–212	35
Boiler scale	At 212	0.5–1.5
Asbestos	100–1000	0.04–0.12
Brick, carborundum	At 1800	5.6
Brick, building	At 70	0.4
Sil-O-Cel	At 1800	0.03
Cork	122–392	0.03
Glass, flint	50–59	0.3–0.6
Infusorial earth, 12.5 lb per cu ft	At 122	0.05
Magnesia insulation, 85%	68–310	0.04
Rubber	At 220	0.01
Fire brick	At 2300 F	1.0
Rock wool	At 212	0.023
Wood, pine	At 70	0.1–0.2
Plaster on wood	At 212	0.208
Glass wool	At 212	0.023
Concrete work	At 212	1.0
Lubricating oil	At 86	0.08
Water	At 167	0.372
Air	At 32	0.0137
	At 212	0.0174
Steam	At 32	0.0095
	At 212	0.0129
Carbon monoxide	At 32	0.131
Carbon dioxide	At 32	0.00804
Oxygen	At 32	0.0138
Hydrogen	At 32	0.092
Methane	At 32	0.0174

Q15-1. What is the heat flow per hour through a brick and mortar wall 9 in. thick if coefficient of thermal conductivity has been determined as 0.4 and the wall is 10 ft high by 6 ft wide, the temperature on one side of the surface being 330°F and on the other, 130°F?

ANSWER. Using Eq. (15-1), we obtain the following:

$$Q = 0.4\,[(10 \times 6) \times (330 - 130)]/(9/12) = 6{,}400 \text{ Btu per hr}$$

Note that the solution has not taken into account the fact that there is an air film on either side of the wall.

15-4. Conductors of Nonuniform Cross Section. Where Eq. (15-1) holds for constant cross-section bodies, the heat loss from an insulated pipe takes place radially from the metal surfaces through a constantly increasing area until it reaches the maximum area of the surface of the insulation. On the other hand, heat flow from a room into cold brine flowing inside a pipe covered with cork or hairfelt insulation starts with a large outside surface and passes radially inward through a constantly decreasing area until it reaches the pipe. In either case, the A term in Eq. (15-1) must be expressed as an average of the section. This is the arithmetic mean average and is expressed as:

$$A_a = \frac{A_1 + A_2}{2} \tag{15-2}$$

15-5. Logarithmic Mean Average of Area. In the relatively few cases in which the maximum area is very large in comparison with the minimum, that is, over twice as great, a considerable error is introduced in using the arithmetic mean as the average area. It then becomes necessary to use the logarithmic mean (log mean). Note for both Eqs. (15-2) and (15-3) to follow, the maximum area is A_1 and the minimum area is A_2.

$$A_m = \frac{A_1 - A_2}{2.3 \log \times A_1/A_2} \tag{15-3}$$

15-6. Conductors in Series. When heat flows through several conductors in series, it is convenient to add the resistances of all conductors and then to use the resistance type of formula. For any conductor

$$Q = \frac{\text{driving force}}{\text{resistance}} = \frac{\Delta t}{R} \tag{15-4}$$

Also

$$R = \frac{L}{k \times A} \tag{15-5}$$

Using the units already given for these symbols, the resistances for three conductors in series are

$$R_1 = \frac{L_1}{k_1 \times A_1} \qquad R_2 = \frac{L_2}{k_2 \times A_2} \qquad R_3 = \frac{L_3}{k_3 \times A_3}$$

Then the total resistance is the sum of all three $R = R_1 + R_2 + R_3$. And the heat transferred through the series group is

$$Q = \frac{\Delta t}{R_1 + R_2 + R_3} \tag{15-6}$$

Note that the Δt in Eq. (15-6) is the temperature difference across the entire series and not the individual components. The temperature drop between each individual conductor is $\Delta t = Q_1L_1/k_1A_1$ and so on for each. But we know that for constant rate of transfer $Q = Q_1 = Q_2 = Q_3$ and that $\Delta t = \Delta t_1 + \Delta t_2 + \Delta t_3$. Figure 15-1 shows three conductors in series.

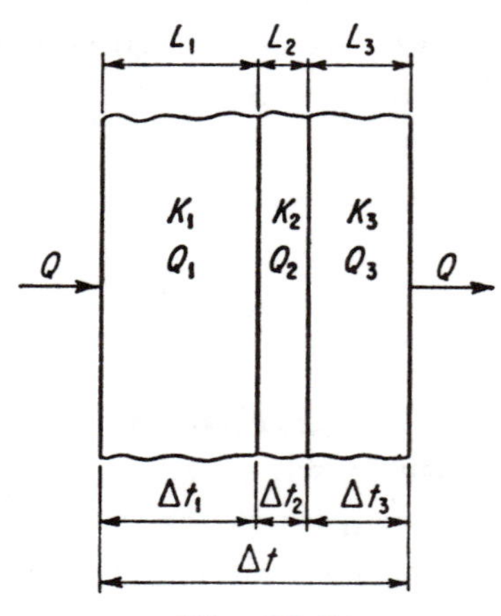

Fig. 15-1

Q15-2. A pipe with an outside diameter of 2½ in. is insulated with a 2-in. layer of asbestos ($k = 0.12$), followed by a layer of cork 1½ in. thick ($k = 0.03$). If the temperature of the outer surface of the pipe is 290°F and that of the outer surface of the cork is 90°F, calculate (*a*) heat loss per hour per 100 ft of insulated pipe, (*b*) temperature at the interface of asbestos and cork, and (*c*) percentage of total resistance due to asbestos and to cork.

ANSWER.

(*a*) Since this problem involves an over-all temperature difference and a series of two resistances, it may be solved with the use of Eq. (15-6).

Inside surface of asbestos $= \pi \times 2.5/12 \times 100 = 65.4$ sq ft
Outside surface of asbestos $= 65.4 \times 6.5/2.5 = 171$ sq ft

This 171 sq ft is also the inside surface of the cork covering.

$$\text{Outside surface of cork} = 65.4 \times 9.5/2.5 = 249 \text{ sq ft}$$

For the asbestos, the average area must be calculated as the log mean because the larger area is greater than twice the smaller.

$$A_m = \frac{171 - 65.4}{2.3 \log (171/65.4)} = 111 \text{ sq ft}$$

For the cork the arithmetic average may be used:

$$A_a = \frac{171 + 249}{2} = 210 \text{ sq ft}$$

The distance through the asbestos is L_1 equal to 2/12, or 0.167 ft. Then

$$R_1 = 0.167/(0.12 \times 111) = 0.0125 \text{ for asbestos}$$

For the cork,

$$R_2 = 0.125/(0.03 \times 210) = 0.0198$$

Then, since $\Delta t = 290 - 90 = 200°\text{F}$,

$$Q = 200/(0.0125 + 0.0198) = \mathbf{6{,}190\ Btu/hr}$$

(*b*) $$\Delta t_1 = QR_1 = 6{,}190 \times 0.0125 = 77.4°\text{F}$$

and the temperature at interface is $290 - 77.4 =$ **212.6°F.**

(*c*) $R_1/(R_1 + R_2) = 0.0125/(0.0125 + 0.0198) \times 100 = 38.8$ per cent, or **38.8 per cent** of the resistance **due to asbestos** and 100 − 38.8, or **61.2 per cent due to cork.**

Q15-3. A furnace wall consists of 9-in. fire brick, 4½-in. Sil-O-Cel brick, 4-in. red brick, and ¼-in. transite board. The k values are: 0.82 at 1800°F for fire brick; 0.125 at 1800°F for Sil-O-Cel; 0.52 at 500°F for transite. Inside wall of furnace is at 1800°F and outside wall is at 200°F. Calculate the heat lost per hour through a wall surface of 10 sq ft.

ANSWER. Let us take as the basis 1 sq ft of wall surface. The surface is the same throughout and is left out of the computations. Use Eqs. (15-4) and (15-5).

$$Q = \frac{1{,}800 - 200}{[(1/0.82)(9/12)] + [(1/0.125)(4.5/12)] + [(1/0.52)(4/12)] + [(1/0.23)(0.25/12)]}$$

$$= 1{,}600/4.646 = 344 \text{ Btu per hr per sq ft, or } \mathbf{3{,}440\ for\ 10\ sq\ ft}$$

Although not required, the temperature within the wall may be computed from the following:

$$\frac{\Delta t}{\Delta t_1} = \frac{R}{R_1} \tag{15-7}$$

For example, between fire brick and Sil-O-Cel

$$\frac{1600}{4.646} = \frac{\Delta t_1}{0.915} \qquad \text{from which } \Delta t_1 = 315$$

Interface temperature is equal to $1800 - 315 = 1485°F$.

Thermal resistances in series form a combination very often met in practice. In many instances the areas of the different resistances are nearly equal. Then Eq. (15-6) will take on the form

$$Q = A\,\Delta t \frac{1}{L_1/k_1 + L_2/k_2 + L_3/k_3} \tag{15-8}$$

For most engineering problems in heat transfer Eq. (15-8) will give answers greatly in excess of the actual realized. What causes this discrepancy? How can this be evaluated?

15-7. Film Concept and Convection. In convection, not only are solids involved but we also have fluid films in series with these solids. Actually, it is of controlling importance in most cases of transfer between solids and fluids (liquids or gases).

Although there are a great many factors that influence the rate at which heat flows by convection to or from a solid in contact with a hotter or colder fluid, the most important of these is the observation that there is always a very thin film of practically stationary fluid on the surface of the solid. Through this film heat must flow by conduction. As you can see from Table 15-1, the value of k for liquids is quite low and for gases is exceedingly low. In the main body of the fluid, heated particles move freely and rapidly, but since the particles of the film are substantially stationary, *the film always offers the major resistance* between the solid and the main body of the fluid.

Consequently, efforts to improve heat transfer by convection are always directed toward reducing the thickness of the highly insulating film, that is, by increasing the velocity of the moving fluid. This will take the flow out of the viscous range and into the turbulent range where greatly improved rates of transfer can be realized.

However, once there, further increase in flow velocity will not reflect itself in much greater transfer rates.

Since the thickness of the fluid film on a solid is unknown and variable, and since the conductivity of most liquids is likewise unknown, it is not practicable to use the heat transfer equation in the form given in Eq. (15-1). Instead, for calculating heat transferred by convection, the expression used is

$$Q = UA\,\Delta t \tag{15-9}$$

where U is known as the over-all heat-transfer coefficient expressed as Btu/(hr)(°F)(sq ft) of surface. It has also been shown that

$$U = \frac{1}{\text{summation of resistances to heat flow}} \tag{15-10}$$

Equation (15-9) may be shown to have been derived from the equation

$$Q = A\,\Delta t\,\frac{1}{1/h_1 + L_2/k_2 + 1/h_3} \tag{15-11}$$

where we have a solid wall resistance (L_2/k_2) and film resistances of $1/h_1$ and $1/h_3$. The ratio within the brackets is the over-all coefficient U. It is well to note that Eq. (15-11) holds for thin-solid walls.

Q15-4. Consider a tube of 16 gauge copper (thickness 0.065 in.), with a difference of 1°F between its inner and outer surfaces. On one side there is a stagnant film of water 0.01 in. thick and a stagnant film of condensed steam on the other. The tube is so thin-walled that a length sufficient to give 1 sq ft of surface on the inside will give practically 1 sq ft of surface on the outside. Calculate the heat transferred by use of Eqs. (15-1) and (15-11) and compare.

ANSWER.

$$Q = (220 \times 1 \times 1)/(0.065/12) = \textbf{40,600 Btu per hr}$$

$$= 1 \times 1 \times \frac{1}{\dfrac{0.01/12}{0.417} + \dfrac{0.065/12}{220} + \dfrac{0.01/12}{0.417}}$$

$$Q = 1 \times 1 \times \frac{1}{0.00199 + 0.00002 + 0.00199} = \textbf{250 Btu per hr}$$

The first answer is extremely out of line as compared with the second and more reasonable rate which might be found in practice under the conditions stated. It is also well to note that the resistance of the copper wall is so small as compared to the film resistances that it may as well be left out with little apparent effect.

Q15-5. Assuming the same conditions as in Q15-4 above, except that air is flowing on the inside of the tube (conductivity of air = 0.0174), what will be the rate of heat transfer?

ANSWER.

$$Q = \frac{1}{0.00163 + 0.00002 + 0.0477} = \textbf{20 Btu per hr}$$

In practice the thickness of the film cannot be measured directly, but can be predicted by an indirect method. As we shall see later, values of the film coefficient h can be computed and Eq. (15-11) may be applied.

15-8. Thick-walled Tubes. Our entire discussion so far has been based on the assumption that the film resistances apply to surfaces that are essentially constant. This is true for flat surfaces, surfaces with only slight curvature, or for thin-walled tubes. Where the confining walls are substantially thick, the following formula ensues.

$$Q = \frac{\Delta t}{1/h_1A_1 + L_2/k_2A_2 + 1/h_3A_3} \qquad (15\text{-}12)$$

The mean area of the tube A_2 must be defined by determining the log mean radius and then A_2 by its use in the standard cylinder area equation. The true mean radius for use in the heat-flow calculations is determined from

$$r_m = \frac{r_1 - r_2}{\ln (r_1/r_2)} \qquad (15\text{-}13)$$

where r_m = mean radius
r_1 = outside radius
r_2 = inside radius

For rigorous accuracy, A_2 should be the area calculated using r_m for thick-walled tubes.

The log mean r_m is not too convenient in engineering calculations. If the arithmetic mean of r_1 and r_2 is used instead of r_m, the results will be in error by 10 per cent or more if r_1/r_2 is 3.4 or greater. If

r_1/r_2 is 1.5 or less, the results will be in error by less than 1 per cent. Now since thick-walled tubes are seldom used in heat-transfer apparatus and since an accuracy of 5 per cent is most likely as much as can be expected from any heat-transfer calculation, the arithmetic mean is under ordinary conditions satisfactory.

In actual heat-transfer calculations for best results take A_1, A_2, and A_3 as the inside, mean, and outside areas per running foot of the tube, respectively. Use Eq. (15-12) to obtain the heat transfer per foot of tube. Then obtain the total number of feet of tube from the total loading.

We previously mentioned that the film resistances were in most cases greater than that through the tube wall, so that the middle term of the denominator is too small to be of practical significance. Then the results in many cases depend on the film resistances. If h_1 and h_3 in Eq. (15-12) are of the same order of magnitude, then use the entire equation; but if one of the film coefficients h_1 or h_3 is very much smaller than the other, then the over-all coefficient U will approach the value of the smaller film coefficient. Thus, the smaller film coefficient will control and use that area coinciding therewith, i.e., if h_1 is small compared to h_3, then the term involving h_1 is the term which largely determines the value U, and, therefore, A_1 is the significant area. Simply, if the film coefficient on the inside of the tube is low, the inside surface should be used in the calculation.

15-9. Estimating Film Coefficients. It is impossible to measure the thickness of films close to the surfaces of tubes and other areas. It is this stagnant film that needs to be given special treatment by means of calculation. A great deal has been accomplished in the field of heat transmission, but there is still much more to be done to cover all cases. The student is referred to the standard texts on heat transfer for a more elaborate treatment of the subject. Here we shall concentrate on questions that have appeared in examinations together with additional material likely to appear in the future.

Most methods used to calculate film coefficients have an accuracy of no more than ± 5 per cent. Nevertheless, the engineer's ability to estimate, approximate as it may be, will prove most beneficial.

There are convection currents which exist across films which cause its thickness to vary downward, i.e., become thinner. Forced circulation of the fluid past a transfer surface has the same effect. Some of the other factors which affect film thickness are viscosity, density, velocity in main body of fluid, conductivity, turbulence, noncondensibles in condensing vapors, corrosion products, and temperature. The transition from viscous to turbulent flow is accompanied by a considerable increase in the film coefficient, but once this state has been achieved and in the turbulent range, further increases are negligible.

Q15-6. Air at a pressure of 30 in. Hg and an average temperature of 100°F is flowing at an average velocity of 30 fps through a horizontal tube of copper 1 in. in diameter (OD) and 0.065 in. wall. The pipe is surrounded with steam at an average temperature of 212°F. What is the over-all rate of heat transfer?

ANSWER. For the air film the Dittus and Boelter equation may be used.

$$\frac{h_i D}{k} = 0.0225 \left(\frac{DG}{Z}\right)^{0.8} \left(\frac{CZ}{k}\right)^{0.4}$$

Here the quantities in the equation may be evaluated as follows:

$$D = (1 - 0.130)/12 = 0.0725 \text{ ft}$$
$$k = 0.0151$$
$$G = u\rho, \text{ where } u = \text{velocity, fps}$$
$$u = 30 \times 3{,}600 = 108{,}000 \text{ fph}$$
$$\rho = \frac{29}{379} \times \frac{460 + 60}{460 + 100} = 0.0728 \text{ lb per cu ft}$$
$$G = u\rho = 108{,}000 \times 0.0728 = 7{,}865 \text{ psf per hr}$$
$$Z = \text{centipoises} \times 2.42 = 0.019 \times 2.42 = 0.0460$$
$$c_p = [6.76 + (0.000606 \times 100) + (0.00000013 \times 100^2)]/29 = 0.236$$

Then by substitution in the Dittus and Boelter equation

$$\frac{h_i \times 0.0725}{0.0151} = 0.0225 \left(\frac{0.0725 \times 7{,}865}{0.0460}\right)^{0.8} \times \left(\frac{0.236 \times 0.0460}{0.0151}\right)^{0.4}$$
$$h_i = \mathbf{7.73\ Btu/(sq\ ft)(hr)(°F)}$$

The film coefficient for the steam side will be very high by comparison, so much higher than the air film that the temperature drop through the steam film will be negligible. Let us now evaluate the terms to be inserted into the Nusselt equation for steam condensing on the outside of a horizontal tube.

$$h_0 = 0.725\left(\frac{r\rho^2k^3g}{ZD\,\Delta t}\right)^{1/4}$$

where r = 970 (from steam tables), latent heat of condensation
ρ = 59.85 lb per cu ft density of condensate
k = 0.416 thermal conductivity of condensate (per ft)
g = 4.18×10^8 acceleration of gravity, ft per hr per hr
Z = $0.286 \times 2.42 = 0.406$ lb-ft-hr units viscosity of condensate
D = 0.0833 ft outside diameter of pipe
Δt = 1°F mean temperature difference between condensing vapor and metal

Then the steam-film coefficient is

$$h_0 = 0.725\left(\frac{970 \times 59.85^2 \times 0.416^3 \times 4.18 \times 10^8}{0.406 \times 0.0833 \times 1.00}\right)^{1/4} = 5{,}400$$

Since the gas-film coefficient controls, the area of the air-film controls, and therefore use Eq. (15-11).

$$U = \frac{1}{1/7.73 + (0.065/12)/222 + 1/5{,}400}$$
$$= \frac{1}{0.1293 + 0.000024 + 0.000185} = 7.72$$

In a problem such as this where h_1 is so greatly different from h_3, it is obviously unnecessary to calculate h_3. Then the over-all coefficient U may be taken as equal to h_1. Let us examine the factors herein. The resistance of the metal wall and the steam film has only changed the coefficient by 1 part in 700, or 0.14 per cent, whereas, it is doubtful if h_1 itself is accurate to more than ± 10 per cent. Thus, we can say that *if one film is a permanent gas* (air in this case) *and the other a liquid or condensing vapor film, the over-all coefficient may be considered equal to the gas film coefficient.*

Q15-7. Crude oil at an average temperature of 150°F is flowing inside standard 3-in. iron pipe. At this temperature its density is 50 lb per cu ft, its viscosity is 1.5 centipoises, its specific heat is

0.5, its thermal conductivity is 0.078, and its velocity is 2 fps. If the steam-film coefficient is 2,500, what is the over-all rate of heat transfer?

ANSWER. The equation of Ulsamer for fluids outside of pipes is used.

$$\frac{h_0 D}{k} = K\left(\frac{DG}{Z}\right)^m \left(\frac{CZ}{k}\right)^n$$

where for values of DG/Z from 0.1 to 50, $k = 0.91$, $m = 0.31$, $n = 0.385$. And for values of DG/Z from 50 to 10,000, $k = 0.60$, $m = 0.31$, $n = 0.50$. The units are the same as for the Dittus and Boelter equation.

$$D = 3.068 \text{ in., or } 0.256 \text{ ft}$$
$$k = 0.078$$
$$u = 2 \times 3{,}600 = 7{,}200 \text{ ft per hr}$$
$$\rho = 50$$
$$G = u\rho = 7{,}200 \times 50 = 360{,}000$$
$$Z = 1.5 \times 2.42 = 3.63$$
$$c_p = 0.50$$

Then h_1 is found to be equal to 80.6 from the following:

$$\frac{h_i \times 0.256}{0.078} = 0.0225\left(\frac{0.256 \times 360{,}000}{3.63}\right)^{0.8}\left(\frac{0.50 \times 3.63}{0.078}\right)^{0.4}$$

$$U = \frac{1}{1/80.6 + (0.216/12)/35 + 1/2{,}500}$$
$$= \frac{1}{0.01240 + 0.00051 + 0.00040} = \mathbf{75.1}$$

Note that the metal wall has a resistance about equal to the steam film, but the two together hardly affect the rate of heat transfer by an appreciable amount.

Q15-8. A certain heater has 1¼-in. standard iron pipe for its heating surface. It is operating with a liquid-film coefficient of 250 and a steam-film coefficient of 2,000. What will be the relative effect of (*a*) increasing the liquid velocity so that the liquid film coefficient becomes 300, (*b*) venting the steam space more thoroughly so that the steam-film coefficient becomes 3,000, or (*c*) substituting copper tubes 1.25 in. ID, 0.067 in. wall?

ANSWER. The existing over-all coefficient of heat transfer is

$$U = \frac{1}{1/250 + (0.140/12 \times 35) + 1/2{,}000}$$

$$= \frac{1}{0.004 + 0.00033 + 0.0005} = 207$$

(*a*) By increasing the liquid-film coefficient to 300 gives

$$U = \frac{1}{0.00333 + 0.00033 + 0.0005} = 240$$

(*b*) By increasing the steam-film coefficient to 3,000 gives

$$U = \frac{1}{0.004 + 0.00033 + 0.00033} = 214$$

(*c*) Changing to copper tubes gives

$$U = \frac{1}{0.004 + (0.067)/(12 \times 220) + 0.0005}$$

$$= \frac{1}{0.004 + 0.000025 + 0.0005} = 221$$

We see that increasing the steam-film coefficient by 50 per cent or changing to copper tubes increases the over-all coefficient only by 12 in 200, while increasing the liquid-film coefficient by 20 per cent increases the over-all coefficient 17 per cent.

The nature and thickness of the metal wall is not always negligible. For instance, if in the above problem the liquid-film coefficient were 1,500, changing from iron pipe to copper tubes would increase the over-all coefficient from 685 to 835, a decided improvement. In general, whenever *both* the film coefficients are high, a change from a heavy metal wall of poor conductivity to a thin wall of high conductivity will be advantageous; but if *either* film coefficient is low, a change in metal would be of slight significance.

One exception may be made to this statement. A film of rust or oxide may result in a greatly thickened stagnant film. The difference between a polished steel tube and a rusty one may change the liquid-film coefficient as much as 1,000 per cent. Thus, in some cases the substitution of copper tubes for iron pipe may greatly improve the rate of heat transfer. This is so not because of the increased conductivity of the metal (copper in this case), but because the copper stays bright and gives a thin stagnant film on the liquid side, while the iron rusts and the rust results in a thicker film, resulting in an increased resistance on the liquid side.

15-10. Pipe Insulation. In the case of pipe insulation, the area through which heat is transferred is not constant. If the thickness of the insulation is small, compared with the diameter, the arithmetic average of the larger area and the smaller area may be used. The arithmetic mean area may be used for all cylindrical vessels and for pipe sizes down to about 2 in., if standard insulation is used and if an error of 4 per cent is permitted. If greater accuracy is required, and the thickness of insulation is great compared with the diameter, the logarithmic mean area must be used.

$$\text{Logarithmic mean area} = \frac{A_2(\text{large}) - A_1(\text{small})}{2.3 \log (A_2/A_1)} \qquad (15\text{-}14)$$

When the ratio of A_2/A_1 is 2 or less, use the arithmetic mean area $(A_2 + A_1)/2$. But it is much simpler to use the radii as previously indicated under Sec. 15-8, Thick-walled Tubes (see also Fig. 15-2).

$$r_m = \frac{r_2 - r_1}{2.3 \log (r_2/r_1)} \qquad (15\text{-}13)$$

The arithmetic mean is

$$r_a = \frac{r_2 + r_1}{2} \qquad (15\text{-}15)$$

Q15-9. A tube 2.5 in. in diameter is lagged with a 2-in. layer of asbestos (conductivity 0.12) which is followed by a 1.5-in. layer of cork (conductivity 0.03). If the skin temperature of the pipe is 290°F and the canvas temperature is 90°F, calculate the loss in Btu/(hr)(ft) of lagged pipe. The various conductivities are expressed per foot units.

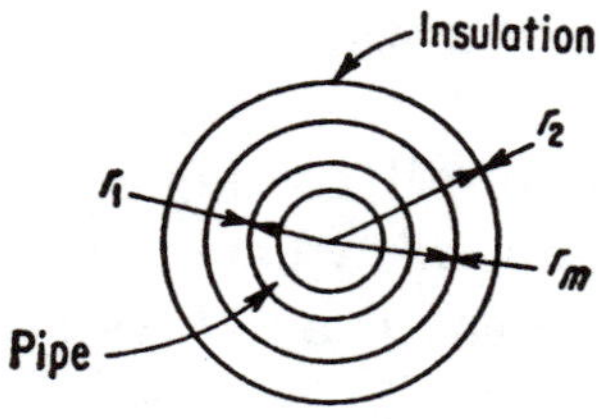

Fig. 15-2

ANSWER. Refer to Fig. 15-2. The layers are too thick to use the arithmetic mean so that we must use the log mean. The loss in Btu per hr is determined

$$Q = \frac{\Delta t}{\Sigma R} = \frac{\Delta t}{R_1 + R_2} = \frac{290 - 90}{R_1 + R_2} = \text{loss in Btu per hr}$$

$$R_1 = \frac{L_1}{k_1 A_1} \quad \text{for asbestos layer}$$

$$R_2 = \frac{L_2}{k_2 A_2} \quad \text{for cork layer}$$

The resistance of each layer is determined by using the mean radius in each case.

For the asbestos,

$$r_m = \frac{3.25 - 1.25}{2.3 \log (3.25/1.25)} = 2.09 \text{ in.}$$

For the cork,

$$r_m = \frac{4.75 - 3.25}{2.3 \log (4.75/3.25)} = 3.95 \text{ in.}$$

$$R_1 = \frac{2/12}{0.12 \times 2\pi(2.09/12) \times 1} = 1.270$$

$$R_2 = \frac{1.5/12}{0.03 \times 2\pi \times (3.95/12) \times 1} = 2.015$$

$$Q = \frac{290 - 90}{1.27 + 2.015} = \mathbf{60.9\ Btu/(hr)(ft)\ of\ lagged\ pipe}$$

Note that the air film outside the pipe was not considered. Now if the air film resistance is equal to 1/6.0, then the heat loss would be 58, a negligible difference.

Q15-10. Heat is being transferred from a gas through the walls of a standard 2-in. steel pipe and into water flowing on the inside of the pipe. Calculate the clean over-all coefficient of heat transfer. Let $h_0 = 6$, $h_i = 500$,

ANSWER.

$$U = \frac{1}{\frac{1}{6} + (0.154)/(12 \times 25) + \frac{1}{500}}$$

$$= \frac{1}{0.1667 + 0.00051 + 0.002} = \mathbf{5.91}$$

Note that here the thickness is in inches and conductivity is in Btu/(sq ft)(hr)(°F)(in.) thickness. The resistance of the metal wall and the water film could be neglected with an error of only 1.52 per cent. If the tube were copper ($k = 220$), the over-all coefficient would be 5.93.

15-11. Fouling Factors. So far we have calculated over-all coefficients that are clean, i.e., no consideration has been taken to include the effects of operation and the build-up of scale or corrosion products on the tube walls. Now we shall take into account these fouling effects.

So that the heat transfer surface in heat exchangers, engine coolers, and the like will have sufficient excess tube surface to maintain satisfactory performance during normal operation and with reasonable service time between cleanings, fouling factors must be applied as shown below. Under special conditions of fouling higher factors may be used.

Fouling factors represent the resistance r to heat transfer caused by the layer of foreign substance deposited on the surfaces of the heat transfer surfaces. Fouling resistance varies directly with the thickness of the film and inversely with the conductivity of the film k. The general equation for the over-all coefficient taking into account all factors is shown below.

$$U = \frac{1}{1/h_0 + r_0 + r_w + r_i(A_0/A_i) + (1/h_iA_i/A_0)} \qquad (15\text{-}16)$$

where U = over-all coefficient of heat transfer, Btu/(hr)(°F)(sq ft) outside surface
h_0 = film coefficient of fluid medium on outside of tubing
h_i = film coefficient of fluid medium on inside of tubing
r_0 = fouling resistance on outside of tubing
r_i = fouling resistance on inside of tubing
r_w = resistance of the tube wall (L/k)
A_0/A_i = ratio of outside tube surface to inside tube surface

Table 15-2 lists minimum fouling factors to apply in Eq. (15-16).

TABLE 15-2*

Fuel oil	0.005
Machinery and transformer oils	0.001
Vegetable oils	0.003
Coke-oven gas	0.01
Diesel-engine exhaust gas	0.01
Organic vapors	0.0005
Steam (nonoil bearing)	0.0
Alcohol vapors	0.0
Steam, exhaust (oil-bearing from reciprocating engines)	0.001
Refrigerating vapors (condensing from reciprocating compressors)	0.002
Air	0.002
Organic liquids	0.001
Brine (cooling)	0.001

* Table taken from Standards of Tubular Exchanger Manufacturers Association (TEMA).

TABLE 15-3. FOULING FACTORS—WATER*

Temperature of heating medium	Up to 240°F		240–400°F	
Temperature of water	125°F or less		Over 125°F	
Types of water	Water velocity, fps		Water velocity, fps	
	3 ft and less	Over 3 ft	3 ft and less	Over 3 ft
Sea water	0.0005	0.0005	0.001	0.001
Brackish water	0.002	0.001	0.003	0.002
Cooling tower and artificial spray pond:				
Treated make-up	0.001	0.001	0.002	0.002
Untreated make-up	0.003	0.003	0.005	0.004
City or well water	0.001	0.001	0.002	0.002
Great Lakes	0.001	0.001	0.002	0.002
River water				
Minimum	0.002	0.001	0.003	0.022
Mississippi	0.003	0.002	0.004	0.003
Delaware, Schuylkill	0.003	0.002	0.004	0.003
East River and New York Bay	0.003	0.002	0.004	0.003
Chicago Sanitary Canal	0.008	0.006	0.010	0.008
Muddy or silty	0.003	0.002	0.004	0.003
Hard (over 15 gr per gal)	0.003	0.003	0.005	0.005
Engine jacket	0.001	0.001	0.001	0.001
Distilled	0.0005	0.0005	0.0005	0.0005
Treated boiler feed water	0.001	0.0005	0.001	0.001
Boiler blowdown	0.002	0.002	0.002	0.002

* Table taken from Standards of Tubular Exchanger Manufacturers Association (TEMA).

15-12. Mean Temperature Difference. In our discussion so far the assumption has been that the hot fluid remains at a constant temperature and the cold fluid does likewise. In actual heat transfer equipment either one or the other, or both, are changing in temperature throughout the apparatus. In Fig. 15-3 the most common arrangements of heat transfer are shown. The arrow indicates the direction of flow of the fluids with respect to each other. Almost all commercial and industrial equipment is designed for *countercurrent flow*, or at least for conditions approaching this (Fig. 15-3*a*). Parallel flow (Fig. 15-3*b*) is rarely used, while Fig. 15-3*c* and *d* are simulated in all refrigeration coolers and steam condensers and wherever a fluid is condensing on one side and the other fluid is being either cooled or heated on the other. Parallel flow gives poor heat-transfer results in that it would require a greater amount of surface

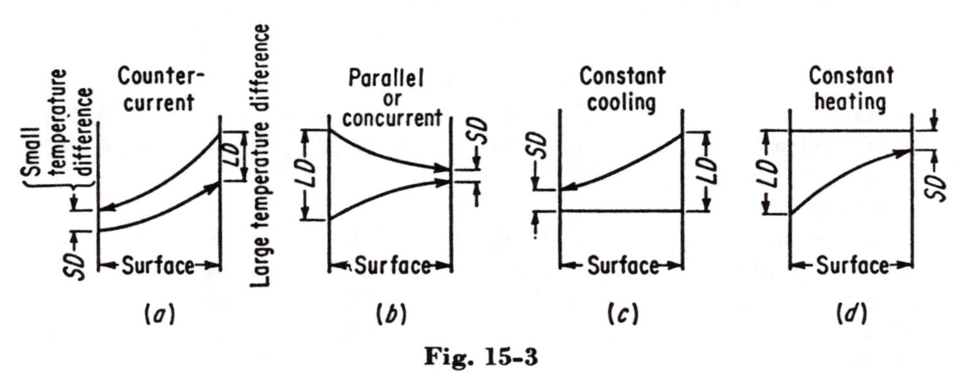

Fig. 15-3

to accomplish the same heat transfer load as compared to countercurrent flow.

The temperature difference for a fluid system is an average quantity. If the equipment is perfectly insulated and the over-all coefficient is a constant quantity, specific heat and weight rate are constant unless the process is that of evaporation or condensation, parallel path flow is taking place. Then the logarithm mean temperature is used in the transfer equation

$$Q = UAt_m \tag{15-17}$$

And the logarithm mean temperature is determined from the following equation:

$$t_m = \frac{\text{large temp. diff.} - \text{small temp. diff.}}{2.3 \log (\text{large diff./small diff.})} \tag{15-18}$$

If the ratio LD/SD is less than or equal to 2, the arithmetic mean temperature may be used with an error of only 4 per cent; if it is greater than 2, the log mean temperature difference must be used.

The results of using Fig. 15-4 are correct when heat-transfer equipment is of the true countercurrent type. However, when the heat exchanger is of the multipass type and the flow is not strictly countercurrent, the proper t_m is less than that obtained from Fig. 15-3. Factors that can be used to correct for this may be obtained directly from charts appearing in the Standards of Tubular Exchanger Manufacturers Association (TEMA).

Q15-11. There is to be heated to 170°F in a tubular heater, by means of saturated steam condensing at 220°F outside the tubes,

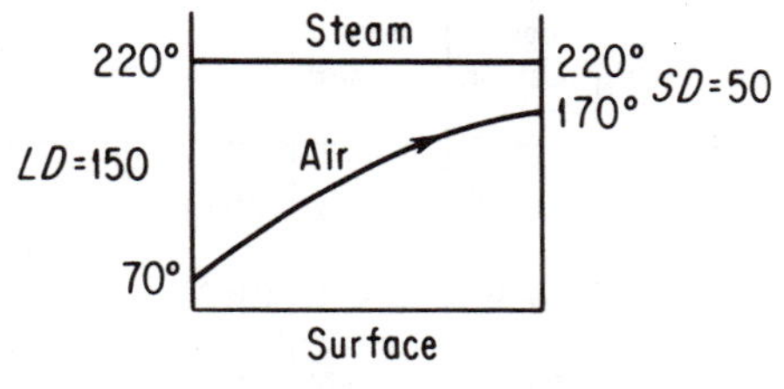

Fig. 15-4

10,000 cfm of dry air at 70°F and normal barometric pressure. The air is to be blown through a number of cold-drawn steel tubes 2.0 actual inside diameter, horizontal, arranged in parallel, the tubes being expanded into suitable tube sheets at the two ends of the heater. Assuming that the initial air velocity in the tubes is 26 fps, calculate (*a*) the number of tubes in parallel, and (*b*) the length of each tube.

ANSWER. Refer to Fig. 15-4 showing the thermodynamic process going on within the heat exchanger. Air is through the tubes with steam condensing on the outside. Assume clean surface with no fouling. The flow rate of air is first found simply. Dry air at 70°F and standard barometer has a density of 0.075 lb per cu ft. Then the weight rate is

$$10{,}000 \times 0.075 = 750 \text{ lb per min}$$

The log mean temperature from Fig. 15-4 is found to be 90°F. This may also be calculated, but it is simpler to use the graphic method. The heat transferred is determined from

$$Q = 750 \times 0.25 \times (170 - 70) = 18{,}800 \text{ Btu per min}$$

The heat transferred Q is also equal to that expressed by Eq. (15-23).

$$Q = 18{,}800 = UAt_m$$

TABLE 15-4. RANGE OF MISCELLANEOUS OVER-ALL COEFFICIENTS*
(Heat transfer in Btu per hr per sq ft of surface per degree difference in temperature between two fluids)

Type of heat exchanger	Controlling resistance		Typical fluid	Typical apparatus
	U, free convection	U, forced convection		
Liquid to liquid	25–60	150–300	Water	Liquid-to-liquid heat exchanger
Liquid to liquid	5–10	20–50	Oil	
Liquid to gas (atm pressure)	1–3	2–10	....	Hot-water radiators
Liquid to boiling liquid	20–60	50–150	Water	Brine coolers
Liquid to boiling liquid	5–20	25–60	Oil	
Gas (atm pressure) to liquid	1–3	2–10	....	Air coolers, economizers
Gas (atm pressure) to gas	0.6–2	2–6	....	Steam superheaters
Gas (atm pressure) to boiling liquid	1–3	2–10	....	Steam boilers
Condensing vapor to liquid	50–200	150–800	Steam-water	Liquid heaters and condensers
Condensing vapor to liquid	10–30	20–60	Steam-oil	
Condensing vapor to liquid	40–80	60–150	Organic vapor-water	
Condensing vapor to liquid	....	15–300	Steam-gas mixture	
Condensing vapor to gas (atm pressure)	1–2	2–10	....	Steam pipes in air; air heaters
Condensing vapor to boiling liquid	40–100	....	....	Scale-forming evaporators
Condensing vapor to boiling liquid	300–800	....	Steam-water	
Condensing vapor to boiling liquid	50–150	....	Steam-oil	

* From a tabulation by A. P. Colburn in Perry, "Chemical Engineer's Handbook," 3d ed., McGraw-Hill Book Company, Inc., New York, 1950.

Refer to Stoever.* The film coefficient is indicated as Btu/(sq ft)(hr)(°F). With a mass flow rate G equal to ($^{750}/_{60}$) per sq ft of tube cross section, or taking on another equivalent form

$$26 \text{ fps} \times 0.075 = 1.95$$

from which the basic value of the film coefficient is taken directly

* Herman J. Stoever, "Applied Heat Transmission," pp. 74, 75, 84, McGraw-Hill Book Company, New York, 1941.

from the curve given by Stoever (p. 79) to be 6.8. With the proper suggested correction factors

$$6.8 \times 1.00 \times 0.87 = 5.91$$

The tube cross-sectional area is $(750/60)(1/1.95) = 6.4$ sq ft. Now since the cross-sectional area of one 2-in. tube is 3.355 sq in., the **number of tubes** (*a*) is

$$(6.4/3.355)(144) = \mathbf{275}$$

(*b*) The total area of heat transfer is now calculated from a rearrangement of Eq. (15-17):

$$A = \frac{18{,}800 \times 60}{5.91 \times 90} = 2{,}120 \text{ sq ft}$$

Since the area per foot of 2-in. tube of thin construction is 1.608 sq ft, the total length of tubing would be (2,120/1.608), or 1,320 ft. We calculated the number of tubes to be 275. Thus, each tube would be (1,320/275), or **4.8 ft.** This would be the length inside the tube sheets. Of course, an additional length would be needed to accommodate the thickness of the tube sheets. As a check refer to Table 15-4 and we see that for condensing vapor to gas (air) at atmospheric pressure and under forced convection the over-all coefficient given as 2 to 10 is in line with the 5.91 calculated from Stoever.

Q15-12. A very long steel cylinder 3 ft in diameter, initially at 1000°F, is suddenly immersed in a fluid, and the surface temperature is instantaneously changed to 500°F. Estimate the core temperature at the end of 3 hr, using the values of physical properties tabulated below:

	Fluid	*Steel*
Thermal conductivity, Btuh-ft-°F	0.1	23.0
Specific heat, Btu/(lb)(°F)	0.8	0.13
Density, lb per cu ft	60	485

ANSWER. The statement "very long steel cylinder" suggests infinite extent in the axial direction. We will assume an infinite cylinder. Surface temperature is instantaneously changed from 1000°F to 500°F. Properties of fluid are therefore not needed because film resistance may be neglected. The problem will be solved using the cylinders series. Now refer to Fig. 15-5. Thermal diffusivity

$$\alpha = \frac{k}{C\rho} = \frac{23.0}{(0.13)(485.0)} = 0.365 \text{ sq ft/hr}$$

Fourier modulus:

$$\theta = \frac{\alpha t}{(r_m)^2} = \frac{0.365(3)}{(3/2)^2} = 0.486$$

$$C\theta = \frac{T_0 - T_1}{T_i - T_1} = \frac{T_0 - 500}{1000 - 500} = 0.0632$$

Core temperature $T_0 = 0.0632(500) + 500 = 31.6 + 500 =$ **531.6°F**

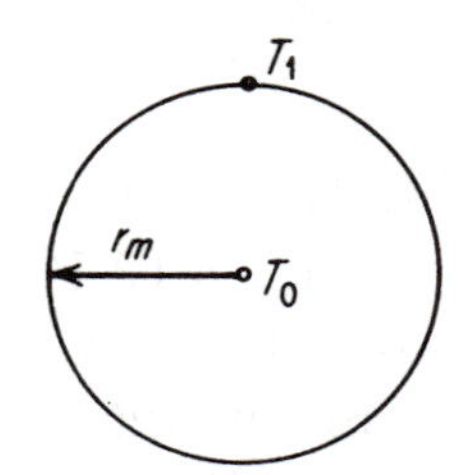

Fig. 15-5

Q15-13. A water line is to be buried underground. It is possible that for long periods of time there will be no flow through the pipe, but the pipe will not be drained. The soil in which the pipe will be buried is dry and has an assumed initial temperature of 40°F. The soil has a density of 40 lb per cu ft, a thermal conductivity of 0.2 Btuh-ft-°F, and a specific heat of 0.44 Btu/(lb)(°F). It is desired to design for a 36-hr period at the beginning of which the soil surface temperature suddenly drops to 0°F. Determine the minimum earth cover needed above the water pipe to prevent the possibility of freezing during the 36-hr cold spell. It is assumed that no flow occurs through the pipe during this period.

ANSWER. This problem considers a solid of infinite extent in the direction of the heat flow. We must find the thickness of earth cover that will keep the pipe at a temperature no lower than 32°F. Now refer to Fig. 15-6.

$$\text{erf}\left(\frac{x}{2\sqrt{\alpha t}}\right) = \frac{T - T_s}{T_0 - T_s} = \frac{32 - 0}{40 - 0} = 0.8$$

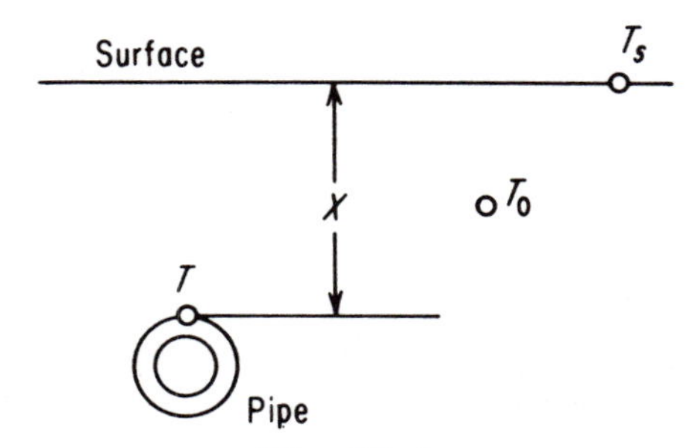

Fig. 15-6

The error factor (erf) may be obtained from the reference *Conduction Heat Transfer,* by P. J. Schneider (Addison-Wesley, 1955). Then thermal diffusivity from given properties is given by

$$\alpha = \frac{k}{C\rho} = \frac{0.2}{0.44 \times 40} = 0.01136 \text{ sq ft per hr}$$

If
$$\text{erf}\left(\frac{x}{2\sqrt{\alpha t}}\right) = 0.8$$

then from Schneider, Table A = 10, p. 382,

$$\left(\frac{x}{2\sqrt{\alpha t}}\right) = 0.906$$

Earth cover for time $t = 36$ hr:

$$\begin{aligned} x &= 0.906(2\sqrt{\alpha t}) \\ &= 0.906[2\sqrt{0.0136 \times 36}] \\ &= 0.906[2\sqrt{0.409}] \\ &= 0.906[2(0.64)] \\ &= \mathbf{1.16\ ft} \end{aligned}$$

Chapter **16**

REFRIGERATION

16-1. General. The natural flow of heat from a hot body to a cold body will be rapid if the difference in temperature is great, if the cooling or heating surface is large, and if the resistance to heat flow is small. When these conditions are not favorable, the flow will be correspondingly slow. We know that heat will not flow from a cold to a hot body, nor between bodies of equal temperature.

Refrigeration has to do with the artificial means for removing heat when conditions are unfavorable to natural or rapid flow. It has to do not only with producing low temperatures where desired, but also with accelerating the natural flow of heat at normal temperatures. Refrigeration is accomplished by providing a substance that is colder than the substance to be refrigerated.

16-2. Refrigerants. Many liquids boil at temperatures low enough for refrigeration, but comparatively few are suitable for refrigeration purposes. Those which have practical usefulness are called refrigerants. The boiling points of some of the more important refrigerants at atmospheric pressure are given in Table 16-1. Increased pressure on any of these liquids raises its boiling point. Decreased pressure has the reverse effect upon the refrigerant and lowers the boiling temperature. The term "boiling point" is generally understood to mean the temperature at which vaporization takes place under atmospheric pressure at sea level. In this presentation the term "boiling temperature" will be used in referring to temperatures of vaporization at other pressures than atmospheric. The candidate for licensure should take with him into the examination tables of refrigerants usually appearing in handbooks or other sources.

16-3. Refrigerating Terms Explained. The following terms are to be understood so that problems appearing on the examination can be handled with facility.

TABLE 16-1. BOILING POINTS OF SOME REFRIGERANTS AT ATMOSPHERIC PRESSURE

Refrigerant	Chemical symbol	Temp., °F
Dichloromethane (Carrene)	CH_2Cl_2	103.6
Ethyl chloride	C_2H_5Cl	55.0
Sulfur dioxide	SO_2	14.0
Isobutane	C_4H_{10}	10.0
Methyl chloride	CH_3Cl	−10.0
Dichlorodifluoromethane (Freon or F-12)	CCl_2F_2	−21.7
Ammonia	NH_3	−28.0
Propane	C_3H_8	−48.1
Carbon dioxide	CO_2	−109.3
Ethane	C_2H_6	−126.9
Ethylene	C_2H_4	−154.7

Refrigerating effect. This is the amount of heat absorbed in the evaporator. This is also the amount of heat removed from space to be cooled. It is measured by subtracting heat content (enthalpy) of one pound of liquid refrigerant as it enters the expansion valve

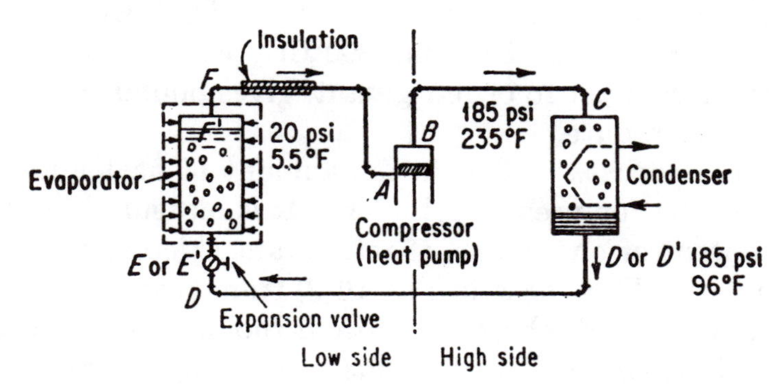

Fig. 16-1

from the heat content of the same pound as it enters the compressor, i.e., refer to Figs. 16-1 and 16-2; from heat content at F or F' subtract heat content at either condition D' (saturated liquid) or D (subcooled liquid).

Ton of refrigeration. When the boiling refrigerant removes sensible heat from the environment of the evaporator at a rate equivalent to the melting of 1 ton (2,000 lb) of water ice in 24 hr, the rate of heat removal is a ton of refrigeration. This is equivalent to the removal at a rate equal to

$$2{,}000 \times {}^{144}\!/_{24} = 12{,}000 \text{ Btu per hr}$$

where 144 Btu per lb is the heat of melting 1 lb of water ice. Heat of sublimation of 1 lb of dry ice (CO_2) is equal to 275 Btu. To say

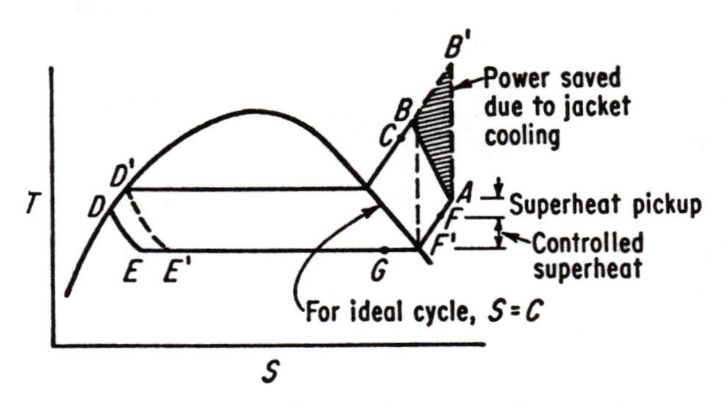

Fig. 16-2

that a refrigeration machine has a capacity of 10 tons is to say that the rate of refrigeration is 10 × 200 = 2000 Btu per min. Note that 1 ton of refrigeration is equal to a rate of 200 Btu per min.

Refrigerant circulated. Dividing 200 Btu per min by the refrigerating effect, in Btu per lb of refrigerant, gives pounds of refrigerant circulated each minute.

Work of compression. This is the amount of heat added to the refrigerant in the compressor cylinder. It is measured by subtracting heat content of one pound of refrigerant at compressor suction conditions (point F', F, or A in Fig. 16-2) from heat content of same pound at compressor discharge conditions (point B or B' in Fig. 16-2).

Theoretical horsepower requirements. Multiplying work of compression in Btu per lb by pounds of refrigerant circulated in an hour, and dividing this product by 2545 Btu per hp-hr gives theoretical horsepower requirements.

$$\frac{\text{Work of compression} \times \text{refrigerant circulated}}{2{,}545} \qquad (16\text{-}1)$$

Coefficient of performance. This is the ratio of refrigerating effect to work of compression. A high coefficient of performance (COP) means high efficiency. Theoretical COP ranges from about 2.5 to more than 5.

$$\frac{\text{Refrigerating effect}}{\text{Work of compression}} \tag{16-2}$$

Horsepower per ton of refrigeration. The mechanical input in horsepower divided by tons of refrigeration effect produced provides the answer to this quantity. If COP is known, horsepower per ton of refrigeration (TR) can be figured directly.

$$\frac{12{,}000 \text{ Btu per hr}}{2{,}545 \times \text{COP}} \tag{16-3}$$

Standard-ton conditions. An evaporating temperature of 5°F, a condensing temperature of 86°F, liquid before expansion valve at 77°F, and suction-gas temperature at 14°F are the necessary conditions for the standard TR. Refrigerating machines are often rated under these conditions.

Head pressure. This is the pressure at the discharge of compressor or in condenser. This is also known as "high-side" pressure.

Suction pressure. This is the pressure at the compressor suction or at outlet of evaporator. Also known as "low-side" pressure.

16-4. Compression Refrigeration Cycle. Refer to Figs. 16-1 and 16-2. In Fig. 16-1 the pipelines are considered short so that pressure drop becomes negligible. The cycle shown is for ammonia. Check the conditions for pressure and temperature. If the condenser is clean and there is just enough surface, then the liquid refrigerant will leave the condenser at condition D' as saturated liquid. If the surface is clean and a large amount of cooling water at low temperature is used, the liquid is subcooled to condition D. The pipe loses sensible heat from B to C. The refrigerant picks up heat from F to A through pipe insulation. Saturated vapor at F' inside the evaporator is superheated in evaporator from F to F' to ensure dry vapor to the compressor to ensure against wet compression and resulting compressor destruction.

Starting with liquid at D, the refrigerant expands through the expansion valve along constant-enthalpy line with a large pressure drop and a small increase in volume, as indicated by line *DE*. The saturated liquid that has not already flashed into vapor now boils in the evaporator, changing its state from a saturated liquid to a saturated vapor at the downstream pressure condition within the evaporator. This refrigerating effect is accompanied by a relatively small pressure drop and a large increase in volume, as along path *EF′*. This is where each pound of refrigerant does its work of cooling. Vapor is superheated and reaches the compressor suction where its volume is reduced and pressure increased along path *F′FAB* under normal operating conditions. In the compressor work is done on the refrigerant to raise its pressure to cause flow and also to raise its temperature above the cooling medium (water or air) so that heat can leave the refrigerant and cool it.

In the condenser the refrigerant loses its heat and superheat of compression and condenses out as liquid along path *BCD′D*. Then the cycle starts anew. Note that shaded area in the compression cycle is power saved due to jacket cooling of the compressor. If the compression were adiabatic and reversible, then path of compression would follow *AB′*. But since there is real cooling either by air or water, the compression path would lie along *AB* with the shaded area in Fig. 16-2 indicative of some degree of isothermal cooling. Cool and large quantities of water increase shaded area and reduce power consumption accordingly.

16-5. Refrigerant Circulated per Ton of Refrigeration. First determine how much refrigerating effect we can expect from each pound of refrigerant. This is the difference between the enthalpy at *D* or *D′* and *F′*. It can also be expressed as the difference between the heat removed by the condenser water and the work done by the compressor. Refer to Fig. 16-2. Enthalpy of saturated liquid at 185 psi is 150.5 Btu per lb and enthalpy of saturated vapor at 20 psi is 613.5 Btu per lb. The difference is 462.6 Btu and each pound of circulating ammonia removes that amount of heat in the evaporator. Then

$$200/462.6 = 0.432 \text{ lb ammonia per min per TR}$$

16-6. Horsepower per Ton of Refrigeration. This is a common method of expressing the actual efficiency of a compression system. Its relation to coefficient of performance is

$$\frac{\text{hp}}{\text{TR}} = \frac{4.71}{\text{COP}} \qquad (16\text{-}4)$$

This expression may be applied to either actual or ideal cycles.

16-7. Wet and Dry Compression. The state point of the refrigerant vapor as it enters the compressor suction may be in the wet region (point G, Fig. 16-2), saturated vapor point (point F'), or in the superheat region (point F or A). If compression occurs with the vapor wet (point G), the process is known as *wet compression*. If at either points F', F or A, it is known as *dry compression*. You will note that under wet compression the cycle approaches the ideal or Carnot cycle. Naturally we would expect wet compression to be more efficient than gas refrigeration. This is in accord with the Joule-Thomson effect.

For the ideal refrigeration machine wet compression would give a large coefficient of performance between two temperature limits. However, wet compression in the actual machine produces low volumetric efficiencies due to vaporization of liquid particles in the compressor cylinder. Actually, the mechanical efficiency of the machine for dry compression is much better all around, and dry compression is used universally in order to protect the machine against damage.

16-8. Displacement of the Vapor-compression Machine. For a particular capacity the size of the compressor depends upon the number of pounds of refrigerant that is removed from the evaporator and circulated per unit time in the closed cycle.

If a refrigeration plant is to have a capacity of N tons of refrigeration, then the rate is $N \times 200$ Btu per min. From this and the refrigerating effect per pound the weight of refrigerant circulated per minute may be determined. For 100 per cent volumetric efficiency the displacement V_d is

$$V_d = (\text{sp vol at suction})(200 \times N/\text{refrigerating effect}) = \text{cfm} \qquad (16\text{-}5)$$

The displacement at volumetric efficiency E_v is obtained by dividing the displacement at 100 per cent by E_v as a decimal less than unity. Volumetric efficiencies will fall between 65 and 85 per cent as limits in actual compressors. The approximate relative displacements required per ton of refrigeration for carbon dioxide, ammonia, and sulfur dioxide are 0.25, 1.0, and 2.5, respectively. The piston speeds for ammonia compressors vary from about 150 fpm in sizes up to 15 tons of refrigeration to 400 rpm in sizes above 200 tons. The ratio stroke to piston diameter is usually from 1¼ to 2. Radial compressors operate at much higher speeds.

Q16-1. Calculate the displacement in an ammonia compressor for a 50-ton refrigeration machine when operating with ammonia at 0°F in the expansion coils (evaporator). At this temperature the heat absorbed by the evaporation of 1 lb of ammonia is 500 Btu (refrigerating effect), and the specific volume is 9 cu ft per lb. The vapor enters the compressor saturated. If speed is 180 rpm and stroke is 1.2 × bore, what is bore and stroke, the compressor being single-acting?

ANSWER. The heat to be absorbed is equal to 50 × 200 = 10,000 Btu per min. The compressor displacement is

$$10{,}000/500 \times 9 = \mathbf{180} \text{ cfm}$$

Now let N be compressor speed (rpm), D be bore (ft), L be stroke (ft), V be piston displacement (cu ft per stroke). Then

$$V = 0.785D^2L \qquad N \times V = 180 \text{ cfm}$$
$$= 180/N = 180/180 = 1 \text{ cu ft}$$
$$L = 1.2D \qquad V = 0.785D^2 \times 1.2D = 1 \text{ cu ft}$$

Rearranging and transposing, we see that

$$D = (1/1.2)(0.785)^{1/3} = \mathbf{1.02 \text{ ft}}$$

also $$L = 1.2D = 1.2 \times 1.02 = \mathbf{1.224 \text{ ft}}$$

Q16-2. The weight of ammonia circulated in a machine is found to be 21.8 lb per hr. If the vapor enters the compressor with a specific volume of 9.6 cu ft per lb, calculate the piston displacement necessary for this machine, assuming 80 per cent volumetric efficiency.

ANSWER.

$$E_v = \frac{\text{actual volume of charge drawn into the compressor}}{\text{piston displacement}}$$
$$0.80 = 21.8 \times 9.6 = 209.28 \text{ cu ft of piston displacement}$$
$$\text{PD} = 209.28 \times 1/0.80 = \mathbf{262 \text{ cu ft per hr}}$$

Q16-3. A single-stage ammonia compressor is producing 10 tons of refrigeration and the power consumed is 15 bhp. Suction pressure is 25 psi, condensing pressure 180 psi. Brine temperature is 20°F off brine cooler. Determine actual coefficient of performance (COP) and amount of ammonia circulated.

ANSWER.

$$\text{COP} = \frac{\text{refrigerating effect}}{\text{power consumed}}$$
$$= \frac{12{,}000 \times 10 \times 24}{15 \times 2{,}545 \times 24} = \mathbf{3.15}$$

Assume no subcooling and that saturated vapor enters compressor. Then

$$\frac{288{,}000}{615 - 149} = \mathbf{620\ lb\ per\ TR}$$

16-9. Carnot Cycle for Refrigeration. This is a reversed cycle of the Carnot type, as shown in Fig. 16-3. The refrigerant is isentropically compressed along *ab* from a cold temperature T_1 in the evaporator to T_2 above that of some naturally available heat sink. The refrigerant then discharges heat at constant temperature T_2 along *bc*. Heat-sink temperature is T_0. At some point *c*, an isentropic expansion *cd* lowers the temperature to T_1, which is below that of the environment, and heat flows from the environment into the evaporator at T_1 along *da* and the cycle is repeated.

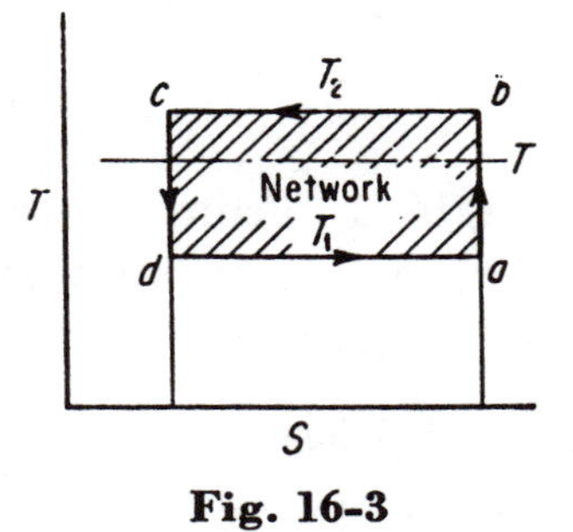

Fig. 16-3

The COP of this cycle may now be expressed as for the Carnot cycle

$$\frac{T_1}{T_2 - T_1} \qquad (16\text{-}6)$$

This is the highest possible COP for all cycles operating between the temperature limits T_1 and T_2 and serves as the standard of comparison for other cycles which more nearly approach the natural events.

16-10. Conclusions Drawn from the Carnot Cycle. It is desirable that work expended be a minimum since it is paid for. Work will be reduced as T_2 is lowered. Steps taken to keep this temperature down are important. T_0 is the lowest temperature attainable by a natural coolant (well water, air). There is a definite limit of improvement here. Work will be reduced as the evaporator temperature T_1 is increased. There are limits, however. To freeze water 32°F and below is necessary. To cool air, higher temperatures may be used, i.e., 50°F.

Q16-4. In an ammonia condensing machine (compressor plus condenser) the water used for condensing is at 55°F and the evaporator is at 15°F. (*a*) Calculate the ideal COP. (*b*) If 1.5 hp per TR is required, what is actual COP? Mechanical efficiency may be taken as 90 per cent.

ANSWER.

(*a*) Ideal COP:

$$(460 + 15)/(55 - 15) = \mathbf{11.885}$$

(*b*) Actual COP:

$$(12{,}000 \times 0.9)/(1.5 \times 2{,}545) = \mathbf{2.84}$$

Q16-5. From the following data calculate the bore and stroke of a single-acting ammonia compressor necessary to produce 10 tons of refrigeration per 24 hr.

Head pressure	160 psia
Suction pressure (at compressor suction)	28 psia
Temperature of ammonia in expansion coils	0°F
Pressure of ammonia in expansion coils	30.42 psia
Temperature at compressor suction	15°F
Latent heat of vaporization of ammonia at 30.42 psia	568.9 Btu per lb
Assume no subcooling	
Thermal capacity of liquid ammonia at 160 psia	134.99 Btu per lb
Thermal capacity of liquid ammonia at 30.42 psia	42.9 Btu per lb
Compressor speed	200 rpm
Sp vol of saturated ammonia vapor at 28 psia	9.86 cu ft per lb

Saturation temperature at 28 psia............	−3°F
All pressures are absolute	
Bore equals stroke	
Volumetric efficiency......................	80 per cent
Pressure in expansion coils..................	30.42 psia
Heat of saturated vapor at 30.42 psia.........	611.8 Btu per lb

ANSWER. Refer to Fig. 16-4. Point A is at compressor suction; B is at discharge; D is at point entering expansion valve; E is at

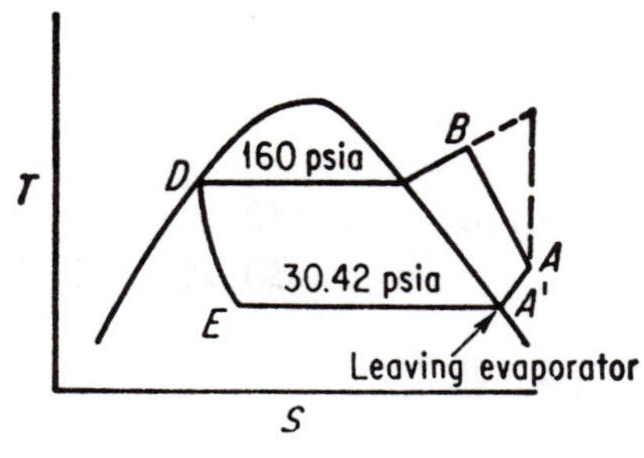

Fig. 16-4

point leaving expansion valve and entering evaporator.

$$\text{Refrigerant circulated} = \frac{20 \times 10}{H_{A'} - H_D} = \frac{2{,}000}{611.8 - 134.99}$$

$$= 4.22 \text{ lb per min}$$

Effect of wiredrawing (expansion) of liquid through expansion valve: Basis is heat of liquid at 160 psia = heat in vapor at 30.42 psia + heat in liquid at 30.42 psia

$$134.99 = 611.8(1 - x) + 42.9(x)$$

where x is weight of unflashed liquid. Solve for x; this is found to be equal to 0.84 lb of liquid at expanded pressure.

$$\mathbf{1 - x = 0.16 \text{ lb of saturated vapor at expanded pressure}}$$

The 0.84 lb of ammonia liquid fed to the evaporator develops refrigerating effect.

Volume of ammonia gas at compressor suction conditions is

$$4.22 \times 9.86 \times \frac{460 + 15}{460 - 3} = 43.4 \text{ cfm}$$

The vapor leaving the evaporator has been superheated:

$$3 + 15 = 18°F$$

Piston displacement × volumetric efficiency = volume of actual gas, or

$$\text{Piston displacement} \times 0.80 = 43.4$$
$$= 43.4/0.80 = 54.3 \text{ cfm}$$

$$\text{Piston displacement} = \text{piston face area} \times \text{stroke} \times \text{strokes per min}$$
$$= 0.785 \times D^2 \times L \times N$$

But D equals L (given). Thus

$$\text{Piston displacement} = 0.785 \times D^2 \times D \times N$$
$$= 0.785 \times D^3 \times 200 = 54.3$$

Rearranging and solving for D,

$$D^3 = \frac{54.3}{0.785 \times 200} \qquad D = \mathbf{0.7\ ft} \text{ and } L = \mathbf{0.7\ ft}$$

Q16-6. Show by appropriate diagram how the horsepower per ton of refrigeration changes with evaporator temperature. Indicate compressor loss curve, compressor indicated horsepower per ton curve, volumetric efficiency relation, per cent of capacity rated at 5°F evaporator temperature, and per cent theoretical liquid feed to suction gas to keep discharge temperature below 250°F. Assume refrigerant to be ammonia.

ANSWER. Refer to Fig. 16-5.

Q16-7. Discuss the determination of a refrigeration cooling load and give an example to show how this is calculated.

ANSWER. The quantity of heat that must be removed from a cooling place (refrigerator, icebox, air-cooled space) depends on (*a*) insulating-loss load (heat loss through insulated walls, floor, and ceiling), (*b*) service load (heat admitted when doors are opened, infiltration), (*c*) special loads (heat added by apparatus such as lamps and motors), (*d*) product load (heat added by the product to be cooled), and (*e*) temperature surrounding outside surface of the space.

For example, assume we wish to store beef at 35°F in a room 10 by 20 by 10 ft high with walls of 4-in. corkboard. Ambient temperature outside of storage refrigerator is 90°F. Determine the load in tons of refrigeration.

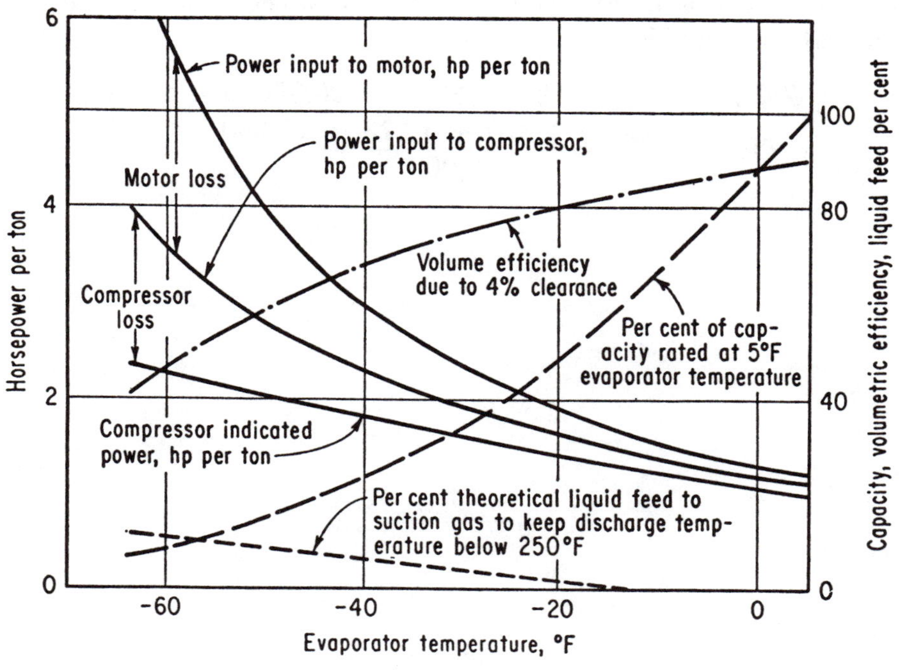

Fig. 16-5

Area of room walls, floor, and ceiling is easily calculated as equal to 1,000 sq ft. The heat loss through these surfaces is found in the usual manner to be

$$1{,}000 \times 0.07(\text{4-in. corkboard } U) \times (90 - 35) = 3{,}850 \text{ Btu per hr}$$

Service load. Heat admitted through doors depends, of course, on how frequently they are opened. Assume that the air is changed completely once each hour (60 min air change). This represents cooling 2,000 cu ft of air from 90 to 35°F.

$$2{,}000 \times 0.0709 \times 0.24 \times (90 - 35) = 1870 \text{ Btu per hr}$$

Special load. Assume two $\frac{1}{4}$-hp motors operating continuously in cooling units; 1 hp is equal to 2545 Btu per hr, including all motor losses.

$$0.25 \times 2 \times 2{,}545 = 1273 \text{ Btu per hr}$$

Product load. Suppose that 500 lb of beef is taken in and out of the room every hour and that when it is placed in storage it has a temperature of 90°F.

$$500 \times 0.74(\text{sp ht beef}) \times 55 = 20{,}350 \text{ Btu per hr}$$

$$\text{TR} = \frac{3{,}850 + 1{,}870 + 1{,}273 + 20{,}350}{12{,}000} = \mathbf{2.27}$$

Chapter 17

HEATING AND VENTILATING

17-1. General Considerations. The concepts of heat transmission we reviewed in Chap. 15 may be applied to the study of the material in this chapter and to the next on air conditioning. Over-all heat-transfer coefficients are determined in the same fashion in order to calculate building heat losses. The working substance is the air around us and carries the heat into every corner of the building. The flow of heat takes place at constant atmospheric pressure conditions.

Heating and ventilating problems usually involve maintenance of conditions of temperature and humidity within a building. Essentially a building is a shell for keeping out the weather and permitting a satisfactory indoor climate to be maintained. How much heat must be added to an enclosure depends on the construction and tightness. Light construction increases heat losses; substantial construction, while costing more initially, keeps down the costs of operation and fuel. Flimsy and loose construction encourages air infiltration and results in high heat-load needs. Different types of construction are used in the northern and southern parts of the country, and difference in climate is the governing factor.

HEATING

Heat passes from a region of higher temperature to lower temperature. In cold weather heat is lost by a warm building by radiation, by convection, and by the escape of warm air replaced by the colder outside air.

17-2. Over-all Coefficient of Heat Transmission. In actual calculations standard handbooks list over-all coefficients of heat transmission for walls having actual thickness, the material involved, and the surface resistances (film resistances). Such an over-all coefficient is called U and represents the number of Btu passing through a square foot of surface of wall construction in one hour, for one degree temperature difference between inside and outside air temperatures. As mentioned before, there is assumed to be still air on the inside and 15 mph wind on the outside. Over-all coefficients can be found by directly testing complete walls or by calculating from individual coefficients. For a simple brick wall the formula used in practice is

$$\frac{1}{U} = \frac{1}{f_i} + \frac{x}{k} + \frac{1}{f_0} \qquad (17\text{-}1)$$

where f_i = film conductance for still inside air equal to 1.65
f_0 = film conductance for outside 15-mph wind on wall surface equal to 6.0
k = conductivity of the brick (or any other material)
x = thickness, in.

Q17-1. Calculate the over-all coefficient of heat transmission through an 8-in. common brick wall with still air inside and 15-mph wind outside.

ANSWER. From Eq. (17-1),

$$\frac{1}{U} = \frac{1}{1.65} + \frac{8}{5} + \frac{1}{6.0} = 0.98$$

$$U = 1/0.98 = \mathbf{1.02\ Btu/(sq\ ft)(hr)(°F)}$$

Under actual conditions walls are more complicated, having any number of materials in series. In many cases an air space is included for its insulating value. For single air space and simple wall:

$$U = \frac{1}{1/f_i + x/k_1 + 1/a + x_2/k_2 + 1/f_0} \qquad (17\text{-}2)$$

For a simple wall with several air spaces add $1/a$ for each space. For more detailed data and instructions the student is referred to the Guide of the American Society of Heating, Refrigerating and Air-Conditioning Engineers. This excellent reference source and other handbooks give conductivity values for various materials, and over-all coefficients for common wall, floor, and roof constructions.

Q17-2. Compute the coefficient of heat transmission for a wood frame wall with 1 in. fir sheathing, building paper and yellow pine siding on outside of studs, air space and wood lath and plaster on inside of studs.

ANSWER. Refer to the ASHRAE Guide for conductivities. Then

$$U = \frac{1}{\frac{1}{1.65} + \frac{1}{1.10} + \frac{1}{0.50} + \frac{1}{2.50} + \frac{1}{6}} = \mathbf{0.25}$$

17-3. Wind Velocity Effects on U. As we know, heat-transmission coefficients used in heating and air-conditioning load calculations are based on outside wind velocity of 15 mph and still air (50 fpm) on the inside surface. Equations (17-1) and (17-2) give us the U value for these conditions. For any one wall type the only change will occur in f_0. For low values of U occurring in multiple and air-space types of construction, a change in f_0 is not as effective as for higher values of U occurring in simple walls or single wall materials.

In order to give a quick visual indication of what happens when wind velocity changes, Fig. 17-1 has been worked. Coefficients U based on an average wind velocity of 15 mph are plotted against correction factors for various wind velocities designated in the Guide. Above an average wind velocity of 30 mph, infiltration takes on a more important role in heating and air-conditioning calculations. Once knowing the 15-mph coefficient U, read up to the proper wind-velocity curve, and then left to read the multiplier. Then use it thus:

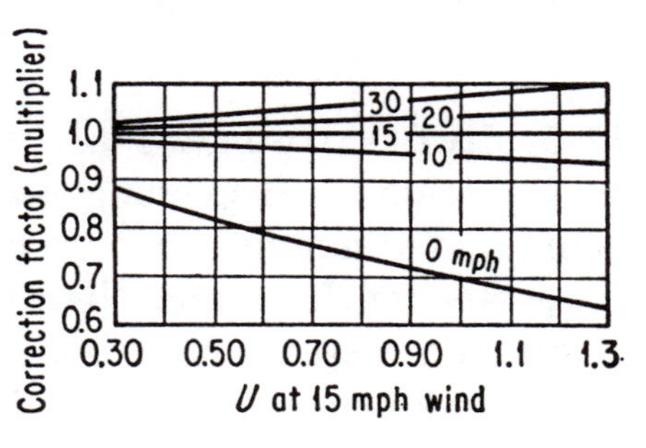

Fig. 17-1

$$\text{Corrected } U = \text{multiplier (from chart)} \times U \tag{17-3}$$

Tables giving conductivities and over-all coefficients for heating, ventilating, and air-conditioning are readily available in the Guide.

Heat flows through an air space by conduction, radiation, and by convection currents. It has been shown that there is always a certain amount of air movement even in "theoretical dead-air spaces." The resulting mechanism is rather complex, but experimental work shows that from about ¾ to 1½ in. the conductance of an air space is more or less constant; below ¾ in., the conductance increases rapidly as the space gets narrower. For air spaces wider than 1½ in., figure the surface conductance on each side of the space instead of using a single conductance value for the air space.

17-4. Insulation. The heat loss through walls, floors, and roofs can be reduced by using insulating materials having high heat-flow resistance. Most insulating materials depend on extremely low conductivity for their effectiveness due to the retention of air within tiny pores. Reflective insulations, such as aluminum foil, feature low emissivity and high reflectivity, and stop effectively transfer by radiation.

Choice of insulation for a particular installation depends on many factors, including insulating value, cost, ease of installation, chemical and physical stability, resistance to fire, vermin, etc. The Guide lists conductivities for various insulating materials. Suppliers of insulating materials also include over-all coefficients for their products, but care should be exercised in evaluating the U's published in their literature.

17-5. Infiltration Losses. Heat losses through infiltration of cold outside air displacing warm inside air through cracks around doors and windows can be very telling. In some cases, infiltration losses are much greater than transmission losses. This movement of air and displacement is due to wind pressure and temperature differences between inside and outdoors.

17-6. Figuring Heating Loads. A series of such calculations is rather simple, but does involve experienced judgment in the choice of coefficients for certain conditions.

There are two kinds of heat-loss calculations. In the first, they are made to determine the maximum heat loss with which to size the heating plant. In the second, they are made to determine the total heat loss for a heating season or part of a season, so as to check fuel consumption or to estimate heating costs. In both cases, they begin with measurement of wall, floor, and roof surfaces, and determination of infiltration flows. The same coefficients of heat transmission are used in both cases. Deviation takes place when temperature differences are introduced. For design calculations, the maximum design temperature difference is used, while for seasonal figures the average difference (say January average) is used.

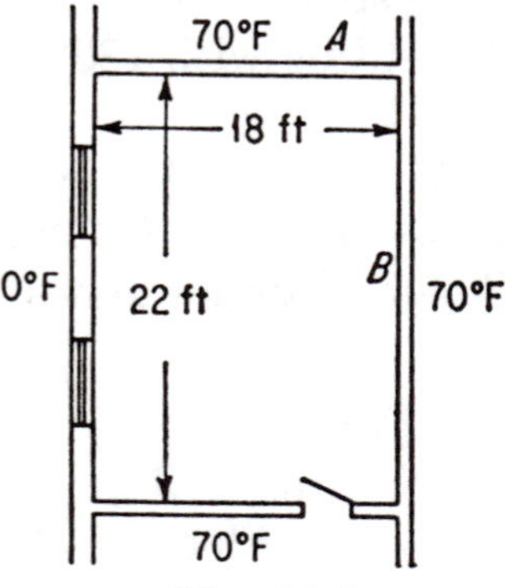

Fig. 17-2

Q17-3. Find the heat in Btu per hr that is to be supplied to a room 18 by 22 by 9 ft high to maintain a temperature of 70°F with the outside temperature of 0°F. One outside wall is of brick 12 in. thick, furred and plastered on wood lath, and two single glazed weather-stripped windows each 4 by 6 ft. Room above, below, and on three sides are also maintained at 70°F. Wind velocity is 15 mph. Refer to Fig. 17-2.

ANSWER.

Areas. Window glass: $2 \times 4 \times 6 = 48$ sq ft. Exposed wall (gross): $9 \times 22 = 198$ sq ft. Exposed wall (net): $198 - 48 = 150$ sq ft.

Coefficients. From tables here or in the Guide for 12-in. brick walls furred and plastered on wood lath, $U = 0.24$. U for single glazed window = 1.13.

Transmission losses. Use the equation $Q = U \times A \times (t_i - t_0)$. (17-4)

Through wall: $Q = 150 \times 0.24 \times 70 = 2{,}520$ Btu per hr
Through glass: $Q = 48 \times 1.13 \times 70 = 3{,}797$ Btu per hr

Infiltration losses.

$$Q = \frac{\text{cu ft per hr} \times (t_i - t_0)}{55.2} \qquad (17\text{-}5)$$

Length of crack; two windows $= 2[(3 \times 4) + (2 \times 6)] = 48$. For 15-mph wind, from the Guide, infiltration = 22.9 cu ft per hr per ft of crack. Thus,

$$Q = \frac{48 \times 22.9 \times 70}{55.2} = 1436 \text{ Btu per hr}$$

Total losses from room and total heat required to balance these losses = 2,520 + 3797 + 1,436 = **7753 Btu per hr.** For heating with steam, the equivalent direct radiation is 7,753/240 = 32.3 EDR. If heating with 180°F hot water, the equivalent is 7,753/150 = 51.6 EDR.

If the rooms adjoining were unheated, say at 35°F and a cellar below at 32°F, U for the partitions = 0.34 and for the floor = 0.24. The additional losses would then be:

$$\text{Partition } A = 162 \times 0.34 \times (70 - 35) = 1928 \text{ Btu per hr}$$
$$\text{Partition } B = 198 \times 0.34 \times (70 - 35) = 2356 \text{ Btu per hr}$$
$$\text{Floor} = 396 \times 0.24 \times (70 - 32) = 3612 \text{ Btu per hr}$$

Many designers add 15 to 20 per cent of total losses for rooms exposed to prevailing winds.

Q17-4. By means of insulation, the loss in heat through a roof per square foot is reduced from 0.40 to 0.18 Btu per hr for each degree Fahrenheit difference between inside and outside temperatures. The area of the roof is 10,000 sq ft and the average difference between inside and outside temperatures is 35°F during the heating season of 5,000 hr. If the heating value of coal is 13,000 Btu per lb and the efficiency of the heating plant is 60 per cent, find the value of the coal saved per heating season at $15 per ton of 2,000 lb.

ANSWER.

$$(0.4 - 0.18) \times 10{,}000 \times 35 \times 5{,}000 \times \frac{1}{0.6} \times \frac{1}{13{,}000} \times \frac{1}{2{,}000} \times 15 = \mathbf{\$370.13}$$

Q17-5. A building having a volume of 500,000 cu ft is to be heated in zero weather to 70°F. The wall and roof surfaces aggregate 28,000 sq ft and the glass surface aggregates 7,000 sq ft. The air is changed three times every hour (a 20-min. air change). Allowing transmission coefficients of 0.25 for the wall and roof surfaces and 1.13 for the single-glass-paned windows, calculate the square feet of steam radiation required if each square foot emits 240 Btu per hour.

ANSWER. The heating load consists of:

$$\text{Wall and roof losses} = 0.25 \times 28{,}000 \times 70 = 490{,}000 \text{ Btu per hr}$$
$$\text{Glass loss} = 1.13 \times 7{,}000 \times 70 = 555{,}000 \text{ Btu per hr}$$
$$\text{Ventilation load} = 500{,}000 \times 3 \times 0.075 = 112{,}500 \text{ lb per hr}$$
$$112{,}500 \times 0.24 \times 70 = 1.89 \times 10^6 \text{ Btu per hr}$$
$$\frac{(490 + 555 + 1.89 \times 10^3) \times 1{,}000}{240} = \mathbf{12{,}200 \text{ sq ft EDR}}$$

If hot-water radiation were used, 150 instead of 240 would be the emission in Btu per hr, and more radiation would be required to do the same job. The fuel-rate consumption may be determined by dividing the total heating load by the system efficiency and the result again by the gross heating value of the fuel as fired. It is convenient to remember that 4 sq ft of steam radiation are equivalent to the condensation rate of 1 lb steam per hour for low-pressure heating steam.

Q17-6. A control room for an oil refinery unit is to be heated and ventilated by means of a central duct system. Ventilation is to be at a rate of 3 cfm of outside air per square foot of floor area. Room to be ventilated measures 40 by 60 ft, and is to be pressurized to keep out hazardous gases. Outside design temperature is −10°F and inside is to be maintained at 75°F. Determine steam consumption rate for maximum design conditions with the use of 5 psig saturated steam.

ANSWER. The rate of outside air to be handled by the heating and ventilating unit is 40 × 60 × 3 = 7,200 cfm. The heating load is found to be

$$Q = 7{,}200 \times 1.08 \times (75 + 10) = 660{,}000 \text{ Btu per hr}$$

where factor 1.08 = sp ht air (0.24) × min per hr (60)/sp vol air (13.3). Saturated steam at a pressure of 5 psig has heat of condensation equal to 960 Btu per lb. Then steam rate required is simply

$$660{,}000/960 = \mathbf{690 \text{ lb per hr}}$$

VENTILATING

17-7. Ventilating Terms and Definitions. The following terms are commonly used in ventilation work.

(*a*) Cubical contents is the contents of the space to be ventilated expressed in cubic feet. Length × width × height = cubical contents. No deduction is made for equipment, tables, etc., within the space.

(*b*) Cubic feet per minute (cfm) is the rate of air flow.

(*c*) Capacity is the volume of air handled by a fan, group of fans, or by a ventilation system usually expressed in cubic feet per minute.

(*d*) Fan rating is a statement of fan performance for one condition of operation and includes fan size, speed, capacity, pressure and horsepower.

(*e*) Fan performance is a statement of capacity, pressure, speed, and horsepower input.

(*f*) Fan characteristic is a graphical presentation of fan performance throughout the full range from free delivery to no delivery at constant speed for any given fan.

(*g*) Resistance pressure (RP). The resistance pressure of any ventilating system is the total of the various resistance factors that oppose the flow of air in the system stated in inches water gauge (WG).

(*h*) Static pressure (SP) is the force exerted by the fan to force air through the ventilating system. If exerted on the discharge side, it is said to be positive and if on the inlet side, it is negative or suction pressure. The total static pressure which the fan exerts is the sum of these two pressure readings. In any system the SP exerted by the fan is equal to the RP of the system. The SP is measured by an inclined draft gauge in inches of water.

(*i*) Velocity is the speed at which the air is traveling expressed in lineal feet per minute. It is measured by use of a velometer or anemometer. Average velocities may be calculated for a fan by dividing the cfm by the square-foot area of the fan discharge. Average velocities in any part of the system may be calculated by dividing the cfm flowing by the cross-sectional area of the duct.

(*j*) Velocity pressure (VP) is a measure of the kinetic energy of horsepower in the moving air. It is measured directly by the use of a Pitot tube and a draft gauge. It can be calculated from the velocity from the formula

$$\text{VP} = (\text{velocity, fpm}/4{,}005)^2 \qquad (17\text{-}6)$$

Figure 17-3 shows a chart for determining velocity of any gas within a duct from manometer readings in inches of water. First determine relative density of the flowing gas, remembering that density of standard air at 70°F and atmospheric pressure is 0.075 lb per

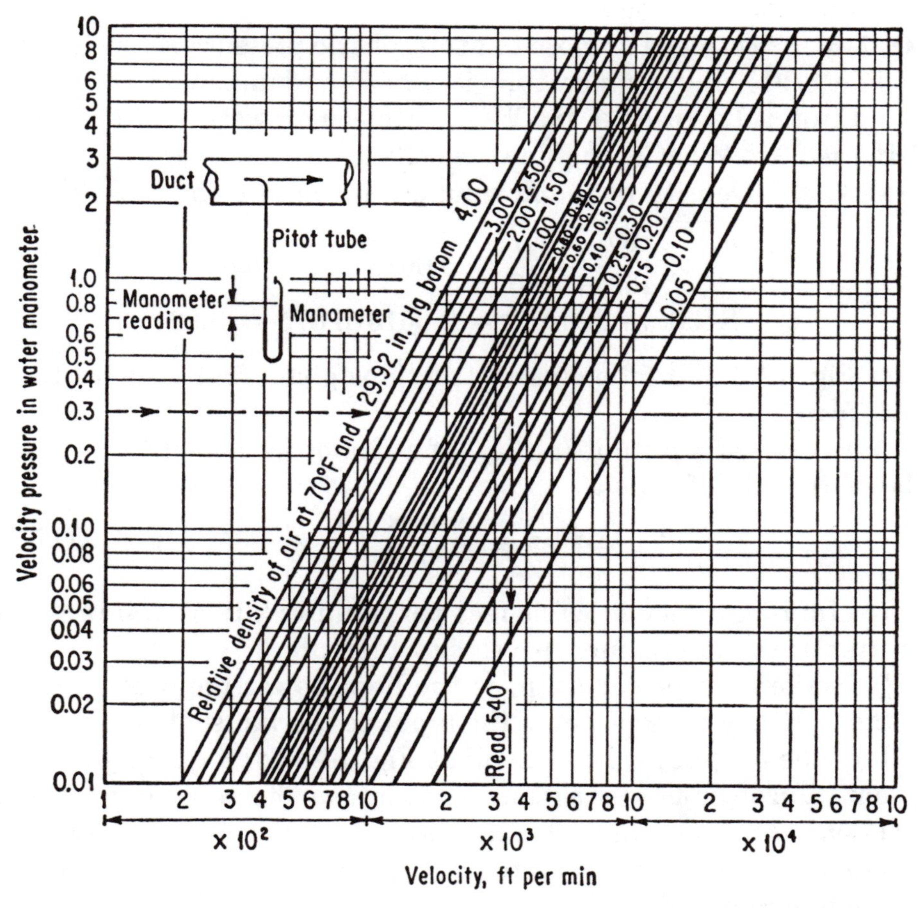

Fig. 17-3. Velocity vs. velocity pressure, gases at various densities.

cu ft. Then use manometer reading and read right to relative density line and down to actual velocity in feet per minute.

Q17-7. Air at 100°F is flowing in a duct. Pitot tube reading for velocity pressure is ½ in. WG. What is the actual velocity of flow?

ANSWER. Air at 100°F has a density of 0.075 × (460 + 70)/(460 + 100), or 0.071 lb per cu ft. Relative density is

$$0.071/0.075 = 0.946$$

Now refer to Fig. 17-3. From the left-hand side start at ½ in. and read right along this line to the sloping relative density line of 0.946. An interpolation must be made between 1.0 and 0.90. Read down to velocity of close to **2,900 fpm.** Normally for such temperatures no real deviation is made.

Q17-8. Acetylene at close to atmospheric pressure is flowing in a pipeline. Flowing temperature is 120°F and manometer reading from Pitot tube reads 1 in. WG. Molecular weight of acetylene (C_2H_2) is 26. What is the flow velocity?

ANSWER. Density of the flowing gas is found to be as follows:

$$\frac{26}{379} \times \frac{460 + 60}{460 + 120} = 0.0615 \text{ lb per cu ft}$$

$$\text{Relative density} = 0.0615/0.075 = 0.82$$

From Fig. 17-3 and following instructions given find velocity to be **4,600 fpm.**

(*k*) Horsepower output (air hp) of a fan, or air horsepower, is calculated from the formula

$$\text{Air hp} = \frac{\text{cfm} \times \text{TP}}{6{,}356} \qquad (17\text{-}7)$$

where cfm is capacity in cubic feet per minute; TP is total pressure of water or static pressure + velocity pressure.

(*l*) Brake horsepower (bhp) is the horsepower required to drive the fan. Brake horsepower is the input to the fan shaft required to produce the output air hp. Look at Fig. 14-4 for a quick graphical presentation of air-flow–resistance pressure–air-horsepower relations.

Q17-9. Air flows through a duct system at the rate of 20,000 cfm. Resistance pressure as calculated is 1 in. WG. Estimate the brake horsepower required for the fan.

ANSWER. Refer to Fig. 14-4. Follow arrowed line. Read air horsepower to be 3.2. Normally fan efficiencies range from 50 to 75 per cent, but let us assume an efficiency of 50 per cent. Then the brake horsepower is said to be 3.2/0.5 = **6.4.** Use a **7½-hp motor.**

(*m*) Mechanical efficiency of a fan is the ratio of horsepower output to horsepower input. Therefore,

$$\text{Mechanical efficiency} = \frac{\text{air hp}}{\text{bhp}}, \text{ expressed as a percentage} \qquad (17\text{-}8)$$

(*n*) Fan discharge or outlet is the place provided for receiving a duct through which air leaves the fan.

(*o*) Fan inlet is the place provided for receiving a duct through which air enters the fan.

17-8. General Ventilation. Where little or no ductwork is required to ventilate a space, the application is known as general ventilation. In most cases, exhaust fans high in the side walls or in the roof are used with general movement of air through windows or louvers across the space and out the fans. The air movement is caused to flow throughout the space to remove smoke, fumes, gases, excess moisture, heat, odors, or dust or simply to provide a constant inflow of fresh, outside air by the removal of foul, stale air.

To select the fans properly for the solution of general ventilation problems, the cubical contents of the space to be ventilated should be determined by multiplying the length by the width by the average height. All dimensions used must be in feet to give volume (cubage) in cubic feet.

The next step is to select the rate of air change required to give satisfactory ventilation. The ASHRAE Guide or manufacturers' catalogues contain much useful data and should be used to solve examination questions.

17-9. How to Measure CO_2 Build-up. In many industrial installations it is important to know how long it will take for a space to become uninhabitable where there is a negligible amount of air, or no outside air at all, entering the area by any means, natural or mechanical.

Atmosphere containing less than 12 per cent oxygen or more than 5 per cent carbon dioxide by volume is considered dangerous to occupants. The following formula is often used to determine the time for carbon dioxide to build up to 3 per cent with a safety factor.

$$T = 0.04 \frac{V}{P} \tag{17-9}$$

where T = time to vitiate space air, hr

V = *net* room volume, cu ft (gross volume less volume of equipment, etc.)

P = number of persons occupying space

This formula takes into account a minimum condition of activity of occupants and no generation of industrial fumes. Figure 17-4 gives the answer to this yardstick at once.

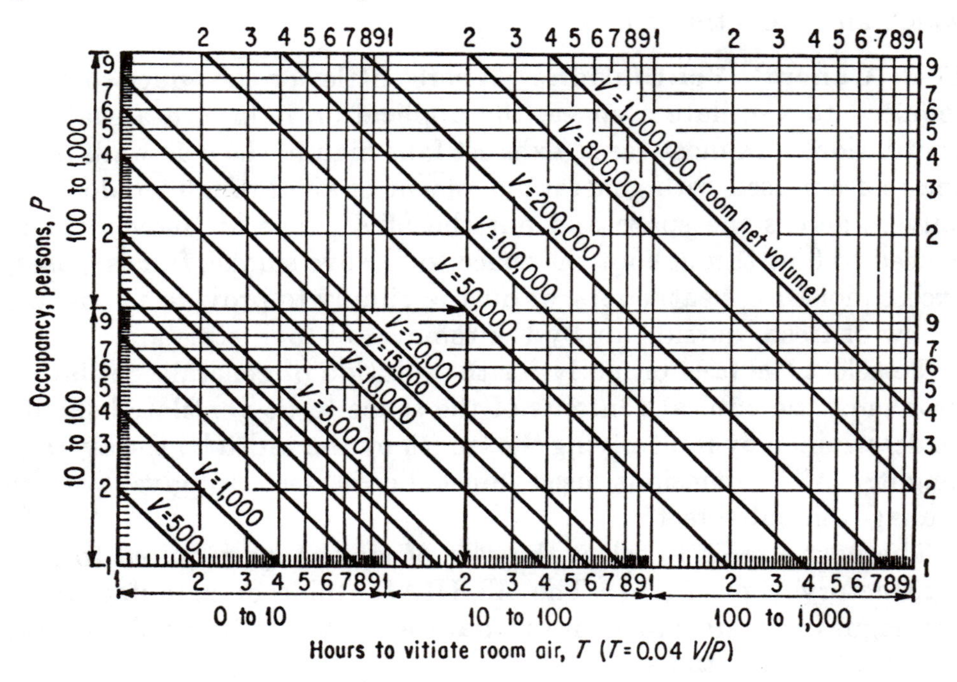

Fig. 17-4. (*Courtesy of Power Engineering.*)

Q17-10. Given a space of 50,000 net cubage with an occupancy of 100 employees with all outside air supply cut off. How long will it take to render the space uninhabitable?

ANSWER. Use Fig. 17-4. Follow arrows on chart and **20 hr.** During this time the oxygen content will have been reduced from a nominal 21 per cent to 17 per cent by volume. It is a good rule to consider that after 5 hr, or one-quarter of the calculated time of 20 hr, the air would become stale and affect worker efficiency.

Where reduced air supply is needed for production purposes, this graph is a useful guidepost to find time between airings. A 12-hr time limit is a good figure to use as one that comes close to the "uninhabitable" time. It is suggested, however, that a check be made with state or local code groups.

The graph may also be used for storage tanks where fumes are not involved in any way, rooms, auditoriums, and other places used for large gatherings of people. For determinations not covered by the graph, use the formula.

17-10. Specific or Local Ventilation. Throughout industry there is hardly an industrial plant that does not have at least one and usually many operations, machines, or processes that require special or, as it is known in industry, specific attention. By this we mean a system of ductwork and fans designed to prevent release or spread of smoke, fumes, odors, dusts, vapors or excess heat into other working areas. This means the control of the problem at its source. This local ventilation requires hoods well designed or special collecting systems and a thorough knowledge of air-flow principles and laws governing the behavior of gases and vapors Since it is beyond the scope of this presentation to cover all the detailed data required by the infinite variety of local exhaust problems, we shall review with the help of fundamental charts and tables that which has been of paramount value.

The best ventilating system can be rendered almost useless unless there is selected balance between fan and duct system. Care should also be exercised to be sure that the fumes, gases, or vapors are collected as close to the source of generation as possible. The closer the better, and the less will be the ventilation needs with attendant reduction in heating load for outside air make-up. Follow these simple rules and no real trouble will ensue:

(*a*) Make all duct runs as short as possible with direct connections.

(*b*) Make area of inlet and outlet ducts equal to outside diameter of fan for lowest friction loss.

(*c*) Where hoods requiring a large area are required, use baffles to give higher edge velocities and reduce air volume required.

(*d*) Do not forget that you cannot take more air out of a room than you are putting into it. Location and sizing of even gravity air inlets are frequently as important as the design of the exhaust system.

For exhaust systems the quantity of air to be moved should be determined by the selection of a suitable face velocity at hood entrance or by other considerations such as heat or moisture absorption, cubic feet per minute of liquid surface, by duct velocities required to convey the material, or combinations of the above. For supply systems the air flow usually is already determined by the volume being exhausted or by other considerations similar to those listed for the exhaust system. But let us digress for a moment to look into the matter of evaporation of water from tanks.

17-11. Evaporation of Moisture from Open Tanks. Where steam or moisture is released to the working space from open tanks or similar vessels, with resultant high humidity conditions, the problem often is to estimate rate of evaporation and provide for a reduction in moisture content of the room air.

In the case of steam, this escape rate is not a difficult problem, but

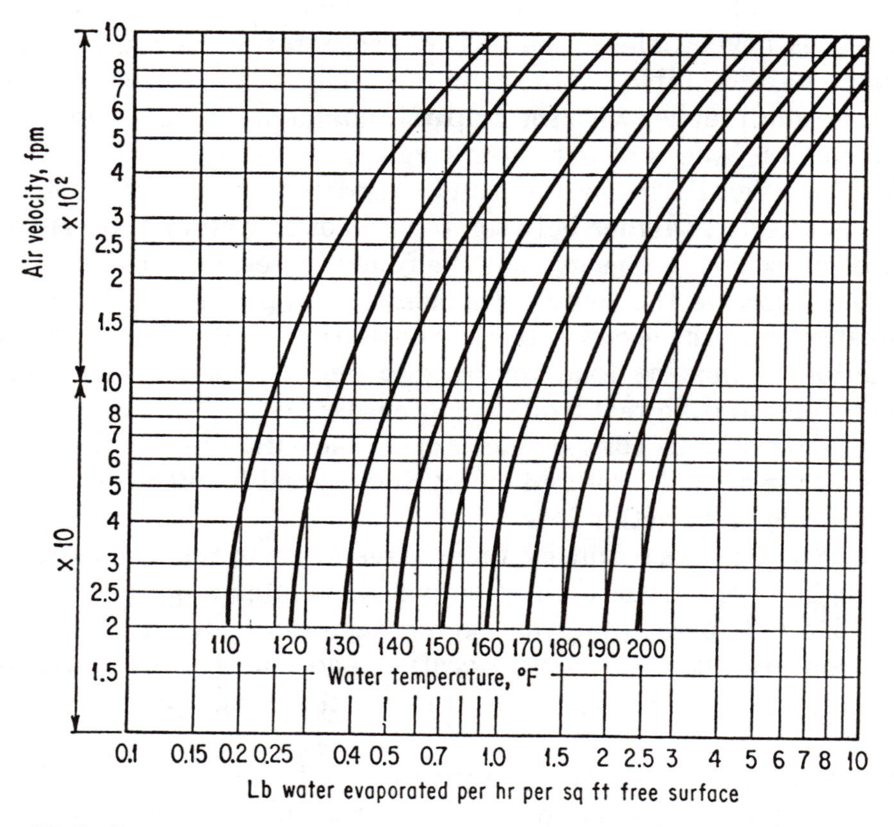

Fig. 17-5. Graph gives quick results under normal operating and design conditions. Given air velocity and tank temperatures, rate of evaporation can be picked off graph. Dark band covers normal room conditions. (*Courtesy of Power Engineering*.)

for hot water there has not been too much information disclosed in the technical literature. High humidity affects worker comfort, and presents quite a condensation problem with dripping from walls and roofs. Typical industries in which this "wet" heat is found are:

Textile industry: Elimination of fog and condensation in dye houses, bleacheries, and finishing departments.

Paper and pulp mills: Elimination of high vapor in machine, beater, and grinder rooms.

Steel and metal goods: Elimination of high vapor in pickling rooms.

Food industries: Control of high vapor in kettle, canning, blanching, bottle washing, and filling areas.

Process industries: Control of high vapor in electroplating, coating, and chemical processes.

However, since time is of the essence and the examinee would find it time-consuming to evaluate conditions, Fig. 17-5 gives quick results. Once air velocity and tank temperature are known, the rate of evaporation can simply be picked off the graph.

Q17-11. A paper-mill machine room produces 50 tons of finished paper per day and it is found that approximately 1½ lb of water must be evaporated for every pound of finished paper as the paper goes over the dryer rolls. What exhaust cfm is needed if room conditions are 100°F and 40 per cent relative humidity and tempered air enters the room at 70°F and 50 per cent relative humidity?

ANSWER. The weight of water evaporated per minute is calculated as follows: 50 tons per 24 hr equals 100,000 lb or

$$\frac{100{,}000}{24 \times 60} = 68 \text{ lb per min of paper}$$

Water is evaporated at the rate of 68 × 1.5, or 102 lb per minute. Now assume entering air is at 70°F and 50 per cent relative humidity. At this condition each 100 cfm already has 0.059 lb moisture. If exhausted at a temperature of 100°F and 40 per cent relative humidity, the moisture content would be increased to 0.117 lb per 100 cfm. The moisture picked up from the room per 100 cfm equals

$$0.117 - 0.059 = 0.058$$

And the cfm required is

$$(102/0.058) \times 100 = \mathbf{177{,}000\ cfm}$$

If room temperature remained at 100°F but it was found possible to boost exhaust humidity to 60 per cent, from the table the water content is found to be 0.175 lb. Then the new difference is

$$0.175 - 0.059 = 0.116 \text{ lb per 100 cfm}$$

Finally,

$$(102/0.116) \times 100 = \mathbf{88{,}000\ cfm}$$

Chapter **18**

AIR CONDITIONING

18-1. Equations of State. The water vapor present in the atmosphere is usually in the form of superheated steam as an invisible gas. The air is then "clear." If the atmosphere is cooled below the dew point, the excess vapor is condensed out in the form of minute drops of water or crystals of ice, so minute at first that they float as fog or cloud. If the droplets coalesce to form large drops, they fall to earth as rain. The maximum water vapor which can be held in the atmosphere increases greatly as the temperature is increased. At any temperature and pressure the quantity of water vapor can vary practically from none to maximum amount for that temperature. Absolute or "bone dry" air is never found in nature, but the amount of water vapor may be so low that it is difficult to measure it.

For practical air-conditioning purposes, it is frequently assumed that air and water vapor in the atmosphere follow the general gas equation for ideal gases, and that their mixture obeys Dalton's law of partial pressures.

To determine the maximum per cent of water vapor for gauge pressure:

$$\frac{\text{Partial pressure of water (from steam tables), psi}}{14.7 + \text{gauge pressure, psi}} \qquad (18\text{-}1)$$

Gauge pressure in denominator is for the entire system. Actual per cent of water vapor by volume is

$$\text{Maximum per cent of water vapor} \times \text{relative humidity} \qquad (18\text{-}2)$$

Q18-1. Find the mass of air contained in a room 25 by 30 by 10 ft at atmospheric pressure and 65°F.

ANSWER.

$$PV = 0.37WT = 14.7 \times (25 \times 30 \times 10) = 0.37 \times W \times (460 + 65)$$

from which weight W is calculated to be **568 lb.**

Exactly, the relative humidity of an air mixture is the ratio of the water-vapor partial pressures at the two different temperatures.

$$\frac{\text{Partial pressure actual at dew-point temp.}}{\text{Partial pressure saturated at dry-bulb temp.}} \times 100 \qquad (18\text{-}3)$$

Q18-2. Find relative humidity of air at 80°F dry bulb and 60°F.

ANSWER. From Eq. (18-3) and the steam tables

$$\frac{0.2561 \text{ psia at } 60°\text{F}}{0.5067 \text{ psia at } 80°\text{F}} \times 100 = \textbf{50.6 per cent}$$

Of course, the relative humidity may be more quickly determined from the psychrometric chart.

The term relative humidity is often misused in a way to mean a feeling of dryness of the air. However, as a matter of fact, air with a relative humidity of 60 per cent when the dry-bulb thermometer is 40°F is drier than air with a relative humidity of 10 per cent and a dry-bulb temperature of 100°F. Relative humidity is the relationship of water vapor in the air at the dew point temperature to the amount that would be in the air if the air were saturated at the dry-bulb temperature. In reality it can be considered as a thermal condition of steam (or water vapor) at one pressure compared with steam (or water vapor) at a higher pressure.

Percentage humidity is not to be confused with relative humidity. Percentage humidity is defined as the percentage ratio of the existing weight of water vapor per unit weight of vapor-free air to the weight of water vapor which would exist per unit weight of vapor-free air if the mixture were saturated at the existing temperature and pressure. Percentage humidity and relative humidity approach equality when the vapor concentrations are low. *Percentage humidity is always somewhat smaller than relative humidity.*

Q18-3. The temperature in a room is 72°F dry bulb and the relative humidity is 40 per cent. Barometric pressure is 29.92 in. Hg. Find (*a*) the partial pressure of the water vapor in the air, (*b*) the specific humidity, and (*c*) the percentage humidity.

ANSWER.

(*a*) The pressure of saturated vapor at 72°F can be found in the steam tables as 0.791 in. Hg. For a relative humidity of 40 per cent, we find the partial pressure of the water vapor to be

$$p_w = \phi \times p_s = 0.40 \times 0.791 = \mathbf{0.3164\ in.\ Hg}$$

(*b*) The specific humidity is next found to be same as previously discussed in the chapter on thermodynamics (Chap. 6).

$$\frac{p_w}{p_a} \times 0.622 = \frac{0.3164}{29.92 - 0.3164} \times 0.622$$
$$= \mathbf{0.665 \times 10^{-2}\ lb\ water\ per\ lb\ dry\ air}$$

(*c*) The percentage humidity is μ:

$$\mu = \frac{0.665 \times 10^{-2}}{1.697 \times 10^{-2}} \times 100 = \mathbf{39.2\ per\ cent}$$

where 1.697×10^{-2} is the weight of water held in the saturated vapor at 72°F. This may be found from calculation or table. From calculation

$$\frac{0.791}{29.92 - 0.791} \times 0.622 = 1.697 \times 10^{-2} \text{ lb vapor per lb dry air}$$

18-2. The Psychrometric Chart. This chart is well known and will not be reproduced here. Its use is of such importance that one should be taken along to the examination.

18-3. Latent Heat Load in Air Conditioning. Latent heat is defined as that which is added to or subtracted from a substance in changing its physical state from a solid to a liquid (latent heat of fusion) or from a liquid to a vapor (latent heat of vaporization). In air-conditioning work the latent heat refers to the change of liquid water to water vapor in humidifying or the change of water vapor to liquid water in dehumidifying.

The latent heat load in air-conditioning work is the sum of all items that produce water vapor. Such items as people, infiltration of outside air, coffee urns, and certain products give off water vapor which is latent heat. For a person at rest the latent heat given off is 180 Btu per hr. The average hourly total heat per person at rest is taken as 400 Btu per hr, of which 180 Btu is latent heat and 220 Btu is sensible heat. These values vary with degree of activity and environmental temperature and humidity. Reference should be made to the Guide.

The quantity of water vapor from infiltration of outside air is determined from the difference between grains of moisture at the outside conditions of dew point and the room dew point. The latent heat value for water vapor is taken at the outside dew point temperature. Manufacturers of air-conditioning equipment usually provide calculation forms complete with factors so that it becomes a simple matter to determine air-conditioning loads.

The quantity of water vapor (latent heat) given up in a space is more difficult to establish accurately than is the sensible heat load. The total sum of water vapor once established is converted from water vapor to heat by using an average value of 1040 Btu per lb water vapor. If the amount of water vapor is 50 lb per hr, the latent heat load would be 50 × 1,040, or 52,000 Btu per hr.

Table 18-1 lists sources of internal heat. More elaborate tables are contained in the Guide.

TABLE 18-1. SOURCES OF INTERNAL HEAT BTU PER HR

1 Horsepower..........2546.0	1 cu ft producer gas......150	60 gr moisture per cu ft
1 kilowatt.............3415.0	1 cu ft illumin. gas...550–700	450 gr moisture per cu ft
100-watt lamp.......... 341.5	1 cu ft natural gas......1000	675 gr moisture per cu ft

(Welsbach Burner averages 3 cu ft per hr; and Fishtail burner, 5 cu ft per hr)

	Heat load of person normally clothed at rest, Btu per hr*								
Dry bulb	90°	85°	80°	75°	70°	65°	60°	55°	50°
Latent heat	241	197	153	118	87	66	61	61	61
Sensible heat	96	144	193	232	267	293	324	354	385
Total heat	337	341	346	350	354	359	385	415	446
Gr. mois. evap. per hr	1589	1300	1011	779	578	433	405	405	405

* Add 25% for man at work.

18-4. Air Required to Pick Up Latent Heat. Since air is the medium of absorbing latent heat (moisture pickup), it is necessary to put the air in condition to be able to absorb moisture. This is accomplished similarly to the absorption of sensible heat in that the absolute humidity (dew point) is reduced sufficiently to make the air capable of taking on moisture. As previously explained, the pickup of sensible heat in the conditioned room is accomplished by the quantity of air circulated at a reduced dry-bulb temperature. The pickup of latent heat is accomplished by the same air circulated at a reduced entering dew-point temperature. Thus, if we circulate 9,333 cfm to pick up 150,000 Btu sensible heat, this same quantity of air can have its dew point reduced sufficiently to accomplish the latent heat pickup. If the latent heat load is 52,000 Btu, this is equal to

$$\frac{52{,}000 \text{ Btu per hr}}{1{,}040 \text{ Btu per lb}} = 50 \text{ lb per hr}$$

and $$50 \times 7{,}000 \text{ grains per lb} = 350{,}000 \text{ grains per hr}$$

The quantity of water vapor to be picked up per cubic foot of air circulated is

$$\frac{350{,}000}{60 \text{ min} \times 9{,}333} = 0.625 \text{ grains per cu ft}$$

If the average room conditions are to be 80°F dry bulb and 50 per cent relative humidity, the dew-point temperature is 59.5°F and the air contains 75.85 grains per lb, or 5.699 grains per cu ft, the air must be supplied to the room at a condition of 5.699 − 0.625, or 5.074 grains per cu ft. Referring to air-property tables, we find that 5.074 grains per cu ft corresponds to approximately 56°F dew point (5.06 grains per cu ft). Thus, the air entering dew point must not be higher than 56°F, or 3½°F lower than the expected room dew point so as to maintain the conditions established of 80°F dry bulb and 50 per cent relative humidity.

Since the calculations were based on maintaining an average condition in the room, the return air is probably the best indication of the average conditions so that in most cases the thermostats and hygrostats are placed in the return air duct or near the return air grille. This is especially true if a number of rooms are connected to the air-conditioning apparatus. As we previously explained, the pickup of sensible heat in a conditioned room is accomplished by the quantity of air circulated at a reduced dry-bulb temperature and the pickup of latent heat is accomplished by the same air circulated at a reduced entering dew-point temperature.

18-5. Total Heat Load. For air-conditioning calculations, total heat is defined as the sum of all the sensible heat loads and all the latent heat loads, and is known as the *room total heat.* The room total heat does not establish the refrigeration load since it does not include the cooling load of outside air. The room total heat does, however, give us an easy means of establishing the dew point of the air entering the conditioned space.

Since the wet-bulb temperature determines the total heat in air, the total heat absorbed (sensible plus latent) by a quantity of air is measured by the difference between the respective wet-bulb temperatures. If H_1 is the enthalpy at the wet-bulb temperature of the air entering a space and H_2 at the wet-bulb temperature of the air leaving the space (return air), the total heat picked up in Btu per lb (W) of air is

$$\text{Total heat, Btu} = W(H_2 - H_1) \tag{18-4}$$

If the sensible heat formula is divided by the total heat formula, the following relation exists

$$\frac{\text{Sensible heat}}{\text{Total heat}} = \frac{W(T_2 - T_1)c_p}{W(H_2 - H_1)} = \frac{H_s}{H_t} \tag{18-5}$$

In the previous examples, the room condition was taken as 80°F dry-bulb temperature, 50 per cent relative humidity, 59.5°F dew point, and 66.7°F wet-bulb temperature. H_s equals 150,000 Btu per hr, H_L equals latent heat equals 52,000 Btu per hr.

$$H_t = H_s + H_L = 150{,}000 + 52{,}000 = 202{,}000 \text{ Btu per hr} \tag{18-6}$$

The dry-bulb temperature was taken as a differential of 15°F, i.e., $T_2 - T_1 = 80 - 65 = 15$°F, and the total heat corresponding to the room wet bulb of 66.7°F is 31.276 Btu per lb.

Substituting values in the formula, Eq. (18-5),

$$\frac{150{,}000}{202{,}000} = \frac{15 \times 0.24}{31.276 - H_1} = 0.7426 = \frac{3.6}{31.276 - H_1}$$

from which H_1 is found to be equal to 26.42 Btu per lb.

By referring to tables of properties of air, 26.42 Btu per lb is seen to correspond to a wet-bulb temperature of about 60°F. Thus, the temperature of the air entering the conditioned area is 65°F dry bulb and 60°F wet bulb, which by referring to the psychrometric chart, gives a dew point of 57°F and a relative humidity of 75 per cent. This is an easy method of determining the entering room dew point which will necessarily vary with the quantity of air circulated.

A line drawn through the room conditions on the psychrometric chart and the air entering conditions and extended through the saturation curve will give a locus of all dew points for any given dry-bulb differential (see chart). Note that the lowest dew point of entering air would be about 55°F. Any lower dew point would cause lower relative humidities. This condition would not be objectionable unless it drops too low to cause the space to feel uncomfortably cool.

Q18-4. Ninety pounds of water vapor are released into the atmosphere of a room every hour. An air supply of 8,250 cfm is provided. Find supply air dew point to maintain a room dew point of 59°F.

ANSWER. Moisture released to room is equal to

$$90 \times 7{,}000 = 630{,}000 \text{ grains per hr}$$

Required change in grains per pound of air is

$$630{,}000/(4.5 \times 8{,}250) = 17$$

From psychrometric chart at 59°F dew-point air has 74.6 grains per lb. Then the required initial moisture content of supply is 74.6 − 17, or 57.6. Dew point for this condition is found to be **52°F** from chart.

Q18-5. The atmosphere of a room having a latent heat gain of 60,000 Btu per hr is to be maintained at a dew point of 58°F. If the air supply to the room is 8,000 cfm, find the required dew point of the air supply.

ANSWER. Required change in latent heat content of air supply is

$$60{,}000/(4.5 \times 8{,}000) = 1.67 \text{ Btu per lb}$$

Latent heat content of room air from psychrometric chart is 71 grains per lb dry air and is equivalent to

$$71/7{,}000 \times 1{,}040 = 10.65 \text{ Btu}$$

Latent heat content of air supply is 10.65 − 1.67, or 8.98 Btu. This is equivalent to $(8.98 \times 7{,}000)/1{,}040 = 59.9$ grains per lb dry air with an equivalent dew point from chart of **53.2°F.**

Q18-6. A room having 80 per cent of its total heat gain as sensible heat is to have its dry-bulb temperature maintained at 76°F and its wet-bulb temperature at 64°F. If air is to be supplied at a dry-bulb temperature of 55°F, find the required wet-bulb temperature of the air supply.

ANSWER. Air supply will heat up 76 − 55, or 21°F. Then, total heat × 0.80 is equal to 0.24 × 21, or

$$\text{Change in total heat} = 0.24 \times 21/0.80 = 6.3 \text{ Btu per lb}$$

From psychrometric chart total heat of air (enthalpy) at 64°F wet-bulb temperature is 29.3. Required total heat of air supply is 29.3 − 6.3, or 23 Btu per lb. This is equivalent to a wet-bulb temperature of **54.4°F.**

Q18-7. With outside conditions of 95°F dry-bulb and 78°F wet-bulb temperature and inside conditions of 75°F dry-bulb and 57.5°F wet-bulb temperature determine the dew point, relative humidity, specific volume, grains of moisture per pound of dry air and enthalpy, using the psychrometric chart.

ANSWER.

Conditions	Outside air	Inside air
Dry bulb	95°F	75°F
Wet bulb	78°F	57.5°F
Dew point	71°F	43°F
Relative humidity	46 per cent	32 per cent
Specific volume	14.4 cu ft per lb	13.6 cu ft per lb
Grains per lb	114.4	40.9
Enthalpy (total heat)	40.64 Btu per lb	25.54 Btu per lb

18-6. Calculation of Cooling Load. Much of the work in calculating summer cooling load is identical with the calculation of winter heating load except that the heat flows into the building instead of out. There is one important new factor which has not been discussed, that of *solar heat* or radiant heat from the sun, which has a tremendous effect on cooling requirements when involving buildings more so than for process cooling.

Ordinary heat transmission takes place even when there is no sun effect. This transmission heat load is calculated as for heating. For the inside temperature that of 80°F dry-bulb temperature and 50 per cent relative humidity are most common although there may be slight deviations depending on a particular job. Outside design temperatures and wet-bulb temperatures for a number of locations are listed in Table 18-2. For a more complete list, refer to latest edition of the Guide.

TABLE 18-2

State	City	Design dry bulb, °F	Design wet bulb, °F
Alabama	Mobile	95	80
Arizona	Tucson	105	72
California	Los Angeles	90	70
Connecticut	New Haven	95	75
Delaware	Wilmington	95	78
District of Columbia	Washington	95	78
Florida	Miami	91	79
Illinois	Chicago	95	75
New Jersey	Trenton	95	78
New York	New York	95	75
Pennsylvania	Philadelphia	95	78
Texas	Houston	95	78
Utah	Salt Lake	95	65
Vermont	Burlington	90	73
Washington	Seattle	85	65
Wisconsin	Madison	95	75

You will recall that in calculating winter heating loads we choose an outdoor temperature 10 to 15°F above the lowest outdoor temperature ever recorded by the Weather Bureau. The choice of summer outdoor dry-bulb design temperature is not quite this simple. Wet-bulb temperature is just as important as dry-bulb temperature. The conditions given in Table 18-2 are those which will not be exceeded more than 5 to 8 per cent of the time during an average summer cooling season (June through September) of one-hundred and twenty 10-hr days, or 1,200 hr. In choosing outdoor design conditions for a location where this information is not known, it is necessary to obtain Weather Bureau or other records of conditions during past years. Select values which have not been exceeded more than 5 per cent of the time in previous average summers. Be sure to use daytime data and *not nighttime data.*

18-7. Cooling Load Due to Solar Radiation. The solar heat load is made up of that from sunlit walls and glass exposed to the rays of the sun. The Guide goes into this matter rather completely and the candidate is referred to that source for coverage. In its proper place we shall devote ourselves to the question and answer approach to focus attention.

18-8. Heat from Machinery and Appliances. When we calculate a winter heating load, we do not deduct the heat generated by all heat-producing equipment within the space to be heated because some of this equipment might be idle and thus not contributing heat. In summer, for which we are now calculating a cooling load, we have the opposite situation. We must consider any heat-generating machinery or appliance which is likely to be operating at the same time that cooling is needed. If there is at all a question as to whether an appliance is operating, we shall consider that it will be operating. Table 18-3 shows the heat added by miscellaneous electric, gas, and steam-heated equipment. Considerable judgment factor is required in using values for equipment where the sensible and latent heat values in Table 18-3 are marked with an asterisk. Motor heat will be accounted for in the short form for figuring the cooling load. Electric lights are figured for full wattage whether incandescent or fluorescent.

18-9. Maximum Cooling Load. When does maximum cooling load occur? So far we have reviewed the procedure and the nature of the calculations involved in calculating summer cooling load. Let us see how we would go about determining the cooling load on a building of certain characteristics. If the top floor had a number of skylights and electric lighting, we could assume that on sunny days the lights would not be in use. We would only calculate the sun effect and exclude the lights. The sun effect on the east and west walls would not be added to the total sun load, for just one side or the other (whichever is the greater) would be included. At noon just the roof and skylight loads would be considered. Then the

TABLE 18-3. HEAT GAIN FROM VARIOUS SOURCES

Source	Btu per hr		
	Sensible	Latent	Total
Electric-heating Equipment			
Electrical equipment—dry heat—no evaporated water	100%	0%	100%
Electric oven—baking	80%	20%	100%
Electric equipment—heating water—stewing, boiling, etc	50%	50%	100%
Electric lights and appliances per watt (dry heat)	3.4	0	3.4
Electric lights and appliances per kilowatt (dry heat)	3413	0	3413
Electric motors per horsepower	2546	0	2546
Electric toasters or electric griddles	90%	10%	100%
Coffee urn—large, 18-in. diameter—single drum	2000	2000	4000
Coffee urn—small, 12-in. diameter—single drum	1200	1200	2400
Coffee urn—approx. connected load per gallon of capacity	600	600	1200
Electric range—small burner	*	*	3400
Electric range—large burner	*	*	7500
Electric range—oven	8000	2000	10000
Electric range—warming compartment	1025	*	1025
Steam table—per square foot of top surface	300	800	1100
Plate warmer—per cubic foot of volume	850	0	850
Baker's oven—per cubic foot of volume	3200	1300	4500
Frying griddles—per square foot of top surface	*	*	4600
Hot plates—per square foot of top surface	*	*	9000
Hair dryer in beauty parlor—600 watts	2050	0	2050
Permanent-wave machine in beauty parlor—24–25 watt units	2050	0	2050
Gas-burning Equipment			
Gas equipment—dry heat—no water evaporated	90%	10%	100%
Gas-heated oven—baking	67%	33%	100%
Gas equipment—heating water—stewing, boiling, etc	50%	50%	100%
Stove, domestic type—no water evaporated—per medium-size burner	9000	1000	10000
Gas-heated oven—domestic type	12000	6000	18000
Stove, domestic type—heating water—per medium size burner	5000	5000	10000

TABLE 18-3. HEAT GAIN FROM VARIOUS SOURCES (*Continued*)

Source	Btu per hr		
	Sensible	Latent	Total
Residence gas range—giant burner (about 5½ in. diameter)	*	*	12000
Residence gas range—medium burner (about 4 in. diameter)	*	*	10000
Residence gas range—double oven (total size 18 × 18 × 22 in. high)	*	*	18000
Residence gas range—pilot	*	*	250
Restaurant range—4 burners and oven	*	*	100000
Cast-iron burner—low flame—per hole	*	*	100
Cast-iron burner—high flame—per hole	*	*	250
Simmering burner	*	*	2500
Coffee urn—large, 18 in. diameter—single drum	5000	5000	10000
Coffee urn—small, 12 in. diameter—single drum	3000	3000	6000
Coffee urn—per gallon of rated capacity	500	500	1000
Egg boiler—per egg compartment	2500	2500	5000
Steam table or serving table—per square foot of top surface	400	900	1300
Dish warmer—per square foot of shelf	540	60	600
Cigar lighter—continuous flame type	2250	250	2500
Curling iron heater	2250	250	2500
Bunsen type burner—large—natural gas	*	*	5000
Bunsen type burner—large—artificial gas	*	*	3000
Bunsen type burner—small—natural gas	*	*	3000
Bunsen type burner—small—artificial gas	*	*	1800
Welsbach burner—natural gas	*	*	3000
Welsbach burner—artificial gas	*	*	1800
Fish-tail burner—natural gas	*	*	5000
Fish-tail burner—artificial gas	*	*	3000
Lighting fixture outlet—large, 3 mantle 480 c.p.	4500	500	5000
Lighting fixture outlet—small, 1 mantle 160 c.p.	2250	250	2500
One cu ft of natural gas generates	900	100	1000
One cu ft of artificial gas generates	540	60	600
One cu ft of producer gas generates	135	15	150

Steam-heated Equipment

Source	Sensible	Latent	Total
Steam-heated surface not polished—per square foot of surface	330	0	330
Steam-heated surface polished—per square foot of surface	130	0	130

TABLE 18-3. HEAT GAIN FROM VARIOUS SOURCES (*Continued*)

Source	Btu per hr		
	Sensible	Latent	Total
Insulated surface, per square foot	80	0	80
Bare pipes, not polished per square foot of surface	400	0	400
Bare pipes, polished per square foot of surface	220	0	220
Insulated pipes, per square foot	110	0	110
Coffee urn—large, 18-in. diameter—single drum	2000	2000	4000
Coffee urn—small, 12-in. diameter—single drum	1200	1200	2400
Egg boiler—per egg compartment	2500	2500	5000
Steam table—per square foot of top surface	300	800	1100
Miscellaneous			
Heat liberated by food per person, as in a restaurant	30	30	60
Heat liberated from hot water used direct and on towels per hour—barber shops	100	200	300

* Per cent sensible and latent heat depends upon use of equipment; dry heat, baking, or boiling.

largest combination of them all is included. For parts of the country in the north latitude the maximum sun effect on the east wall occurs about 8 A.M., at which time there is zero sun effect on the west wall, and almost zero sun effect on the south wall. It may be shown that the sun effect on the east and west walls occurs at totally different times.

Time lag plays an important role in solar-heat gain. Sun will pass immediately through glass and will have an immediate effect on cooling load. However, the same sun impinging on walls will suffer a time lag in having its effect felt. Table 18-4 lists representative time lags for several materials.

As we can see from the table that portion of the sun effect may not reach the inside and have its effect until several hours later. By the time this heat reaches the inside of the conditioned space, the sun effect on the south wall is at a maximum and the portion striking the windows of the south wall is entering the conditioned space immediately. Hence it is likely that the maximum solar-heat effect *inside* this room might occur sometime between 11 A.M. and 1 P.M. This maximum would *not* include the solar heat striking the solid portion of the south wall, because time lag would delay this heat and it would finally enter the conditioned space between 3 and 4 P.M. By this time most of the sun effect on the south windows would have disappeared, and since the room does not have an exposed west wall, the solar heat on the room would diminish as the afternoon progressed.

TABLE 18-4. TIME LAG IN TRANSMISSION OF SOLAR RADIATION THROUGH WALLS AND ROOFS

Type and thickness of wall or roof	*Time lag, hr*
2-in. pine	1.5
6-in. concrete	3
4-in. gypsum	2.5
3-in. concrete and 1-in. cork	2
2-in. iron and cork	2.5
4-in. iron and cork	7.25
8-in. iron and cork	19
22-in. brick and tile wall	10

By combining the solar effects on glass and wall and by use of available extensive tables, a definite and orderly procedure may be established for determining the time of day at which the maximum effect of solar heat effect is actually entering the room.

Q18-8. A building as illustrated in Fig. 18-1 is to be maintained at 75°F dry-bulb and 64.4°F wet-bulb temperatures. This building is situated between two similar units which are not cooled. There is a second-floor office above and a basement below. The south wall, containing 45 sq ft of glass area, has a southern exposure. On the north side there are two show windows which are ventilated to the outside and are at outside conditions. Between the show windows is located the doorway. This will be assumed as normally closed but it is opened quite frequently and allows an average of 600 cfm of outside air to be admitted. This will cause slightly more than two air changes per hour in the building. Number of persons is 35. Lighting is 1,100 watts on a sunny day. Basement temperature is 80°F assumed. Maximum outside conditions for design purposes

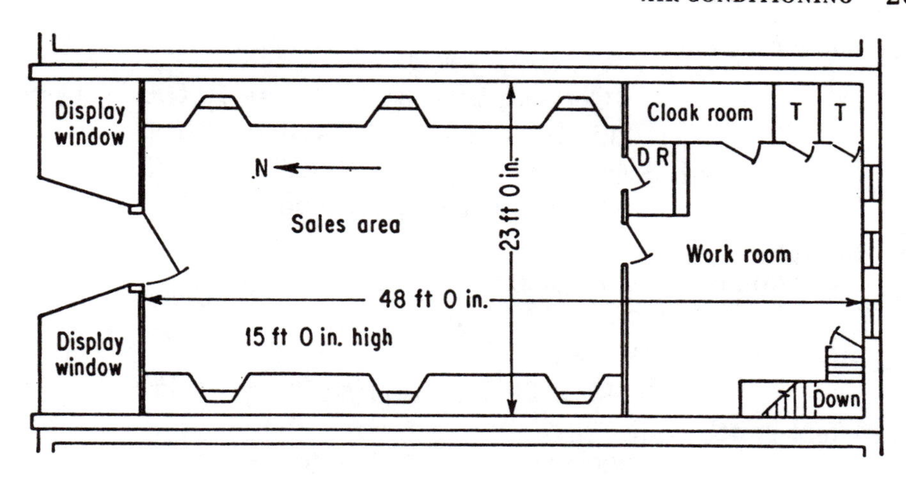

Fig. 18-1

are 95°F dry-bulb and 78°F wet-bulb temperature. Determine the refrigeration load.

ANSWER.

Calculations of heat load using U coefficients:	U
East and west walls (24 in. brick, plaster one side)	0.16
North partition (1¼ in. tongue-and-groove wood)	0.60
Plate-glass door	1.0
South wall (13 in. brick, plaster one side)	0.25
Windows (single thickness)	1.13
Floor (1 in. wood, paper, 1 in. wood over joists)	0.21
Ceiling (2 in. wood on joists, lath and plaster)	0.14

Temperature difference walls and ceiling: 95 − 75 = 20°F

Temperature difference floor: 80 − 75 = 5°F

Heat leakage:

East and west wall	= 0.16 × 1,440 × 20	=	4,610 Btu per hr
North partition	= 0.6 × 299 × 20	=	3,590
Glass door	= 1.0 × 35 × 20	=	700
South wall	= 0.25 × 300 × 20	=	1,500
Windows 45 + 46	= 1.13 × 91 × 20	=	2,060
Floor	= 0.21 × 1,104 × 5	=	1,160
Ceiling	= 0.14 × 1,104 × 20	=	3,090
	Total	=	16,710 Btu per hr

Sensible heat load.

$$\text{Occupants} = 35 \times 300 = 10{,}500 \text{ Btu per hr}$$
$$\text{Lights} = 1{,}100 \times 3.413 = 3{,}760$$
$$\text{Motor horsepower} = 0.5 \times 2{,}546 = 1{,}275$$
$$\text{Heat leakage above} = 16{,}710$$

Air leakage

$$= \frac{36{,}000 \text{ cu ft per hr} \times 0.24 \times (95 - 75)}{13.70 \text{ cu ft per lb}} = 12{,}600$$

Sun effect glass south wall

$$= 45 \times 30 \text{ Btu/(hr)(sq ft)} = 1{,}350$$

Sun effect south wall

$$= 300 \times 0.25 \times (120 - 95) = 1{,}875$$
$$\text{Grand total} = 48{,}070 \text{ Btu per hr}$$
$$\text{Dry tons of refrigeration} = 48{,}070/12{,}000 = 4.01 \text{ tons}$$

Latent heat load (moisture):

Air leakage:

Conditions	Outside air	Inside air
Dry bulb	95°F	75°F
Wet bulb	78°F	64.4°F
Per cent relative humidity	47	55
Dew point	71°F	58°F
Gr per lb	114.4	71.9
Sp vol	.....	13.7 cu ft per lb

First determine the pounds per hour of air.

$$\frac{36{,}000 \times (114.4 - 71.9)}{13.7 \times 7{,}000} = 15.95 \text{ lb per hr}$$
$$\text{Air latent heat} = 15.95 \times 1{,}040 = 16{,}900 \text{ Btu per hr}$$
$$\text{Person load} = 35 \times 100 = 3{,}500 \text{ Btu per hr}$$
$$\text{Total latent} = 16{,}900 + 3{,}500 = 20{,}400 \text{ Btu per hr, or } 20{,}400/12{,}000 = 1.70 \text{ tons}$$
$$\text{Total load} = 4.01 \text{ sensible} + 1.70 \text{ latent} = \mathbf{5.71 \text{ total tons}}$$

18-10. Adiabatic Cooling. Let us examine a simple problem in cooling air adiabatically. Follow the process on the psychrometric chart. The water is merely recirculated and only a small amount of make-up is added due to loss by humidification; neither is water-heated or water-cooled before passing through the washer. For

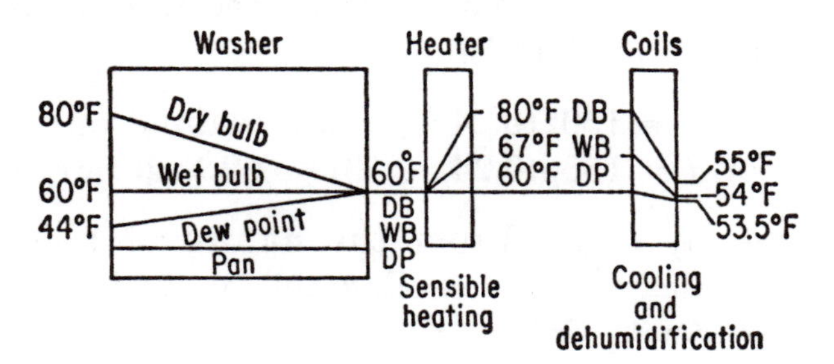

Fig. 18-2

convenience let us superimpose temperature scales on this washer. The condition of entering air is 80°F dry bulb, 44°F dew point, 60°F wet bulb. With appropriate conditions and perfect internal mixing of water and air, the air leaves at 60°F and saturated (see Fig. 18-2). Then, as we said before, all three temperatures become identical. Now since the wet bulb does not vary since enthalpy remains constant, we have true adiabatic cooling, for the dry-bulb temperature has dropped. However, dew point has risen, for humidification has taken place. The heat released by the drop in dry bulb has been used in evaporating the water.

Now let us consider a heater placed in the air stream as shown. This adds only sensible heat to the air. With no moisture additions or subtractions, the dew point would remain the same as it is independent of sensible heat. But with the addition of sensible heat the total heat of the air and water vapor rises and thus the wet-bulb temperature is also raised. Also, there is a rise in dry-bulb temperature to some point above the wet-bulb temperature.

With the addition in line of a dehumidifier coil or washer, all three temperatures suffer a drop.

Q18-9. A room is to be cooled by air from an adiabatic air washer, and has no moisture load, but does have a sensible heat load of 150 Btu per min. The conditions to be maintained are 80°F dry-bulb temperature and 60 per cent relative humidity. Find the required entering wet-bulb temperature to the washer and the quantity of air that must be washed each minute.

ANSWER. The wet-bulb temperature corresponding to the room conditions is **70°F.** This is the entering wet-bulb temperature required. Then the air to be sensibly heated from 70 to 80°F with 150 Btu per min is found as follows:

lb per min = 150/(0.24 × temperature drop)
lb per min = 150/(0.24 × 10) = **62.5 lb per min**

18-11. Bypass System. From the nature of air-conditioning apparatus the air leaving the cooling coil or washer is practically saturated. The dew point so resulting is necessary to maintain proper conditions in the room. When the air-conditioning requirements are reduced due to lower outdoor temperatures, fewer people than calculated, less lighting, etc., the temperature leaving the apparatus is so low that the air entering the space will cause drafts and chilled environment. In order to avoid this condition, the leaving air must be sensibly heated to the proper dry-bulb temperature before being admitted to the space. Normally the difference (diffusion) between dry-bulb temperature to be maintained within the space and that of the entering air is from 15 to 20°F, depending on the manner of air distribution and kind of space.

In the bypass system, a portion of the return air is mixed with the outside air which has passed through the conditioner together with the remainder of the return air. The quantity of return air being automatically controlled warms the saturated air leaving the conditioner to correct dry-bulb temperature. Since only a portion of the recirculated air that passes through the apparatus is refrigerated (the balance being passed around the apparatus), the refrigeration is reduced proportionately.

Q18-10. Air is to be delivered to a space at the outlet grille at 64°F dry bulb. The dehumidifier leaving air temperature (apparatus dew point) is 52°F. If the return air at 80°F dry bulb is available, what proportion of conditioned air and bypass air must be used?

ANSWER. Let x be proportion of conditioned air necessary and y be proportion of by-passed air necessary.

Then

$$52x + 80y = 64(x + y)$$

from which may be cleared

$$12x = 16y$$

Therefore x = **57.15 per cent conditioned air**
y = **42.85 per cent bypassed air**

Q18-11. A space to be conditioned has a sensible heat load of 10,000 Btu per min and a moisture load of 26,400 grains per min; 2,300 lb of air are to be introduced each minute to this space for its conditioning to 80°F dry bulb and 50 per cent relative humidity. Draw a sketch of this problem and work out the amount of air to be bypassed and the amount and temperature of air leaving the dehumidifier.

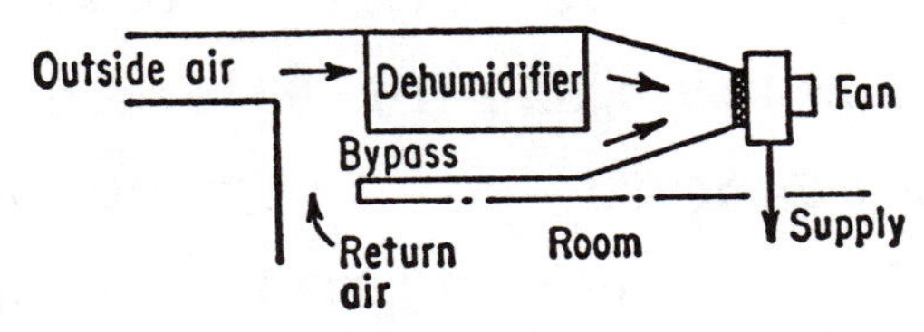

Fig. 18-3

ANSWER. Refer to Fig. 18-3. Now set up the following table of conditions.

	Room conditions	Air leaving dehumidifier
Dry bulb	80°F	54°F
Wet bulb	67°F	54°F
Dew point	60°F	54°F
Relative humidity	50 per cent	100 per cent
Total heat	31.15	22.54
Grains per lb	77.3	62.1
Sp vol	13.84	13.13
Lb per min	2,300	1,610

Moisture load = (26,400/7,000) × 1,040 = 3,920 Btu per min
Sensible load = 10,000 Btu per min
Total load = 13,920 Btu per min

Solve by trial and error, assuming 53°F air leaving dehumidifier.

Total heat at 80°F and 50 per cent rel. humidity = 31.15
Total heat at 53°F and 100 per cent rel. humidity = 21.87
Difference is the pickup = 9.28 Btu

On basis of first trial, air circulated is 13,920/9.28 = 1,500 lb per min. Check this figure as follows:

$$1{,}500 \times 0.24 \times (80 - 53) = 9{,}700 \text{ Btu per min}$$

This is not enough. By trial and error again, assume 54°F air leaving dehumidifier.

Total heat at 80°F and 50 per cent rel. humidity = 31.15
Total heat at 54°F and 100 per cent rel. humidity = 22.54
Difference is the pickup = 8.70 Btu

Air circulated is now 13,920/8.70 = 1,600 lb per min. Now check,

$$1{,}600 \times 0.24 \times (80 - 54) = 9{,}970 \text{ Btu per min}$$

This figure is slightly short of the 10,000 Btu per min required, but will do. Thus,

Total air leaving dehumidifier = **1,600 lb per min**
Air bypassed = 2,300 − 1,600 = **700 lb per min**
Temperature air leaving dehumidifier = **54°F**

Q18-12. Air enters a drier at 70°F dbt and 20 per cent relative humidity and leaves the drier at 180°F dbt and a relative humidity of 50 per cent. The drier operates at atmospheric pressure of 14.3 psia. Determine the number of cubic feet of air entering per minute needed to evaporate 2.0 lb water per min from the material being processed.

ANSWER. At 70°F, saturation pressure is 0.3631 psia from steam tables. At 20 percent relative humidity, vapor pressure is 0.20 × 0.3631 = 0.073 psia.

$$t_1 = 70°\text{F} \qquad \text{RH}_1 = 20\%$$

Using humidity ratio relations, pounds of water per pound of dry air is

$$0.622 \times (0.073/14.227) = 0.00319 \text{ lb per lb for entering air}$$

Note that $14.3 - 0.073 = 14.227$ psia. Then for leaving air,

$$0.622 \times (3.755/10.545) = 0.2214 \text{ lb water per lb dry air}$$

Thus water removed from system $= 0.2214 - 0.00319 = 0.2182$ lb water per lb dry air circulated.

$$\frac{2.0 \text{ lb water to be removed}}{0.2182 \text{ lb water removed per lb dry air}} = 9.17 \text{ lb per min air to be circulated}$$

To convert to cfm, first calculate air density at entering conditions.

$$(29/379) \times (14.227/14.7) \times (520/530) = 0.073 \text{ lb per cu ft}$$

Finally, air cfm $= 9.17/0.073 =$ **125 cfm.** Note application of humidity ratio.

Q18-13. An underground space 15 ft wide by 20 ft long by 10 ft high has a wall surface temperature of 55°F. Outside air conditions are 70°F dbt and 60 per cent relative humidity, and the space is naturally ventilated to the outdoors. Will surface condensation take place? If so, what will be the dehumidification load if one air change per hour is assumed? The following data are to be used in the solution: moisture content in outside air is 4.788 grains per cu ft, moisture content to be maintained in conditioned space is 1.223 grains per cu ft. In the second part of the problem, to prevent condensation on the walls and ceilings, space air must be dehumidified to a low dry-bulb temperature of 50°F and a relative humidity of 30 per cent.

ANSWER. **Use the graph in Fig. 18-4 to determine the possibility** of condensation.

(*a*) **Will surface condensation take place? Refer to Fig. 18-4;** 70°F dbt and 60 per cent relative humidity intersect at a ground temperature of 55°F. Since this is surface temperature (dew point), **condensation will take place.**

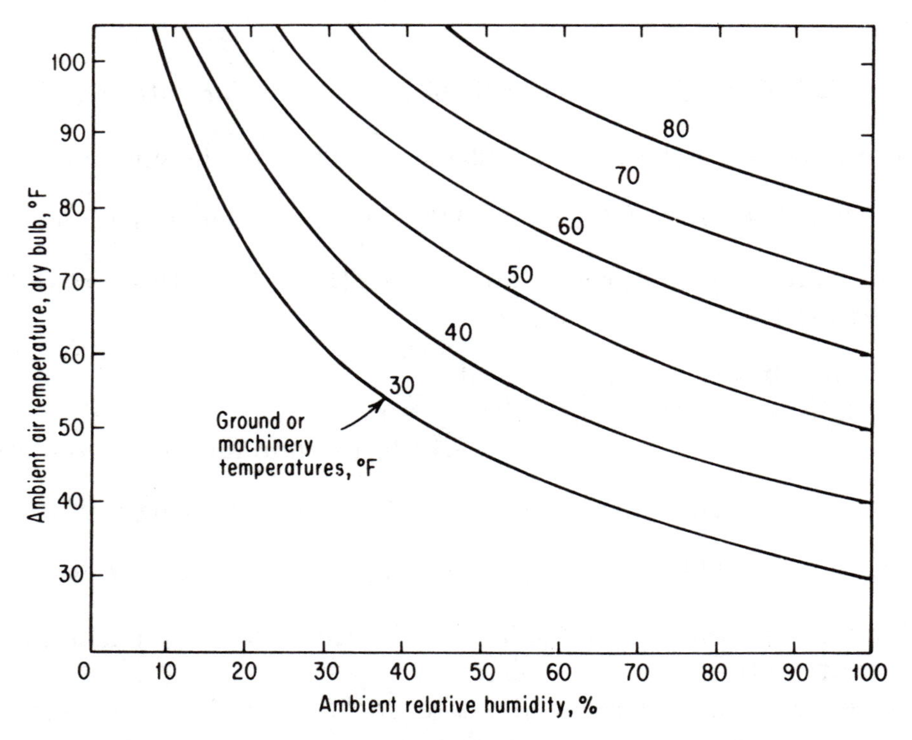

Fig. 18-4

(*b*) The dehumidification load is

$$\frac{\text{Room volume} \times (4.788 - 1.223)}{7{,}000}$$

$$= \frac{(15 \times 20 \times 10) \times (4.788 - 1.223)}{7{,}000}$$

$$= 1.528 \text{ lb condensate per hour}$$

Since the weight of 1 pint of water is about 1 lb, the dehumification load is 1.528 × 24 = **37 pints per 24 hr.**

Supplement A

MECHANICS

Questions and Answers

A-1. A projectile is fired at an angle of 37° with the horizontal at an initial velocity of 2,000 fps across terrain as shown in Fig. A-1. Find the time it takes to reach point L on the plateau.

ANSWER.

$$t = \frac{1}{2}\left(\frac{2v_0}{g}\sin\theta \pm \sqrt{\frac{4v_0^2}{g^2}\sin^2\theta - 4\frac{2h}{g}}\right)$$
$$= \frac{1}{2}\left(\frac{2 \times 2{,}000}{32.2} \times 0.6 \pm \sqrt{\frac{4 \times 2{,}000^2}{32.2^2} \times 0.36 - 4 \times \frac{2 \times 100}{32.2}}\right)$$
$$= 37.5 \pm 37.4 = \mathbf{74.9\ sec}$$

The other is imaginary for the problem.

The first term $(2v_0/g)\sin\theta$ is the time the projectile takes to reach the top of the trajectory. The first term under the square root sign

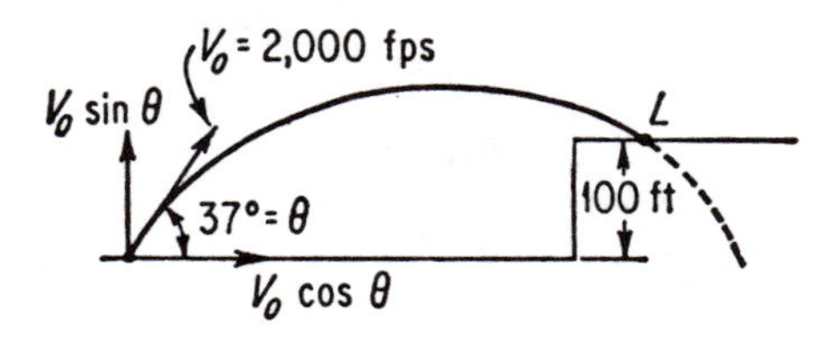

Fig. A-1

is the same, since it takes the projectile just as long to come down to the same level as it takes to reach the top of the trajectory, if the square root is taken. The second term under the square root sign is a minus because the plateau has shortened the time of flight by that amount. If there were a drop below the initial level and the time of flight were longer as a result, the minus (−) sign would become a plus (+).

A-2. A sphere weighs 1,500 lb. Find the tension in the wire and the direction and magnitude of the reaction at the hinge at A. Refer to Fig. A-2.

ANSWER.

$$Ab = 2'0''$$
$$P_1 \sin 30 = 3{,}000 \times 0.5 = 1{,}500 \text{ lb}$$
$$P_1 = 3{,}000 \text{ lb}$$
$$BC \times 3 = 3{,}000 \times 2 = 6{,}000$$

Thus $$BC = 2{,}000 \text{ lb}$$

Reaction at A is **1,000 lb** at 30° with the horizontal.

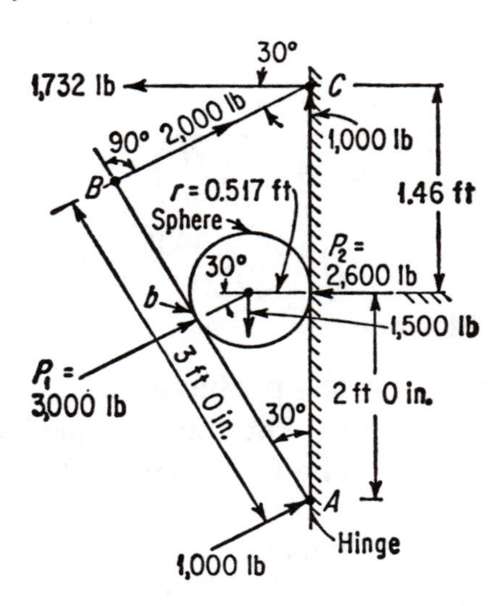

Fig. A-2

Note. AB was given as 3′0″.

Point A to center line of sphere was given as 2′0″.
Angle ABC was given as 90°.
Angle BCA was given as 30°.

A-3. A rocket bomb weighs 14 tons, of which 6 are fuel plus oxygen. In order to simplify the calculations, suppose the rocket to be projected vertically upward and the stream of gas ejected to be at a constant rate of 260 lb per sec. The velocity of the gas stream is 6,400 fps. Neglect air resistance and the decrease of rocket weight. (*a*) What maximum velocity will the rocket attain? (*b*) To what height will it rise? (*c*) With what velocity will it strike the earth in falling?

ANSWER.

(*a*) Propelling force $F = W$, rate of gas stream $\times$ gas velocity/g. This is rate of momentum of gas stream. Therefore,

$$F = 260/32.2 \times 6{,}400 = 52{,}000 \text{ lb}$$

Time during which gas stream acts until fuel is exhausted is

$$t = 6 \times 2{,}000/260 = 46.2 \text{ sec}$$

Resulting force acting on rocket is

$$F_r = F - W_r = 52{,}000 - 28{,}000 = 24{,}000 \text{ lb}$$

Impulse force of rocket during entire interval gas stream acts is same as momentum of rocket at end of impulse. Therefore,

Rocket impulse $\times\ t$ = weight of rocket $\times$ max. rocket velocity/g

$$24{,}000 \times 46.2 = 14 \times 2{,}000 \times v_r/32.2$$

Rearrange and solve for v_r, which is found to be equal to **1,265 fps.** This is maximum rocket velocity upward.

(*b*) Velocity on striking the earth. Then rise during impulse is

$$s_1 = \frac{v_r - v_0}{2}t = \frac{1{,}265 - 0}{2} \times 46.2 = 29{,}200 \text{ ft}$$

Rise after impulse is s_2:

$$s_2 = \frac{(v_r)^2}{2g} = \frac{1{,}265^2}{64.4} = 25{,}000 \text{ ft}$$

Total rise is $29{,}200 + 25{,}000 =$ **54,200 ft.**

(*c*) Velocity on striking the earth is

$$v_f = (2gs_f)^{1/2} = (64.4 \times 54{,}200)^{1/2} = \mathbf{1{,}860\ fps}$$

A-4. A single flat surface (vertical plate) is carried at a speed of 10 fps toward a nozzle that discharges water at 8 cfm at 50 fps.

What force is required to carry the plate if it is maintained directly toward the nozzle? How much water hits the plate each minute?

ANSWER.

$$2\left[\frac{W}{g}(v_1 - v_2)\right] = 2\left[\frac{8 \times 62.4}{60} \times \frac{1}{32.2} \times (50 + 10)\right] = \mathbf{31\ lb}$$

$$8 \times {}^{60}\!/_{50} = \mathbf{9.6\ cfm\ of\ water}$$

A-5. A tank truck full of fresh water is proceeding along the road at 40 mph. The tank may be considered to be a cylinder with the diameter of 6 ft and a length of 20 ft. If the truck is suddenly stopped, what will be the longitudinal force in pounds per inch of circumference in the front end of the cylinder wall? What will be the lateral force per inch of length in the tank wall? Neglect static pressure due to water head.

ANSWER. Total weight of water in the tank is calculated out to be 35,300 lb. The truck has a speed equal to ${}^{40}\!/_{60} \times 88 = 59$ fps. Now, assume it takes 1 sec to effect a dead stop. Then the deceleration is 59 ft per sec² and the force exerted by the water on the front end is

$$F = 35{,}300/32.2 \times 59 = 64{,}600 \text{ lb}$$

The longitudinal force in the cylinder wall is given by

$$f_1 = \frac{64{,}600}{12\pi \times 6} = \mathbf{286\ lb\ per\ in.}$$

The lateral force in the cylinder wall is given by

$$f_2 = 2 \times 286 = \mathbf{572\ lb\ per\ in.}$$

A-6. The position of a body of 70-lb mass is given by the equation $x = 3t^2 + 2t + 4$, where x is in feet and t is in seconds. (*a*) Compute the velocity when x is equal to 70; (*b*) what force is required for the motion to take place?

ANSWER.

(*a*)

$$70 = 3t^2 + 2t + 4$$
$$t^2 + {}^{2}\!/_{3}t - 22 = 0$$
$$t = -{}^{1}\!/_{3} \pm \sqrt{{}^{1}\!/_{9} + 22} = 4.4 \text{ sec}$$

when x is equal to 70 the velocity is $(6 \times 4.4) + 2 =$ **28.4 fps.** Note that this is obtained from the equation $x = 6t + 2$ derived from the equation given in the problem.

(*b*) Force required is equal to $M\ddot{x} = (70/32.2) \times 6 =$ **13 lb.**

A-7. In the hand-operated press shown in Fig. A-4 a force F produces a compressive reaction on the body B placed between the movable platform and the fixed anvil. If $F = 20$ lb, $a = 1$ in.,

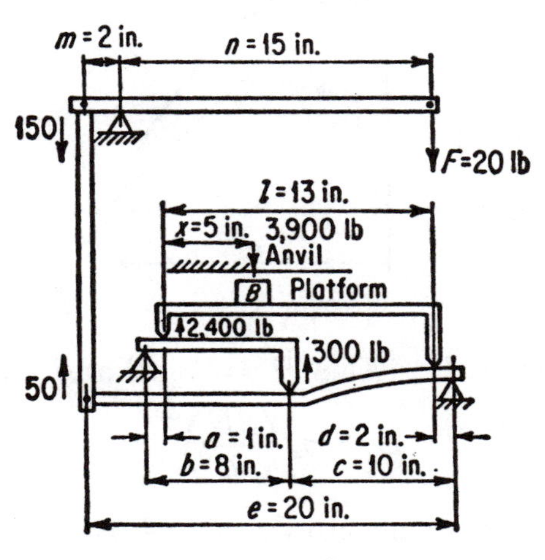

Fig. A-3

$b = 8$ in., $c = 10$ in., $d = 2$ in., $e = 20$ in., $m = 2$ in., $n = 15$ in., and $x = 5$ in., what will be the compressive reaction on body B?

ANSWER. The forces shown in Fig. A-3 have been added in accordance with simple calculations involving moments. Thus, in order to find the reaction at point m,

$$\frac{20 \text{ lb} \times 15 \text{ in.}}{2 \text{ in.}} = 150 \text{ lb}$$

$$\frac{e \times 150}{c} = \frac{20 \text{ in.} \times 150 \text{ lb}}{10 \text{ in.}} = 300 \text{ lb}$$

$$\frac{b \times 300}{a} = \frac{8 \text{ in.} \times 300 \text{ lb}}{1 \text{ in.}} = 2{,}400 \text{ lb}$$

$$\frac{1 \times 2{,}400}{13 \text{ in.} - 5 \text{ in.}} = \frac{13 \text{ in.} \times 2{,}400}{8 \text{ in.}} = \mathbf{3{,}900 \text{ lb}}$$

A-8. A bomber is flying with a constant velocity v at an elevation H toward its objective, which is being sighted by an observer in the plane. At what angle of sight with respect to the vertical should a bomb be released so that it would strike the objective? Neglect air resistance.

ANSWER. Refer to Fig. A-4. Let $L = v_x t$, since velocity is constant. We know it will take just as long for the bomb to fall through

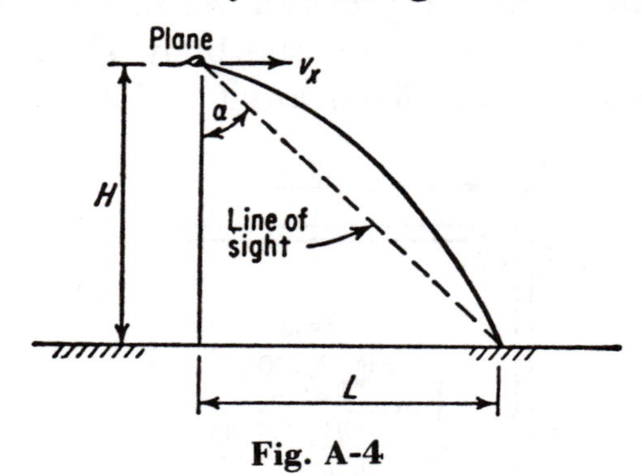

Fig. A-4

H as to cover distance L. Also $H = \frac{1}{2}gt^2$ and by substitution

$$L = v_x \left(\frac{2H}{g}\right)^{1/2}$$

The angle of sight α may be obtained in terms of tan α so that

$$\tan \alpha = \frac{L}{H} = \frac{v_x}{H} \times \left(\frac{2H}{g}\right)^{1/2}$$

$$\tan \alpha = v_x \times 0.25 \left(\frac{1}{H}\right)^{1/2}$$

and

$$\alpha = \mathbf{tan^{-1}\ 0.25 \times \frac{v}{H^{1/2}}}$$

A-9. An automobile weighing 2,800 lb is traveling at 30 mph when it hits a depression in the road which has a radius of curvature of 50 ft. What is the total force to which the springs are subjected?

ANSWER. The acceleration into the curve is

$$a = \frac{v^2}{\text{radius of curvature}} = \frac{(30/60 \times 88)^2}{50} = 38.7 \text{ ft per sec}^2$$

Let us now equate the vertical forces to obtain the following equation of motion

$$F - 2{,}800 = \frac{W}{g} 38.7 = \frac{2{,}800}{32.2} 38.7 \quad \text{or} \quad F = \mathbf{6{,}170\ lb}$$

A-10. A rocket missile of weight W is to leave the earth's surface and be projected out into space to infinite height. Determine the escape velocity, i.e., the initial velocity required to accomplish this feat.

ANSWER. The missile must pass out beyond the earth's force of gravity or gravitational pull. We remember that the earth's attraction force F is proportional to the square of the distance x away from the center of the earth to that point of force where it becomes zero. For this purpose it is equal to

$$F = \frac{WR^2}{x^2}$$

where R is the earth's radius. The work required to go from $x = R$ to $x = \alpha$ is by integration of $\int_R^a F\,dx$ is WR. There is a degeneration of initial kinetic energy to zero kinetic energy and this equals the work expended. Initially, the kinetic energy is $\frac{1}{2}(W/g)v_0^2$, and finally it is zero. Then

$$WR = \frac{1}{2}\frac{W}{g}v_0^2$$

where v_0 is the escape velocity in fps. By clearing both sides of the equation and solving for v_0,

$$v_0 = (\mathbf{2gR})^{1/2}$$

A-11. A ship is propelled by a constant thrust $p \times m$, where m is the mass of the ship. The water exerts a resistance $c \times m \times v^2$, where p and c are constants.

(*a*) What is the terminal velocity?

(*b*) Starting from rest, what is the distance traveled? Express the solution as a function of time.

(*c*) Repeat part (*b*) and express the solution as a function of velocity.

ANSWER. **Refer to Fig. A-5. $\Sigma F = m \times a$. Then from Newton's second law of motion, at terminal velocity, V = constant and therefore $a = 0$.**

(*a*) $p \times m - c \times m \times v^2 = 0$ and terminal velocity $v = \sqrt{p/c}$.

Fig. A-5

Answer.

(*b*) The differential equation of motion is

$$\Sigma F = m \times \frac{d^2x}{dt^2} = m \times \frac{dV}{dt}$$

$$p \times m - c \times m \times V^2 = m \times \frac{dV}{dt} \qquad \text{and} \qquad p - c \times V^2 = \frac{dV}{dt}$$

$$\frac{dV}{dt} + c \times V^2 - p = 0$$

Separate variables so that $(dV/p - cV^2) = dt$.

$$\int \frac{dV}{p - cV^2} = \int dt$$

From tables,

$$\frac{1}{2pc} \ln \frac{p + V\sqrt{pc}}{p - V\sqrt{pc}} + C_{\text{constant}} = t$$

When $t = 0$, $V = 0$. Then

$$\frac{1}{2\sqrt{pc}} \ln \frac{p}{p} + c = 0$$

Then since ln (1) = 0, $c = 0$.

$$\ln \frac{p + V\sqrt{pc}}{p - V\sqrt{pc}} = 2\sqrt{pc}\,(t)$$

$$e^{2\sqrt{pc}(t)} = p + V\sqrt{\frac{pc}{p}} - V\sqrt{pc}$$

$$V = \sqrt{\frac{p}{c}}\,\frac{e^{2\sqrt{pct}} - 1}{e^{2\sqrt{pct}} + 1}$$

$$V = v\,\frac{1 - e^{2\sqrt{pct}}}{1 + e^{2\sqrt{pct}}}$$

where V = terminal velocity

Since $p = dx/dt$,

$$\int dx = \int V \frac{1 - e^{2\sqrt{pct}}}{1 + e^{2\sqrt{pct}}}\, dt$$

$$\int dx = V \int \frac{1}{1 + e^{2\sqrt{pct}}}\, dt - V \int \frac{e^{2\sqrt{pct}}}{1 + e^{2\sqrt{pct}}}\, dt$$

$$\int dx = V \int \frac{dt}{1 + e^{2\sqrt{pct}}} - V \int \frac{dt}{1 + e^{2\sqrt{pct}}}$$

$$\frac{x}{V} = \frac{1}{2\sqrt{pc}} [2\sqrt{pct} - \ln(1 + e^{2\sqrt{pct}})]$$

$$= -\frac{1}{-2pc} [-2\sqrt{pct} - \ln(1 + e^{-2\sqrt{pct}}] + c$$

where c = constant of integration

$$\frac{x}{V} = \frac{1}{2\sqrt{pc}} [-\ln(1 + e^{2\sqrt{pct}}) - \ln(1 + e^{-2\sqrt{pct}})] + c$$

When $t = 0$, $x = 0$, and

$$0 = \frac{1}{2\sqrt{pc}} (-\ln e - \ln 2) + c$$

Therefore

$$c = \frac{2 \ln 2}{2\sqrt{pc}} = \frac{1.386}{2\sqrt{pc}}$$

Finally

$$x = \frac{V}{2\sqrt{pc}} [1.386 - \ln(1 + e^{2\sqrt{pct}}) - \ln(1 + e^{-2\sqrt{pct}})] \quad \textbf{Answer.}$$

(*c*) Now going back: $dV/dt + cV^2 - p = 0$. And dividing through by dx/dt,

$$\frac{dV}{dt} + \frac{cV^2}{dx/dt} - \frac{p}{dx/dt} = 0$$

and since $dx/dt = V$,

$$\frac{dV}{dx} + cV - \frac{p}{V} = 0$$

Now separate variables:

$$\int \frac{dV}{-cV + p/V} = \int dx$$

Let $V^2 = \mu$, then $2V\,dV = d\mu$.

$$\int dx = \tfrac{1}{2} \int \frac{2V\,dV}{p - cV^2} = \tfrac{1}{2} \int \frac{d\mu}{p - c\mu}$$

$$x = \tfrac{1}{2} \left[\frac{1}{-c} \ln (p - c\mu) \right] + c$$

And therefore,

$$x = \tfrac{1}{2} \left[\frac{1}{-c\,(p - cV^2)} \right] + \text{constant} \cdot$$

When $x = 0$, $V = 0$, so that

$$0 = \tfrac{1}{2} \left[\frac{-1}{c} \ln p \right] + c$$

$$c = \frac{1}{2c} \ln p$$

and substituting, we have

$$x = \frac{1}{2c} [\ln p - \ln (p - cV^2)]$$ **Answer.**

A-12. A curved section of railroad track has a radius of 2,500 ft. As shown in Fig. A-6, track gage is 4 ft 8½ in. and the superelevation of the outside rail is 4 in. What is the maximum speed in miles per hour at which a train may negotiate the curve without having wheel flanges exert side thrust on the rails?

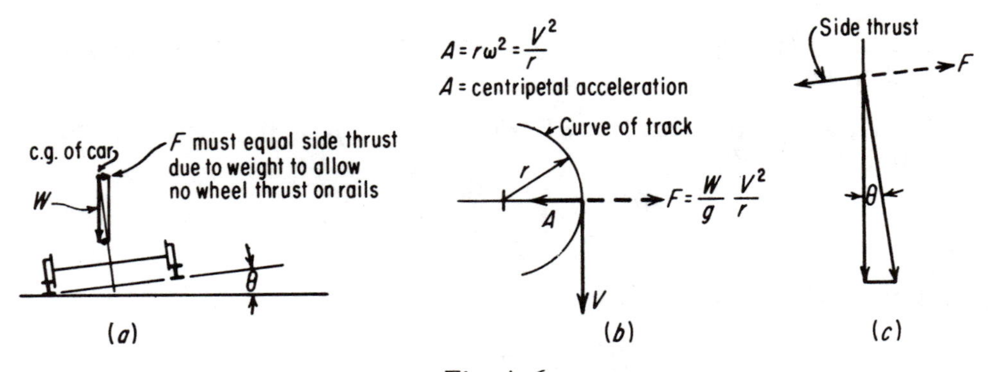

Fig. A-6

ANSWER. Refer to Fig. A-8*b* and *c*. Centripetal force F must equal side thrust due to weight to allow no wheel thrust on rails. Track gage expressed in inches = 56.5 in.

$$\theta = \arcsin 4/56.5 = \arcsin 0.0708$$

From vehicle dynamics,

$$\theta = \arctan \frac{F}{W} = \arctan \frac{W}{g} \times \frac{V^2/r}{W} = \arctan \frac{V^2}{gr}$$

For small angles tan = sin. Therefore, let $V^2/gr = 4/56.5$ and $V^2 = 5{,}700\ \text{ft}^2/\text{sec}^2$.

$$V = 75.5 \text{ ft per sec or } \mathbf{51.5\ mph}$$

A-13. A hanger for an equipment platform is to carry a load of 175 kips. Design an eyebar of A-440 steel.

ANSWER. Refer to Fig. A-7 for the notational system. Let subscripts 1 and 2 refer to cross sections through the body of the bar and through the center of the pinhole, respectively. From the AISC *Manual* for A-440 steel:

If $t \lesseqgtr 0.75$ in, $f_y = 50$ kips/in².

If $0.75 < t \lesseqgtr 1.5$ in, $f_y = 46$ kips/in².

If $1.5 < t \lesseqgtr 4$ in, $f_y = 42$ kips/in².

(*a*) Design the body of the member, using a trial thickness.

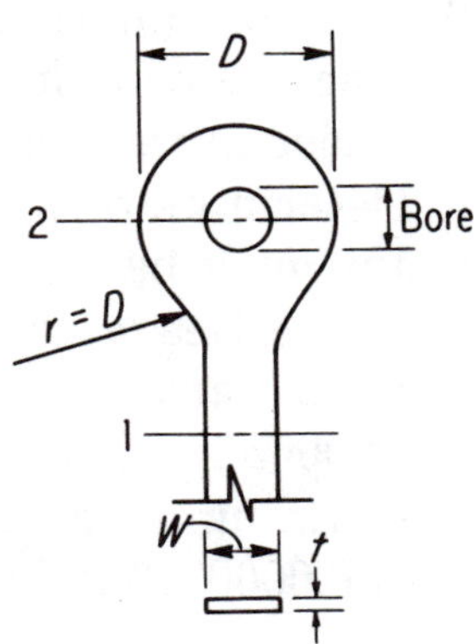

Fig. A-7

The specification restricts the ratio w/t to a value of 8. Compute the capacity P of a ¾-in eyebar of maximum width. Thus, $w = (8)(¾) = 6$ in; $f = 0.6\ (50) = 30$ kips/in²; $P = (6)(0.75)(30) = 135$ kips. This is not acceptable because the desired capacity is 175 kips. Hence, the required thickness exceeds the trial value of ¾ in. With t greater than ¾ in, the allowable stress at 1 is $0.60f_y$, or 0.60(46 kips/in²) = 27.6 kips/in², say 27.5 for design purposes. At 2 the allowable stress is 0.45(46) = 20.7 kips/in², say 20.5 kips/in² for design purposes.

To determine the required area at 1, use the relation

$$A_1 = \frac{P}{f}$$

where f = allowable stress as computed above. Then

$$A_1 = \frac{175}{27.5} = 6.36 \text{ in}^2$$

in which $A_1 = 6.5$ in². Use a plate 6½ × 1 in.

(*b*) Design the section through the pinhole. The ASCI specification limits the pin diameter to a minimum value of 7 ($w/8$). Select a pin diameter of 6 in. The bore will then be 6$\frac{1}{32}$ in in diameter. The net width required will be $P/ft = 175/(20.5 \times 1.0) = 8.54$ in. And $D_{min} = 6.03 + 8.54 = 14.57$ in. Set D = 14¾ in; $A_2 = (1.0)(14.75 - 6.03) = 8.72$ in². $A_2/A_1 = 1.34$. This result is OK, because the ratio A_2/A_1 must lie between 1.33 and 1.50.

Determine the transition radius r. In accordance with the specification, set $r = D$ = 14¾ in.

A-14. A rigid horizontal slab of uniform construction weighing 120,000 lb is first centrally supported by a 12-in-high steel column. Two aluminum columns 11.90 in high are symmetrically placed, one on each side of the steel column. All three columns are subjected to a rise in temperature to pick up the entire 120,000-lb load.

Determine to what minimum rise in temperature the three columns must be subjected to pick up the *entire 120,000-lb* load.

Properties of columns:

Steel—area = 10 in²; $E_s = 30 \times 10^6$; coefficient of expansion 0.0000065 in/(in) (°F).

Aluminum—area = 20 in²; $E_a = 9.6 \times 10^6$; coefficient of expansion 0.0000240 in/(in) (°F).

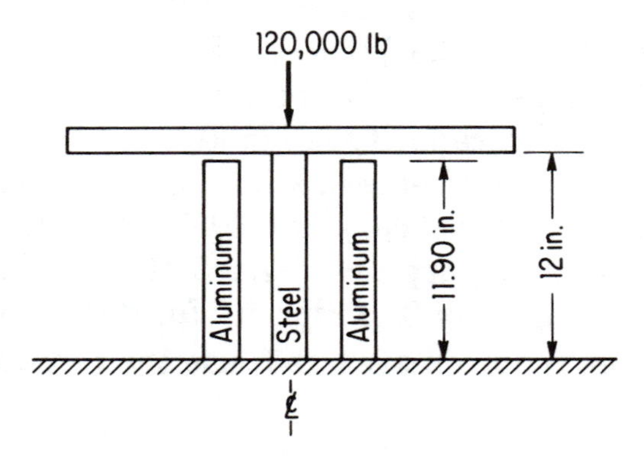

Fig. A-8

ANSWER. Refer to Fig. A-8. Aluminum columns will carry full load when their shortening under load just equals the differential thermal expansion over the steel column. Then, shortening of aluminum columns

$$\Delta_T = \frac{Pl}{AE} = \frac{120 \times 12}{40 \times 9600} = 3.75 \times 10^{-3} \text{ in}$$

Differential thermal expansion

$$\Delta_T = \Delta\alpha\Delta T\, l = (24 - 6.5)\, 10^6 \times 12\, \Delta T = 2.1 \times 10^{-4}\, \Delta T \text{ in}$$

Therefore, $\Delta_P = \Delta_T - 0.1$ in or $3.75 \times 10^{-3} = 2.1 \times 10^{-4}\, \Delta T - 0.1$.

$$\Delta T = \frac{0.1 + 3.75 \times 10^{-3}}{2.1 \times 10^{-4}} = \frac{1000 + 3.75}{2.1} = 494°\text{F} \qquad \textbf{Ans.}$$

Supplement **B**

MACHINE DESIGN

Questions and Answers

B-1. Calculate the torsional stress in the shaft of a blower from the following data: shaft diameter 2 in., speed 200 rpm, total head of air equivalent to 2 in. WG, volume discharged 30,000 cfm, and blower efficiency 60 per cent.

ANSWER. First determine the shaft horsepower, and then the torque, and finally the stress.

$$\text{Shaft horsepower} = (30{,}000/13.8 \times 2/12 \times 62.4/0.075)/(33{,}000 \times 0.6) = 15.8$$

where specific volume of air is 13.8 cu ft per lb and the term 62.4/0.075 converts feet of water to feet of air, the fluid flowing.

$$\text{Torque} = T = (15.8 \times 33{,}000)/(2\pi \times 200) = 415 \text{ lb-ft, or } 4{,}980 \text{ lb-in.}$$

Then,

$$\text{Stress} = S_s = \frac{T}{Z_p} = \frac{4{,}980}{\pi \times 0.5} = \mathbf{3{,}180\ psi}$$

B-2. A truck body lowers 5 in. when a load of 25,000 lb is placed on it. What is the frequency of vibration of the truck if the total spring-borne weight is 40,000 lb?

ANSWER. The frequency is $1/T = W/2\pi$ and the acceleration of the weight is

$$a = -W^2x$$

but $F = Ma$ and the force acting on the mass is $-kx$.

Then
$$-kx = -W^2Mx$$

and
$$W = \sqrt{\frac{k}{M}} = \sqrt{\frac{kg}{W}}$$

Now we see that frequency f is determined from the following:

$$f = \frac{W}{2\pi} = \frac{1}{2\pi} \times \sqrt{\frac{kg}{W}}$$

where k is the spring constant and is equal to 25,000/5 in. = 5,000 lb per in.

$$f = \frac{1}{2\pi}\sqrt{\frac{5{,}000 \times 12 \times 32.2}{40{,}000}} = \textbf{1.105 oscillations per sec}$$

B-3. A mass weighing 25 lb falls a distance of 5 ft upon the top of a helical spring having a spring constant k of 20 lb per in. (*a*) What will be the velocity of the mass after it has compressed the spring 8 in.? (*b*) Calculate the maximum compression of the spring.

ANSWER. The fall of the weight will be 5 ft *plus* 8 in. when it has compressed the spring 8 in. At that time the spring will have absorbed energy equal to $\frac{1}{2}ks^2 = \frac{1}{2} \times 20 \times 8^2 = 640$ lb-in., or 53.3 lb-ft.

The change in potential energy is equal to

$$\Delta\text{PE} = W \times h = 25(5 + 0.67) = 142 \text{ lb-ft}$$

Part of this energy has been absorbed by the spring, the remainder 142 − 53.3, or 88.7 lb-ft, must remain as kinetic energy of the mass equal to $\frac{1}{2}Mv^2$. The velocity of the mass after it has compressed the spring is

$$v = \sqrt{\frac{88.7 \times 2}{M}} = \sqrt{\frac{88.7 \times 2g}{25}} = \textbf{15.1 fps}$$

When the spring has been fully compressed, the velocity of the mass will become zero and all the change in potential energy of the mass will have been converted into potential energy of the spring.

$$W(5 \times 12 + s) = \tfrac{1}{2} \times 20 \times s^2 = 60 \times 25 + 25s = 10s^2$$

Rearranging and dividing by 5 throughout, we obtain

$$2s^2 - 5s - 300 = 0$$

By means of the binomial theorem,

$$s = \frac{5 \pm \sqrt{25 + 2{,}400}}{4} = \frac{5 \pm 49.3}{4} = \textbf{13.6 in.}$$

B-4. Investigate the acceptability of the fan and motor drive shown from the standpoint of possible excessive vibration. Motor: a-c 110 volts, 60 cycles, 2,400 rpm. Armature: 40 lb. Radius of gyration: 5 in. Three-blanded fan: wt 10 lb; radius of gyration 9 in. Shaft is of steel. If design is not acceptable, indicate what changes should be made.

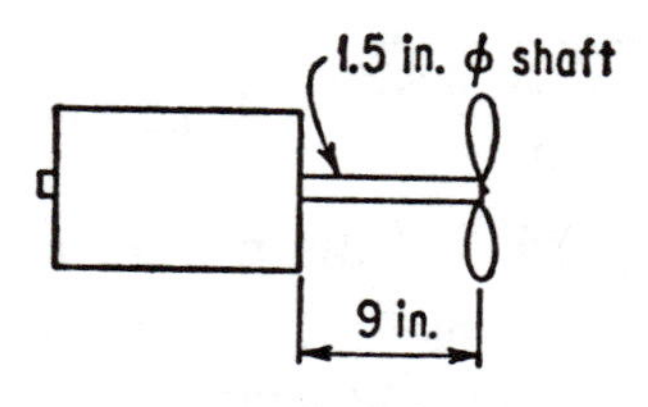

Fig. B-1

ANSWER. Refer to Fig. B-1. The frequency of the fan assembly is

$$f = \frac{1}{2\pi}\sqrt{\frac{(I_f + I_m)k}{I_f I_m}}$$

For the fan:

$$I_f = (10 \text{ lb}/32.2)(1/12)(9^2) = 2.1 \qquad \text{(moment of inertia)}$$

For the motor:

$$I_m = (40 \text{ lb}/32.2)(1/12)(5^2) = 2.68 \qquad \text{(moment of inertia)}$$

Also

$$k = 11.5 \times 10^6/9 \times 0.495 = 6.3 \times 10^5 \qquad \text{(torsional constant)}$$

$$f = \frac{1}{2\pi}\sqrt{\frac{(2.1 + 2.68)(6.3 \times 10^5)}{2.1 \times 2.68}} = 116 \text{ cps}$$

Motor frequency is 2,400/60 = 40 cps. We may get excessive vibration when the system starts or is shut down if the rate of increase (or decrease) in speed is small, say around 40 cps.

In order to design the system without danger of excessive vibration, we should increase radius of gyration of both fan and motor armature and increase shaft diameter or decrease its length until frequency of entire system is below 40 cps.

B-5. A 22-coil squared and ground spring is designed to fire a 10-lb projectile into the air. The spring has a 6-in. diameter coil with $\frac{3}{4}$-in. diameter wire; it has a free length of 26 in. and is compressed 8 in. when loaded or set. The shear elastic limit for the spring material is 85,000 psi. The spring constant Wahl factor is 1.18. The shear modulus of elasticity of the spring material is

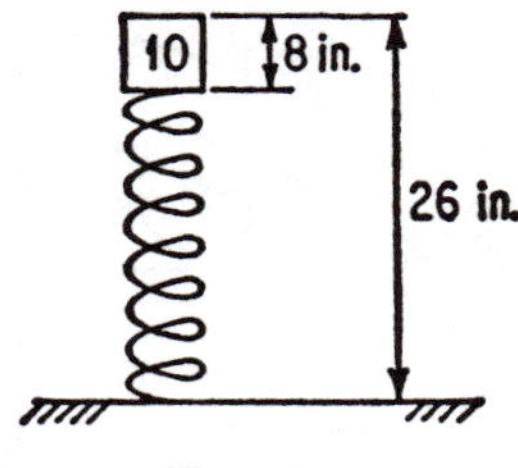

Fig. B-2

12,000,000. (*a*) Determine the height to which the projectile will be fired. (*b*) Determine the safety factor for this spring (see Fig. B-2).

ANSWER. Let the following be understood:

n = 22 coils
r = 3 in.
d = 0.75 in.
k = 1.18
$G = 12 \times 10^6$ psi
$\bar{\imath}$ = 85,000 psi

(*a*) The force to which the spring is stressed when loaded is calculated to be

$$P = 0.1963 \frac{0.75^3 \times 34{,}100}{3 \times 1.18} = 800 \text{ lb}$$

$$10 \text{ lb wt} \times \text{distance fired} = \frac{P \times \text{deflection}}{2}$$

$$10 \times s = (800 \times 8)/2 \qquad \text{and} \qquad s = \frac{800 \times 8}{2 \times 10} = \mathbf{320\ in.}$$

The actual stress is found to be

$$\frac{8 \text{ in.} \times 12 \times 10^6 \times 0.75 \text{ in.} \times 1.18}{4 \times 22 \times \pi \times 3 \text{ in.}^2} = 34{,}100 \text{ psi}$$

(*b*) The factor of safety is

$$\frac{\text{Stress at elastic limit}}{\text{Actual stress}} = \frac{85{,}000}{34{,}100} = \mathbf{2.5}$$

B-6. A rod shown in Fig. B-3 forms a circular path *ABC* for a slider *S*. It is connected to *A* by a spring normally 2.5 ft long with a spring constant of 1 in. per lb force applied. The block is held at

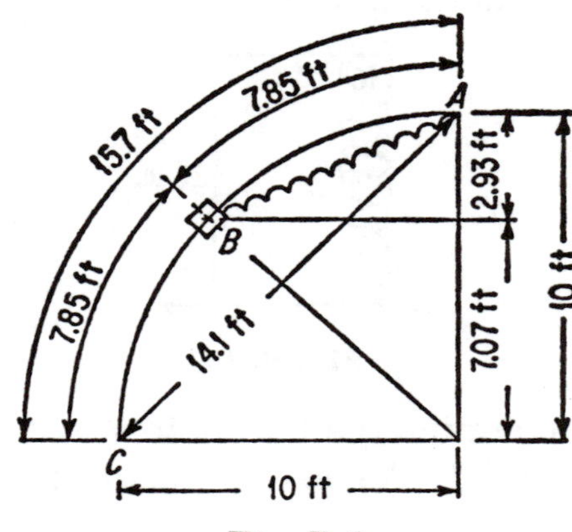

Fig. B-3

C and then released. What work has been done on the block when it reaches *B*, midway between *C* and *A*?

ANSWER. With slider in position *C* the force pulling is F_1. This is equal to

$$F_1 = (14.1 - 2.5)12 = 139 \text{ lb}$$

The horizontal and vertical components of the force are F_{1x} and F_{1y}, respectively.

$$F_{1x} = 139 \times 0.707 = 98.3 \text{ lb} = F_{1y}$$

With slider in position *B* as required,

$$F_2 = (7.65 - 2.5)12 = 62 \text{ lb}$$

where 7.65 is chord distance, in feet.

$$F_{2x} = 62 \times (7.07/7.65) = 57.3 \text{ lb}$$
$$F_{2y} = 62 \times (2.93/7.65) = 23.8 \text{ lb}$$

Then the work done is

$$\{[(98.3 + 57.3)2.93] + [(98.3 + 23.8)7.07]\}\tfrac{1}{2} = \mathbf{660\ in.\text{-}lb}$$

B-7. It is known that a certain piece of machinery weighing 2 tons creates a disturbing force of 1,200 lb at a frequency of 2,000 cpm. The machine is to be supported on six springs, each taking an equal share of the load in such a way that the force transmitted to the building is not to exceed 20 lb. Assume that the machine is guided so that it may move only in a vertical direction. Calculate the required scale of each spring.

ANSWER. The natural frequency of free vibration of the system in question is given by

$$f = \frac{1}{2\pi}\sqrt{\frac{kg}{W}}$$

where k = spring modulus, lb per in.
g = gravitational constant, in. per sec²
W = weight on each spring, lb

The amplitude of the forced vibration is given by A in the following equation:

$$A = x_s \frac{1}{1 - (f_1/f)^2}$$

where x_s = displacement of spring caused by steady-state vibration
f_1 = $W/2\pi$, or frequency of exciting force, cps
f = natural frequency of system, cps

The ratio f_1/f is called the frequency ratio, and the ratio $1/[1 - (f_1/f)^2]$ may be interpreted as the *magnification factor*. Now, because of proportionality factors, we can say

$$\frac{20}{1{,}200} = \frac{1}{1 - \left[\dfrac{(2{,}000/60)^2}{f^2}\right]}$$

For a reduction in the disturbing force which we are after, the ratio A/x_s is negative. Now solve for f and this is found to be equal to 4.28 cps. Finally, substitute f in the first equation for natural frequency and

$$4.28 = \frac{1}{2\pi}\sqrt{\frac{k \times 386}{4{,}000/6}}$$

and k is found to be equal to **1,240 lb per in.**

B-8. An automobile has main helical springs that are compressed 6 in. by the weight of the car body. If the axles of the wheels of the auto are clamped to a test platform and the platform is given a vertical harmonic motion having an amplitude of 1 in. and a frequency of 1 cps, determine the amplitude of the motion of the body of the car. Assume that there are no shock absorbers and thus the vibration takes place without damping. What is the maximum shortening of the spring?

ANSWER. The frequency of the free vibration is f in accordance with

$$f = \frac{1}{2\pi}\sqrt{\frac{g}{6}} = \frac{1}{2\pi}\sqrt{\frac{386}{6}} = 1.28 \text{ cps}$$

Since x_s is 1 in. and f_1 is 1 cps, the amplitude A of the forced vibration is

$$A = x_s \frac{1}{1 - (f_1/f)^2} = 1 \times \frac{1}{1 - (1/1.28)}\, 2 = \mathbf{2.56 \text{ in.}}$$

Since the frequency of the impressed motion is below the resonant frequency, the motion of the axles is in phase with the motion of the body of the car. Hence, the change in the length of the spring is 2.56 − 1, or 1.56 in. The maximum shortening of the spring, therefore, is 1.56 + 6 = **7.56 in.**

B-9. A light artillery gun is mounted on a wheeled carriage; combined weight of gun and carriage is 3,000 lb. The gun fires a 60-lb projectile which has a velocity of 1,500 fps just as it leaves the gun. The angle at which the projectile is fired is 30° above the horizontal. Neglecting the recoil due to powder acceleration, (*a*) determine the velocity of the gun and carriage just after the projectile leaves the gun, and (*b*) determine the required modulus of a spring which is to be used as a "bumper" to bring the recoiling carriage to rest. The spring is to deflect no more than 18 in.

ANSWER. The momentum of the gun and carriage backward must be equal to the momentum of the projectile forward. Thus,

$$M_p v_p \cos\theta + M_g v_g = 0$$

where subscripts p and g are for projectile and gun (plus carriage).

(*a*) Velocity of gun (and carriage) is found by rearranging the equation and

$$v_g = \frac{60 \times 1{,}500 \times 0.8660}{3{,}000} = \textbf{26 fps in a backward direction}$$

B-10. The impeller of a slow-speed centrifugal pump has a mass moment of inertia of 5 in.-lb-sec². The pump is driven by an electric motor that has a mass moment of inertia of 10 in.-lb-sec². A steel shaft whose diameter is 1 in. and whose length is 12 in. connects pump and motor. At a constant speed of 120 rpm, 5 hp are transmitted through the shaft with no measurable vibration. A steel rod enters the impeller eye along with the liquid and jams the impeller, causing the rotor to stop rotating instantly. At the same time a circuit breaker opens the electrical circuit to the motor and the motor stops instantly. What is the resulting maximum shear stress in the shaft?

ANSWER. The resulting stress is the sum of the normal running stress and that due to torsional impact because of the steel rod. The shearing stress during normal operation is

$$f_s = \frac{\text{Torque} \times 16}{\pi d^3}$$

The torque must first be determined from the equation

$$\frac{\text{Hp} \times 33{,}000}{2\pi N} \times 12$$

Then
$$T = \frac{5 \times 33{,}000 \times 12}{2\pi \times 120} = 2{,}650 \text{ in.-lb}$$

$$f_s = \frac{2{,}650 \times 16}{\pi 1^3} = 13{,}500$$

On impact, all the kinetic energy of the rotating mass is instantly converted into potential energy which would tend to disrupt the shaft. We may neglect the kinetic energy of the shaft and the mass moment of inertia of the impeller to simplify the calculations. Now, the kinetic energy of the motor is

$$E_m = \tfrac{1}{2}I\omega^2 = \tfrac{1}{2} \times 10 \times (4\pi)^2 = 788 \text{ in.-lb}$$

where the angular velocity is ω equal to $2\pi \times (^{120}\!/_{60})$, or 4π radians per sec. And I is equal to the mass moment of inertia of motor as 10 in.-lb-sec.² The shearing stress on the shaft is found by use of

$$S_s = \sqrt{\frac{4E_sE_k}{V_{\text{shaft}}}} = \sqrt{\frac{4(12 \times 10^6)788}{9.45 \text{ cu in.}}} = 63{,}200 \text{ psi}$$

Finally, the maximum shear stress in the 1-in. shaft is

$$13{,}500 + 63{,}200 = \mathbf{76{,}700\ psi}$$

B-11. It has been estimated that 4,000 ft-lb of energy must be supplied at the drive shaft of a belt-driven punch press to blank satisfactorily a hole in a sheet of 115-lb coke tin. Of the total energy required, 800 ft-lb are supplied by the belt and the remainder by the flywheel. The weight of the flywheel must be such that its speed will not vary more than 10 per cent from its maximum speed of 200 rpm. Assuming the radius of gyration to be 18 in., calculate the weight of the flywheel.

ANSWER. Energy supplied by flywheel is 4,000 ft-lb total minus 800 ft-lb supplied by belt, leaving 3,200 ft-lb to be supplied by flywheel. The energy given up by the flywheel is dependent upon its change in rpm.

$$3{,}200 = \frac{W}{2g} \times \rho^2 \times (\omega_1^2 - \omega_2^2)$$

Angular velocity $\omega_1 = 200 \times 2\pi/60 = 200 \times 2\pi/60 = 21$ radians per sec. Angular velocity ω_2 is similarly found to be equal to 18.9 radians per sec. Then substituting in the equation

$$3{,}200 = \frac{W}{64.4} \times \left(\frac{18}{12}\right)^2 (21^2 - 18.9^2)$$

from which W is found to be equal to **1,090 lb.**

B-12. One of the routine steps in preliminary design of a piece of turbomachinery is to compare the critical and operating speeds. Such a routine check disclosed that a centrifugal fan would be operating at 98 per cent of its first critical speed.

(*a*) What would be the effect on critical speed caused by decreasing bearing span 10 per cent?

(*b*) What would be the effect on critical speed caused by increasing shaft diameter 10 per cent?

(*c*) What would be the effect on critical speed caused by increasing rotor weight 10 per cent?

Approximate quantitative results are desired.

ANSWER. Speed of rotation of a shaft and the rotor attached to it at which resonance occurs is frequently called the critical speed for the shaft. In general, it is important to know the critical speed of a rotating member so that speed of operation can be maintained either considerably above or below this dangerous (critical) speed, or so that the member can be designed for an operating speed that will not be too near to the critical speed. To calculate the critical speed, first determine equivalent spring modulus of the shaft with a concentrated load W (fan wheel) at the center. Spring modulus $k = W/\Delta$, for which Δ is static deflection for a beam with concentrated load *at center*,

$$\Delta = \frac{Wl^3}{48EI} \text{ in.}$$

where l is distance between bearings, in.; E is modulus of elasticity (for steel it's 30×10^6 psi); and I is moment of inertia for the circular cross-sectional area of diameter d, that is, $\pi d^4/64$. Spring modulus may be found to be equal to $k = 48EI/l^3$, lb per in. If we neglect weight of shaft, the natural frequency or critical speed is

$$S_c = \frac{60}{2\pi}\sqrt{\frac{kg}{W}} = \frac{60}{2\pi}\sqrt{\frac{48EI}{l^2} \times \frac{386}{W}}$$

(*a*) By decreasing span length 10 per cent, with all other items constant, we can apply a correction factor to the original S_{c1}. Thus,

$$S_{c2} = S_{c1}\sqrt{\frac{1}{l^3}} = S_{c1}\sqrt{\frac{1}{(1 - 0.1)^3}} = S_{c1} \times 1.16$$

Critical speed will be **increased 16 per cent.**

(*b*) By increasing shaft diameter 10 per cent

$$S_{c2} = S_{c1}\sqrt{d^4} = S_{c1}\sqrt{1.1^4} = S_{c1} \times 1.22$$

Critical speed will be **increased 22 per cent.**

(*c*) By increasing rotor weight 10 per cent

$$S_{c2} = S_{c1}\sqrt{\frac{1}{W}} = S_{c1}\sqrt{\frac{1}{1.1}} = S_{c1} \times 0.953$$

Critical speed will be decreased **4.7 per cent.**

B-13. A motor weighing 100 lb is mounted on four springs, each having a spring constant k equal to 15 lb per in. What is the natural frequency and period of the motor as it vibrates?

ANSWER. Each spring carries 25 lb, assuming uniform distribution.

$$f = \frac{1}{2\pi}\sqrt{\frac{kg}{W}} = \frac{1}{2\pi}\sqrt{\frac{15 \times 386 \text{ in. per sec}^2}{25}} = \frac{15}{6.28} = \mathbf{2.4\ cps}$$

The period is $1/f = 1/2.4 =$ **0.417 sec.**

B-14. *What is the natural frequency of a shaft* of weight W simply supported and of length l? How can this approach be applied to similar systems?

ANSWER. This is an example of an evenly distributed weight that vibrates because of the elasticity of the shaft rather than that of a spring. To solve for the frequency f, which equals $(1/2\pi)\sqrt{g/\Delta}$, it is necessary to determine the static deflection Δ of the system. We know that a simply supported shaft (or beam) has $\Delta = (5Wl^3)/(386EI)$. Where E is the tensile modulus of elasticity and I is the moment of inertia of the cross-sectional area about the neutral axis. Finally

$$f = \frac{1}{2\pi}\sqrt{\frac{386 \times E \times I}{5Wl^3}} \qquad \text{cps}$$

Note 386 is gravitational constant g, in. per sec². Thus, for similar systems merely substitute for Δ the static deflection from tables or calculated values.

B-15. A variable-speed motor is mounted on an elastic beam at the center of the beam span. The motor weighs 23 lb and drives a shaft which has a 2.0-lb eccentric weight located 1.0 ft from the shaft center. When the motor is not running, the combined motor and eccentric weight causes a 0.50-in. deflection in the supporting beam. (*a*) Determine the speed of the system at resonance. (*b*) Determine the amplitude of the forced vibration when the motor is running at 360 rpm.

ANSWER. Refer to Fig. B-4. Rotor revolves with angular velocity ω (radians per sec) and unbalanced mass M at a distance r (ft) from axis of rotation, causes a rotating unbalance (centrifugal) force equal to $Mr^2\omega = P_0$. In a forced vibration, if the period of the impressed force is the same as that of the free or natural period of vibration of the system, the theoretical amplitude of the vibration

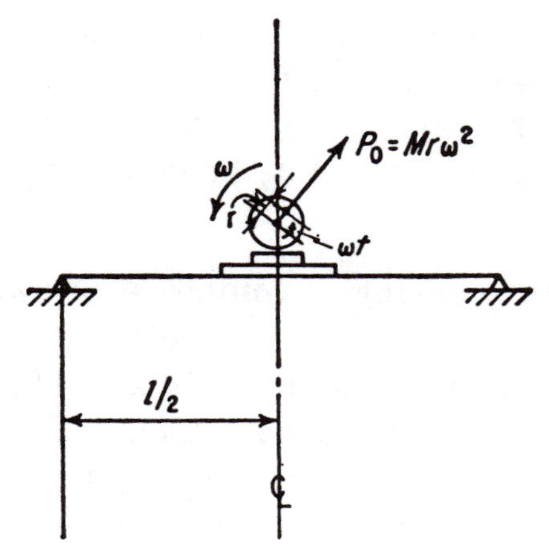

Fig. B-4

becomes exceedingly large. This condition is known as *resonance* and is, of course, to be avoided in parts of machines and structures, i.e., natural frequency equals resonant frequency, $f_n = f_r$. Thus

$$f_r = f_n = \frac{1}{2\pi}\sqrt{\frac{kg}{W}} = \frac{1}{2\pi}\sqrt{\frac{g}{\Delta}} = \frac{1}{2\pi}\sqrt{\frac{386}{0.5}} = 4.42 \text{ cps}$$

(*a*) Rotating speed at resonance $= f_r \times 60 =$ **265.2 rpm**

(*b*) The amplitude of the forced vibration is the product of the static deflection caused by the rotating unbalanced force and the *amplification factor*

$$\frac{P_0}{k}\left[\frac{1}{1-(f_1/f_n)^2}\right] \text{ in.}$$

where f_1/f_n is frequency ratio. Continuing

$$P_0 = \frac{W}{g}r\omega^2 = \frac{2}{32.2} \times 1 \times \left(\frac{2\pi \text{ rpm}}{60}\right)^2 = 89 \text{ lb}$$

Then
$$f_1 = \frac{\omega}{2\pi} = \frac{12\pi}{2\pi} = 6 \text{ cps}$$

Also
$$f_1 = (\text{rpm}/60) = {}^{360}\!/_{60} = 6 \text{ cps}$$

Finally

$$\frac{89}{W/\Delta}\left[\frac{1}{1-(6/4.42)^2}\right] = \frac{89}{(25/0.5)}\left[\frac{1}{1-1.85}\right]$$
$$= 1.78(-1.178) = \mathbf{-2.1 \text{ in.}}$$

The value of the amplification factor can be positive (+) or negative (−) depending on whether $f_1 < f_n$ or not. At resonance ($f_1 = f_n$) the amplitude is theoretically infinite. Actually, however, the damping, which is always present, holds the amplitude to a finite amount.

B-16. A heavy-duty spring is composed of two concentric coils. The outside coil, with an outside diameter of 9 in., has six active coils of 1½-in.-diameter round bar stock. The inner spring has nine active coils of 1-in. round bar stock. The outside diameter of the inner coils is 5½ in. The free height of the outer spring is ¾ in. greater than that of the inner spring. For a total load of 20,000 lb, find the deflection of each spring and the load carried by inner and outer coils.

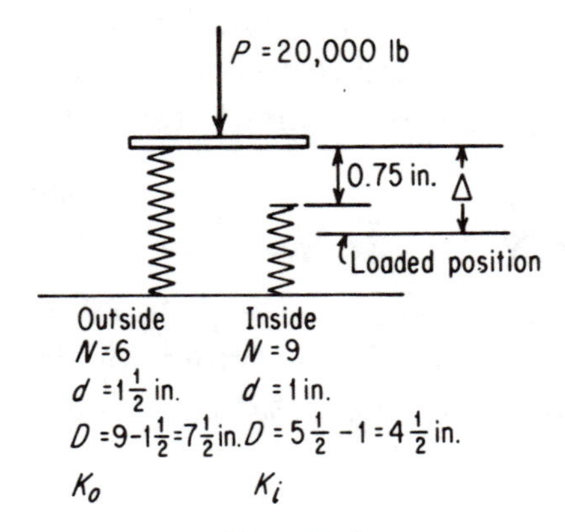

Fig. B-5

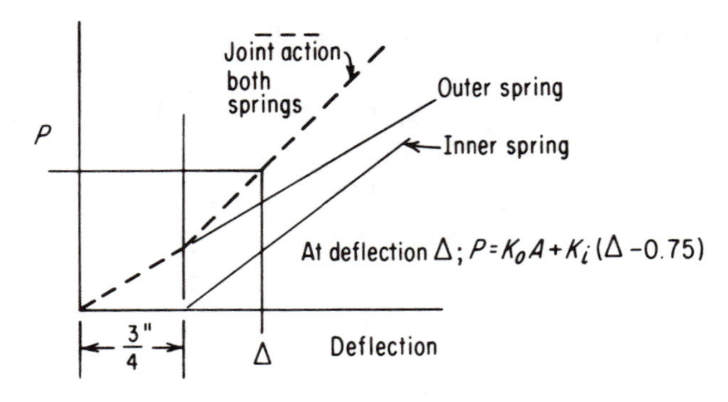

Fig. B-6

ANSWER. Refer to Figs. B-5 and B-6. In general, $K = d^4G/8D^3N$ for spring constant. G for carbon steel is 11.5×10^6 psi. Therefore,

$$K_0 = \frac{(1.5)^4 \times (11.5 \times 10^6)}{8 \times 7.5^3 \times 6} = 2{,}870 \text{ lb per in.}$$

$$K_i = \frac{(1^4) \times (11.5 \times 10^6)}{8 \times 4.5^3 \times 9} = 1{,}750 \text{ lb per in.}$$

Since $P = K_0\Delta + K_i(\Delta - 0.75)$, then

$$20{,}000 = 2{,}870\Delta + 1{,}750(\Delta - 0.75)$$

$$\Delta = 21{,}310/4{,}620 = \mathbf{4.6\ in.}$$

Outer spring $\Delta_0 = 4.6$ Inner spring $\Delta_i = 4.6 - 0.75 = 3.85$

Therefore,

$$P_0 = K_0\Delta_0 = \mathbf{13{,}200\ lb} \qquad P_i = K_iA_i = \mathbf{6{,}720\ lb}$$

B-17. A flexible coupling has an active element in the form of a torsional helical spring. At 900 rpm, 25 hp is transmitted. To what radius of helix must the spring be wound if it is made of ¾-in.-square wire; and if the permissible stress is 30,000 psi, how many active coils must there be if the torsional deflection is not to exceed 5 degrees?

ANSWER.

$$T = \frac{63{,}000 \times \text{hp}}{N} = \frac{63{,}000 \times 25}{900} = 1{,}750 \text{ in.-lb}$$

$$Ks = \frac{6M}{bh^2}\,\frac{3c_1^2 - c_1 - 0.8}{3c_1(c_1 - 1)}$$

$$30{,}000 = \frac{6 \times 1{,}750}{0.75^3}\,\frac{3c_1^2 - c_1 - 0.8}{3c_1(c_1 - 1)}$$

$c_1^2 - 4.254c_1 + 1.302 = 0 \qquad c_1 = 3.922$

$$R = \frac{c_1d}{2} = \frac{3.922 \times 0.75}{2} = \mathbf{1.47\ in.}$$

$$\phi = \frac{Ml}{EI} \qquad l = \frac{\phi EI}{M} = \frac{\phi Ebh^3}{12M} = \frac{5\pi}{180} \times \frac{30 \times 10^6 \times 0.75^4}{12 \times 1{,}750} = 39.4 \text{ in.}$$

$$N = \frac{l}{2\pi R} = \frac{39.4}{2\pi 1.47} = \mathbf{4.27\ coils}$$

where c_1 = spring index
d = diameter of wire
E = modulus of elasticity
I = moment of inertia
K = spring rate
N = number of active coils
R = mean radius of helix
T = torque

Reference: M. F. Spotts, Design of Machine Elements, 3d ed. Chapter 4, "Springs," Prentice-Hall (1961).

Supplement C

GEARING

Questions and Answers

C-1. A 750-hp 500-rpm motor geared at 1:10 ratio to a hoist having a cylindrical drum 8 ft in diameter is used to raise coal vertically from a mine. At what percentage of full load is the motor operating when raising a 6-ton load at uniform speed? Neglect frictional losses in the hoist.

ANSWER. The drum shaft speed is $n = N/10 = 500/10 = 50$ rpm. The vertical load speed is v.

$$v = \pi n D = \pi \times 50 \times 8 = 1{,}258 \text{ fpm}$$

The horsepower to raise the load $= Wv/33{,}000 = 6 \times 2{,}000 \times 1{,}258/33{,}000 = 457$.

Per cent of full load on motor is $457/750$, or **61 per cent.**

C-2. Show the arrangement of gears in a simple lathe for cutting screw threads, and calculate proper gear sizes to cut 6 threads per inch, the lead screw having a pitch of 1/8 in.

ANSWER. Let the following nomenclature apply:

N_c = turns in stud gear
T_c = number of teeth in stud gear
N_s = turns in screw gear
T_c = number of teeth in screw gear

For the tool to advance 1 in., the screw must turn 1/(1/8), or 8 times. But for 8 turns of the screw, the spindle must make 6 turns, or

$$\frac{N_c}{N_s} = \frac{6}{8} = \frac{T_s}{T_c}$$

If we assume T_s is 60, then T_c is equal to $(8/6)T_c$, or 80 teeth. If a 42-tooth gear is available, T_c is equal to 56 teeth. The idler may have any number of teeth, but will be selected as to keep the line of gear centers (A-A) straight (see also Fig. C-1).

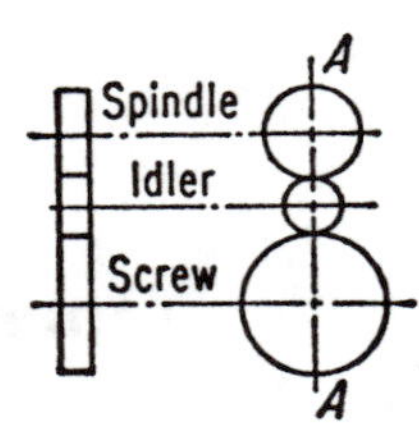

Fig. C-1

C-3. An automobile motor has a torque of 230 lb-ft driving the rear axles through a differential having a ratio of 4.11 and an efficiency of 97 per cent. Calculate the tractive effort of the rear wheels, their diameters being 30 in. What rear-wheel horsepower is developed when speed is 40 mph?

ANSWER. Let the following nomenclature apply:

N_m = motor rpm
N_a = rear-axle rpm
T_m = motor torque, lb-ft
T_a = rear-axle torque, lb-ft
P = tractive effort, lb
R = rear-wheel radius, ft

Input to differential $= 2\pi N_m T_m$
Differential output $= 2\pi N_a T_a$
Differential efficiency is therefore

$$\frac{2\pi N_a T_a}{2\pi N_m T_m} = 0.97$$

Canceling, $N_a/N_m \times (T_a/T_m) = 0.97$. And now rearranging and solving for T_a since $N_m/N_a = 4.11$,

$$T_a = T_m\left(\frac{N_m}{N_a}\right)0.97 = 230 \times 4.11 \times 0.97 = 916 \text{ lb-ft for both wheels}$$

For one wheel:

$P = (916/2)(1/30)(2)(12) = 366$ lb, or **732 lb both wheels**

Rear-wheel horsepower:

$$\frac{2\pi N_a T_a}{33{,}000}$$

But we must first find N_a.

$$N_a = (40/60)(5{,}280/\pi)(1/30)(12) = 448 \text{ rpm}$$
$$2\pi(448)(916)/(33{,}000) = \mathbf{78.2\ rear\text{-}wheel\ hp}$$

C-4. A tensile-testing machine applies the tension through two screws actuating a drawhead or nut, the maximum force on each one being 50,000 lb. The length of the nut in which the screw revolves is 6 in. and the pitch of the screw is ½ in. The depth of the flat thread on the screw is ¼ in. The pressure on the lubricant between threads is 2,000 psi, this assumption being the basis of the calculation. Calculate the mean and root diameters of the screw and the stress on it.

ANSWER: Let the following nomenclature apply:

W = maximum load on each screw, lb.
h = length of nut, in.
d_0, d_r, d_m = outside, root, and mean diameter of thread, in.
L = lead of screw, in.
n = number of threads per in.
w = pressure between threads, psi
μ = coefficient of friction
T = twisting moment on screw, lb-in.
S_t = tensile stress in screw, psi

Then

$$W = wnh(\pi/4)(d_0^2 - d_r^2)$$

By rearrangement

$$(d_0^2 - d_r^2) = \frac{4W}{wnh\pi} = \frac{4 \times 50{,}000}{2{,}000 \times 12 \times 3.14} = 2.65$$

From the problem

$$d_0 = d_r + 0.5 \qquad \text{also} \qquad d_r = 2.65 - 0.25 = \mathbf{2.4\ in.}$$
$$d_0 = 2.4 + 0.5 = 2.9 \text{ in.}$$
$$d_m = \frac{d_0 + d_r}{2} = \frac{2.9 + 2.4}{2} = \mathbf{2.65\ in.}$$

Assuming μ to be 0.15, we have the twisting moment T to be equal to

$$T = Wr_m \frac{1 + \mu\pi d_m}{\pi d_m - L\mu}$$

$$T = 50{,}000 \times (2.65/2)[(1 + 0.15 \times 3.14 \times 2.65)/(3.14 \times 2.65 - 0.5 \times 0.15)]$$

$$= 10{,}080 \text{ lb-in.} \qquad \text{where } r_m \text{ is mean radius}$$

$$S_s = \frac{16T}{\pi d_m^3} = \frac{16 \times 10{,}080}{3.14 \times 2.65^3} = 2{,}750 \text{ psi}$$

$$S_t = \frac{4W}{\pi d_r^2} = 50{,}000 \times \frac{4}{3.14 \times 2.4^2} = 11{,}050 \text{ psi}$$

Maximum shear stress:

$$\tfrac{1}{2}(S_t^2 + 4S_s^2)^{1/2} = \tfrac{1}{2}(11{,}050^2 + 4 \times 2{,}750^2)^{1/2} = \mathbf{6{,}175 \text{ psi}}$$

Maximum tensile stress:

$$\tfrac{1}{2}[S_t + (S_t^2 + 4S_s^2)^{1/2}] = \tfrac{1}{2}[11{,}050 + (11{,}050^2 + 4 \times 2{,}750^2)^{1/2}] = \mathbf{11{,}700 \text{ psi}}$$

C-5. As shown, a horizontal beam is supported by two rods: one of aluminum 50 in. long and ¾ in. in diameter, the second of steel 36 in. long and ⅜ in. diameter. Rods are 50 in. apart. For a load of 3,000 lb applied at a distance x from the left rod, find, if the beam is to remain horizontal, (*a*) distance x and (*b*) stress induced in each rod.

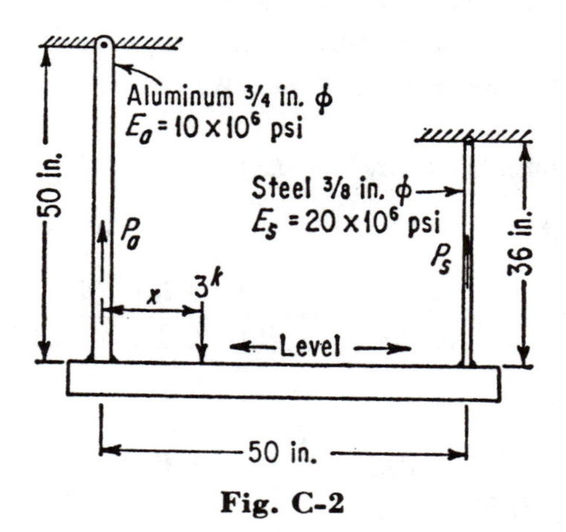

Fig. C-2

ANSWER. See Fig. C-2. For the beam to remain horizontal the total change in length of each rod Δl under load must be the same, i.e., Δl_a equals Δl_s. The following relation will apply

$$\Delta l_a = \frac{P_a L_a}{A_a E_a} = \Delta l_s = \frac{P_s L_s}{A_s E_s}$$

where P_a = load on aluminum rod, lb
L_a = original length of aluminum rod, in.
A_a = cross section of aluminum rod, sq in.
E_a = modulus of elasticity for aluminum, psi
P_s, L_s, A_s, E_s are similarly for the steel rod

(a) Neglect weight of beam and establish relation between P_a and P_s.

$$\frac{P_a L_a}{A_a E_a} = \frac{P_s L_s}{A_s E_s}$$

$$\frac{P_a \times 50}{0.785(\frac{3}{4})^2 \times 10 \times 10^6} = \frac{P_s \times 36}{0.785(\frac{3}{8})^2 \times 20 \times 10^6}$$

from which $P_a = 1.42P_s$. From the problem we also know that $P_a + P_s = 3{,}000$ lb, and combining both relations of P_a and P_s,

$$1.42P_s + P_s = 3{,}000 \qquad P_s = 3{,}000/2.42 = 1{,}240 \text{ lb}$$
$$P_a = 1.42P_s = 1.42 \times 1{,}240 = 1{,}760 \text{ lb}$$

In order to establish dimension of x, refer to Fig. C-2. Taking summation of moments about P_a equal to zero

$$3{,}000 \times x - P_s \times 50 = 0$$
$$3{,}000 \times x - 1{,}240 \times 50 = 0$$
$$x = 62{,}000/3{,}000 = \mathbf{20.66 \text{ in.}}$$

(b) $S_a = \dfrac{P_a}{A_a} = \dfrac{1{,}760}{0.785(\frac{3}{4})^2} =$ **4,000 psi** for the **aluminum**

$S_s = \dfrac{P_s}{A_s} = \dfrac{1{,}240}{0.785(\frac{3}{8})^2} =$ **11,200 psi** for the **steel**

C-6. What is the minimum length a cooling conveyor must be to take a production of 180 pieces per hour, in a single row, allowing each piece 10 min for cooling, if the articles manufactured are 15 in. in diameter and spaced at 3-in. intervals? What sprocket ratio must be used between 7.2 rpm output gear head and 12-in.-diameter conveyor drive pulley?

ANSWER. The problem is to determine the number of articles to be handled every 10 min and with the size and spacing, the minimum conveyor length can be found. The sprocket ratio can be found from the conveyor speed and drive-pulley rpm. 180 pieces per hour is equivalent to 30 pieces every 10 min that the conveyor can handle. If the articles are 15 in. in diameter and 3 in. apart, each article requires 18 in., or 1.5 ft. The minimum effective length of conveyor without excess loading and unloading return or overlap should be $30 \times 1.5 = 45$ ft. Total length with 12-in. pulleys and allowing 6.68 ft for loading and unloading will probably be

$$2 \times 45 + 1 \times 3.14 + 6.86 = \mathbf{100\ ft}$$

The conveyor speed, which is also the periphery speed of the drive pulley, is

$$45/10 = 4.5 \text{ fpm}$$

Since the drive pulley diameter is 12 in., its rpm would be

$$(4.5)/(1 \times 3.14) = 1.43 \text{ rpm}$$

$$\text{Sprocket ratio} = \frac{\text{rpm of output gear head}}{\text{rpm of conveyor drive pulley}} = \frac{7.2}{1.43} = \mathbf{5\ to\ 1}$$

C-7. A horizontal shaft 75 in. long is supported in bearings at each end. At each bearing a hole is drilled completely through bearing and shaft. Cylindrical pins, similar to cotter pins, are then inserted to keep the shaft from moving. A pure torque of 15,000 in.-lb is then applied. If pins and shaft are made of SAE 1045 steel, what pin diameter is indicated?

ANSWER. First determine shaft diameter indicated. From "Machinery's Handbook" 13th ed., pages 511 to 514, the formula $B = \sqrt[3]{D^2} \times 5.2$ is given, where B is maximum distance (ft) between shaft bearings and D is shaft diameter, in. Then

$$75/12 = \sqrt[3]{D^2} \times 5.2 = 6.25$$

from which D is found to be equal to 1.32 in. Now, general proportions indicates the use of a 2-in.-diameter shaft. Then from page 368 of the same reference source, i.e., strength of taper pin

$$d = 1.13 \sqrt{\frac{\text{turning moment, in.-lb}}{D \times \text{safe unit stress, psi}}}$$

$$d = 1.13 \sqrt{\frac{15{,}000/2}{2 \times 10{,}000}} = 1.13 \sqrt{0.375} = 1.13 \times 0.611 = 0.69 \text{ in.}$$

Use a **3/4-in.-diameter pin.**

Supplement **D**

MECHANISM

Questions and Answers

D-1. Describe and illustrate by sketches three different mechanisms that could be used to provide a rotational velocity ratio of 12,500 between driving and driven shafts. Each must be a compact, practical mechanism, suitable for transmitting small amounts of power. Determine the proportions affecting the velocity ratio in each case.

ANSWER.

Case 1. A spur gear train, consisting of three 10-tooth pinions driving 100-tooth gears and one 10-tooth pinion driving a 125-tooth

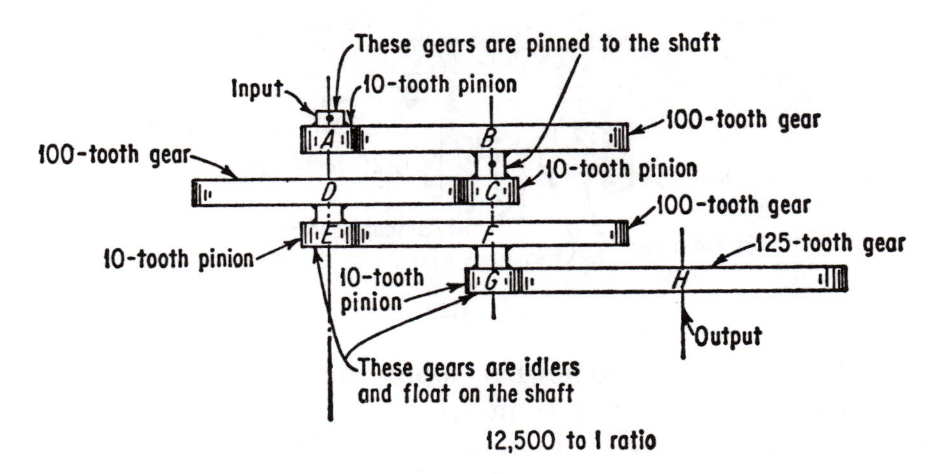

Fig. D-1

gear. The reason for using a 10-tooth pinion is to obtain a satisfactory contact ratio (see Fig. D-1). Pinions with fewer teeth are not recommended. This gear train is very efficient, capable of an over-all efficiency of about 95 per cent or better depending on type of bearing used. This gear train is also reversible.

Case 2. A compound worm-gear drive. This is used where high reductions are required and power consumption is no object (see Fig. D-2). It has very low efficiency and is not reversible. Its

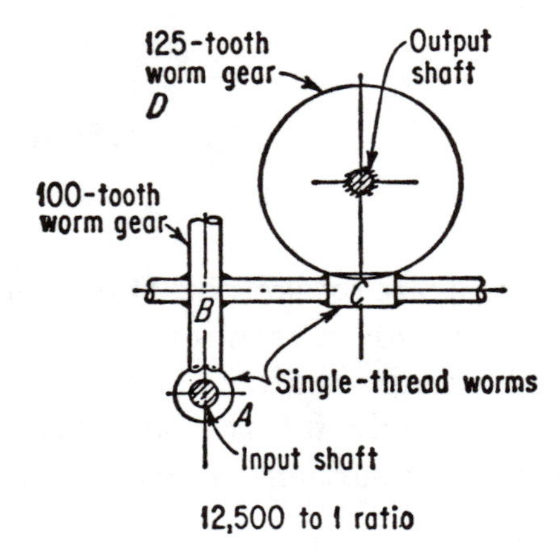

Fig. D-2

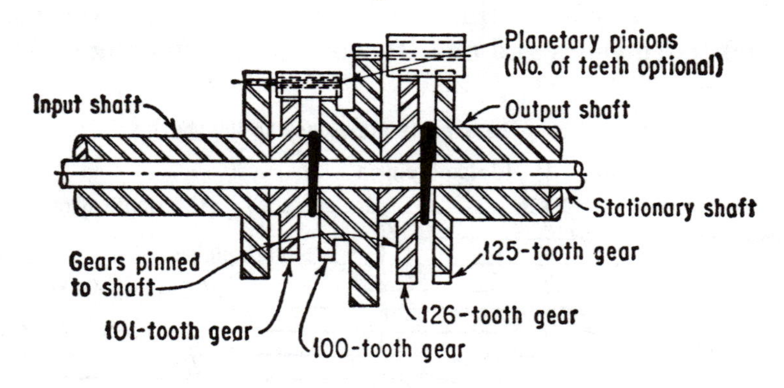

Fig. D-3

chief advantages are: somewhat more compactness and greater accuracy of angular motion from driver to driven.

Case 3. Epicylic gear reduction. This is based on the principle of obtaining a speed differential between two sun gears engaging with the same planet pinion (see Fig. D-3). Both gears have the same pitch diameter, but have a different number of teeth. The efficiency of this gear train is almost as high as that of the spur gear train.

Determination of proportions or number of teeth, where N = rpm; t = number of teeth.

Case 1. $N_a/N_b = 12{,}500/N_b = t_b/t_a = {}^{100}/_{10}$. Thus

$$N_b = 1{,}250 \text{ rpm}; N_b = N_c$$

$N_c/N_d = 1{,}250/N_d = t_d/t_c = {}^{100}/_{10}$. Thus $N_d = 125$ rpm;

$$N_d = N_e$$

$N_e/N_f = 125/N_f = t_f/t_e = {}^{100}/_{10}$. Thus $N_f = 12.5$ rpm;

$$N_f = N_g$$

$N_g/N_h = 12.5/N_h = t_h/t_g = {}^{125}/_{10}$. Thus $N_h = 1$ rpm; therefore, N_a/N_h is equal to 12,500 to 1.

Case 2. $N_a/N_b = 12{,}500/N_b = t_b/t_a = 100$. Thus $N_b = 125$; $N_b = N_c$.

$N_c/N_d = 125/N_d = t_d/t_c = 125$. Thus $N_d = 1$ rpm; therefore, N_a/N_h is equal to 12,500 to 1.

Case 3. The input shaft rotating at 12,500 rpm will cause the 100-tooth sun gear to rotate at 125 rpm by a reduction of 100 to (101 − 100) or 100 to 1. Likewise, the output shaft will have an rpm of 1 by a reduction of 125 to (126 − 125), or 125 to 1.

D-2. An eccentric is to be placed on a shaft 3 in. in diameter and to impart a 2-in. stroke to a pump. If the minimum thickness of the

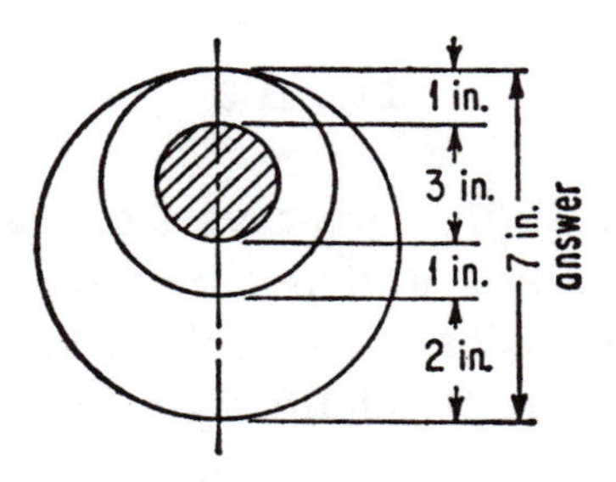

Fig. D-4

eccentric from the shaft to its outer periphery is 1 in., calculate the outside diameter of the eccentric.

ANSWER. See Fig. D-4.

D-3. Show how to construct a cam driving a follower in a line of motion which is offset 1 in. from the cam center. The lowest position of the follower is 1 in. to the right and 1 in. above the cam center. The rise of the follower is 2 in. The follower ascends during the first half-turn with harmonic motion and descends during the remaining half-turn with uniformly accelerated and retarded motion.

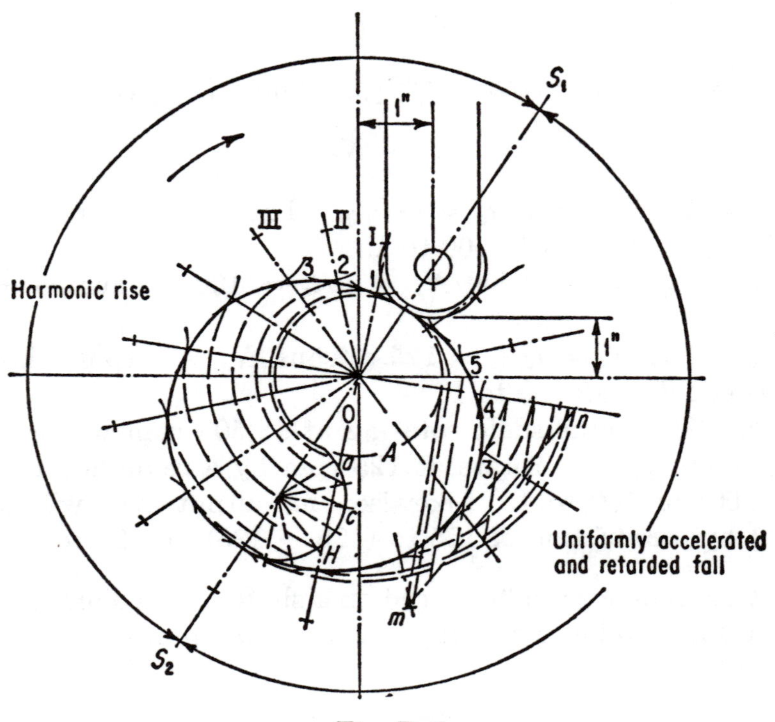

Fig. D-5

ANSWER. See Fig. D-5. Draw vertical center line of follower 1 in. from cam center line and position bottom of follower 1 in. above center line. With O as center, draw shaft circle A tangent to follower. Divide circle in two halfs by means of diameter S_1 and S_2 passing through centers of cam and follower.

For a rise with harmonic motion divide the left half of the cam, beginning at S_1 and ending at S_2, into two equal parts, say 8. Draw semicircle H, having a diameter equal to the rise, namely, 2 in., and divide it also into 8 parts, projecting points a, b, c, etc., on the rise. With O as center, describe arcs which intersect the radial lines at

1, 2, 3, etc. If the follower were a point, a line drawn through points 1, 2, 3, etc. would give the required cam profile. With a roller follower it is necessary to shift the profile in such a manner that centers I, II, III, etc., describe arcs passing through points 1, 2, 3, etc., and equal to the follower radius. The accurate cam profile is now drawn tangent to these arcs.

To obtain a fall with uniformly accelerated and retarded motion, the right-hand half of the cam, again beginning at S_1, is divided into 8 equal arcs as before. If it is assumed that the acceleration is 1 unit, the rise is divided into 8 parts in the ratios of 1:2:3:4:4:3:2:1. To do this, draw a line *mn* at such an angle that it is easily divided by 20 (the sum of $1 + 2 + 3 + 4 + 4 + 3 + 2 + 1$), and lay off the required intervals projecting these points so as to divide the rise in the same ratios. Then with center at O draw arcs intersecting the radial lines at 3, 4, 5, etc. Next, proceed as for the harmonic rise to locate the accurate cam profile.

D-4. A rocker 10 in. long is actuated by a crank and connecting link, the crank and rocker shaft centers being 20 in. apart. The

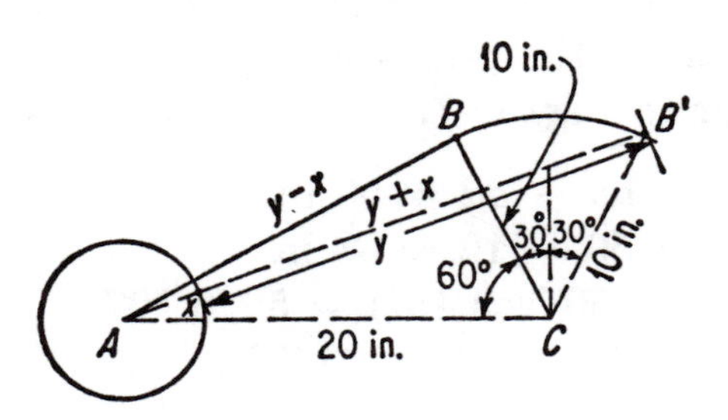

Fig. D-6

rocker moves 30° both sides of a line perpendicular to the line of shaft centers. Calculate the lengths of the crank and connecting link.

ANSWER. Refer to Fig. D-6. By the law of cosines in triangle ABC

$$y - x = \sqrt{10^2 + 20^2 - 2 \times 10 \times 20 \cos 60°} = \sqrt{300} = 17.32$$

In triangle $AB'C$ by the law of cosines

$$y + x = \sqrt{10^2 + 20^2 - 2 \times 10 \times 20 \times \cos 120°} = \sqrt{700} = 26.47$$

Therefore, $2y = 43.79$ and $2x = 9.14$. Finally, $\boldsymbol{y}$ **= 21.90 in. connecting link.** Also $\boldsymbol{x}$ **= 4.57 in. crank.**

D-5. Two rockers are connected at their free ends by a link. One rocker is 12 in. long and is perpendicular to the line of centers which is 50 in. long. The other rocker is 6 in. long and at an angle of 45° with the line of centers. A force of 100 lb is applied at the end of

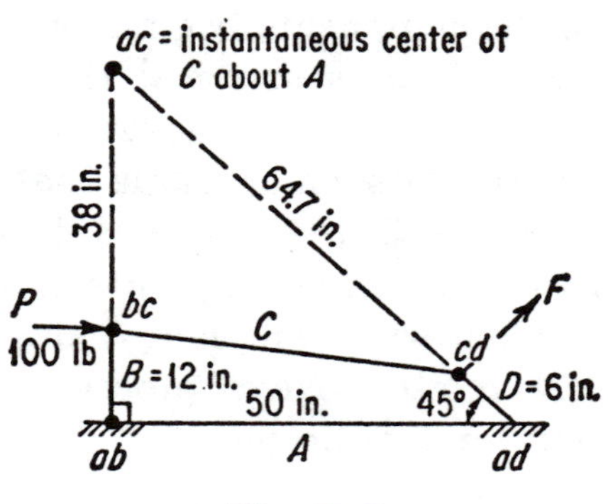

Fig. D-7

the 12-in. rocker and directed perpendicular to it. What torque is exerted on the shaft of the 6-in. rocker?

ANSWER. Refer to Fig. D-7. With crank D at 45°.

$$ac - ab = ab - ad = 50 \text{ in.}$$
$$ac - bc = 50 - 12 = 38 \text{ in.}$$
$$ac - ad = (50/\cos 45°) = 50/0.707 = 70.7 \text{ in.}$$
$$ac - cd = 70.7 - 6 = 64.7 \text{ in.}$$

Since P and F are the only forces acting on link C, their respective moments about the instantaneous center of $C - ac$ have to be equal. That is to say,

$$P \times 38 \text{ in.} = F \times 64.7 \text{ in.}$$
$$F = \frac{100 \times 38}{64.7} = 58.7 \text{ lb}$$

Torque on shaft at ad = 58.7 × 6 in. = **352.2 in.-lb**

D-6. Show how four-bar linkages can be applied to typical industrial applications.

ANSWER. All mechanisms can be broken down into equivalent

four-bar linkages. They can be thought of as the basic mechanism and are useful in many mechanical operations as shown in Figs. D-8 to D-31.

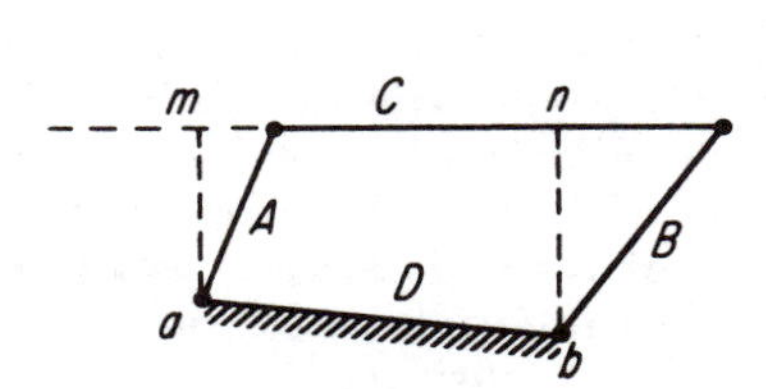

Fig. D-8. Four-bar linkage—two cranks, a connecting rod, and a line between the fixed centers of the cranks made up the basic four-bar linkage. Cranks can be rotated if *A* is smaller than *B* or *C* or *D*. Link motion can be predicted.

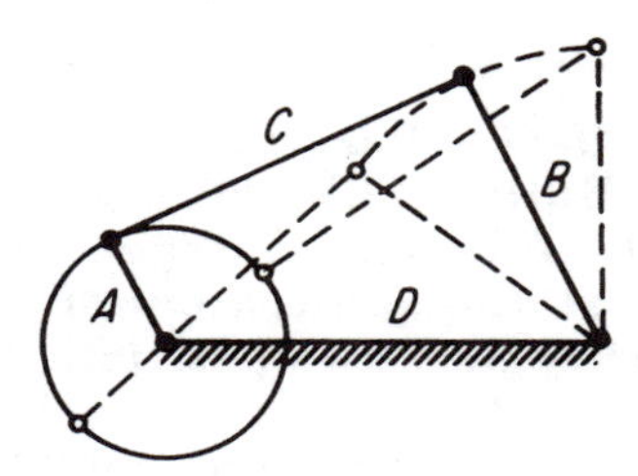

Fig. D-9. Crank and rocker—The following operations must hold for operation: $A + B + C > D$; $A + D + B > C$; $A + C - B < D$, and $C - A + B > D$.

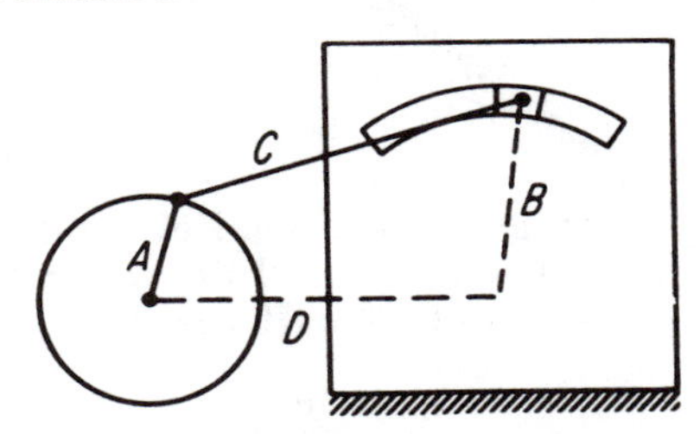

Fig. D-10. Four-bar link with sliding member—One crank replaced by circular slot with effective crank distance at *B*.

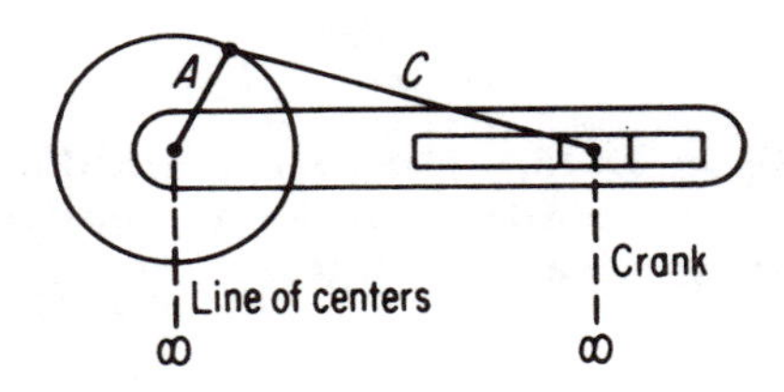

Fig. D-11. Straight sliding link—This is the form in which a slide is usually used to replace a link. The line of centers and the crank *B* are both of infinite length.

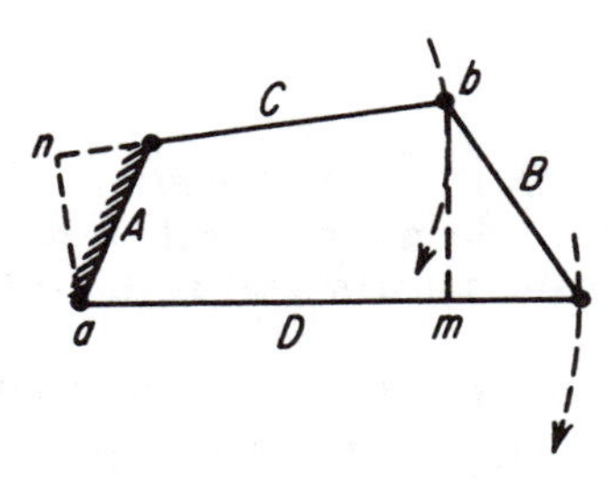

Fig. D-12. Drag link—This linkage is used as the drive for slotter machines. For complete rotation, $B > A + D - C$ and $B < D + C - A$.

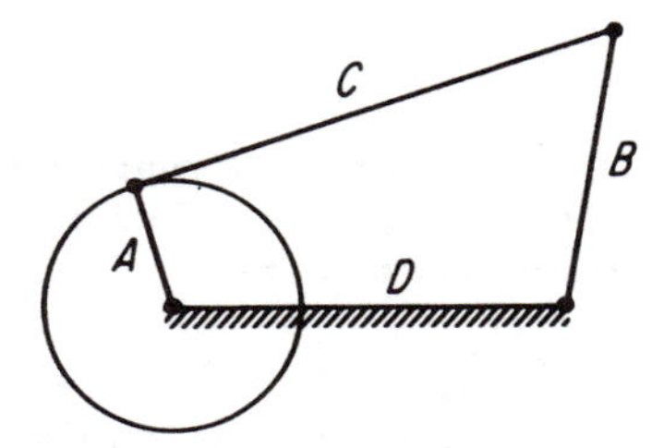

Fig. D-13. Rotating crank mechanism—This linkage is frequently used to change a rotary motion to a swinging movement.

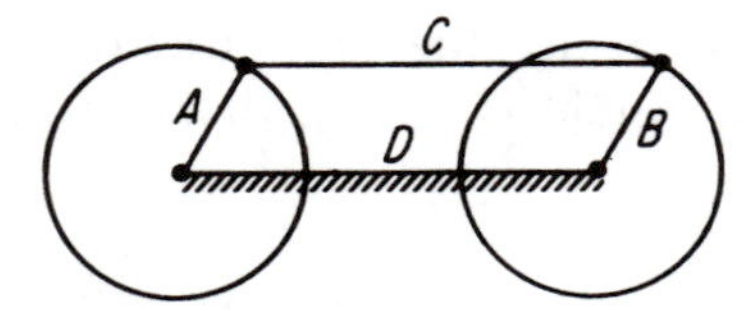

Fig. D-14. Parallel crank four-bar—Both cranks of the parallel crank four-bar linkage always turn at the same angular speed, but they have two positions where the crank cannot be effective. They are used on locomotive drivers.

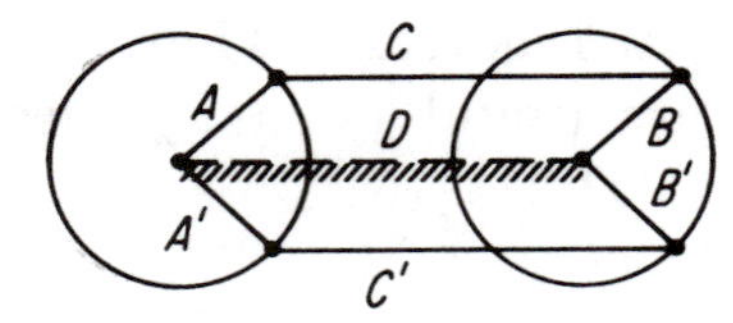

Fig. D-15. Double parallel crank—This mechanism avoids dead center position by having two sets of cranks at 90° advancement. Connecting rods are always parallel. Sometimes used on driving wheels of locomotives.

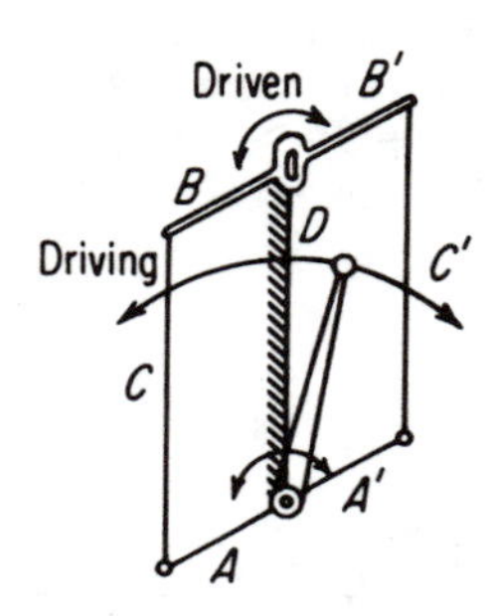

Fig. D-16. Parallel cranks—Steam control linkage assures equal valve openings.

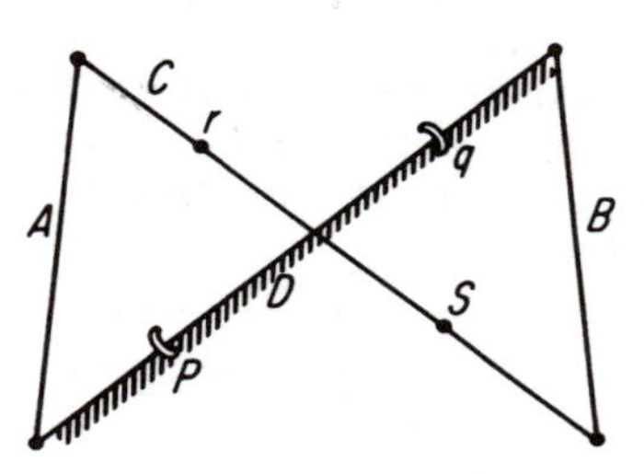

Fig. D-17. Nonparallel equal crank—If crank *A* has uniform angular speed, *B* will vary.

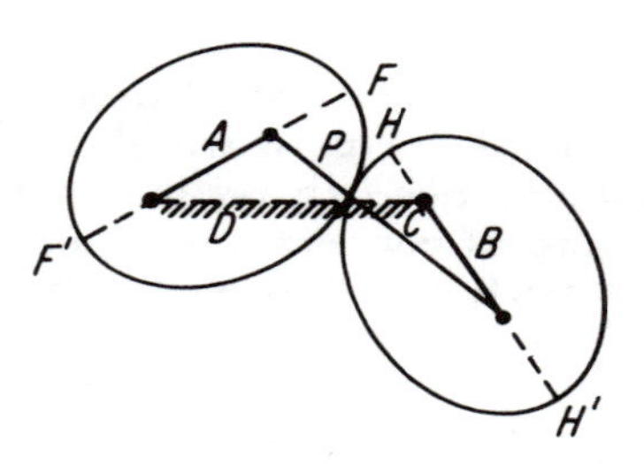

Fig. D-18. Elliptical gears—They produce the same motion as nonparallel equal cranks.

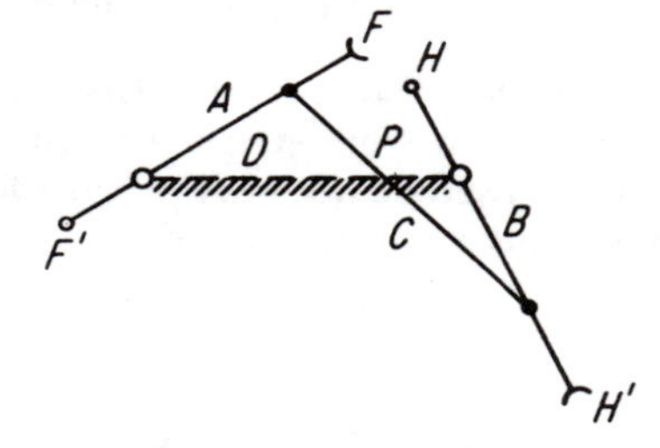

Fig. D-19. Nonparallel equal cranks—Same as first but with crossover points and link ends.

D-7. A rigid bar of length L slides with its ends constrained to move in slots that make an angle δ with each other. Find the following in terms of L, θ, and the angular velocity ω of the bar (see Fig. D-32).

(*a*) When the bar is perpendicular to one slot, find the magnitude and direction of the velocity and acceleration of the point in contact with the other slot.

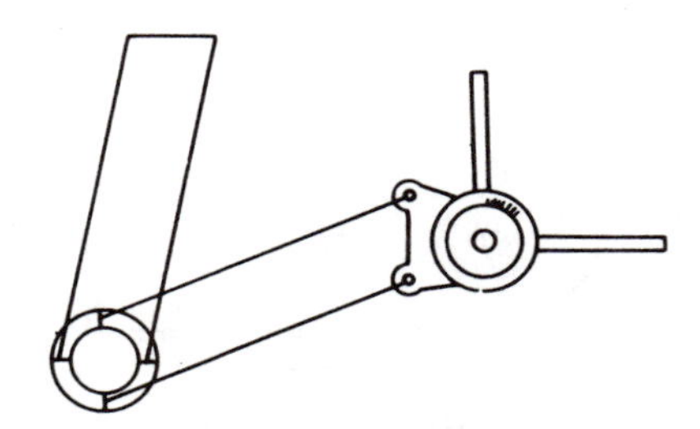

Fig. D-20. Double parallel crank mechanism—This mechanism forms the basis for the universal drafting machine.

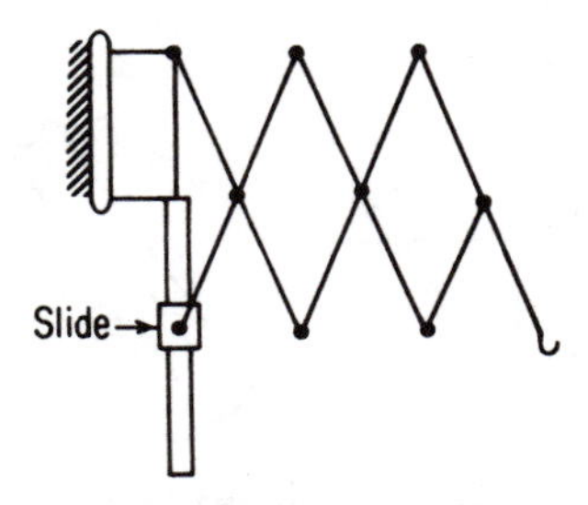

Fig. D-21. Isosceles drag lines—"Lazy-Tong" device made of several isosceles links; used for movable lamp support.

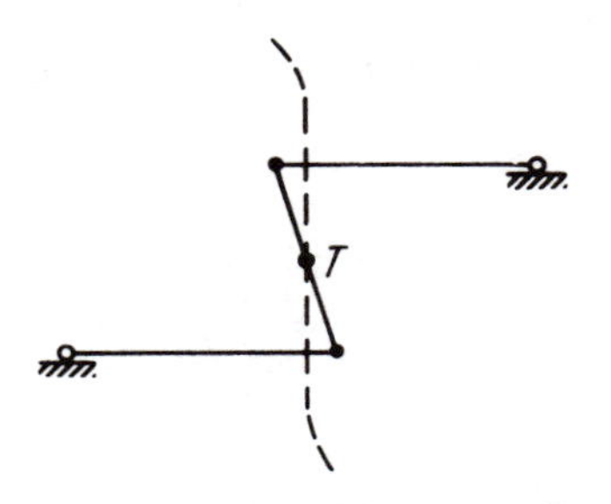

Fig. D-22. Watt's straight-line mechanism—Point *T* describes a line perpendicular to parallel position of cranks.

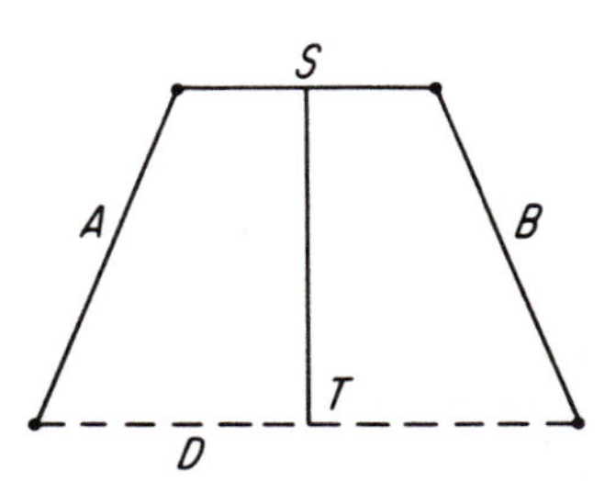

Fig. D-23. Robert's straight-line mechanism—The lengths of cranks *A* and *B* should not be less than 0.6*D*; *C* is one-half *D*.

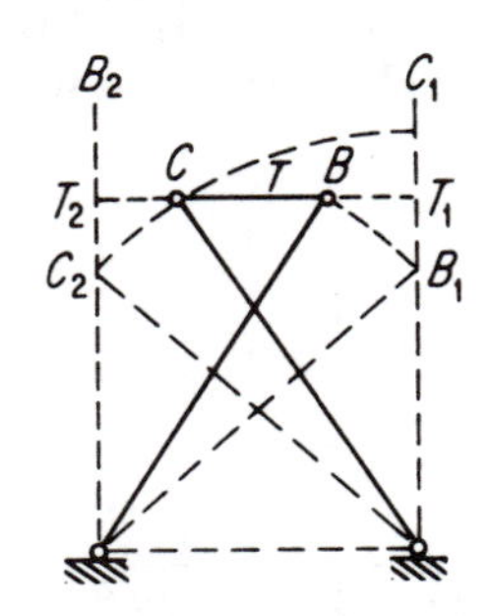

Fig. D-24. Tchebicheff's links—made in proportion: *AB* = *CD* = 20; *AD* = 16; *BC* = 8.

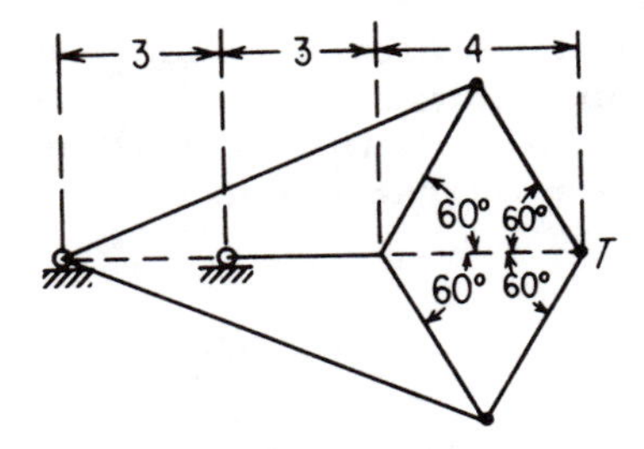

Fig. D-25. Peucellier's cell—When proportioned as shown, the tracing point *T* forms a straight line perpendicular to the axis.

(*b*) If the bar makes equal angles with the slots, find the speeds of the two ends.

ANSWER. The general position is shown in Fig. D-33. Assume ω is constant, i.e., $\alpha = 0$.

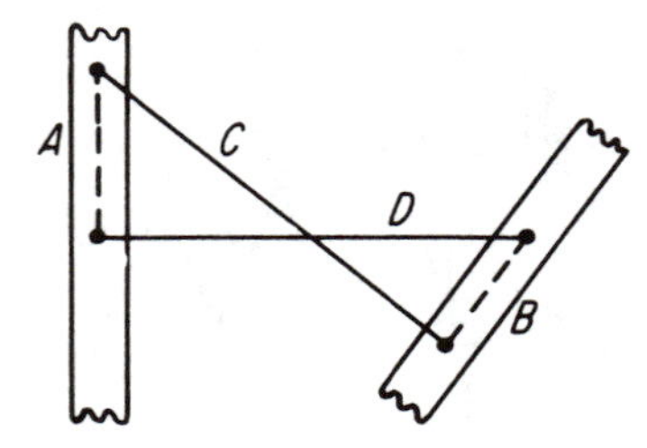

Fig. D-26. Nonparallel equal crank—The centrodes are formed as gears for passing dead center and can replace ellipticals.

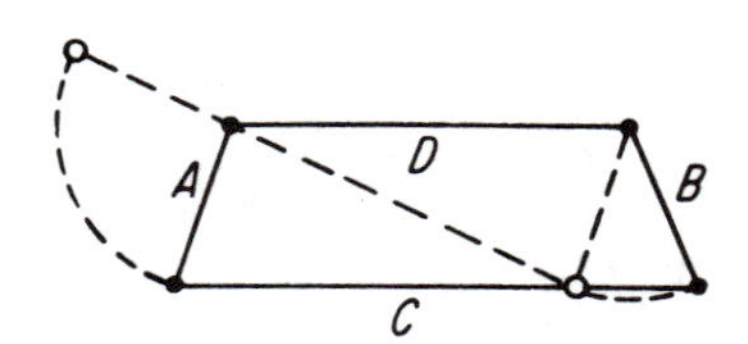

Fig. D-27. Slow-motion line—As crank *A* is rotated upward, it imparts motion to crank *B*. When *A* reaches dead center position, the angular velocity of crank *B* decreases to zero. This mechanism is also used on the Corliss valve.

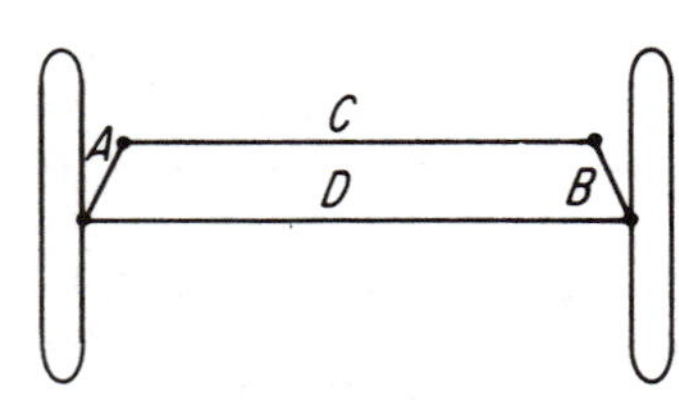

Fig. D-28. Trapezoidal linkage—This linkage is not used for complete rotation but can be used for special control. Inside moves through larger angle than outside with normals intersecting on extension of rear axle in cars.

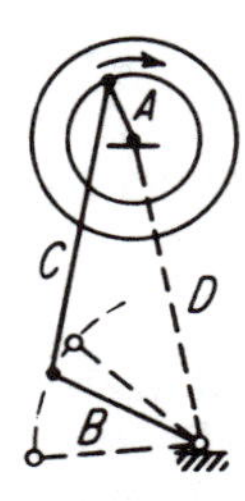

Fig. D-29. Treadle drive—This four-bar linkage is used in driving grind-wheels and sewing machines.

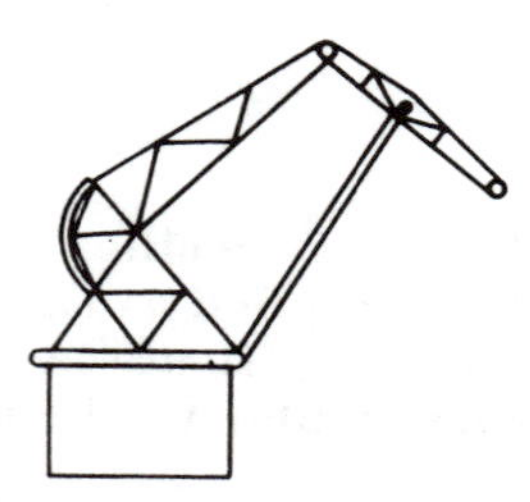

Fig. D-30. Double-lever mechanism—Slewing crane can move load in horizontal direction by using D-shaped portion of top curve.

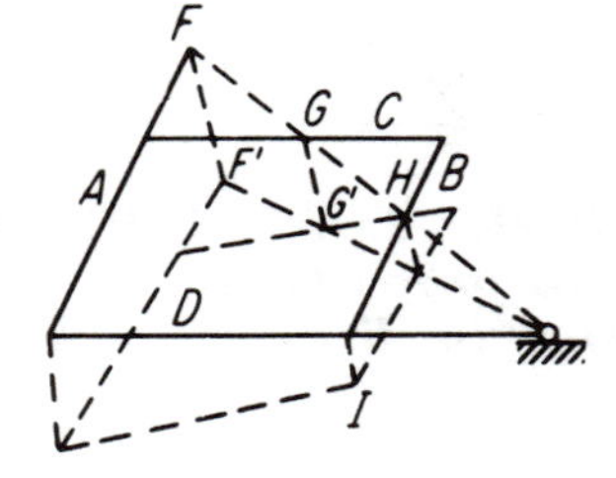

Fig. D-31. Pantograph—The pantograph is a parallelogram in which lines through *F*, *G*, and *H* must always intersect at a common point.

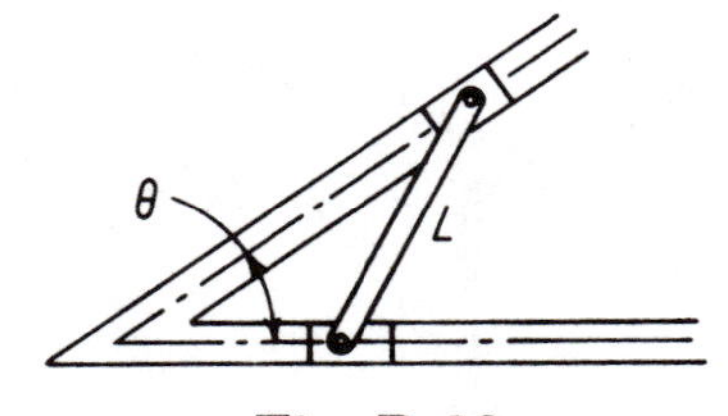

Fig. D-32

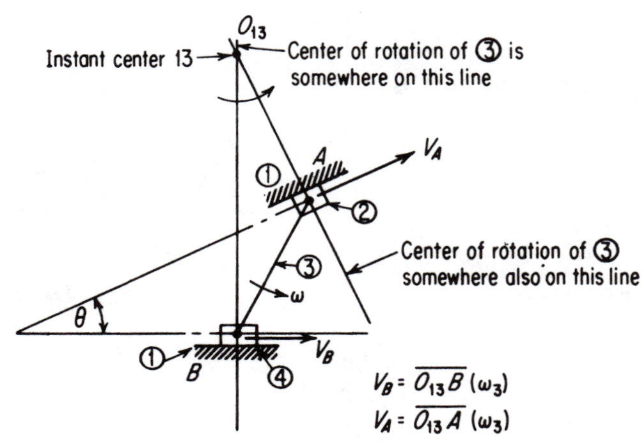

Fig. D-33

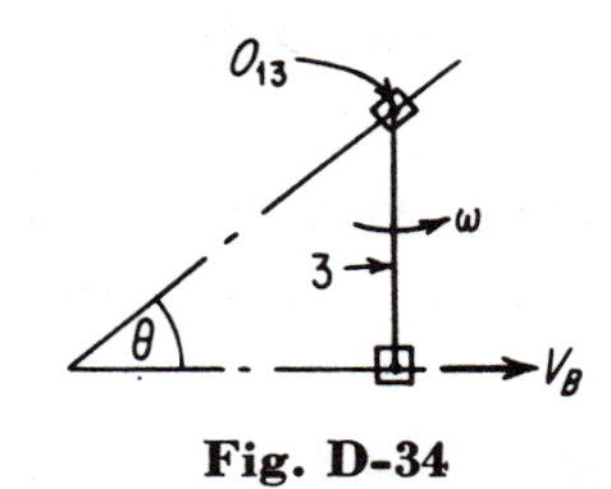

Fig. D-34

(*a*) Bar perpendicular to one slot (see Fig. D-34). The velocity analysis is: Since the instant center between link 3 and ground 1 is coincident with A, point A cannot have any velocity at this instant. From the above general expression, since $\overline{O_{13}A} = 0$, $V_A = \overline{O_{13}A}(\omega)$, and therefore, $V_A = 0$.

Acceleration analysis (see Fig. D-35). The relative acceleration equation is $\bar{a}_A = \bar{a}_B + \bar{a}_{AB}$ (see Fig. D-36). Therefore,

$$a_A = L\omega^2/\sin\theta \qquad \textbf{Ans.}$$

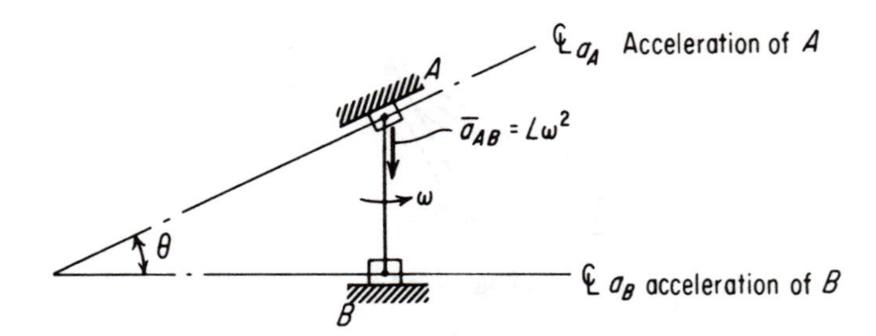

Fig. D-35

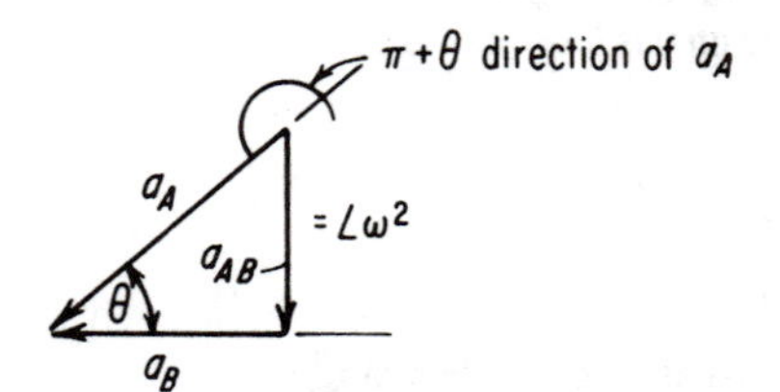

Fig. D-36

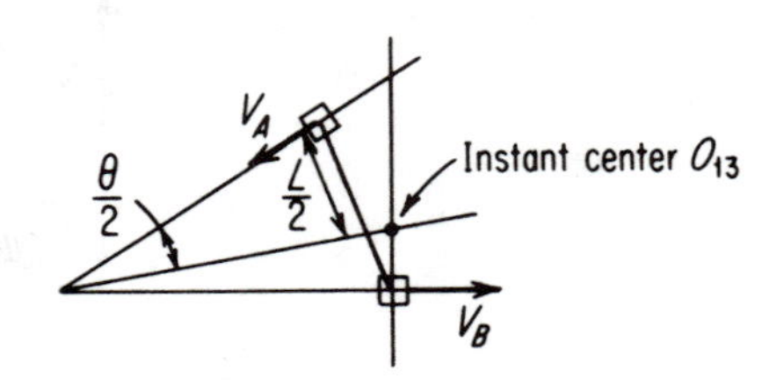

Fig. D-37

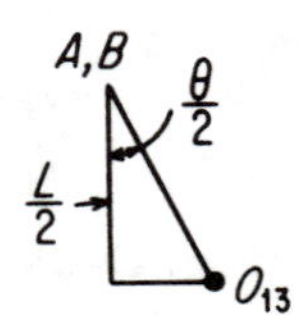

Fig. D-38

(*b*) Bar at equal angles with slots (see Fig. D-37). $V_A = O_{13}A \times \omega$ and $V_B = O_{13}B \times \omega$. Now refer to Fig. D-38 and conclude that

$$\overline{O_{13}A} = \overline{O_{13}B} = \frac{L/2}{\cos \theta/2} = \frac{L}{2 \cos \theta/2}$$

Finally,

$$V_A = \frac{L\omega}{2 \cos \theta/2} \qquad \textbf{Ans.}$$

$$V_B = \frac{L\omega}{2 \cos \theta/2} \qquad \textbf{Ans.}$$

Supplement **E**

HYDRAULICS AND FLUID MECHANICS

Questions and Answers

E-1. An elevated water tank consists of a cylindrical section 16 ft in diameter and 20 ft high. Below this is a hemispherical bottom. With water level 2 ft below the top a 12-in.-diameter circular orifice with discharge coefficient of 0.60 is opened in the bottom at the center. Compute the time required to draw the water level down 22 ft.

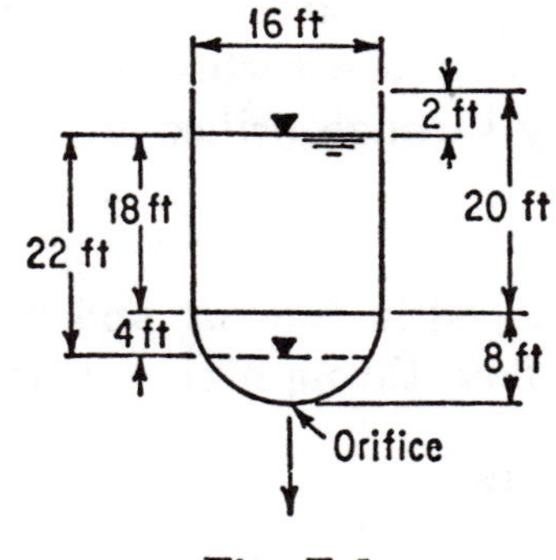

Fig. E-1

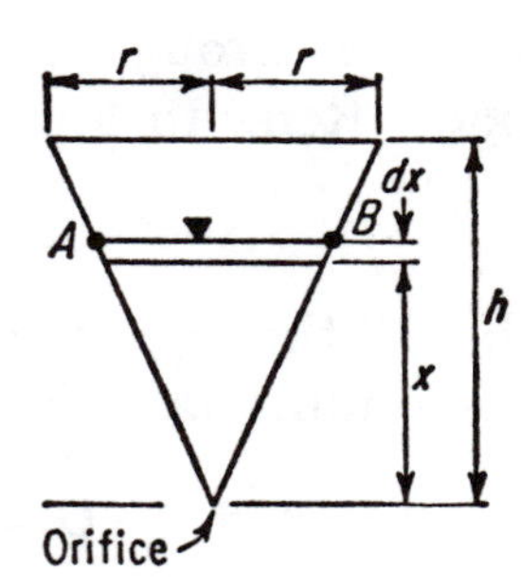

Fig. E-2

ANSWER. The total time to lower the level may be broken up into two sections: in the cylindrical section and in the hemispherical section. Refer to Fig. E-1. The time for the lowering in the straight section is

$$t_1 = \frac{2A}{CA_0\sqrt{2g}}(\sqrt{h_1} - \sqrt{h_2})$$

$$= \frac{2 \times 0.785 \times 16^2}{0.60 \times 0.785 \times \sqrt{64.4}}(\sqrt{26} - \sqrt{8}) = 242 \text{ sec}$$

Time to lower water in an hemispherical container is given by the general formula

$$t_2 = \frac{\pi}{CA_0\sqrt{2g}}\left[\frac{4}{3}Rh^{3/2} - 0.4h^{5/2}\right]_{h_2}^{h_1}$$

where R is radius of hemispherical container and h_1 and h_2 are the respective levels. Then for the problem at hand

$$t_2 = \frac{\pi}{CA_0\sqrt{2g}}\left[\frac{4}{3} \times 8(8^{3/2} - 4^{3/2}) - 0.4(8^{5/2} - 4^{5/2})\right]$$

$$= \frac{\pi}{0.60 \times 0.785 \times 64.4}$$

$$[10.67(22.6 - 8) - 0.4(181 - 32)]$$

$$= 0.819(156 - 59.5) = 0.819 \times 96.5 = 79 \text{ sec}$$

Total time $t_1 + t_2 = 242 + 79 = 321$ sec, or **5.4 min**

E-2. A vessel in the form of a right circular cone is filled with water. If h is its height and r the radius of its base, what time will it require to empty itself through an orifice of area a at the vertex?

ANSWER. Refer to Fig. E-2. Velocity through orifice is

$$v_0 = \sqrt{2gx}$$

Denote by dQ the volume of water discharged in time dt, and by dx the corresponding fall of surface. The flow through the orifice is

$$Q = a\sqrt{2gx} \qquad \text{cfs}$$

being measured as a right cylinder of area of base a and altitude. Therefore, in time increment dt

$$dQ = a\sqrt{2gx}\,dt \tag{1}$$

Denoting S the area of the surface of the water when depth is x, we have, from geometry,

$$\frac{S}{\pi r^2} = \frac{x^2}{h^2} \qquad \text{or} \qquad S = \frac{\pi r^2 x^2}{h^2}$$

But the volume of water discharged in time dt may also be considered as the volume of a cylinder AB of area of base S and altitude dx;

hence

$$dQ = S\,dx = \frac{\pi r^2 x^2\,dx}{h^2} \tag{2}$$

Equating (1) and (2) and solving for dt,

$$dt = \frac{\pi r^2 x^2\,dx}{ah^2\sqrt{2gx}}$$

$$t = \frac{\pi r^2 x^2\,dx}{ah^2\sqrt{2gx}} = \frac{2\pi r^2\sqrt{h}}{5a\sqrt{2g}} \quad \textbf{Ans.}$$

E-3. Two sources of pressure, M and N, are connected by a water-mercury differential gauge, as shown in Fig. E-3. What is the difference in pressure between M and N in psi?

ANSWER.

P_m + pressure equivalent to head over $A = P_n$ + pressure equivalent to head over B + pressure equivalent of mercury head (AB)
Then $P_m + (5 \times 62.4) = P_n + (2 \times 62.4) + (1 \times 62.4 \times 13.6)$

$$P_m - P_n = 662.8 \text{ psf, or } 662.8/144 = \textbf{4.6 psi}$$

E-4. It is necessary to pump 3 cfs of water through a pipeline 4,000 ft long to a reservoir 200 ft above the pumps. What minimum size of cast-iron pipe (new) should be installed if the lost head is not to exceed 10 per cent of the static lift?

ANSWER. The following relation and equation are involved.

$$\text{Velocity} = \frac{Q}{A} = \frac{Q}{0.785d^2} = \text{fps} = \frac{Q}{\pi d^2/4}$$

$$h_f = f_1 \frac{L}{d}\frac{v^2}{2g} = f_1 \frac{L}{d}\left(\frac{4Q}{4d^2\pi}\right)^2 \times \frac{1}{2g}$$

Then

$$d^5 = \frac{8f_1 LQ^2}{\pi^2 g h_f} = \frac{8 \times 0.02 \times 4{,}000 \times 3^2}{\pi^2 \times 32.2 \times (200 \times 0.10)} = 0.907 \text{ ft}$$

$$d = 11.77$$

Use **12-in.** cast-iron pipe.

E-5. Water enters a turbine through a 4-in.-diameter pipe at 150 psia and leaves through an 8-in.-diameter pipe at a point 3 ft lower under a pressure of 5 psig. Flow is 2 cfs. Turbine efficiency is 85 per cent. Compute horsepower output of turbine.

ANSWER. See Fig. E-4. The flow velocity through the 4-in. pipe may be found to be 23 fps; that through the 8-in. exit pipe, 5.75 fps. Now apply Bernoulli's equation.

$$Z_A + \frac{v_A^2}{2g} + \frac{P_A}{w_A} + h_f + \text{TDH} = Z_B + \frac{v_B^2}{2g} + \frac{P_B}{w_B}$$

$$0 + (23)^2/64.4 + (150 \times 144)/62.4 + 0 + \text{THD} = -3 + (5.75)^2/64.4 + (5 \times 144)/62.4$$

$$0 + 8.24 + 346 + \text{THD} = -3 + 0.513 + 11.55$$

$$\text{TDH} = 8.24 + 346 + 3 - 0.513 - 11.55 = 345.18 \text{ ft}$$

$$\text{hp output} = \frac{2 \times 60 \times 62.4 \times 345.18}{33{,}000} \times 0.85 = \mathbf{66.5}$$

The static head of 3 ft was taken as negative because datum was passed through the supply line. The horsepower equation above is in the form

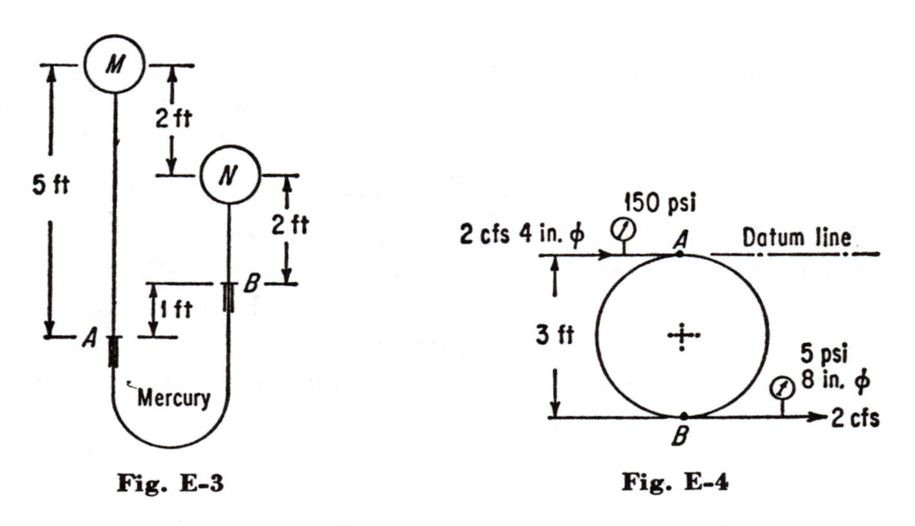

Fig. E-3

Fig. E-4

$$\text{Horsepower} = \frac{\text{lb per min} \times \text{total dynamic head (TDH)}}{33{,}000}$$

E-6. A pipe 30 in. in diameter discharges water at 20 cfs. At point *A* in the pipe the pressure is 30.5 psig and the elevation is 100 ft. At point *B* in the pipe, which is 5,000 ft along the pipe from point *A* the pressure is 50.2 psia and the elevation is 80 ft. Compute the value of the friction factor.

ANSWER. See Fig. E-5. Again, this is a "natural" for Bernoulli's equation. Equation may then be modified to suit as

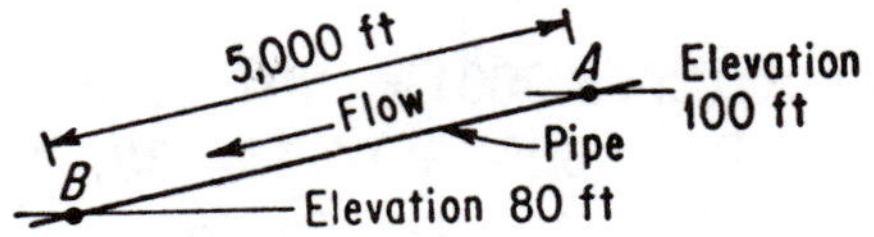

Fig. E-5

follows:

$$h_f = \frac{P_A - P_B}{w} + (Z_A - Z_B) + \frac{v_A^2 - v_B^2}{2g}$$

Since flow areas are the same with constant pipe size,

$$v_A = v_B = \frac{Q}{(30^2 \times 0.785)/144} = \frac{20}{4.9} = 4.08 \text{ fps}$$

$$P_A = (30.5 \times 144) = 4{,}390 \text{ psf} \qquad P_B = (50.2 - 14.7)144 = 5{,}110 \text{ psf}$$

Therefore

$$h_f = \frac{4{,}390 - 5{,}110}{62.4} + (100 - 80) = -11.5 + 20 = 8.5 \text{ ft}$$

From the Darcy equation

$$f_1 = \frac{h_f}{L/D \times v_A^2/2g} = \frac{h_f \times D \times 2g}{L \times v_A^2} = \frac{8.5 \times (30/12) \times 64.4}{5{,}000 \times 4.08^2} = \mathbf{0.0164}$$

E-7. A hydraulic test on a needle valve having a 7/32-in.-diameter seat showed that a flow of 1.485 gpm of water at 60°F caused a drop in pressure through the valve of 2.4 psi. It is intended that this valve be used for compressed-air service and the drop anticipated was to be about 5 psi when connected to the air. What would be the pressure drop in psi if the anticipated air flow is 2,710 scfh (14.7 psia and 60°F) when at an internal pressure of 325 psig and 70°F upstream of the needle valve?

ANSWER. Use equation $Q = CA\sqrt{2g\,\Delta h}$. The discharge coefficient C applies to all fluids for all practical intents and purposes. For the water

$$\frac{1.485}{60 \times 7.5} = C\,\frac{(7/32)^2 \times 0.785}{12^2}\sqrt{64.4 \times 2.4 \times 2.31}$$

Solve for C. It is found to be equal to 0.677, a realistic figure. This value of C is also good for the air flow here. However, flow formula for air must be checked for type of flow, retarded or unretarded.

$$G = (2{,}710/379) \times (29/3{,}600) = 207/3{,}600 = 0.58 \text{ lb per sec}$$
$$P_c = 0.53 \times P_1 \qquad P_1 = 325 + 14.7 = 339.7 \text{ psia}$$
$$P_2 = 339.7 - 5 \text{ (assumed)} = 334.7 \text{ psia}$$

Note: The specification of a 5-psi drop through the valve was given the valve manufacturer to select the valve. The entire purpose of this solution is to check the water-flow and pressure-drop figures submitted by the manufacturer.

$$P_c = 0.53 \times 339.7 = 180 \text{ psia} \qquad \text{thus, } P_2 > P_c, \text{ retarded flow}$$

Using the equation of flow that applies for retarded flow, the flow rate at actual operating conditions is found to be

$$G = \frac{2.056}{\sqrt{460 + 70}} \times \frac{0.785 \times 0.218^2}{144} \sqrt{\left(\frac{334.7}{339.7}\right)^{1.43} - \left(\frac{334.7}{339.7}\right)^{1.71}} \times 0.677 = 0.054$$

This flow rate of 0.054 lb per sec is based on a 5 psi drop. Since the actual flow rate is equivalent to 0.058 lb per sec and flow is considered to be turbulent, the anticipated drop is found as follows applying the laws of affinity.

$$5/x = (0.054/0.058)^2 = 0.885 \qquad x = 5/0.885 = \mathbf{5.67\ psi}$$

Valve is acceptable.

E-8. Steam at an average pressure of 2,000 psia and a temperature of 1000°F flows in a pipeline whose inside diameter (ID) is 16 in. If the flow rate is 1,500,000 lb per hr, calculate the pressure drop per 100 ft of pipe. Note that many of the simplified formulas are not applicable at an elevated temperature and pressure. The solution must take into account various friction parameters such as viscosity and Reynolds number.

ANSWER. First step in any problem involving friction loss is to determine the Reynolds number. See appropriate section of Chap. 5.

$$Re = \frac{Dv\rho}{\mu} = \frac{(16/12) \times 118 \times 2.54}{0.03 \times 0.000672} = 19.7 \times 10^6$$

where

$$\text{Velocity of steam} = \frac{Q}{A} = 164/1.39 = \frac{(0.394 \times 1{,}500{,}000)/3{,}600}{(\pi \times 1.33^2)/4} = 118 \text{ fps}$$

$$\text{Density of steam} = 1/\text{sp vol} = 1/0.394 = 2.54 \text{ lb per cu ft}$$

Now we see that flow is definitely turbulent. From the Fanning correlation the friction factor is found to be 0.0035. Then the pressure drop is

$$\Delta P/100 \text{ ft} = \frac{2 \times 0.0035 \times 100 \times 2.54 \times 118^2}{144 \times 32.2 \times 16/12} = \mathbf{4\ psi}$$

The viscosity of steam was found in Marks, "Mechanical Engineers' Handbook" (6th ed., pp. 4-66). Specific volume of steam was obtained from the steam tables.

E-9. A pipe carrying oil having a specific gravity of 0.875 changes in size from 6 in. internal diameter at section A to 15 in. internal diameter at section B. Section A is 10 ft lower than section B. The pressure in the pipe is 12.00 psi at A and 8.50 psi at B. The discharge is 4.50 cfs. Determine whether the flow is upward or downward.

ANSWER. See Fig. E-6. Head at

$$B = 8.5 \times 2.31/0.875 = 22.4 \text{ ft}$$

The velocity head at B is calculated as 0.2 ft. Then the total energy head at B is if datum is taken through B

$$0 + 0.2 + 22.4 = 22.6 \text{ ft}$$

Head equivalent to 12 psi is $12 \times 2.31/0.875$, or 31.8 ft. The velocity head at the same point may be calculated as 8.25 ft. Then the total energy at A is

$$-10 + 8.25 + 31.8 = 30.05 \text{ ft}$$

Thus we see that the head at A is greater than at B. Therefore, flow is **upward.**

E-10. An oil pipe has a horizontal 60° bend which also reduces the pipe size from 18 to 12 in. What is the magnitude and direction of the total resultant force on the bend when 20 cfs of oil, having a specific gravity of 0.85, is entering the bend per second at a pressure of 8 psi?

ANSWER. See Fig. E-7. By standard calculations for 20 cfs the velocity v_1 in the 18-in. line is 11.31 fps; that in the 12-in. line, 25.5

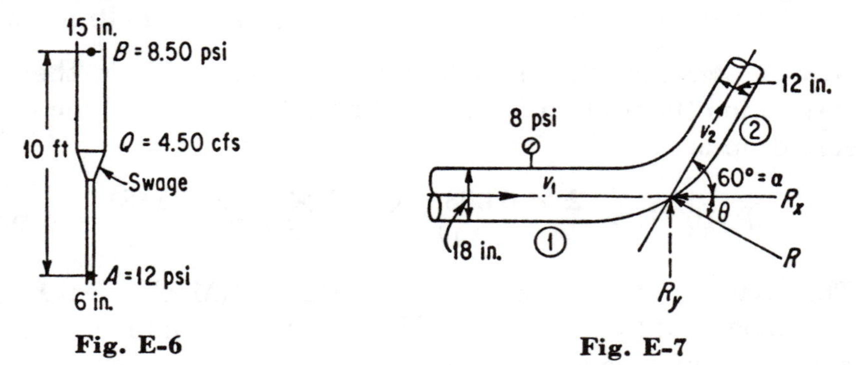

Fig. E-6 Fig. E-7

fps as v_2. Velocity heads: 18-in. pipe $(11.31)^2/2g = 1.99$ ft oil; 12-in. pipe $(25.5)^2/2g = 10.1$ ft oil. Pressure head at point 1 is

$$(8/62.4 \times 0.85)144 = 21.7 \text{ ft oil}$$

From Bernoulli's equation

$$\frac{P_2}{w} = \frac{P_1}{w} + \frac{v_1^2}{2g} - \frac{v_2^2}{2g}$$

$$21.7 + 1.99 - 10.1 = 13.6 \text{ ft oil}$$

This is so because friction head is negligible and potential heads equate each other because bend is in the horizontal plane. The pressure in the 12-in. end of the bend is P_2 which we shall now determine.

$$P_2 = 13.6 \times 62.4 \times 0.85 = 722 \text{ psf}$$

The axial component $R_x = \dfrac{Qw}{g}(v_1 - v_2 \cos \alpha) + (P_1A_1 - P_2A_2 \cos \alpha)$

$$R_x = (20 \times 62.4 \times 0.85/32.2)(11.31 - 25.5 \times 0.5) + (1{,}152 \times 0.785 \times 18^2/12^2) - (722 \times 0.785 \times 0.5) = 1{,}710 \text{ lb}$$

$$R_y = \left(\frac{Qwv_2}{g} + P_2A_2\right) \sin \alpha$$

$$[(20 \times 62.4 \times 0.85 \times 25.5/32.2) + (722 \times 0.785 \times 1^2)]0.866 = 1{,}217 \text{ lb}$$

$$R = \sqrt{R_x^2 + R_y^2} = \sqrt{1{,}710^2 + 1{,}217^2} = \mathbf{2{,}100 \text{ lb}}$$

Direction in which R operates: $\theta = \arctan (R_y/R_x) = \mathbf{35°25'}$

E-11. At what rate is the surface of the water rising in a vessel whose form is that of an inverted right circular cone when the water is 20 ft deep and is flowing in at a uniform rate of 40 cfm?

ANSWER. Volume of vessel: $\frac{1}{3} \times H \times \pi H^2 = \pi H^3/3$. Then rate of change of volume with respect to height: $dV/dH = (3\pi H^2)/3 = \pi H^2$. Given: $dV/dH = 40$. Now, $dV/dt = dV/dH \times dH/dt$. Or, $40 = \pi H^2 \times dH/dt$. When H is 20 ft, $dH/dt = (40)/(\pi \times 400) =$ **0.03183 fpm,** rate of rising of water level.

E-12. Under what conditions may the effect of restrictions such as valves and elbows be neglected in a pipeline?

ANSWER. Whenever the ratio of pipe length (ft) to pipe diameter (ft) is equal to or greater than 1,000, the effect of valves, etc., may be neglected.

E-13. Water is being pumped up a 5 per cent grade through a 12-in. cast iron pipe at 5 cfs. If a pressure gauge at a certain point in the pipe reads 50 psi, what will the pressure reading be 200 pipeline ft further upstream?

ANSWER. Vertical lift is $(5/100)(200)$, or 10 ft. Velocity is 6.35 fps calculated. Darcy friction factor f_1 may be taken as 0.02. Then,

$$h_f = 0.02 \times 200 \times 6.35^2/64.4 = 2.51 \text{ ft head loss}$$

From Bernoulli's equation, $P_A/w_A = 10 + (P_B/w_B) + 2.51 = 115.5$. Finally,

$$\frac{P_w}{w_B} = 115.5 - 10 - 2.51 = 103 \text{ ft of water}$$

Finally, $103/2.31 =$ **44.5 psi**

E-14. The accompanying sketch shows an oil-speed indicator. The relation between speed and head h is given by: rpm $= C \times h^n$. Determine the calibration constants C and n.

ANSWER. See Fig. E-8. Write energy equation at points A and B. The datum plane is at point C.

$$\frac{1.75}{12} + h + 0 = \frac{1.75}{12} - \frac{v^2}{2g} + \frac{P_A}{w}$$

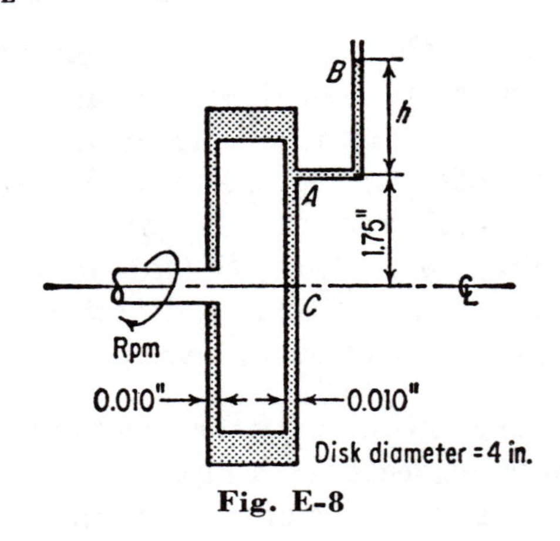

Fig. E-8

Since P_A/w is small, it may be neglected. Then

$$h = \frac{v^2}{2g} \qquad \text{or} \qquad v = \sqrt{2gh}$$

$$v = 2\pi r \times \text{rpm}/60 \qquad \text{with } r = 1.75/12$$

Hence $\text{rpm} = [(60 \times 12)/(2\pi \times 1.75)] \times \sqrt{64.4} \times \sqrt{h}$

$$\text{rpm} = 65.5 \times 8.03\sqrt{h} = 526\sqrt{h}$$

Or since $\text{rpm} = C \times h^n$, $\mathbf{C} = \mathbf{526}$ and $\mathbf{n} = \mathbf{1/2}$

E-15. Saturated hot water at 2,000 psia flows through a single-seat valve with a back pressure of 1,000 psia. For a valve-flow coefficient $C_v = 400$, determine the flow through the valve under these conditions.

(*a*) Assume no "flashing" in the valve, and determine the flow Q through the valve.

(*b*) Assume flashing in the valve, and determine the flow Q in the line. The flow through the valve can be considered adiabatic. The effect of flashing is to change the specific gravity of the fluid flowing. The general equation for flow through a valve is $Q = C_v \sqrt{\Delta P \times G}$, where

Q = flow rate through the valve at inlet conditions
C_v = valve flow coefficient = gpm/$\sqrt{\text{psi}}$
$\Delta P = P_1 - P_2$, pressure drop across valve
G = specific gravity of fluid flowing through valve

ANSWER.

(*a*) No flashing flow. If there is no flashing through the valve, the specific gravity of fluid passing through the valve is constant. For hot saturated water at 2,000 psia,

$$G = \frac{1}{62.4(v_f)}$$

where v_f = specific volume of water at 2,000 psia

$$G = \frac{1}{62.4 \times 0.0257}$$

$$\Delta P = 1{,}000 \text{ psi}$$

$$C_v = 400$$

$$Q = C_v \sqrt{\frac{\Delta P}{62.4 \times 0.0257}} \simeq \mathbf{10{,}000\ gpm}$$

(*b*) With flashing. Due to flashing some water changes to vapor, and the specific gravity of fluid flowing through the valve will be that of the saturated water and vapor mixture. Let h_1 be the enthalpy of the fluid upstream of the valve and h_2 be the enthalpy of the fluid downstream of the valve. Then

$$h_1 = h_2 = h_{f2} + xh_{fg2}$$

where x is the fraction of the water flashed, h_1 is the enthalpy of saturated water upstream at point 1, h_{f2} is the enthalpy of saturated liquid at the valve outlet, and h_{fg2} is heat of vaporization at point 2 downstream of the valve. From steam tables, $h_1 = 671.1$, $h_{f2} = 542.4$, and $h_{fg2} = 649.4$. Then

$$x = \frac{671.1 - 542.4}{649.4} = 0.2$$

Let W = mass flow rate through the valve. Through the valve 20 per cent will "flash" into vapor. Therefore, the total volume of liquid is given by $0.8 \times W(v_{l2})$.

$$V_{l2} = 0.8 \times 0.0216 \times W$$

$$V_{g2} = \text{total volume of vapor} = 0.2 \times W \times v_{g2} = 0.2 \times W \times 0.4456$$

$$V = \text{total volume of fluid through the valve} = V_{l2} + V_{g2}$$

$$V = W(0.8 \times 0.0216 + 0.2 \times 0.4456) = 0.1064W$$

G', the specific gravity of fluid flowing through the valve, is found next by

$$\frac{W}{(62.4)V} = \frac{W}{(62.4 \times 0.1064)W} = \frac{1}{62.4 \times 0.1064}$$

Therefore,

$$\text{Flow } Q = C_v \sqrt{\frac{\Delta P}{G'}} = 400 \sqrt{\frac{1{,}000}{62.4 \times 0.1064}}$$
$$= 400 \times 12.27 \text{ or approx. } \mathbf{4{,}908 \text{ gpm}}$$

E-16. Discuss two-phase flow of a flashing mixture of steam and water and show by example how calculations may be resolved.

ANSWER. The accompanying chart, Fig. E-9, entitled "Capacity and Outlet Pressure of Pipes Conveying a Flashing Mixture of Steam and Water," can be used to determine the maximum flow capacity of a mixture of water and steam through a pipe of a fixed constant diameter and a fixed equivalent length for the following inlet and outlet conditions:

Pipe inlet: water at saturation pressure P_i and temperature

Pipe outlet: critical pressure P_0 corresponding to maximum flow

The fluid at the outlet is a vapor-liquid mixture. Note that the outlet pressure P_0 indicated on the chart is always critical pressure. The chart is not applicable to cases where the container at the pipe outlet is maintained at a pressure exceeding the critical.

A typical example of such a flow is through a boiler blowoff line where P_i is the pressure at the steam disengaging drum. The chart is based on a formula derived from basic principles assuming an isenthalpic (constant enthalpy) flow and a constant Fanning friction factor of 0.004, following the method outlined by M. W. Benjamin and J. G. Miller in their paper "Flow of a Flashing Mixture of Water and Steam through Pipes," published in the ASME Transactions, vol. 64, pp. 657–669, 1942.

The equation used in the calculations, as adapted for arithmetic integration, is as follows:

$$\left(\frac{W}{a}\right)^2 = \frac{(5.79 \times 10^6) \sum_{\rho_0}^{\rho_i} \rho_w \times AP}{2 \ln \rho_1/\rho_0 + 0.192(L/d)}$$

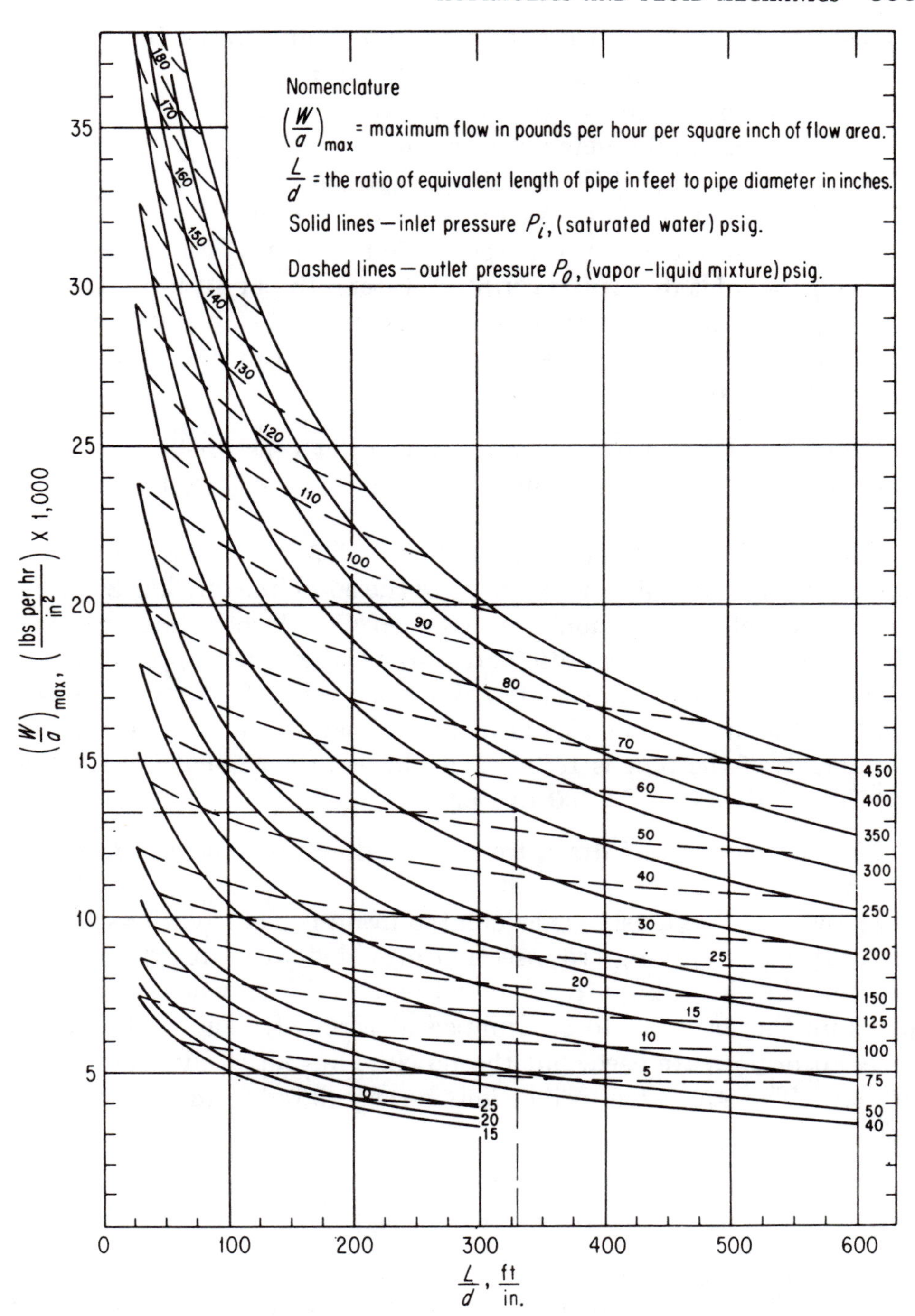

Fig. E-9. Capacity and outlet pressure of pipes conveying a flashing mixture of steam and water.

where W = flow of mixture of water and steam, lb per hr
a = cross-sectional area of pipe, sq in.
L = equivalent length of pipe, ft
d = inside diameter of pipe, in.
ΔP = pressure differential interval, psi
P_1 = inlet pressure (saturated water), psia
P_0 = outlet pressure (vapor-liquid mixture), psia
ρ_i = inlet density (saturated water), lb per cu ft
ρ_0 = density of vapor-liquid mixture at P_0, lb per cu ft
ρ_{av} = average density in any AP interval, lb per cu ft
ln = natural logarithm

Example. Obtain the capacity and outlet pressure for a 1½-in. line, Schedule 80, whose equivalent length is 495 ft, for an inlet pressure of 250 psig.

Refer to the chart, and for an L/d = 495/1.5 = 330, enter the **chart on the L/d scale and move up to the solid line marked 250 psig.** This point of intersection lies between the dashed lines marked 50 and 60 psig, and the outlet pressure is, therefore, approximately 54 psig. Turn and move left from this point of intersection and read on the W/a scale 13,400 lb per hr per sq in. The cross-sectional area of the pipe is 1.767 sq in.; hence, the capacity of the line is 1.767 × 13,400 = **23,700 lb per hr.**

E-17. Discuss vessel drain time when flow takes place through a pipeline.

ANSWER. Previously we calculated drain time for various vessels of constant or varying cross section provided with an orifice at the exit but did not take into account the actual case where a pipeline picks up the drainage to be dumped at a remote point. In such a case no orifices are used, but the pipeline has an equivalency to an orifice. The equivalent orifice may be calculated from the following equation:

$$\frac{d_p^3}{c_0^2[12fL + (c_L + c_E)d_p]} = d_0^4$$

This equation gives d_0, the equivalent orifice diameter, i.e., the diameter of an orifice which will produce the same pressure drop as a pipe of diameter d_p, equivalent length L, and average coefficient of friction f, the coefficients $(c_L + c_E)$ being equal to 1.5. This relationship does not apply to pipes with equivalent length L less than about five pipe diameters. Such short pipes are referred to as short tubes, and special discharge coefficients are to be used. The equation above is solved in graphical form; this graph has been published through the courtesy of *Petroleum Refiner*, where it appeared in the March 1960 issue. The article, "Graphs Find Vessel Drain Time," was written by W. A. Rostafinski. When using the chart, Fig. E-10, enter the base line with equivalent pipe length (straight pipe plus valves and fittings) and rise vertically until you strike the pipe size (in this case 3 in.), then move to the right and read the orifice diameter to be used in the calculation. The chart has a built-in orifice coefficient of 0.6.

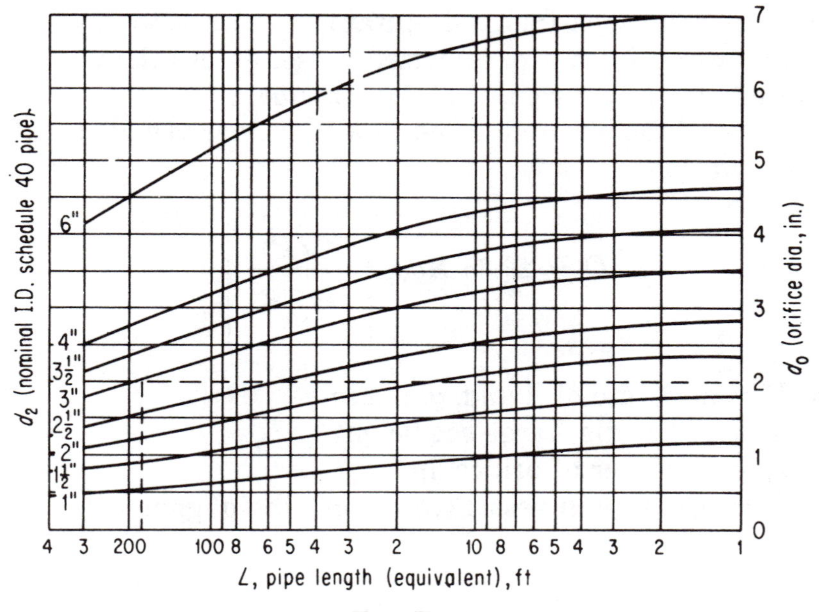

Fig. E-10

E-18. Pressurized vessels are often used in plant practice to control the injection rate of a liquid into a process system. When connected to a source of compressed gas, a pressure vessel with an orifice for liquid discharge will eject liquid at a controlled rate. The rate at which the liquid is forced through the orifice depends on the orifice diameter, gas pressure, and liquid density.

Kerosene is to be injected into a pressurizing system by means of nitrogen gas under a constant pressure of injection. The injection takes place through an orifice. Determine the flow rate of injection.

Data and Assumptions

Sp gr kerosene = 0.8.

Pressure of processing system = 14.7 lb/in^2 abs.

Injection pressure constant at 100 lb/in^2 gauge.

Orifice diameter = 0.5 in.

Energy losses in the vessel, lines, and orifice are negligible.

Expanding gas behaves as an ideal gas in isothermal conditions, i.e., pressure × volume = constant, when its reduced pressure is relatively low.

Liquid is incompressible and contains no dissolved gases.

Liquid flowing from a pressurized vessel enters the process system at ambient pressure, i.e., 14.7 lb/in^2 abs.

Pressures of the gas and liquid in a container are identical.

Basic equation

$$Q = 29.85\, K d_o \left| \frac{\sqrt{P}}{\sqrt{\text{sp gr}}} \right|$$

where Q = flow rate of liquid, gal/min
K = orifice discharge coefficient
d_o = diameter of orifice, in
P = pressure in vessel (constant), lb/in^2 gauge

ANSWER.

$$Q = 29.85 \times 0.6 \times 0.5 \frac{\sqrt{100}}{\sqrt{0.8}} = 100 \text{ gal/min} \qquad \textbf{Ans.}$$

Usually, a specific flow rate is required for a liquid injection. The above empirical formula may be used with the necessary variables. In a single-charge system, gas pressure diminishes and the liquid flow rate also decreases from the initial discharge rate as the liquid leaves the vessel. Here a different equation is used.

$$Kd_0^2 \frac{t}{v_{g1}\sqrt{\text{sp gr}}} = -0.00228\,(P_1 + 14.7)\frac{\sqrt{P_2}}{P_2 + 14.7} - \frac{\sqrt{P_1}}{P_1 + 14.7} + 0.261\tan^{-1}\frac{\sqrt{P_2}}{3.834} - 0.261\tan^{-1}\frac{\sqrt{P_1}}{3.834}$$

where t = time of flow, min
v_{g1} = initial volume of gas, gal
P_1 = initial pressure of gas, lb/in^2 gauge
P_2 = final pressure of gas, lb/in^2 gauge

The quantity of liquid injected during a specific time period is found from $V = v_{g2} - v_{g2}$, gal.

E-19. Water is being pumped from a deep well through an open-ended pipeline to develop the aquifer. Pipeline length from the well pumps to the plant site is 1000 ft, and the flow velocity in the pipeline is 5 ft/s at a pressure of 250 lb/in^2 gauge. Water temperature is 60°F. During the test a plant attendant shut off the flow of water by turning the handle on a plug valve at the end of the open-ended pipeline instantaneously, and as a result the pipe connection back at the pump discharge was ruptured.

(*a*) Determine the pressure rise and the peak line pressure developed in the pipeline by the sudden shutoff.

(*b*) If the valve had been closed gradually, say in 5 s, what would have been the pressure rise and peak line pressure under those conditions? In both cases, neglect the elasticity of water and the pipe.

ANSWER.

(*a*) Effect of sudden shutoff:

$$\text{Pressure rise} = \frac{\rho\, c\, v}{g} = \frac{62.4 \times 4701 \times 5}{32.2 \times 144} = 316 \text{ lb/in}^2 \qquad \textbf{Ans.}$$

$$\text{Peak line pressure} = 250 + 316 = 566 \text{ lb/in}^2 \text{ gauge} \qquad \textbf{Ans.}$$

(*b*) Gradual closure:

$$\text{Pressure rise} = \frac{\rho L}{g}\left(-\frac{dv}{dt}\right)$$

where p = water density = 62.4 lb/ft^3
L = length of pipeline = 1000 ft
$-dv/dt$ = deceleration of the liquid stream = 5/5 = 1 ft/s^2
Assume $-dv/dt$ is constant or $-dv/dt = -1$ ft/s^2. Then

$$\text{Pressure rise} = \frac{62.4 \times 1000 \times 1}{32.2 \times 144} = 13.5 \text{ lb/in}^2 \qquad \textbf{Ans.}$$

$$\text{Peak line pressure} = 250 + 13.5 = 263.5 \text{ lb/in}^2 \text{ gauge} \qquad \textbf{Ans.}$$

Note velocity of sound in water (acoustic velocity) at stated conditions = 4701 ft/s. Thus we see that, had the valve been closed gradually, the pump connection would not have ruptured.

E-20. A siphon piping system is connected to a reservoir as shown in Fig. E-11. Determine the maximum height in feet that can be used for the siphon in a water system if the length of the pipe from the water source to its highest point is 500 ft, the water velocity is 13 ft/s, pipe is 10 in in diameter, the water temperature is 70°F, and the flow is 3200 gal/min

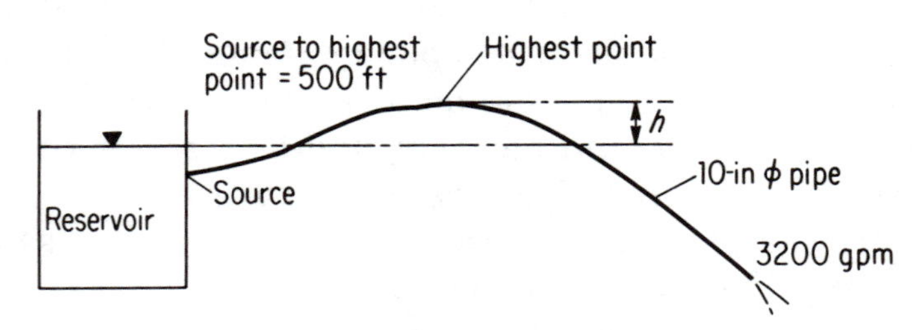

Fig. E-11

ANSWER. Determine vapor pressure of the water by use of steam tables. The vapor pressure of the water at 70°F is found to be P_v = 0.3631 lb/in^2 abs or 52.3 lb/ft. The water's specific volume from the steam tables is found to be 0.01606 ft^3/lb. Converting this to density, 1/0.01606 = 62.2 lb/ft^3. The vapor pressure in feet of 70°F water is f_v = 52.3/62.2 = 0.84 ft of water.

Now determine the friction head loss and the velocity head. From the reservoir to the highest point of the siphon point B in Fig. E-11, the friction head in the pipe must be overcome. Use the Hazen-Williams formula or a pipe friction table to determine the friction head. Use the Hydraulic Institute Pipe Friction Manual,

h_f = 4.59 ft per 100 ft of pipe or (500/100)(4.59) = 22.95 ft. From the same table, velocity head = 2.63 ft.

Finally, for a siphon handling water, the maximum allowable height, h ft, at sea level with atmospheric pressure at 14.7 lb/in^2 abs is found to be equal to (14.7 × 144 in^2/ft^2 per density of water at operating temperature, lb/ft^3) − (vapor pressure of water at operating temperature, ft + 1.5 × velocity head, ft + friction head, ft). Thus,

$$h = (14.7 \times 144/62.2) - (0.84 + 1.5 \times 2.63 + 22.95) = 11.32 \text{ ft}$$

In actual practice, the value of h is taken as 0.75 to 0.80 of the computed value. Thus, 11.32 × 0.75 = 8.5 ft. **Ans.**

E-21. Determine the minimum wall thickness and schedule number for a branch steam pipe operating at 900°F if the internal steam pressure if 1000 lb/in^2 gauge. Use ANSA B31.1 Code for Pressure Piping and the ASME Boiler and Pressure Vessel Code values and equations where they apply. Steam flow rate is 72,000 lb/h.

ANSWER. Assume a steam velocity of 12,000 ft/min, which is a reasonable one. From steam tables, the specific volume v is found to be 0.7604 ft^3/lb. Determine cross-sectional area:

$$a = \frac{2.4(72{,}000)(0.7604)}{12{,}000} = 10.98 \text{ in}^2$$

$$\text{ID} = 2\left(\frac{a}{\pi}\right)^{1/2} = 3.74 \text{ in} \qquad \text{Use a 4-in-diameter pipe}$$

Determine the pipe schedule number. The ANSA Code for Pressure Piping defines schedule number as SN = 1000 P_1/S. Assume for this pipe and operating conditions that a seamless ferritic alloy steel (1 percent Cr, 0.55 percent Mo) pipe is used.

$$\text{SN} = \frac{(1000)(1014.7)}{13{,}100} = 77.5$$

$$= 1000\,\frac{P_1}{S}$$

Use next higher SN, which is 80. **Ans.**

Pipe-wall thickness may be found from piping handbooks. The thickness for a 4-in, SN 80 pipe is 0.337 in.

Note. Use the method given above for any type of pipe—steam, oil, water, gas, or air—in any service—power, refinery, process, commercial, etc. Refer to the proper section of B31.1 Code for Pressure Piping when computing the schedule number, because the allowable stress S varies for different types of service.

E-22. A storage tank containing a flammable liquid is to be provided with a quick-opening valve to an underground sump for draining in the event of a fire. The tank is 20 ft in diameter, and a 20-ft depth of liquid is normally stored. The valve and dumping system is assumed to be equivalent to a 6-in-diameter orifice. The liquid has a density of 50 lb/ft^3. Also assume a constant orifice coefficient. How long will it take to empty the tank?

ANSWER. Assume large Reynolds number

Orifice coefficient = C_o = 0.61

Pressure drop = height of liquid

Orifice diameter/tank diameter = D_o/D_T = much less than 1

$A_o = 0.7854\ (1/2)^2 = 0.1963\ \text{ft}^2$

Flow = $(0.61)(0.1963)\ \sqrt{2gh} = 4.30$

$A_T = \pi\ (20)^2/4 = 314\ \text{ft}^2$

$$\text{And } 314\ dh/dt = -0.958\ \sqrt{h}\ \int_0^{20} dh/\sqrt{h}$$

$$= -(0.985/314) \int_0^t dt$$

$2\sqrt{h} = -0.00305 + C$. Note when $t = 0$, $h = 20$. Therefore, $C = 8.94$.

$$t = \frac{-2\sqrt{h} + 8.94}{0.00305} = \frac{8.94}{3.05} = 2920 \text{ s or } 48.67 \text{ min} \qquad \textbf{Ans.}$$

Time to empty tank from full level to zero level = about 48.67 min.

E-23. An elevated tank supplies water to a closed process vessel, the flow being due to gravity alone. It is desired to double the rate of flow by increasing the size of the pipe used. One engineer states this can be effected by doubling the size of the pipe in diameter. Another claims that the pipe cross section should be doubled. You are asked to settle this dispute. Present your decision, together with supporting calculations. You may assume a constant water level in the tank and in the process vessel, turbulent flow in the pipe, negligible kinetic energy, and negligible entrance and exit losses. Estimate the Fanning friction factor from the equation: $f = (0.0460)/(N_R \times 0.2)$.

ANSWER. See Fig. E-12. And $(P_2/\rho) - Z_1(g/g_c) - h_f = 0$ via Bernoulli. Friction loss through the line must remain constant so

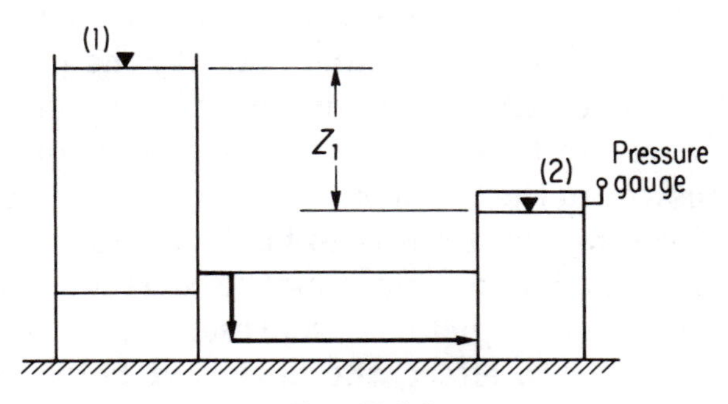

Fig. E-12

long as P_2 and Z_1 do. The Fanning friction factor $f = (h_f\, Dg_c)/(2Lv^2)$ = constant.

$$h_f = 2\frac{0.046/R_e^{0.2}}{Dg_c} Lv^2 = \frac{2(0.046)(L)(v^2)}{(D^{0.2}\rho^{0.2}v^{0.2})\rho^{0.2}} D_c$$

$$h_f = \frac{P_2}{\rho} - Z_1\frac{g}{g_c} \qquad h_f = \frac{0.092Lv^{1.8}\mu^{0.2}}{D^{1.2}\rho^{0.2}g_c}$$

Flow rate $= W = \rho A v$

$$W^{1.8} = \rho^{1.8}A^{1.8}v^{1.8} = \frac{\rho^{1.8}A^{1.8}h_f D^{1.2}\rho^{0.2}g_c}{(0.092)\ L\ \mu^{0.2}}$$

$$W^{1.8} = \frac{\rho^{1.8}\pi^{1.8}D^{3.6}}{4^{1.8}}\,\frac{h_f\rho^{0.2}g_c D^{1.2}}{0.092L\ \mu^{0.2}}$$

Everything is constant except W and D; hence

$$W^{1.8} = BD^{4.8}$$

where B is a constant.

For the original case, $W_1^{1.8} = BD_1^{4.8}$. Double flow rate $(2W_1)^{1.8} = BD_2^{4.8}$. Taking ratios gives

$$\left(\frac{2\ W_1}{W_1}\right)^{1.8} = \frac{BD_2^{4.8}}{BD_1^{4.8}} \text{ or } \frac{D_2}{D_1} = 2 \times \frac{1.8}{4.8} = 2^{0.374} = 1.296 \text{ or } 1.3$$

Therefore, doubling pipe diameter is wrong. Doubling cross-sectional area is increasing diameter by 2 = 1.414; so this is wrong. Use $D_2 = 1.3D_1$ to double the flow.

E-24. Describe how you would proceed to set up a computer program for pressure drop in a pipeline. Write the pipe friction formula in FORTRAN or any other computer language.

ANSWER. There are two stages in preparing a computer for calculating the answer to the problem. The first consists of analyzing the problem, deciding what needs to be done, and adopting a plan. The most-used method of determining the correct sequence of operations of this plan is by pictorial representation, or in computer jargon, a flowchart. In effect, the flowchart helps you analyze a system and determine if it has computer application. The rule here is simple: If you can write down the course of action you follow to get the answer to a problem, the action can be programmed into a computer. The computer's apparent capability is largely due to skillful and imaginative programming, whereby the basic building blocks provided by the computer—its input, storage, control, arithmetic, output, and decision functions—are assembled into an almost limitless variety of configurations.

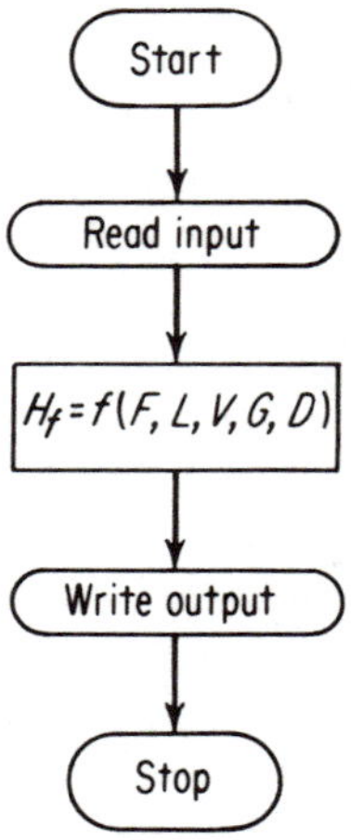

The second stage is to write the action in a form the computer can understand and to decide on a means of identifying the objects at hand. This involves a translation process.

Flow Chart. The computer program is based on the following pipe friction equation by Darcy. See Fig. E-13.

Start

Read input

$H_f = f(F, L, V, G, D)$

Write output

Stop

Flow chart

Fig. E-13

$$H_f = \frac{2FLV^2}{GD}$$

The input data required for the program are listed in the nomenclature:

H_f = friction head loss, ft
F = friction factor (dimensionless)
L = pipe length, ft
V = velocity of flow, ft/s
G = gravitational constant, 32.2 ft/s^2
D = pipe diameter, ft

If so desired, the friction equation may be simplified by combining constants to: $H_f = FLV^2/(16.1 \times D)$. The input data are entered on specially prepared data sheets, using problem-oriented language. After scanning the input data, the computer links to the program. See FORTRAN coding form reproduced in Fig. E-14.

The pipe friction formula in FORTRAN is

$$\mathrm{HF} = \frac{\mathrm{F * L * V * V}}{\mathrm{16.1 * D}}$$

E-25. A 3-in sharp-edged orifice is located in a 6-in-diameter pipe. A manometer connected to the orifice taps reads 1.2 in Hg. The pipe is transporting water at 70°F through 500 ft of pipe to the top of a building. The total static head on the pump is 50 ft. Determine the motor horsepower required, assuming an overall efficiency of 78 percent.

ANSWER.

Pressure drop through orifice $= 1.2 \times 1.133 = 1.36$ ft of water

$Q = A_o C_d\,(2g\Delta h_.)^{1/2}$, where $A_o = 0.7854(3/12)^2 = 0.0491$ ft^2, and $C_d = 0.61$

C_d evaluated from $d/D = 3/6 = 0.5$ ratio. Then,

$Q = 0.0491 \times 0.61(64.4 \times 1.36)^{1/2} = 0.2803$ ft^3/s (actual flow rate)

Also, $Q = A_p V_p$, where $A_p = 0.7854\ (6/12)^2 = 5148$ ft/h

hp $= Q\rho H_t/550$, where $H_t = H_s + H_f + H_v$, and $Q =$ ft^3/s, $\rho =$ 62.27 lb/ft^3

$H_s = 50$ ft given, but H_f will be evaluated from use of Reynolds number to find friction factor, so that $H_f = fLV^2/2gD$, where $L =$ 500 ft.

FORTRAN CODING FORM

PROGRAM HF= $\frac{2FLV^2}{gD}$

PROGRAMMER | DATE | PUNCHING INSTRUCTIONS | GRAPHIC | PUNCH | PAGE OF | CARD ELECTRO NUMBER*

STATEMENT NUMBER (1–5)	CONT. (6)	FORTRAN STATEMENT (7–72)	IDENTIFICATION SEQUENCE (73–80)
		NI=5	
		READ(NI,100)F,EL,V,G,D	
100		FØRMAT(5F10.3)	
C		EL = L	
		HF=2 *F*EL*V*V/(G*D)	
		NØ=6	
		WRITE(NØ,101)HF,F,EL,V,G,D	
101		FØRMAT(1X,VALUES FØR HF,F,L,V,G,D ARE RESP '),6F10.3)	
		STØP	
		END	

*A standard card form, IBM electro 888157, is available for punching statements from this form

Fig. E-14

Reynolds number = $VD\rho/\mu$ = (5184 × 0.5 × 62.27)/(2.37) = 6.76 × 10^4. At this value for Reynolds number the friction factor for smooth pipes is f = 0.021. Then

H_f = (0.021 × 500 × $(1.43)^2$/(64.4 × 0.5) = 0.67 ft of water

H_v = $V^2/2g$ (since all the water must be accelerated to the velocity in the orifice). And $v_o = Q/A_o$ = 0.2803/0.0491 = 5.71 ft/s

H_v = $(5.71)^2$/64.4 = 0.506 ft of water

Therefore, H_t = 50 + 0.67 + 0.506 = 51.2 ft and finally

hp = 0.2803 × 62.27 × 51.18 (550 × 0.78) = 2.082 hp, say, **2.0 hp**

Supplement F

THERMODYNAMICS, HEAT AND POWER

Questions and Answers

F-1. A rocket combustion chamber is supplied with 20 lb per sec of hydrogen and 80 lb per sec of oxygen. Before entering the nozzle all of the oxygen is consumed, the pressure is 20 atmospheres (atm), and the temperature is 4900°F. Specific heat ratio k of the combustion products may be assumed constant and equal to 1.25. Neglecting disassociation and friction, calculate the required throat area of the nozzle.

ANSWER. As in all problems involving gas flow the condition of flow should first be determined, whether retarded or unretarded flow exists. Since the downstream or discharge pressure is atmospheric while that at the throat is 20 atm, the flow is said to be unretarded. Critical flow exists at the nozzle throat. The formula that may be used in the solution of this problem is

$$G = A_t P_1 \left(\frac{2}{k+1}\right)^{1/(k-1)} \left(\frac{2gk}{R_m T_1 (k+1)}\right)^{1/2} \qquad \text{lb per sec}$$

where A_t = nozzle throat area, sq in.
P_1 = upstream pressure, psia, $20 \times 14.7 = 294$ psia
P_c = critical pressure, psia
R_m = gas constant of mixture
T_1 = chamber temperature, °R, $4900 + 460 = 5360$°R

The only unknowns needed to set up the above equation are W and R_m. First solve for W. We know it requires 8 lb of oxygen to burn theoretically 1 lb hydrogen. On this basis, there will be formed from the theoretical chemical equation

$$2H_2 + O_2 = 2H_2O$$

that 32 lb of oxygen are needed to burn 4 lb of hydrogen. On this basis, then, 8 lb of oxygen are needed to burn 1 lb of hydrogen. It appears that only 10 lb of hydrogen are used up to combine with the 80 lb of available oxygen, leaving 10 lb of hydrogen uncombined. In the products of combustion there will be

90 lb of superheated steam + 10 lb excess hydrogen = 100 lb mixture

The gas constant for the mixture R_m may be simply and quickly determined by

$$R_m = 1{,}544/\text{molecular weight of mixture}$$

The molecular weight of the mixture is equal to total weight divided by the total number of moles in the mixture. Thus,

$$M_m = \frac{100}{(90/18) + (10/2)} = 10$$

$$R_m = 1{,}544/10 = 154.4$$

Repeating the original equation for weight flow through nozzle and inserting the values now available,

$$100 = A_t \times 20 \times 14.7 \left(\frac{2}{2.25}\right)^4 \left(\frac{(2 \times 32.2 \times 1.25}{(154.4 \times 5{,}360 \times 2.25)}\right)^{1/2}$$

Solve for A_t. This is found to be equal to **82.8 sq in.**

F-2. An isentropic convergent nozzle is used to evacuate air from a test cell which is maintained at stagnation pressure and temperature of 300 psia and 800°R, respectively. The nozzle has inlet and outlet areas of 2.035 and 1.000 sq ft, respectively; the nozzle is secured to the wall of the test cell with anchor bolts imbedded in the wall. Constant specific heat c_p is to be taken as 0.24; the specific heat ratio is to be taken as 1.4; and the gas constant R is to be taken as 53.35 lb force foot per lb mass °R. Calculate the total tensile load on the anchor bolts.

ANSWER. See Fig. F-1. Also refer to flow through nozzles, page 109. Since back pressure on the nozzle is considered atmospheric the flow is critical or unretarded. The nozzle reaction force is exerted upon the curved portion of the nozzle contour causing tension in the holding bolts. The critical pressure $P_c = 0.53 \times P_1$. Or $0.53 \times 300 = 159$ psia. Upon leaving the nozzle the gas expands rapidly and the inertia of the rapidly expanding air stream sets up a series of expansion and compression waves.

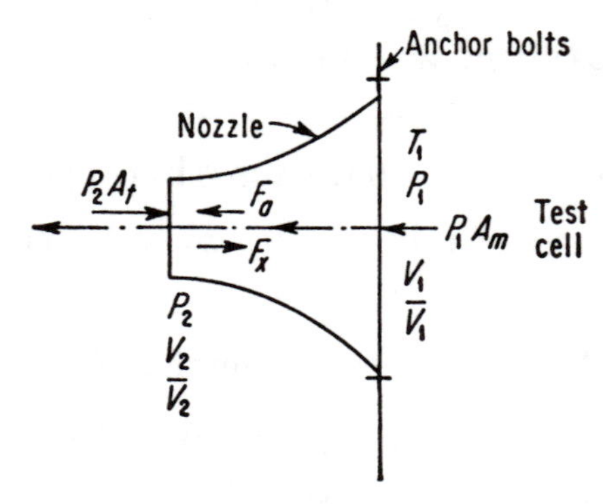

Fig. F-1

Now let the following nomenclature apply:

v_1 = gas velocity at mouth of nozzle, fps
v_2 = gas velocity at nozzle throat, fps
G = gas flow rate, lb per sec
P_1 = pressure at nozzle mouth, psia
P_2 = pressure at nozzle throat P_c, psia
A_m = nozzle mouth area, 2.035 sq ft
A_t = nozzle throat area, 1 sq ft
P_1A_m = force of gas on nozzle mouth, lb
P_2A_t = force of gas on nozzle throat, lb
F_x = axial component of the force F which is exerted upon the curved portion of the nozzle by the expanding gas; this is equal and opposite to the tension forces upon the bolts
F_a = force causing acceleration of the gas, lb

$$F_a = Ma = \frac{G}{g}(v_2 - v_1), \text{ lb}$$

Refer to Fig. F-1, and you will note that the force at the throat is against the nozzle throat. Thus

$$F_a = P_1A_m - P_2A_t - F_x$$

or

$$F_x = P_1A_m - P_2A_t - F_a = P_1A_m - P_2A_t - \frac{G}{g}(v_2 - v_1)$$

Now we must determine the flow G by

$$G = \frac{0.532A_tP_1}{\sqrt{T_1}} = \frac{0.532 \times 1 \times 300 \times 144}{\sqrt{800}} = 810 \text{ lb per sec}$$

In order to calculate the gas velocities at mouth and at throat, we must first find the specific volumes at these two points. Assuming a perfect gas,

$$\bar{v}_1 = \frac{RT_1}{P_1} = \frac{53.35 \times 800}{300 \times 144} = 0.985 \text{ cu ft per lb}$$

For ideal isentropic expansion of gas,

$$\bar{v}_2 = \bar{v}_1 \left(\frac{P_1}{P_2}\right)^{1/k} = 0.985 \left(\frac{300}{159}\right)^{0.714} = 1.574 \text{ cu ft per lb}$$

Throat velocity is

$$v_1 = \frac{G\bar{v}_1}{A_m} = \frac{810 \times 0.985}{2.035} = 392 \text{ fps}$$

$$v_2 = \frac{810 \times 1.574}{1} = 1{,}275 \text{ fps}$$

Finally,

$$F_x = (300 \times 144)2.035 - (159 \times 144)1 - (810/32.2)(1{,}275 - 392)$$
$$= 43{,}200 \times 2.035 - 22{,}900 - 22{,}200 = \mathbf{42{,}700\ lb}$$

F-3. Air at a mixture pressure of 20 psia and a temperature of 240°F with a relative humidity of 2 per cent is used to heat 5,000 lb of solid plastic material prior to a certain manufacturing process. The plastic has a specific heat of 0.36 Btu/(lb)(°F). When placed in the oven, the plastic is at 40°F and when removed, it is at 100°F. The air flow is controlled in such a way that its leaving temperature is always 100°F.

(*a*) Calculate the air quantity in pounds required on the assumption that the oven is perfectly insulated from its surroundings.

(*b*) If the heating process requires 46 minutes, specify the average capacity of the fan in cubic feet per minute.

ANSWER. The heat absorbed by the plastic is released by the air stream. If water vapor were a perfect gas, its enthalpy (total heat) would depend only upon temperature and would be independent of pressure; then, values read from the saturated steam tables could be used without error at pressures other than saturation pressure. Actually, the enthalpy of water vapor varies somewhat with pressure but for most psychrometric calculations the quantity of water vapor is small compared to the quantity of air and a slight error in the enthalpy of water vapor introduces only a negligible error in the enthalpy of air-water vapor mixtures. Thus, for calculations from 32 to 212°F it is sufficiently satisfactory to take the enthalpy of water vapor directly from saturated steam tables, particularly in view of the fact that the saturation pressure of steam is low over this temperature range.

Above 212°F the saturation pressure of steam increases rapidly and at high temperatures becomes many times as great as the total pressure—1 atm—of the air–water-vapor system. Therefore, for temperatures above 212°F, it is preferable to refer to superheated steam tables and to use the enthalpy of superheated steam tables and to use the enthalpy of air–water-vapor mixtures. The enthalpy of the air–water-vapor mixture $(H_m)_t$ at any temperature t can be calculated from

$$(H_m)_t = (H_a)_t + w(H_{wv})_t \qquad \text{Btu per lb dry air}$$

where $(H_a)_t$ = enthalpy of dry air at temperature t, Btu per lb
w = humidity, lb water per lb dry air
$(H_{wv})_t$ = enthalpy of water vapor at temperature t, Btu per lb

Proceeding, the heat released by the air stream is the difference in enthalpies between the condition of air in and air out. There is no change in moisture content of the air stream, so that the amount of water vapor in the air originally remains unchanged.

Enthalpy–Air In. First determine the humidity ratio w_i,

$$w_i = \frac{24.97 \times 0.02}{20 - 0.5} \times 0.622 = 0.016 \text{ lb moisture per lb dry air}$$

From superheat steam tables and dry-air tables in the ASHRAE Guide or any other standard handbook

$$(H_m)_t = (0.24 \times 240) + 0.016(1{,}160.4) = 76.1 \text{ Btu per lb dry air}$$

Here the enthalpy was taken from saturated steam tables because temperature was close to 212°F, acceptable here.

Enthalpy—Air Out. Assume negligible pressure loss of air system. Humidity is the same leaving as entering system.

$$(H_m)_t = (0.24 \times 100) + 0.016(1{,}104.4) = 41.7 \text{ Btu per lb dry air}$$

Heat released to plastic material is by difference:

$$76.1 - 41.7 = 34.4 \text{ Btu per lb dry air.}$$

(*a*) The weight of air required is

$$\text{wt plastic} \times \text{sp ht} \times (100 - 40) = 5{,}000 \times 0.36 \times 60 = 108{,}000 \text{ Btu}$$

$$108{,}000/34.4 = \mathbf{3{,}140 \text{ lb dry air}}$$

(*b*) From molal volume relationships, assume a draw-through fan system with the fan handling air at 100°F. Also neglect affect of humidity at such low-moisture content.

$$3{,}140 \times \frac{379}{29} \times \frac{460 + 70}{460 + 60} \times 1/46 = \mathbf{905 \text{ cfm}}$$

Let us see just what difference in air volume will result if we base our calculations on the dry air with the original moisture in it.

$$\frac{24.97 \times 0.02}{20 - 0.5} = 0.0257 \text{ cu ft vapor per cu ft dry air}$$

Since there is no loss or gain of water vapor, this ratio will be constant. The total volume of vapor plus dry air is

$$(905 \times 46 \times 0.0257) + (905 \times 46) = 1{,}070 + 41{,}600 = 42{,}670 \text{ cu ft}$$

$$42{,}670/46 = \mathbf{928 \text{ cfm}}$$

F-4. Two air streams, *A* and *B*, are mixed in direct contact with each other. These are the pertinent data:

	Streams	
	A	*B*
Mass flow lb dry air per hr	2,000	3,000
Pressure, psia	5	5
Relative humidity, per cent	50	90
Dry-bulb, °F	60	85

Assuming constant pressure conditions, what will be the enthalpy, dry bulb, and relative humidity after mixing?

ANSWER. Applying the same treatment as in Prob. F-3, we shall first find enthalpies of each stream and then combine them.

(*a*) *For stream A*,

$$\frac{0.25614 \times 0.5}{5 - (0.25614 \times 0.5)} \times 0.622 = 0.0164 \text{ lb water per lb dry air}$$

$$\text{Weight of moisture in stream } A = 0.0164 \times 2{,}000 = 32.8 \text{ lb}$$

$$(H_m)_t = 14.48 + (0.0164 \times 1{,}087.2) = 32.30 \text{ Btu per lb dry air}$$

$$\text{Enthalpy of stream } A = 2{,}000 \times 32.30 = 64{,}600 \text{ Btu per hr}$$

For stream B,

$$\frac{0.59588 \times 0.9}{5 - (0.59588 \times 0.9)} \times 0.622 = 0.0746 \text{ lb water per lb dry air}$$

$$\text{Weight of moisture in air stream } B = 0.0746 \times 3{,}000 = 223.8 \text{ lb}$$

$$(H_m)_t = 19.32 + (0.0746 \times 1{,}098.3) = 101.1 \text{ Btu per lb dry air}$$

$$\text{Enthalpy of stream } B = 3{,}000 \times 101.1 = 304{,}000 \text{ Btu per hr}$$

$$\text{Enthalpy of mixture} = \frac{64{,}600 + 304{,}000}{2{,}000 + 3{,}000} = \mathbf{73.8 \text{ Btu per lb dry air}}$$

(*b*) Dry bulb of the mixture, assuming negligible change in specific heat, is

$$\frac{2{,}000}{5{,}000} \times 60 = 24 \qquad \frac{3{,}000}{5{,}000} \times 85 = 51 \qquad 24 + 51 = \mathbf{75\ F}$$

(*c*) Relative humidity of the mixture at 75°F dry-bulb temperature is

$$\frac{32.3 + 223.8}{5{,}000} = \frac{0.42969x}{5 - (0.42969x)} \times 0.622 = 0.0512$$

$$x = 0.885 \qquad \text{relative humidity} = \mathbf{88.5 \text{ per cent}}$$

F-5. Saturated steam at 300 psia flows through a 4-in. standard steel pipe 500 ft long at 1,000 lb per hr. The line is insulated with 2 in. of 85 per cent magnesia. If the pipe is situated in a 20°F atmosphere, what is the quality of the steam at the end of the 500-ft run?

ANSWER. The pressure drop is negligible and we can assume a constant pressure system. Temperature corresponding to 300 psia is 417.33°F. From appropriate tables in the ASHAE Guide the heat loss through a 4-in. pipe insulated as indicated in the problem will lose 0.32 Btu per hr per linear foot of insulated pipe. Then the heat loss will be for the entire length

$$Q = 0.32 \times 500(417.33 - 20) = 63{,}500 \text{ Btu per hr}$$

With heat of condensation at 300 psia equal to 809.3 Btu per lb condensed the weight of steam condensed within the line may be determined as

$$63{,}500/809.3 = 78.5 \text{ lb per hr}$$

Per cent wet = (78.5/1,000)100 = 7.9. Then quality is

$$1 - 0.079 = 0.921, \textbf{ 92.1 per cent}$$

This is a reasonable quality for steam lines operating saturated at source point.

F-6. A 20,000-kw turbogenerator is supplied with steam at a pressure of 300 psia and a temperature of 650°F. The back pressure is 1 in. Hg abs. At best efficiency the combined steam rate is 10 lb per kwhr.

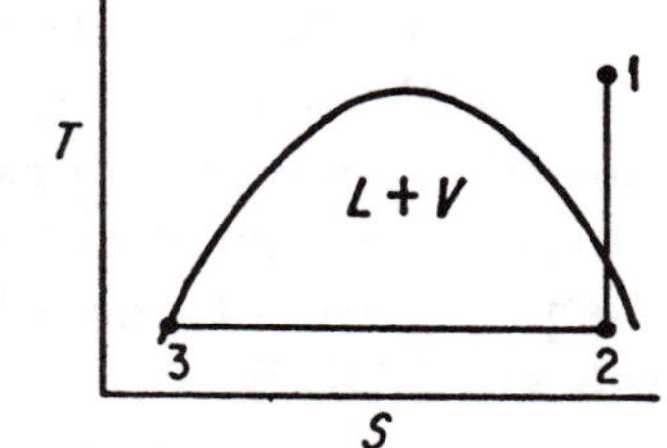

Fig. F-2

(*a*) What is the combined thermal efficiency (CTE)?

(*b*) What is the combined engine efficiency (CEE)?

(*c*) What is the ideal steam rate?

ANSWER. Refer to Fig. F-2. The combined thermal efficiency is given by

$$(a)\ \text{CTE} = \frac{3{,}413}{w_r}\frac{1}{h_1 - h_3}$$
$$= 3{,}413/10 \times 1/(1{,}340.6 - 47.06)$$
$$= 0.264, \text{ or } \textbf{26.4 per cent}$$

$$(b)\ \text{CEE} = \frac{w_i}{w_e} = \frac{\text{wt of steam used by ideal engine}}{\text{wt of steam used by actual engine}}$$

The weights of steam may also be expressed as Btu per lb. Thus, for the ideal engine the value is 3413 Btu per lb; for the actual engine, $H_1 - H_2$. We already know from the steam tables (or Mollier diagram) the value of H_1. Since the steam expands isentropically into the wet region below the dome of the TS diagram we must first determine the quality at point 2 by calculation. It may also be determined directly by use of the TS diagram for steam. By calculation and application of the method of mixtures

$$S_1 = 1.6508 = S_2 = 0.0914 + x_2\,1.9451$$

$$x_2 = \frac{1.6508 - 0.0914}{1.9451} = \frac{1.5594}{1.9451} = 0.80$$

$$H_2 = 47.06 + 0.8 \times 1{,}047.8 = 47.06 + 838 = 885.06 \text{ Btu per lb}$$

$$\text{Ideal steam ram rate } w_i = \frac{3{,}413}{H_i - H_2} = \frac{3{,}413}{1{,}340.6 - 885.06}$$

$$w_i = 3{,}413/455.54 = \mathbf{7.48\ lb\ per\ kwhr} \qquad (Ans.\ c)$$

$$\text{CEE} = 7.48/10 \times 100 = \mathbf{74.8\ per\ cent} \qquad (Ans.\ b)$$

F-7. A turbogenerator is operated on the reheating-regenerative cycle with one reheat and one regenerative feedwater heater. Throttle steam at 400 psia and a total steam temperature of 700°F are used. Exhaust at 2 in. Hg abs. steam is taken from the turbine at a pressure of 63 psia for both reheating and feed-water heating. Reheat to 700°F.

For the ideal turbine working under these conditions, find:

(*a*) Percentage of throttle steam bled for feed-water heating.

(*b*) Heat converted to work per pound of throttle steam.

(*c*) Heat supplied per pound of throttle steam.

(*d*) Ideal thermal efficiency.

(*e*) Draw temperature-entropy diagram showing boiler, turbine, condenser, feed-water heater and piping.

ANSWER. Throttle conditions are designated with subscript 1. From steam tables and the Mollier diagram we have the following:

$$P_1 = 400 \text{ psia}$$
$$t_1 = 700°\text{F}$$
$$H_1 = 1362.2 \text{ Btu per lb}$$
$$S_1 = 1.6396$$
$$H_2 = 1178 \text{ Btu per lb}$$
$$H_3 = 1380.1 \text{ Btu per lb}$$

(*a*) Percentage of throttle steam bled for feed-water heating is

$$\frac{\text{Heat added}}{\text{Heat supplied}} = \frac{H_6 - H_5}{H_2 - H_5} = \frac{265.27 - 69.10}{1{,}178 - 69.10} = \frac{196.17}{1{,}107.9}$$
$$= 0.1771, \text{ or } \mathbf{17.17 \text{ per cent}}$$

(*b*) Heat converted to work per pound of throttle steam is

$$H_1 - H_2 + 0.8229(H_3 - H_4) = 1{,}362.2 - 1{,}178 + 0.8229(1{,}380.1 - 1{,}035.8)$$
$$184.3 + 0.8229(344.3) = 184.3 + 283 = \mathbf{467.3 \text{ Btu per lb}}$$

(*c*) Heat supplied per pound of throttle steam is

$$H_1 - H_6 + (H_3 - H_2) = 1{,}362.3 - 265.27 + 1{,}380.1 - 1{,}178$$
$$1097.03 + 202.1 = \mathbf{1299.13 \text{ Btu per lb}}$$

(*d*) Ideal thermal efficiency is

$$\frac{\text{Heat converted to work}}{\text{Heat supplied}} = \frac{467.3}{1{,}299.13} = 0.361, \text{ or } \mathbf{36.1 \text{ per cent}}$$

(*e*) Refer to Figs. F-3 and F-4.

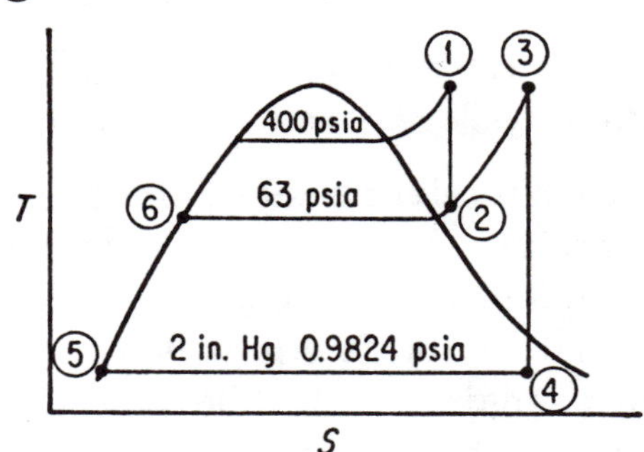

Fig. F-3

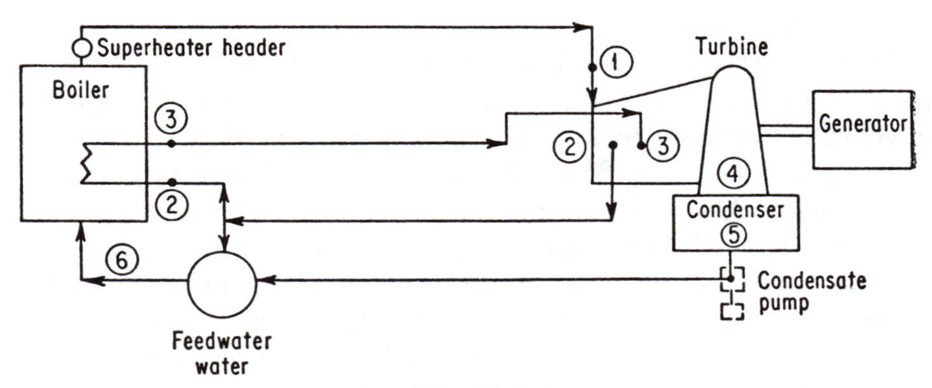

Fig. F-4

F-8. Determine the force acting on a piston 1 in. in diameter if it is pushed 3 in. into an airtight cylinder 4 ft long. Consider that piston is pushed quickly.

ANSWER. The initial pressure is atmospheric and the initial volume is found to be 37.7 cu in. *before* the piston is moved into the cylinder.

$$0.785 \times 1 \times 1 \times (4 \times 12) = 37.7 \text{ cu in.}$$

After moving 3 in. into the cylinder, the new volume becomes 35.3 cu in.

$$0.785 \times 1 \times 1 \times (3.75 \times 12) = 35.3 \text{ cu in.}$$

Assume the compression is polytropic with the compression exponent n equal to 1.35, then set up the following relation with the final pressure P_2:

$$14.7/P_2 = (0.0204/0.0218)^{1.35} \quad \text{or} \quad P_2 = 16 \text{ psia}$$

Pressure generated within the cylinder is $16 - 14.7 = 1.3$ psi. Finally,

$$\text{Force} = 1.3 \times 0.785 \times 1^2 = \mathbf{1.02\ lb}$$

where $0.0204 = 35.3/1{,}728$ and $0.0218 = 37.7/1{,}728$

F-9. A compressed-air receiver is maintained at 300 psia and 80°F. Air is led from the receiver through 2-in.-diameter pipe of Schedule 40 thickness. A complete rupture occurs in the line at a distance from the receiver equal to 100 ft equivalent. Assume compressor capacity is sufficient to maintain the above-stated temperature and pressure, and determine the quantity of air which escapes to atmosphere through the ruptured pipe. Assume air as perfect gas with a specific heat ratio of 1.4.

ANSWER. It is the acoustic velocity which sets the maximum flow rate of a gas in a pipe or other conduit. From the acoustic velocity assumed to take place at point of rupture (2-in.-diameter opening) we can find the maximum gas flow rate in cubic feet per second, giving the greatest possible pressure drop in the 100 ft of pipe. If the pressure drop calculated ensures critical flow, then the flow rate becomes fixed. The resulting pressure drop at acoustic velocity when subtracted from the steady receiver pressure will determine the type of flow (retarded or unretarded) into the atmosphere. The acoustic velocity may be determined in accordance with

$$v_c = \sqrt{kgRT} = \sqrt{1.4 \times 32.2 \times 53.5 \times 540} = 1{,}140 \text{ fps}$$

The flow rate at point of rupture is simply and quickly given by

$$\frac{1{,}140 \times 2^2}{183} = 25 \text{ cfs}$$

To ensure critical flow the minimum pressure upstream is

$$P_1 = 14.7/0.53, \text{ or } 28 \text{ psia}$$

where 14.7 psia is atmospheric pressure downstream of rupture. For unretarded flow, the flow rate is constant for all conditions of critical flow. To ensure critical flow, the pressure drop through the pipeline must not exceed 300 − 28 = 272 psi. For unretarded flow of air the flow rate in pounds per second is found to be

$$G = \frac{0.532A_0P_1}{\sqrt{T_1}} = \frac{0.532 \times 0.785(2/12)^2 \times 28 \times 144}{\sqrt{540}} = 2.1 \text{ lb per sec}$$

This shows the density of the air stream at point of exit to be 2.1/25, or 0.084 lb per cu ft. The free air flow rate is also found to be equal to 2.1/0.075 = 28 cfs. This is also equivalent to 28 × 60, or 1,680 cfm. From pressure-drop charts in standard manuals or standards of the Compressed Air Institute, the pressure drop for the problem at hand and for the conditions given is found to be approximately 8 psi per 100 ft of 2-in. pipe. Thus, the flow remains critical and the flow rate is constant at **2.1 lb per sec.**

F-10. The vacuum in a surface condenser is 28 in. Hg referred to a 30-in. barometer. The temperature in the condenser is 80°F. Find the per cent by weight of air present in this condenser.

ANSWER. At this temperature 1 in. Hg exerts a pressure of 0.4875 psi. Also, if the vacuum as stated above were 30 in. Hg referred to a 30-in. barometer, the absolute pressure in the condenser would be zero. In a condenser, steam (water vapor) is condensing in contact with water and may be taken as saturated.

$$\frac{0.5067}{(2 \times 0.4875) - 0.5067} \times 0.622 = 0.672 \text{ lb water per lb dry air}$$

$$\frac{1}{1 + 0.672} \times 100 = \textbf{60 per cent by weight of air}$$

F-11. When a pressure vessel contains 1 lb of water vapor at 300°F, the pressure gauge reads 10 psi. If 1 lb of dry air is pumped into the vessel with the temperature of the system remaining the same, what will be the new pressure reading?

ANSWER. Problem assumes no water vapor condensation. Assume molecular weight of the air is 29, that of water vapor 18. We will work this solution by finding the vessel volume by assuming the application of the perfect gas law. Add the 1 lb of air to the 1 lb of water vapor, and with the volume and temperature known we can determine the new pressure.

$$p_1V = NKT = (10 + 14.7)V = (1/18)(10.71)(460 + 300)$$

Then $$V = \frac{(1/18)(10.71)(760)}{24.7} = 18.4 \text{ cf vessel volume}$$

The average molecular weight of the mixture of water vapor and dry air is

$$M_m = \frac{2}{(1/18) + (1/29)} = 22.2$$

Set up perfect gas law formula to solve for the new pressure, p_2.

$$p_2 \times 18.4 = (2/22.2)(10.71)(760)$$

$$p_2 = (2/22.2)(10.71)(760)/18.4 = 40 \text{ psia, or } \mathbf{25.3\ psi}$$

F-12. Air discharged from a compressor before cooling to remove moisture is at 150 psig and 150°F. Consider the gas to be saturated with water vapor. Barometer is at 14.0 psia. To what temperature must the gas be cooled at the cooler outlet to prevent condensation in the distribution system where the pressure has dropped to 100 psig and the temperature is 50°F?

ANSWER. From steam tables, at 150°F the saturated vapor pressure is 3.716 psia. At the compressor discharge before cooling, the vapor content is determined

$$3.716/(150 + 14 - 3.716) = 0.0232 \text{ cu ft vapor per cu ft dry air}$$

At remote point where the temperature will be 50°F, the saturated vapor pressure is 0.178 psia, and the vapor content is

$$0.178/(100 + 14 - 0.178) = 0.00156 \text{ cu ft vapor per cu ft dry air}$$

This last figure is the maximum amount of water vapor content at the remote point and this must be the vapor content at the cooler discharge to prevent condensation. The cooler outlet temperature can now be calculated by determining the saturated vapor pressure at remote point:

$$\frac{p}{150 + 14 - p} = 0.00156$$

from which p is found to be 0.2554 psia. Refer to steam tables, saturated steam section. For a pressure of 0.2554 psia the corresponding saturated temperature is approximately **60°F.** This is the cooler outlet temperature desired.

F-13. 100 lb of steam, condensed at 150 psia, is discharged through a trap into the atmosphere without further cooling in the trap discharge line. What percentage of the condensate will "flash" upon discharging? Assume a constant enthalpy expansion and use method of properties of a wet mixture.

ANSWER. Use saturated steam table and check with Mollier diagram. Enthalpy of initial condensate before expansion is equal to the sum of the enthalpies (saturated vapor and saturated liquid) in the final wet mixture.

$$H_1 = H_f + XH_{fg} = 330.53 = 180.07 + X970.3$$

$$X = \frac{330.53 - 180.07}{970.3} = 0.015 \text{ or } \mathbf{1.5\ per\ cent}$$

F-14. An air compressor compresses air from atmospheric pressure to 70 psig at the rate of 400 cfm. The increase in internal energy is 1,200 Btu per min, and 300 Btu per min are rejected during the process through the aftercooler. If the initial specific volume is 13.55 and the final specific volume is 3.53, calculate the work done on the air by the compressor.

ANSWER. Hint: Total work is equal to change in internal energy plus heat rejected plus work of compression. The work done is

$$A(P_1V_1 - P_2V_2) = \frac{1}{778} \times 144[(14 \times 400) - \{(70 + 14 \times 400 \times 3.53/13.55)\}]$$

Work done $= 1{,}200 + 300 + 600 =$ **2,100 Btu**

F-15. Ten horsepower are absorbed during the compression strokes of an engine. During 1 min 100 lb of cooling water flowing through the cylinder jacket has its temperature raised 4.2°F by heat absorbed from the cylinder during the compression strokes. (*a*) Find change in internal energy of the medium being compressed in Btu per minute. (*b*) What kind of a process is this?

ANSWER. (*a*) $10 \times 33{,}000/778 = 424$ Btu per min

Heat absorbed by 100 lb water: $100 \times 1 \times 4.2 = 420$ Btu. This is equivalent to $420 \times 778 = 326{,}900$ ft-lb. Change in internal energy is equal to heat removed minus heat of compression, that is, $420 - 424 =$ **−4 Btu per min**

(*b*) This is a **polytropic process for gases.**

F-16. Superheated steam is generated at 1,350 psia and 950°F. It is to be used in a certain process as saturated steam at 1,000 psia. It is desuperheated in a continuous manner by injecting water at 500°F. How many pounds of saturated steam will be produced per pound of original steam?

ANSWER. Enthalpy steam at 1,350 psia and 950°F (H_1) = 1,465 Btu per lb. Enthalpy saturated steam at 1,000 psia (H_2) = 1,191 Btu per lb. Enthalpy water at 500°F (H_3) = 488 Btu per lb. Now let X equal lb of 500°F water required. Then

$$H_1 + XH_3 = (1 + X)H_2$$

$$X = \frac{H_1 - H_2}{H_2 - H_3} = \frac{1{,}465 - 1{,}191}{1{,}191 - 488} = 0.39$$

Thus, $1 + 0.39 =$ **1.39 lb** saturated steam produced per lb original steam.

F-17. Air at 40°F and 2,800 psia flows through a ½-in.-ID tube at the rate of 75 scfm. It is expanded through a valve into a ½-in. standard pipe (Sch. 40) to atmospheric pressure. Estimate the temperature of the air a short distance beyond the valve, assuming that a negligible amount of heat is transferred in from the surroundings.

ANSWER. Assume a constant enthalpy process. Then

$$H_1 - \frac{v_1^2}{2g(1/778)} = H_2 - \frac{v_2^2}{2g(1/778)}$$

Cfm at condition 1 is $75 \times (^{15}/_{2800}) \times (^{500}/_{520}) = 0.385$

$$v_1 = \frac{0.385 \times 144}{60 \times 0.25 \times 0.785} = 4.7 \text{ fps}$$

Now estimate t_2 equal to $-50°F$. Then

$$v_2 = (^{75}/_{60}) \times (144/0.304) \times (^{410}/_{520}) = 470 \text{ fps}$$

$$H_2 = H_1 - (^1/_{778} \times 1/64.4 \times 470^2) = 102 - 4.4 = 97.6 \text{ Btu per lb}$$

Temperature from enthalpy chart at 97.6 is **−48°F.**

F-18. When nitrogen at 100 atm and 80°F expands adiabatically and continuously to 1 atm through a throttle, the temperature drops to 45°F. A process for liquefying nitrogen involves a continuous flow exchanger system with no moving parts with the gas entering at 100 atm and 80°F unliquefied gas leaving at 70°F and 1 atm and liquid drawn off at 1 atm abs. It is estimated that the system is well enough insulated so that the heat leak is reduced to 25 Btu per lb mole of nitrogen entering. It is claimed that 3.5 per cent of the entering nitrogen can be liquefied on one pass through the apparatus. Do you think this is correct? Show basis for your answer.

ANSWER. Datum is at −460°F vaporization where enthalpy is zero.
Enthalpy (H_1) of nitrogen gas at 100 atm and 80°F = 125 Btu per lb
Enthalpy (H_2) of nitrogen gas at 1 atm and 70°F = 130 Btu per lb

$$Q = {}^{25}/_{28} = \text{heat leak Btu per lb of entering nitrogen}$$

Enthalpy (H_3) of liquid nitrogen at 1 atm and −322°F (77.4 Kelvin)

$$H_3 = 0.24(460 - 322) - (1{,}335 \times 1.8/28) = -53 \text{ Btu per lb}$$
$$Q + H_1 = (1 - X)H_2 + XH_3$$
$$X = \frac{H_2 - H_1 - Q}{H_2 - H_3} = \frac{4.11}{183} = 0.0224, \text{ or } \mathbf{2.24 \text{ per cent}} \text{ liquefied}$$

Only 2.24 per cent of the entering nitrogen can be liquefied.

F-19. A 4-ft-diameter duct carrying air of density 0.0736 lb per cu ft is traversed by a pitot tube using the 10-point method. The readings in in. of water at 72°F from one side of the duct to the other are, respectively, 0.201, 0.216, 0.220, 0.219, 0.220, 0.220, 0.218, 0.219, 0.220, and 0.216. Find the average velocity and the mass flow.

ANSWER. Average the square root of the readings:

$$\frac{\sqrt{0.210} + \sqrt{0.216} + \cdots + \sqrt{0.216}}{10} = 0.4668 = \sqrt{z_m}$$

The densities of water and air are: 62.3 lb per cu ft and 0.0736 lb per cu ft respectively. The mean $z_m = 0.216$ in. H_2O. This represents a head in ft of fluid flowing as

$$H = \frac{z_m(p_m - p_1)}{p_1} = \frac{0.216(62.3 - 0.074)}{12(0.074)} = 15.1 \text{ ft}$$

Finally, the average velocity:

$$V = \sqrt{2gH} = \sqrt{64.4(15.1)} = \mathbf{31.2\ fps}$$

F-20. Air at 60°F and atmospheric pressure is compressed, liquefied and separated by rectification into pure oxygen and pure nitrogen. The two gases are finally compressed into storage cylinders at 2,000 psig and 100°F. Calculate change in enthalpy, entropy, and energy for the whole process per 1,000 cu ft of air treated. Assume ideal gases. Use unit of Btu and degrees F. Assume air to consist only of oxygen and nitrogen.

ANSWER. Change in enthalpy is $H_{air} - H_{02} - H_{n2}$. Thus

$$\Delta H = 125 - (32/39 \times 0.21 \times 109) - (28/29 \times 0.79 \times 128)$$
$$125 - 25.4 - 98.6 = \textbf{Approximately zero}$$

Entropy for air $(0.79 \times 0.95 \times 28) + (21 \times 0.88 \times 32) - 1.987(0.79 \ln 0.79 + 0.21 \ln 0.21) = 28$ Btu/(lb mol)(°F)

Entropy for nitrogen $0.79 \times 0.605 \times 28 = 13.4$

Entropy for oxygen $0.21 \times 0.58 \times 32 = 3.9$

Mole air = 1,000/379 = 2.64

Change in entropy for above $= -2.64(28 - 17.3)$
$= \mathbf{-28.6}$ **Btu/1,000 cu ft air**

Energy change $\Delta H - \Delta PV = -\Delta PV = 2.64R(T_1 - T_2)$
$= 2.64 \times 1.987 \times (60 - 100) = \mathbf{-210}$ **Btu**

Supplement G

FUELS AND COMBUSTION PRODUCTS

Questions and Answers

G-1. Three-hundred pounds of carbon are burned with 65,000 cu ft of air initially at 14.7 psia and 300°F. Assume that the carbon is pure, and calculate the volumetric combustion analysis.

ANSWER. Refer to Chap. 7 and review the general considerations of combustion. Set up the theoretical combustion equation and note relative weights of each constituent. Refer to Table 7-1.

$$C + O_2 \rightarrow CO_2$$
$$12 + 32 \rightarrow 44$$

For theoretical combustion conditions and Table 7-1 the weight of air required to burn 1 lb of carbon is (2.67 + 8.78), or 11.45 lb. Theoretically, then, the total weight of air consumed by the 300 lb of carbon would be 300 × 11.45, or 3,435 lb. Now let us see just what weight of air was injected into the firing chamber.

$$65{,}000 \times \frac{460 + 60}{460 + 300} \text{ (no pressure correction)} = 44{,}500 \text{ cu ft}$$

The weight of this air represents (44,500/379)29 = 3,400 lb. The actual weight of air fed to the combustion chamber is short by 3,435 − 3,400, or 35 lb. Under these conditions incomplete combustion takes place and carbon monoxide appears in the flue gases. Now write out the theoretical combustion equation for CO formation.

$$2C + O_2 = 2CO$$
$$24 + 32 = 56$$

Total weight of carbon in the reaction is 300 lb. This weight must be distributed between that reacted upon to form CO_2 (let us call this part x) and part reacted upon to form CO (let us call this part $300 - x$). Now set up the following relationship in general terms first:

lb air per lb C to form $CO_2 \times x +$ lb air per lb C to form CO $(300 - x) =$ wt of total air

$$(32/12 \times 1/0.2313)x + (32/24 \times 1/0.2313)(300 - x) = 3{,}435$$
$$11.5x + 5.8(300 - x) = 3{,}435$$
$$11.5x + 1{,}740 - 5.8x = 3{,}435$$
$$x = (3{,}435 - 1{,}740)/5.7 = 1{,}695/5.7 = 298 \text{ lb}$$

Remembering that mole per cent is equivalent to volume per cent, calculate constituent breakdown on mole basis.

$$CO_2 = 298 \times \frac{44}{12} \times \frac{1}{44} = 24.8 \text{ moles} \qquad 24.8/119 = \mathbf{20.8 \text{ per cent}}$$

$$CO = 2 \times \frac{56}{24} \times \frac{1}{28} = 0.2 \text{ mole} \qquad 0.2/119 = \mathbf{0.2 \text{ per cent}}$$

$$N_2 = 3{,}435 \times 0.7687 \times 1/28 = 94 \text{ moles} \qquad 94/119 = \mathbf{79 \text{ per cent}}$$

$$\text{Total moles} = 119 \text{ moles}$$

Note that no combustibles were present in the combustion chamber. All the carbon was reacted upon. Some problems involve this complication and the next question will be so written.

G-2. A coal contains 4 per cent moisture, 23 per cent volatile matter, 64 per cent fixed carbon, and 9 per cent ash and has a heating value of 14,100 Btu per lb. Determination of the carbon in the coal shows it to be 79 per cent. The refuse removed from the ash pit of a grate-fired furnace using this coal contains 62 per cent moisture (due to wetting down of the ashes by hose to lay dust), 3 per cent volatile combustible matter, 11 per cent fixed carbon, and 24 per cent ash. Estimate the per cent of the heating value of this coal lost in the furnace as unburnt combustible and the per cent carbon fired which remains in the refuse.

ANSWER. Problem tendered in Q7-4, page 123, gave the ultimate analysis of the coal. The proximate analysis is presented here. Now proceed with solution.

The combustible matter in the refuse is not wholly coked fuel, as is evidenced by the presence of considerable volatile matter in it, neither is it wholly uncoked coal, since the ratio of fixed carbon to volatile in the fuel, $64/23 = 2.78$, is not the same as that, $11/3 = 3.67$, obtained from the refuse. It is fair to assume that some wholly unburnt coal has dropped through the grate, the amount of this being measured by the volatile matter in the refuse. And note, too, that additional coal, coked completely in passing over the grate, has not had all of the carbon burnt out of it.

These are reasonable assumptions, since in the actual coking process there is little loss of volatile matter until a certain temperature is reached. However, when decomposition starts it is completed in a relatively narrow temperature range without much further heat supply and, on a furnace grate, in a relatively short time.

The volatile matter in the refuse is, obviously, a measure and proportional to the uncoked coal. Therefore, the loss in heating value due to uncoked coal is estimated as follows: basis 1 lb coal as fired.

$$0.09 \text{ lb ash} \times \frac{3}{24} \times \frac{100}{23} = 0.049 \text{ lb uncoked coal in refuse}$$

This represents a corresponding loss in heating value of $0.049 \times 14{,}000$, or 690 Btu.

The total fixed carbon in the refuse is made up of that due to uncoked coal as well as that from the coke present. The former, expressed per 100 lb of refuse, is $3 \times 64/23 = 8.35$, whereas—the total being 11—the difference, 2.65, is that corresponding to the coke present. Thus, the pounds of carbon in the refuse are

$$0.09(2.65/24) = 0.0099 \text{ per lb of coal}$$

And the corresponding heating value is

$$(0.0099/12)(97{,}000)1.8 = 145 \text{ Btu}$$

Therefore, the total per cent loss is

$$\frac{690 + 145}{14{,}100} \times 100 = \mathbf{5.92\ per\ cent}$$

The presence of moisture in the refuse presents no difficulties or complications, since ratios alone are used to transform from one basis to another.

If there is 0.049 lb of uncoked coal in the refuse, there is $0.049 \times 0.79 = 0.0387$ lb of carbon in it. The total carbon unburnt is

$$0.0387 + 0.0099 = 0.0486 \text{ lb}$$

This is equal to

$$(0.0486/0.79)(100) = \mathbf{6.15\ per\ cent}$$

of the carbon in the coal.

G-3. Methane, CH_4, is burned in four times the theoretical air quantity, i.e., the excess air is 300 per cent. As a very crude approximation, it may be assumed that the specific heats of air methane, and the combustion gases are equal. If the air is burned at 100°F, estimate the products of combustion temperature if the higher heating value of methane is 21,500 Btu per lb.

ANSWER. Assume that specific heat for all gases in the products of combustion is 0.24 Btu/(lb)(°F), basis is 16 lb of methane, and that mixing is intimate. Heat absorbed is given by the equation: $Q = wc_p\,\Delta t$. Now determine the weights of each of the flue gas components. Set up the theoretical equation for combustion of methane and balance it out.

$$CH_4 + 2O_2 = CO_2 + 2H_2O + 21{,}500 \text{ Btu per lb methane}$$

Products of combustion from the equation per 16 lb methane:

$$\begin{aligned} CO_2 &= 44 \text{ lb} \\ H_2O &= 36 \text{ lb} \\ O_2 &= 64 \times 3 = 192 \text{ lb} \\ N_2 &= 64 \times (0.7687/0.2313) \times 4 = 850 \text{ lb} \end{aligned}$$

Since each component will suffer the same temperature increase while absorbing its share of the total heat of combustion (of 16 lb methane)

$$44 \times 0.24 \times \Delta t + 36 \times 0.24 \times \Delta t + 192 \times 0.24 \times \Delta t + 850 \times 0.24 \times \Delta t = 21{,}500 \times 16$$

$$10.6\,\Delta t + 8.65\,\Delta t + 46\,\Delta t + 204\,\Delta t = 344{,}000 \text{ Btu}$$

$$\Delta t = 344{,}000/269.25 = 1280°\text{F}$$

Final temperature = 1280 + 100 = **1380°F**

G-4. In the combustion of coal, what is the amount of heat liberated per pound of carbon in complete combustion to CO_2 and in combustion to CO?

ANSWER. To answer this question, we may take data on heating values directly from tables. The heating value of a fuel depends upon the combustion conditions: temperatures of fuel and air at moment of ignition, and temperatures of products of combustion when the released Btu's are measured. Various values will be found in the literature, some based upon 60°F, others on 32°F, etc. As a result, for instance, the heating value of CO ranges from 3900 to 4500 Btu.

The heat released by burning carbon to CO cannot be measured directly. It is taken as the difference between complete combustion of carbon to CO_2 and the combustion of CO to CO_2. The American Gas Association has published the following data based on temperatures of 32°F:

Carbon burned to CO_2 yields...........	169,686 Btu per mole C
Carbon monoxide burned to CO_2 yields..	−122,328 Btu per mole C
Difference is carbon burned to CO......	47,358 Btu per mole C

Since the molecular weight of carbon is 12, it requires 169,686/12, or **14,140** Btu per lb of carbon in complete combustion to CO_2. For combustion to CO it liberates 47,358/12, or **3946** Btu per lb carbon.

G-5. A certain coal requires 8 lb of air per lb coal for perfect combustion. If in burning this coal at a rate of 1 ton per hr, 50 per cent excess air is used entering at 70°F and flue-gas temperature of 500°F, what is the heat loss due to this excess air?

ANSWER. Consider specific heat of air a variable expressed by

$$c_p = 0.239 + 4.138 \times 10^{-10}T^2$$

Weight of air involved is $2{,}000 \times 8 \times (^{50}/_{100}) = 8{,}000$ lb per hr excess air. This weight of air will absorb and carry away with it up the stack the amount of heat we are to determine. Also $T_1 = 460 + 70$, and $T_2 = 460 + 530$. Then

$$Q_{1-2} = W \int_{530}^{960} c_p\, dT = 8{,}000 \int_{530}^{960} (0.239 + 4.138 \times 10^{-10})\, dT$$

By integration the answer is found to be **8.25 × 10⁵ Btu per hr**

Supplement **H**

THE STEAM POWER PLANT

Questions and Answers

H-1. An acceptance test on a boiler shows that its induced-draft fan handles 600,000 lb per hr of flue gas at 290°F against a friction of 10 in. WG. The boiler is located afterwards to a spot where the barometric pressure is 24 in. Hg and not the original 30 in. Hg at sea level. With no changes made to the equipment except adjustments to the fan and with gas weights and temperatures the same as before, what are the new volume and suction conditions for the fan design and relocation?

ANSWER. When a fan is required to handle air or gas at conditions other than standard (or any other basis), a correction must be made in the static pressure and horsepower. A fan is essentially a constant-volume machine and at a given speed in a given system the cfm will not materially change, regardless of density. The static pressure, however, changes directly with density. The static pressure must be carefully calculated for the specified conditions. Now for the problem at hand, assume a gas molecular weight of 28. Then the density correction factor is given as

$$24/30 = 0.8$$

There is no temperature correction. Then the new suction condition is

$$10 \times 0.8 = \mathbf{8\ in.\ WG}$$

The new volume condition

$$\frac{600{,}000}{28} \times 379 \times \frac{460 + 290}{520} \times \frac{24}{24 - 8} = \mathbf{17.6 \times 10^6\ cu\ ft\ per\ hr}$$

H-2. The allowable concentration in a certain boiler drum is 2,000 ppm. Pure condensate is fed to the drum at the rate of 85,000 gallons per hour (gph). Make-up, containing 50 grains per gal of sludge-producing impurities, is also delivered to the drum at the rate of 1,500 gph. Calculate the blowdown as a percentage of the boiler steaming capacity.

ANSWER. See also Q8-6. We must first equate the grains per gallon of sludge to ppm to simplify the calculations further on.

$$\text{ppm} = \frac{50}{8.33 \text{ lb per gal} \times 7{,}000 \text{ grains per lb}} = \frac{50}{8.33 \times 7{,}000} = 855$$

To maintain a steady concentration in the boiler drum the sludge-forming impurities rate of feed through the normal arrangement must equate the blow-down rate, or

$$\begin{aligned} \text{ppm to boiler} &= \text{ppm in blowdown} \\ 855 \times \text{gph make-up} &= 2{,}000 \text{ ppm} \times \text{gph blowdown} \\ 855 \times 1{,}500 &= 2{,}000 \text{ ppm} \times \text{gph blowdown} \\ \text{gph blowdown} &= 855 \times 1{,}500/2{,}000 = 640 \end{aligned}$$

Boiler evaporation (steaming capacity)

$$= \text{feed} - \text{blowdown} = 85{,}000 + 1{,}500 - 640 = 85{,}860 \text{ gph}$$

Finally, per cent blowdown is

$$640/85{,}860 \times 100 = \mathbf{0.747 \text{ per cent}}$$

H-3. A process industry is studying the economics of a single-effect vs. a double-effect evaporation. Motive steam is available at 150 psia, dry and saturated; raw water from underground wells is in ample supply at a temperature which may be assumed constant at 60°F. Final evaporator product will be condensed at atmospheric pressure. Calculate the ratio of evaporator product to pounds of motive steam for (*a*) single-effect evaporation and (*b*) for double-effect evaporation.

ANSWER. See Fig. H-1. Use the steam tables. The heat losses are negligible.

(*a*) For single-effect evaporation, heat in minus heat out:

$$1 \times (1{,}194.4 - 330.5) = X(1{,}150.4 - 28)$$

Solving for X, this is found to be equal to **0.77 lb of water evaporated per lb of steam supplied.**

(*b*) For double-effect evaporation, heat in minus heat out:

$$X(1{,}150.4 - 180) = Y(1{,}150.4 - 28)$$

$$Y = \frac{0.77(1{,}150.4 - 180)}{(1{,}150.4 - 28)} = 0.665 \text{ lb water evaporated per lb steam}$$

Total evaporation per lb steam $= X + Y = 0.77 + 0.67$
$= \mathbf{1.44 \text{ lb per lb}}$

The problem as presented does not give the true picture for a double-effect evaporator since the second evaporator is not truly operating

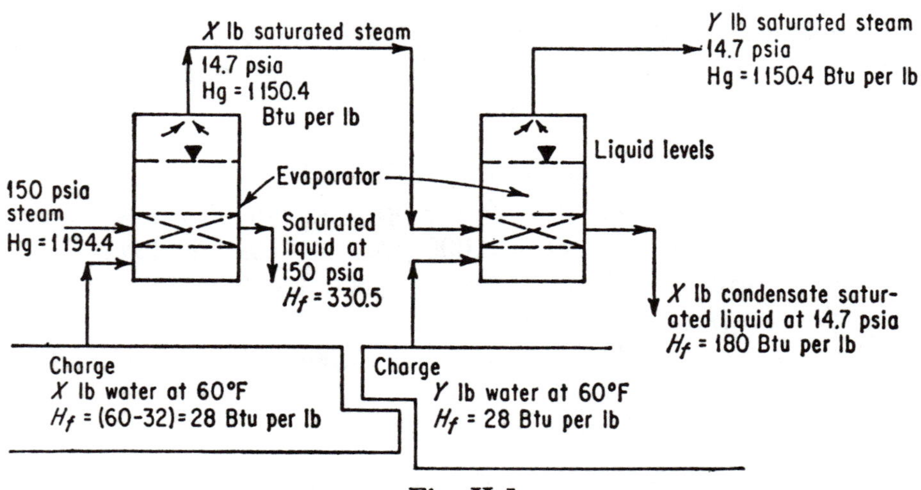

Fig. H-1

under a lower pressure than the first. The vapor coming from the first evaporator contains all the heat (no heat losses), latent in this case, put into it by the steam supplied to the first, and is reused as the steam supply to the second evaporator. In actual practice, the first effect could have been operated at a lower steam pressure, say 15 psia, because at lower pressures the latent heat is greater per pound. The second effect could have been operated at a vacuum of 18 in. Hg, heat being supplied by the vapor coming from the first effect at atmospheric pressure.

H-4. Superheated steam is generated at 1,350 psia and 950°F. It is to be used in a certain process as saturated steam at 1,000 psia. It is desuperheated in a continuous manner by injecting water at

500°F. How many pounds of saturated steam will be produced per pound of original steam?

ANSWER.

Enthalpy of steam at 1,350 psia and 950°F $= H_1 = 1465$ Btu per lb
Enthalpy of saturated steam at 1,000 psia $= H_2 = 1191$ Btu per lb
Enthalpy of water at 500°F $(500 - 32) = H_3 = 488$ Btu per lb

Let X = lb of 500°F water required, then

$$H_1 + XH_3 = (1 + X)H_2 \qquad X = \frac{H_1 - H_2}{H_2 - H_3} = \frac{1{,}465 - 1{,}191}{1{,}191 - 488} = 0.39$$

Finally, $1 + 0.39 =$ **1.39 lb sat. steam produced per lb original steam.**

H-5. By use of a suitable sketch, (*a*) show the major equipment, and indicate the principal circuits associated with a boiling-water nuclear

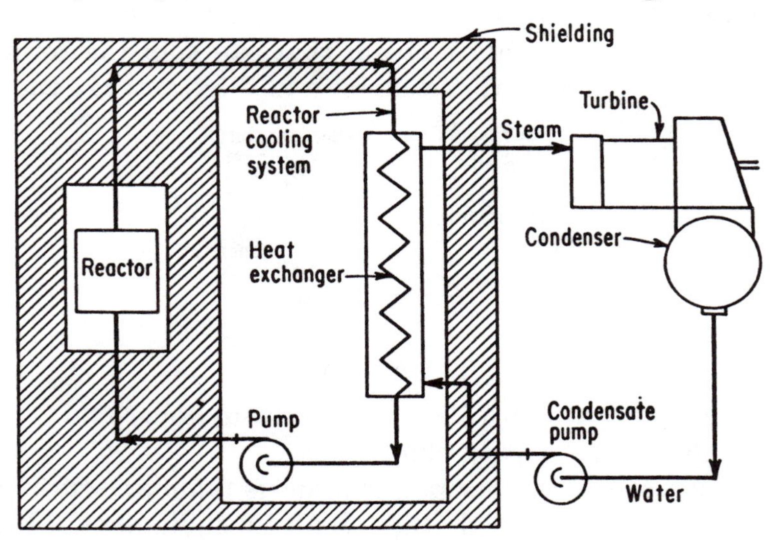

Fig. H-2

reactor power plant. (*b*) Name one advantage and one disadvantage of this cycle. (*c*) What is meant by negative temperature coefficient?

ANSWER.

(*a*) Figure H-2 shows a simplified schematic of a nuclear power-plant cycle. Fissioning (splitting) of uranium atoms in the reactor produces heat that is picked up by the cooling medium. The latter delivers this heat to the heat exchanger to generate steam to the turbine. The reactor and its cooling circuit must be heavily shielded to confine hazardous radiation within the reactor and cooling circuit.

In nuclear power plants, the heat energy comes not from burning fuel but from splitting atoms. This fission process takes place in what used to be called a "pile," because that was a good description of the stack of graphite and uranium bricks under the west stands of Stagg Field at the University of Chicago where the first controlled chain reaction was produced. Now we call the device in which fission occurs a "reactor."

A reactor can be likened to a furnace, not because it resembles one, but because it does the same job, releases heat. The nuclear reaction gives off so much heat that, even if we are not interested in making power, we must keep moving the heat to prevent the reactor from melting down. Heat in the coolant used can be transferred in a heat exchanger to generate steam that can spin the rotor of the turbogenerator. The heat exchanger and the reactor in the nuclear

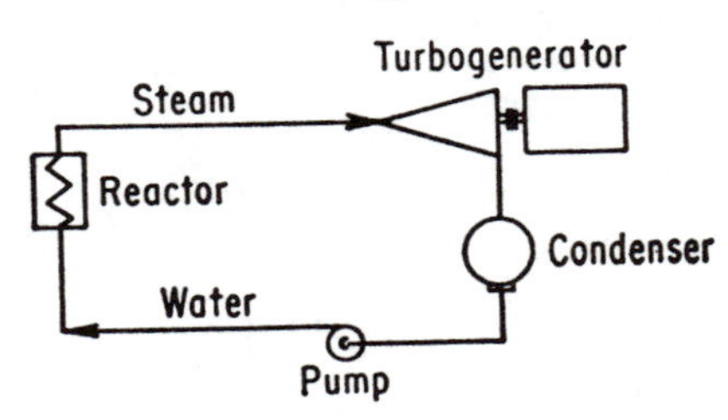

Fig. H-3

power plant are thus equivalent to the steam boiler and its furnace in the familiar steam power plant.

Thus much of the nuclear power plant is familiar—turbinegenerator, condenser, pumps, and other auxiliary equipment. But there are some significant differences—all stemming from the unseen but seething atomic activity in the reactor.

The answer to part (*a*) is shown in Fig. H-3. The coolant (which is also the moderator) picks up heat in the reactor core to form steam bubbles. The bubbles separate from the water at the water level and the steam leaves through the upper reactor tank nozzle. Feed water enters the reactor tank below to pass up through the fuel elements in the core as coolant and moderator. Steam leaving the reactor will be moderately radioactive so that the steam equipment must be shielded.

(*b*)

Advantages	*Disadvantages*
1. Simple pumping needs. Much saving in pumping power over other systems.	1. Possible radioactivity of steam-power equipment and piping may require some shielding. Accessibility for maintenance and repair may be seriously hampered.
2. Note lower fuel temperatures. This is a definite advantage.	2. Radioactive particles in steam may require special shaft seals on turbines, pumps, valves, etc.
3. Simplified equipment layout. Less equipment needed. Note no heat exchanger needed as in Fig. H-2.	3. Noncondensible gases removed from condenser may be radioactive.
4. Lower pressures required; thus their effect on pressure vessel expense.	

(*c*) The great advantage of a boiling-water reactor is direct steam production without an external heat exchanger so that steam of given conditions can be produced with the same temperatures and pressures in the reactor. The reactor can be designed so that it has a *negative temperature coefficient* of reactivity. The formation of steam voids, as the power attempts to go up, will tend to shut down the reactor, or in other words, it has a large negative temperature coefficient.

H-6. It is desired to estimate the fuel requirements for a nuclear reactor which is to operate at a thermal level of 500,000 kw continuously for a two-year period. The fuel is uranium containing 1½ per cent of the 235 isotope; fuel elements will be removed from the reactor for reprocessing when 1.0 per cent of the 235 isotope has fissioned.

Preliminary calculations indicate that 800 megawatt-days of energy result from the fissioning of 1 g of U^{235}. On this basis, and using the enrichment and burnout percentages above, determine the total weight in pounds of uranium (natural plus enriched) that must be put into the core.

ANSWER. For continuous operation, assume a plant factor of unity.

$$\frac{500 \text{ megawatts} \times 365 \text{ days} \times 2 \times 1}{800 \text{ megawatt-days}} = 457 \text{ g } U^{235} \text{ fissioned}$$

$$\text{Total weight} = \frac{457}{0.01 \times 0.015} \times \frac{1}{1{,}000} \times 2.2046 = \mathbf{6{,}700 \text{ lb}}$$

H-7. It is desired to determine by calculation the enthalpy and the per cent of moisture of the steam entering a condenser from a steam

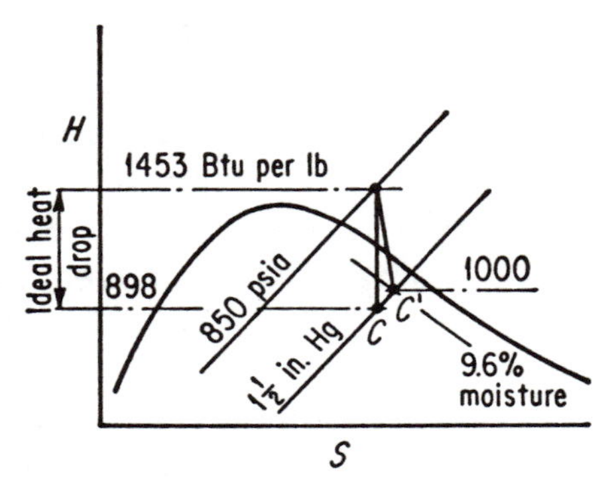

Fig. H-4

turbine. The turbine is delivering 20,000 kw and is supplied with steam at 850 psia and 900°F; the exhaust pressure is 1½ in. Hg abs. The steam rate, when operating straight condensing is 7.70 lb per delivered kwhr and the generator efficiency is 98 per cent.

ANSWER. Refer to Fig. H-4 (Mollier diagram). The engine efficiency is

$$E_e = \frac{3{,}413}{w_s(H_1 - H_c)} = \frac{3{,}413}{7.7 \times (1{,}453 - 898)} = 0.80 \text{ for ideal}$$

The Rankine engine efficiency for the steam turbine concerned is

$$\frac{0.80}{0.98} = 0.817 = \frac{H_1 - H_{c'}}{H_1 - H_c} \qquad \text{and} \qquad H_1 - H_{c'} = 0.817 \times 555$$
$$= 453 \text{ Btu per lb}$$

At the end of the actual expansion of the steam in the turbine $H_{c'}$ is

$$H_{c'} = 1453 - 453 = \mathbf{1000 \text{ Btu per lb enthalpy}}$$

where $H_{c'}$ crosses the pressure line of 1½ in. Hg, moisture per cent may be found to be **9.6 per cent.**

H-8. A 40,000-kw straight-flow condensing steam turbogenerator unit is to be supplied with steam at 800 psia and 800°F and is to exhaust at 3 in. Hg abs. The half-load and full-load throttle steam flows are estimated to be, respectively, 194,000 lb per hr and 356,000 lb per hr. The mechanical efficiency of the turbine is assumed to be 99 per cent and the generator efficiency 98 per cent.

(*a*) Compute the no-load throttle steam flow.

(*b*) Derive an equation for the heat rate of the unit expressed as a function of the output in kilowatts.

(*c*) Compute the internal steam rate of the turbine at 30 per cent of full load.

ANSWER.

(*a*) See Fig. H-5. Assume a straight-line rating characteristic; first determine the difference between full-load and half-load steam

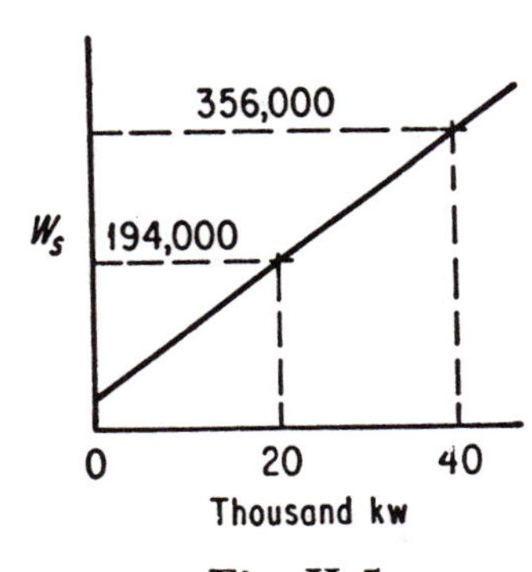

Fig. H-5

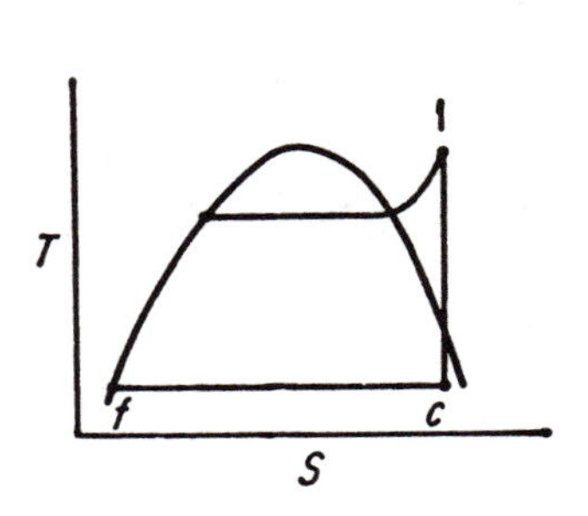

Fig. H-6

rates. This is 356,000 − 194,000 = 162,000. Then

$$194{,}000 - 162{,}000 = \mathbf{32{,}000\ lb\ per\ hr}$$

(*b*) Refer to Fig. H-6. Actual turbine efficiency is

$$E_t = \frac{3{,}413}{w_s(H_1 - H_f)}$$

$$\text{Heat rate} = \frac{3{,}413}{E_t}\ \text{Btu per kwhr} = w_k(H_1 - H_f)$$

Actual steam rate in pounds per kilowatt hour is $w_k = w_s/kw$.

$$w_s = 32{,}000 + \frac{162{,}000}{20{,}000} \times \text{kw} = 32{,}000 + 8.1 \text{ kw}$$

$$w_k = \frac{32{,}000}{\text{kw}} + 8.1$$

Now with $H_1 = 1{,}398$ and $H_f = 83$, then $H_1 - H_f = 1{,}315$. Finally,

$$\text{Heat rate} = \frac{1{,}315 \times 32{,}000}{\text{kw}} + [1{,}315(8.1)]$$

$$= 10{,}650 + \frac{42{,}100{,}000}{\text{kw}}$$

Let us see how the steam rate and heat rate compare for the quarter-load points.

At full load: w_k = 8.9 lb per kwhr   Heat rate = 11,700 Btu per kwhr
At ¾ load: w_k = 9.17 lb per kwhr   Heat rate = 12,080 Btu per kwhr
At ½ load: w_k = 9.7 lb per kwhr   Heat rate = 12,770 Btu per kwhr
At ¼ load: w_k = 11.3 lb per kwhr   Heat rate = 14,870 Btu per kwhr

(c) For engine and turbine combined

$$E_e = \frac{3{,}413}{w_k(H_1 - H_c)}$$

Also

$$H_1 = 1{,}398 \qquad H_c = 912 \qquad H_1 - H_c = 486$$

We saw that

$$w_s \text{ at full-load} = 356{,}000$$
$$w_s \text{ at no-load} = 32{,}000$$

For the full-load range the total change is

$$356{,}000 - 32{,}000 = 324{,}000$$

$$w_k \text{ at 30 per cent load} = \frac{32{,}000 + 0.30(324{,}000)}{0.30 \times 40{,}000} = 10.87 \text{ lb per kwhr}$$

$$E_i = \frac{3{,}413}{10.78 \times 486} = 0.651 \text{ for engine and turbine combined}$$

If we call the internal efficiency of the turbine (not including the friction loss) E_i, then

$$E_i = \frac{2{,}545}{w_a(H_1 - H_c)}$$

Thus $$E_i = \frac{E_e}{\text{mechanical efficiency} \times \text{generator efficiency}}$$

$$E_i = \frac{0.651}{0.99 \times 0.98} = 0.672$$

The actual steam rate in pounds per horsepower is

$$w_a = \frac{2{,}545}{E_i(H_1 - H_c)} = \frac{2{,}545}{0.672 \times 486} = \mathbf{7.8}$$

H-9. The high-pressure cylinder of a turbogenerator unit receives 1,000,000 lb per hr of steam at initial conditions of 1,800 psia and 1050°F. At exit from the cylinder the steam has a pressure of 500 psia and a temperature of 740°F. A portion of that steam is used in a closed feed-water heater to increase the temperature of 1,000,000 lb per hr of 2,000 psia feed water from 350 to 430°F; the balance passes through a reheater in the steam-generating reheater and is admitted to the intermediate-pressure cylinder of the turbine at a pressure of 450 psia and a temperature of 1000°F. The intermediate cylinder operates nonextraction. Steam leaves this cylinder at 200 psia and 500°F. Calculate each of the following:

(*a*) Flow rate to the feed-water heater, assuming no subcooling
(*b*) Work done, in kilowatts, by the high-pressure cylinder
(*c*) Work done, in kilowatts, by the intermediate-pressure cylinder
(*d*) Heat added by the reheater

ANSWER. See Fig. H-7. Use the steam tables and Mollier diagram to find the various enthalpies. Interpolate as required. Neglect pressure drop through feed-water heater in lines and other heat-exchange apparatus except as noted.

(*a*) Steam flow rate to feed-water heater: This is merely a heat balance.

$$\text{Flow to heater } (H_2 - H_7) = \text{water rate } (H_6 - H_5)$$

$$\text{Flow to heater} = \frac{1 \times 10^6(409 - 324.4)}{(1{,}379.3 - 449.4)} = \mathbf{91{,}100\ lb\ per\ hr}$$

(*b*) Work done by high-pressure cylinder:

$$\text{Work done} = \frac{1 \times 10^6(1{,}511.3 - 1{,}379.3)}{3{,}413} = \mathbf{38{,}700\ kw}$$

(*c*) Work done by intermediate-pressure cylinder:

$$\text{Work done} = \frac{(1 \times 10^6 - 91.1 \times 10^3)(1{,}521 - 1{,}269)}{3{,}413} = \mathbf{67{,}150\ kw}$$

(*d*) Heat added by reheater:

$$\text{Heat added} = 908{,}900(1{,}521 - 1{,}379.3) = \mathbf{128.9 \times 10^6\ Btu\ per\ hr}$$

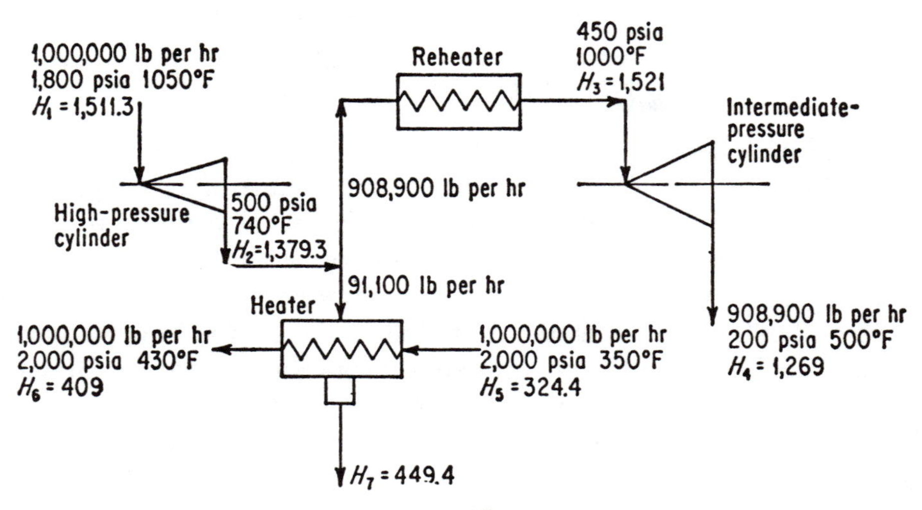

Fig. H-7

H-10. In certain proposed nuclear power plants, the nuclear reactor together with specific accessory apparatus is enclosed in a containment vessel. In the event of rupture in the reactor coolant system, the containment vessel is expected to withstand the sudden increase in pressure.

Describe your approach to the problem of specifying the maximum possible pressure for which the containment vessel should be designed; that is, how would you go about determining the maximum pressure which would develop following rupture of the reactor primary coolant system?

ANSWER. Assume the type of enclosure to be a "reactor only," one not housing the turbogenerator. By housing the turbine in a more nearly conventional building, it renders the plant more amenable to future modifications—an important consideration in a development project such as nuclear power plants are today.

From the nuclear safety standpoint, too, the "reactor only" style of enclosure offers other advantages and simplifies the problem of pressure calculation. The danger of a turbine explosion is minimized with the turbine outside, rotating in a plane not intersecting the containment vessel. Then, too, the danger of damage to the enclosure by explosion of the hydrogen used to cool the generator is eliminated. With less lubricating oil within the enclosure the fire safety is improved.

The enclosure design is usually based on adequacy for pressures that could be developed in severe terminal accidents. However, experience with present installations indicate the possibility of occurrence of any such accident would be extremely remote.

The internal pressure within the containment vessel (enclosure) in the event of a major accident could be created by contributions from three different potential sources of energy:

(*a*) The pressurized hot water in the reactor and in those of its auxiliaries which are not separated from it by a solid barrier. In the Dresden Nuclear Power Station outside Chicago, Illinois, a reactor rupture would expose 188 tons of pressurized hot water at 1,000 psig in a boiling condition. Volume of containment vessel is that from a 190-ft-diameter sphere.

(*b*) A nuclear excursion.

(*c*) Chemical reaction between reactor components, that is, between water and the zirconium cladding in which the uranium oxide fuel is encased.

The hot-water contribution is the most important of these. It is unlikely that any nuclear and chemical contributions that would be significant in comparison are probable.

Accordingly, the design pressure should be determined by the pressure that would be created by release and partial flashing of water to steam of the pressurized hot water contained in the reactor and auxiliaries with no significant nuclear or chemical energy contributions. For the Dresden installation at its rated operating pressure (1,000 psig) and rated thermal level (630 megawatts) the pressure expected to develop is about 25 psig.

Design pressure normally is 30 psig with a 1.25 times design pressure for pneumatic testing.

The basis for design temperature not required in this problem should not go without some word. The equilibrium temperature corresponding to a pressure of 25 psig design pressure plus 100°F brings the design temperature up to slightly over 300°F.

H-11. Water at 300°F is pumped into the boiler feed line which is at a pressure of 1,224 psia and that steam from the boiler leaves at 1,200 psia and a total steam temperature of 700°F. Find the heat added to each pound of water passing through the boiler, assuming no heat losses whatsoever.

ANSWER. Initially the water is a subcooled liquid. The heat added to each pound is the difference in the initial and final state points. The heat equivalent of work done on each pound of liquid by the feed water pump must be included. Enthalpy of the liquid initially is

$$\text{Enthalpy at } 300°\text{F} + [A(P_1 - P_a)]\bar{V}_a$$
$$269.5 + [(1/778)(144)(1{,}224 - 67)](0.0174) = 273.2 \text{ Btu per lb}$$

Subscript a denotes the condition on the saturated liquid line at 67 psia (the saturation pressure for 300°F). The final enthalpy at the superheater outlet is 1,310.6 Btu per lb from steam tables or Mollier diagram. Thus, heat added per lb = 1,310.6 − 273.2 = **1,037.4**

H-12. Determine the stack effect and i.d. fan needs for a fan located at base of a stack. Static pressure of the fan may be consumed in overcoming resistance in the system ahead of the fan. Thus there is no pressure available at fan discharge. Ambient temperature may vary between 85°F and −30°F. Flow of stack gases to range between 175,000 cfm at 180°F to 250,000 cfm at 250°F. Calculation is to be based on summer ambient of 85°F. In cold weather, efficiency of stack improves because of the higher temperature difference. Fan discharges into stack at 45°F up.

ANSWER. Theoretical draft or static at the base of stack calculated from the difference in weight between the column of gas and similar column of ambient air.

$$\begin{aligned}\text{Static pressure} &= (0.256) \times b \times H \times \left(\frac{1}{T_{\text{amb}}} - \frac{1}{T_m}\right)\\ &= (0.256) \times 28.75 \times 80 \times (1/545 - 1/640)\\ &= 0.108 \text{ in. water column}\end{aligned}$$

where b = barometric pressure at 1,100 ft elevation, in. H_g

H = stack height, 80 ft; stack diameter = 11 ft

$T_{\text{amb}} = 460 + 85 = 545R$

$T_m = 460 + 180 = 640R$

The gas flow through the stack depends on the losses which must be overcome, including friction loss through stack itself.

Air flow	175,000 cfm	250,000 cfm
Stack velocity	1,850 fpm	2,640 fpm
Velocity pressure @ 130°F avg	0.201 in. wc	0.422 in. wc
Elbow velocity	3,300 fpm	4,800 fpm
Velocity pressure @ 130°F avg	0.6 in. wc	1.25 in. wc
Stack friction loss	0.045 in. wc	0.075 in. wc
Stack exit loss	Negligible	Negligible
Elbow loss	0.164 in. wc	0.34 in. wc
Stack entrance loss	Negligible	Negligible
Total losses	0.209 in. wc	0.415 in. wc
Plus 10 per cent safety factor	0.021 in. wc	0.042 in. wc
Required stack draft	0.230 in. wc	0.457 in. wc
Available stack draft	0.108 in. wc	0.108 in. wc
Minimum in. wc to be supplied by fan at stack	**0.122 in. wc**	**0.349 in. wc**

Available stack draft alone cannot remove gases from furnace. Fan must be speeded up and larger fan motor provided to give the additional static needed.

H-13. For generation of steam at 90 psig, what is the advantage and percentage gain when boiler feedwater is heated from 50 to 210°F?

ANSWER. There is a definite advantage. A pound of dry saturated steam at 90 psig contains 1,187 Btu above 32°F, and when the feedwater is at the temperature of 50°F, or (50 − 32), 18°F above 32°F, for conversion into steam at the stated pressure each pound must receive 1,187 − 18 or 1,169 Btu. When feedwater is at 210°F, each pound of feedwater evaporated requires 210 − 50 or 160 Btu less, or only (1,169 − 160)/(1,169) × 100 or 86.3 per cent as much heat conversion into steam, and the gain would be **13.7 per cent.**

H-14. Choose a suitable feedwater regulator and combustion control for an industrial boiler serving the following loads: heating, 18,000 lb/h; process, 100,000 lb/h; miscellaneous uses, 12,000 lb/h. The boiler will have a maximum overload of 20 percent, and wide load fluctuations are expected at frequent intervals during operation. Pulverized-coal fuel will be used to fire the boiler.

ANSWER. Determine the required boiler rating by first finding the sum of the individual loads on the boiler, or 18,000 + 100,000 + 12,000 = 130,000 lb/h. With a 20 percent overload, the boiler rating must be 1.2 × 130,000 = 156,000 lb/h. With an additional reserve capacity to provide for unusual loads, the rated boiler capacity should be 1.1 × 156,000 = 171,500 lb/h, say 175,000 lb/h for selection purposes.

Choose the type of feedwater regulator to use. A boiler in the 75,000 to 200,000 lb/h range can use a relay-operated regulator with one or two elements when the load fluctuations are reasonable. With wide load swings the relay-operated three-element regulator is a better choice. In addition, since the boiler will encounter wide load swings, a three-element regulator is a wise choice and a safe one too. **Ans.**

Choose the type of combustion-control system. A stream-flow-airflow type of combustion-control system would probably be the best for the fuel and load conditions in this plant. This is the type of combustion-control system to use.

Any control system selected for a boiler should be checked out by studying the engineering data available from the control system manufacturer. **Ans.**

Excess pressure ahead of the feedwater regulator should be at least 50 lb/in^2 and should be controlled by regulation of the feed pump. Use excess pressure valves only when excess pressure varies more than plus or minus 30 percent. Where drum level is unsteady owing to high solids concentration or boiler feed or other causes, use next higher-class feed regulator. See Kallen, *Handbook of Instrumentation and Controls*, McGraw-Hill.

H-15. A pump valve in our circulating system is provided with a spring having a spring rate of 65 lb/in. During the pumping cycle, the valve opens to its full limit of 1 in. The physical dimensions of the helical spring are: 3.56 in outside diameter, No. 1 W and M gauge (0.283 in) wire diameter, free length of 4.25 in, and 6 total coils with ends squared and ground. When the valve is fully open, determine the total deflection, total load, and corrected maximum stress in the spring.

ANSWER. Initial compression = 100/65 = 1.538 in

Total deflection = 1.538 + 1.0 = 2.538 in **Ans.**

Valve opening load = 65 × 1.0 = 65 lb
Total spring load = P = 100 + 65 = 165 lb **Ans.**
Mean diameter of spring = 3.56 − 0.283 = 3.277 in
Spring index = D/d = 3.277/0.283 = 11.59 and Wahl correction K

where $K = (4C - 1)/(4C - 4) + (0.615)/(C) = 1.124$

$S_{vc} = (8PDX)/(\pi d^3)$

$S_{vc} = (8 \times 165 \times 3.277 \times 1.124)/(3.1416 \times 0.283^3) = 68{,}000$ lb/in² **Ans.**

H-16. A steam turbine with a water rate of 80 takes steam at 150 lb/in² gauge, dry and saturated, and expands it to a backpressure of 25 lb/in² gauge at its outlet. (*a*) Determine the quality of the steam at the backpressure. (*b*) With a reciprocating pump, what would be the steam quality under the same conditions of throttle and backpressure? (*c*) Discuss energy transformations within each type of unit.

ANSWER. Refer to Fig. H-8. This is an H-S diagram.

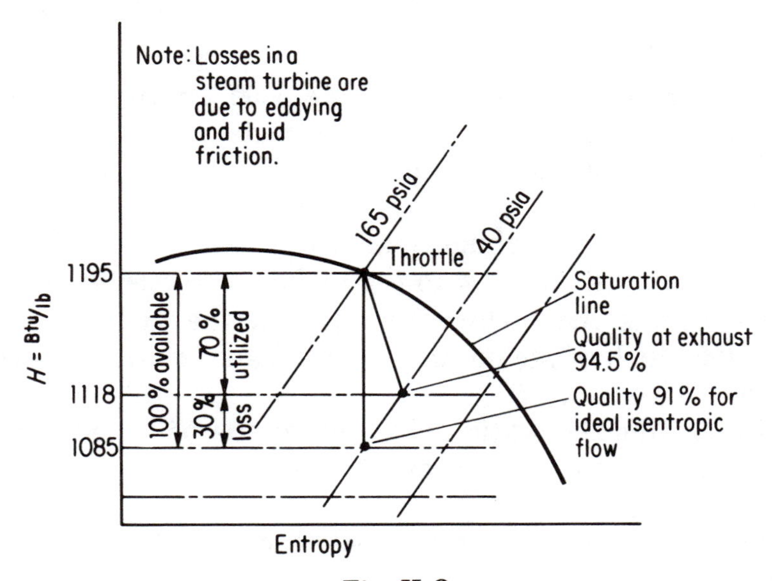

Fig. H-8

(*a*) Steam turbines with water rates of 60 to 80 transform 70 to 75 percent of the heat for energy. Turbines with higher rates transform 80 to 85 percent of the heat for energy. Reciprocating pumps or engines transform 50 percent of the heat available. The above data are based on practical tests.

Now, from Fig. H-8, steam tables, and Mollier diagram and assuming 70 percent utilization of heat

$$(1195 - 1085)(0.70) = 77$$
$$1195 - 77 = 1112 \text{ Btu/lb enthalpy at exhaust}$$

From Mollier diagram as shown, the quality of exhaust steam is found to be 91 percent. Ideal heat drop is equivalent to 100 percent availability of energy.

(*b*) For the reciprocating unit the ideal heat drop is still (1195 − 1085) = 110 Btu/lb steam. Percentage utilized is 50, so that the actual heat drop is (1195 − 1085)(0.50) = 55 Btu/lb steam. Then as before the quality is found to be 94 percent.

(*c*) The higher the thermal efficiency the lower the steam quality after expansion through the equipment. Also the higher the water rate the lower the steam quality at the exhaust. Exhaust from a reciprocating driver is high in steam quality because of wire drawing through the valve ports. Losses in steam turbines are due to eddying and fluid friction.

H-17. A processing plant operates a 5000-kW turbine having an engine efficiency of 73 percent. The initial steam conditions are 600 lb/in^2 abs, a total steam temperature of 600°F, and a backpressure of 1 in Hg.

(*a*) Determine the turbine steam rate.

(*b*) Determine steam flow to the turbine throttle, lb/h.

(*c*) Determine the turbine heat rate if the turbine auxiliaries require 4000 lb/h of steam and their exhaust heats the feedwater to 160°F.

(*d*) Determine the station heat rate if the boiler efficiency is 80 percent and the electrically driven boiler auxiliaries require 100 kW.

(*e*) What is the overall station efficiency?

ANSWER.

(*a*) Draw the T-S diagram for the process. See Fig. H-9.

Let $P_1 = 600$ lb/in^2 abs $p_2 = 1$ in Hg $\eta_e =$ engine efficiency

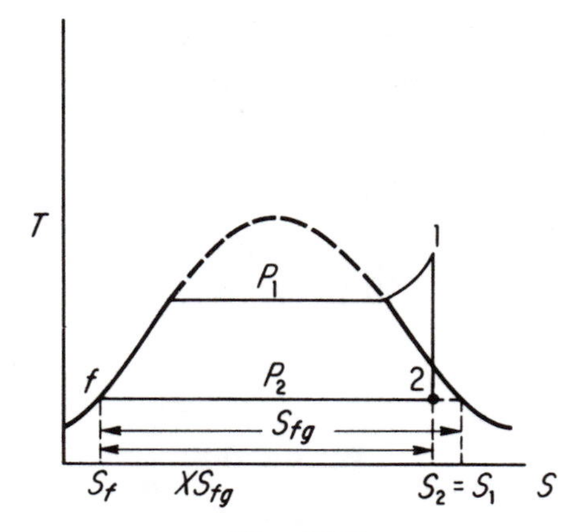

Fig. H-9

S = entropy	t_1 = 600°F	η_b = boiler efficiency
h = Btu/lb steam		X = quality of steam, percent

From steam tables at P_1 = 600 lb/in² abs and t_1 = 600°F

$$h_1 = 1290.9 \qquad S_1 = 1.5334 \qquad S_1 = S_2 \text{ isentropic expansion}$$

At 1 in hg (approximately 0.5 lb/in² abs)

$$h_f = 47.5 \qquad S_f = 0.0924 \qquad t_f = 79°\text{F}$$
$$h_{fg} = 1048 \qquad S_{fg} = 1.9434$$

Now $S_2 = S_f + X\,S_{fg}$ or $1.5334 = 0.0924 + X(19{,}434)$, from which $X = 0.74$, or steam quality at point 2 (fig. H-9) is 74 percent. Then, $h_2 = h_f + X\,h_{fg}$ or $h_2 = 47.5 + 0.74 \times 1048 = 823.5$ Btu/lb steam.

Note. h_2 and percent moisture can be obtained directly from the Mollier diagram. The above arithmetic method is more accurate and is given in case the Mollier diagram is not available.

Then ideal work W_i = ideal heat drop $= h_1 - h_2 = 1290.9 - 823.5 = 467.4$ Btu/lb

$$\text{Rankine steam rate} = \frac{3413}{h_1 - h_2} = \frac{3413}{476.4} = 7.3 \text{ lb/kWh}$$

Actual work $W_a = \eta_e X W_i = 0.73 \times 467.4 = 341$ Btu/lb steam

Turbine steam rate $= 3413/W_a = 3413/341 = 10$ lb/kWh **Ans.**

(*b*) Steam rate to throttle = 10 × 5000 = 50,000 lb/h **Ans.**

(*c*) Input = (throttle steam + auxiliary steam)($h_1 - h_f$). Then from steam tables h_f at 160°F = 127.9.

$$\text{Heat rate} = \frac{(50{,}000 + 4000)(1209.9 - 127.9)}{5000 \text{ kWh}}$$
$$= 12{,}560 \text{ Btu/kWh} \qquad \textbf{Ans.}$$

(*d*) Turbine heat rate per net kWh, where net kWh = 5000 − 100

$$\text{Turbine heat rate} = \frac{62{,}800{,}000}{4900} = 12{,}816 \text{ Btu/kWh}$$

The turbine heat rate per net kWh represents boiler output.

Then the station heat rate = (turbine heat rate/boiler efficiency) = 12,816/0.8 = 16,020 Btu/kWh **Ans.**

(*e*) Station efficiency = (3413/station heat rate) = 3413/16,020 = 0.213 or 21.3 percent **Ans.**

H-18. (*a*) What are the steps in sizing a control valve? (*b*) Size a steam control valve from the following situations: Inlet pressure of dry saturated steam at valve is 30 lb/in² abs. Flow rate is 1000 lb/h, with valve outlet pressure of 20 lb/in² abs. Valve type is straight-through (single-seat) throttling.

ANSWER.

(*a*) Select type of valve in accordance with practice.

Compute the critical pressure P_c.

Determine fluid density ρ using steam tables.

Calculate valve coefficient C_v.

Calculate $C_v/0.8$.

Select valve size, using manufacturer's rating tables.

(*b*) $P_c = P_1 \times 0.58 = 17.4$ lb/in² abs.

$P_2 = 20$ lb/in² abs.

Thus, P_2 is greater than P_c. And therefore flow is noncritical. ρ is based on P_2 conditions. Steam tables give the specific volume of the steam at P_2 = 20 lb/in² abs as 20.11 ft³/lb. Thus, $\rho = (1)/(20.11) = 0.05$ lb/ft³. Calculate C_v from the following standard formula:

$$C_v = \frac{W}{(63.5)(P_1 - P_2)\rho}$$

where C_v = valve flow coefficient
P_1 = valve inlet pressure, lb/in^2 abs
P_2 = valve outlet pressure, lb/in^2 abs
ρ = vapor (or gas) density, lb/ft^3
W = vapor or gas flow rate, lb/h

$$C_v = \frac{1000}{(63.5)(30 - 20)(0.05)} = 31.5$$

From manufacturers' ratings select a 2-in valve straight-through throttling single seat. For normal operation, maximum operating flow should not be greater than 80 percent of the maximum possible flow (C_v for the valve in question). Flow values of less than 10 percent should not be used; that is, calculated flow rate divided by C_v should not be less than 10 percent.

H-19. Industrial gases are often piped from water-sealed gas holders under suction to a processing plant some distance away. The gas is saturated with water vapor and is metered after pumping and the volume is automatically corrected to a base temperature and pipeline pressure as contracted for between gas supplier and customer. However, the water-vapor content is often neglected so that as delivery temperatures vary, actual volumes of gas and water vapor vary. If the customer has contracted for gas at 70°F delivered, its water-vapor content would be smaller than at a delivery temperature of 80°F. Since the customer will be billed for a volume of the saturated mixture at contract conditions, if the correction for water vapor is not made, he will pay for more gas than is actually delivered at the higher temperatures. On the other hand, the gas producer will be penalized for the lower-than-contract-temperature delivery.

Now, here's the situation. Barometric readings averaged at 29.21 in Hg. Average gas pressure as determined by radial planimeter from the flowmeter recording chart was 15.04 lb/in^2 gauge. Average gas temperature was likewise determined as 51.57°F. Uncorrected volume of gas passing through the flowmeter for the period covered by the chart was 349,400 ft^3.

(*a*) Determine the corrected volume of gas passed through the flowmeter and corrected to 60°F and 30 in Hg pressure and saturated with water vapor, i.e., contract conditions.

(*b*) What would be the difference if presence of water vapor were not corrected for?

(*c*) Who would be penalized, the gas producer or the customer?

ANSWER.

(*a*) Average absolute gas pressure

$$15.04 + (29.21 \times 0.4912) = 29.39 \text{ lb/in}^2 \text{ abs}$$

Gas volume reduced to standard conditions

$$V_c = 349{,}400 \times \text{correction factor}$$

$$V_c = 349{,}400 \times \frac{29.39 - 0.1888}{14.69 - 0.256} \times \frac{519.6}{51.57 + 459.6} \tag{H-1}$$

$$V_c = 349{,}400 \times 2.056 = 718{,}366 \text{ ft}^3$$

This is the volume of gas reduced to standard gas saturated with water vapor at 60°F and 30 in Hg and corrected for water vapor.

(*b*) If the presence of water vapor were not corrected for

$$V = 349{,}400 \times \frac{29.39}{14.69} \times \frac{519.6}{511.17} = 709{,}981 \text{ ft}^3$$

or a difference of 718,366 − 709,981 = 8385 ft³.

(*c*) The gas producer would be penalized.

Explanation of Formula (H-1).

$$V_c = V\frac{p - w}{B - W} \times \frac{T}{t}$$

where V = observed flowmeter reading, ft³

V_c = volume of standard gas saturated at 30 in Hg and 60°F

p = observed absolute gas pressure, lb/in² abs

w = vapor pressure of water at t, lb/in² abs

B = standard barometer equivalent, 14.69 lb/in² abs

W = vapor pressure of water at 60°F, lb/in² abs

T = °Rankine

t = observed absolute gas temperature, °R

Supplement I

STEAM ENGINES

Questions and Answers

I-1. A 12- by 16-in. single-acting steam engine takes steam for its full stroke at 100 psia and exhausts against a pressure (back pressure) of 16 psia. What is the net work done per revolution?

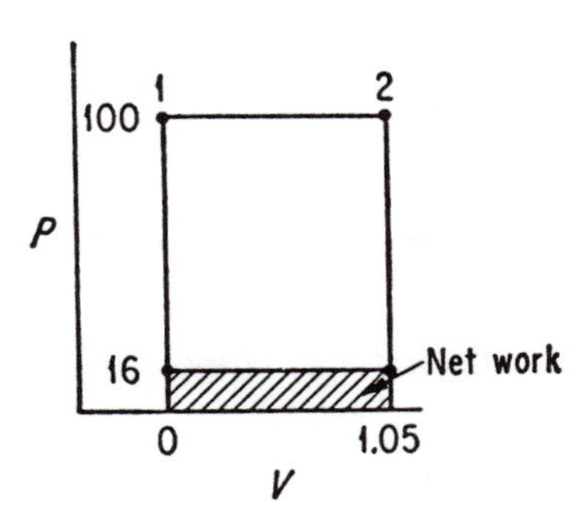

Fig. I-1

ANSWER. See Fig. I-1. V_1 is equal to zero. At the end of the full stroke V_2 is equal to

$$V_2 = 0.785(12/12)^2 \times (16/12) = 1.05 \text{ cu ft}$$

Work on forward stroke is between points 1 and 2. Thus,

$$W_{1-2} = 100 \times 144(1.05 - 0) = 15{,}125 \text{ ft-lb}$$

On the return stroke work is done on the steam; this is W_{2-1}.

$$W_{2-1} = 16 \times 144(0 - 1.05) = -2{,}420 \text{ ft-lb work done on steam}$$

Net work per revolution is 15,125 − 2,420 = **12,705 ft-lb.**

I-2. It is desired to build a compound engine with its low- and high-pressure cylinders double-acting. This engine is to develop 600 indicated hp when using steam having an absolute pressure of 150 psi and an absolute back pressure of 2 psi. The speed of the engine is to be 150 rpm and the piston speed in each cylinder is to be 750 fpm. If the cylinder ratio is to be 4 and the total ratio of expansion 12, find the size of the engine assuming a diagram factor of 0.80.

ANSWER. The length of stroke is 750 fpm/(2 × 150) = 2.5 ft, or **30 in.** The ideal mean effective pressure is

$$P_m = \frac{\text{area of ideal } PV \text{ diagram}}{\text{length of diagram}}$$

This is also

$$P_m = P_1\left(\frac{1 + \ln r_e}{r_e}\right) - P_3$$

Also the ratio of expansion r_e is equal to 1/cutoff. In the problem at hand

$$P_m = 150\left(\frac{1 + \ln 12}{12}\right) - 2 = 150 \times 0.290 - 2 = 41.5 \text{ psia}$$

With the diagram factor being 0.8, the actual mean effective pressure (mep) is

$$\text{Mep} = 41.5 \times 0.80 = 33.2 \text{ psia}$$

Now using the basic equation for hp = $PLAN/33{,}000$ and arrange to solve for A.

$$A = \frac{600 \times 33{,}000}{2 \times 33.2 \times 2.5 \times 150} = 795 \text{ sq in.}$$

This value of 795 sq in. gives a value of **32 in.** for the diameter of the low-pressure cylinder. The diameter of the high-pressure cylinder is solved by $32/\sqrt{4}$ = **16 in.** The machine would be a **16 by 32 by 30 in.** Note diagram factor is ratio of area of actual indicator card to area of ideal indicator card.

I-3. A steam engine develops 50 ihp with dry saturated steam supplied at 150 psia. The consumption is 1,250 lb per hr. Calculate and determine the following: (*a*) Carnot efficiency, (*b*) Rankine efficiency, (*c*) actual thermal efficiency, and (*d*) engine efficiency.

ANSWER. Let the following nomenclature apply:

T_1 = steam temperature at inlet pressure, °R
T_2 = steam temperature at exhaust pressure, °R
H_1 = enthalpy of steam at inlet pressure, Btu per lb
H_2 = enthalpy after isentropic expansion to exhaust (back) pressure, Btu per lb
H_3 = enthalpy of saturated liquid at exhaust pressure, Btu per lb
w_a = actual water rate of engine, lb per ihp-hr

Obtain all values but w_a from steam tables.

(*a*) Carnot efficiency $= \dfrac{T_1 - T_2}{T_1} = \dfrac{(460 + 358) - (460 + 219)}{460 + 358}$, or **17 per cent**

(*b*) Rankine efficiency $= \dfrac{H_1 - H_2}{H_1 - H_3} = \dfrac{1{,}194 - 1{,}034}{1{,}194 - 188}$, or **15.9 per cent**

(*c*) Actual thermal efficiency $= \dfrac{2{,}545}{(H_1 - H_3)(1/w_a)}$

$$= \frac{2{,}545}{(1{,}194 - 188) \times 1/(1{,}250/50)}$$, or **10.1 per cent**

(*d*) Engine efficiency $= \dfrac{\text{actual thermal efficiency}}{\text{Rankine efficiency}} = \dfrac{0.101}{0.159}$

$= 0.635$, or **63.5 per cent**

I-4. Consider a heat engine that operates on the Carnot cycle with steam as the working substance. At the start of the adiabatic expansion, the pressure of the steam is 247.4 psia and the quality is 97 per cent. The lowest temperature in the cycle is 100°F. Find (*a*) heat supplied, (*b*) heat rejected, and (*c*) cycle efficiency.

ANSWER.

(*a*) Heat supplied $= T_1(\Delta S)_t$. But first we must determine ΔS.

$$S_e = S_f + (x_e S_{fg}) = 0.5668 + (0.97 \times 0.9602) = 1.498$$
$$S_d = S_a = S_f = 0.5668$$

Heat supplied $= T_1(\Delta S)_t = (860)(1.498 - 0.567) =$ **801 Btu per lb**

(*b*) Heat rejected $= T_2(\Delta S)_t = (560)(1.498 - 0.567)$

$=$ **521 Btu per lb**

(*c*) Efficiency $= \dfrac{T_1 - T_2}{T_1} = \dfrac{400 - 100}{860} = 0.349$, or **34.9 per cent**

I-5. What would be the increase in power of a 12- by 18-in. steam engine running at 200 rpm noncondensing, if operated condensing with 26 in. Hg vacuum?

ANSWER. A 26-in. Hg vacuum is a pressure of

$$(26 \times 0.491) = 12.76 \text{ psi}$$

less than atmospheric pressure. If the back pressure when operating noncondensing is 2 psi above atmospheric pressure, the reduction of backpressure from operating condensing would be

$$(2 + 12.76) = 14.76 \text{ psia}$$

The power would be increased:

$$\frac{14.76(12 \times 12 \times 0.785)(18)(2)(200)}{12 \times 33{,}000} = \mathbf{30.35\ ihp}$$

Note that engine is double-acting and sectional area of piston rod on crank end was not deducted since complete information was not given in the problem.

I-6. A power plant in a brewery is to generate 19,200 kw of electrical energy and be able to supply 40,000 lb per hr of steam for the unit operations within the plant. The process steam plus 30,000 lb per hr of steam for feedwater heating is extracted from the turbine at a point where the pressure is 25 psia. The turbine throttle conditions are 200 psia and 600°F total steam temperature. Condenser pressure is 2.0 psia. Compute the boiler capacity in pounds per hour of steam needed to serve these purposes at the throttle conditions.

ANSWER. Refer to Fig. I-2.

p_1 = throttle pressure = 200 psia
t_1 = throttle temperature = 600°F
p_2 = extraction pressure = 25 psia
p_3 = 25 psia
p_4 = condenser pressure = 2.0 psia

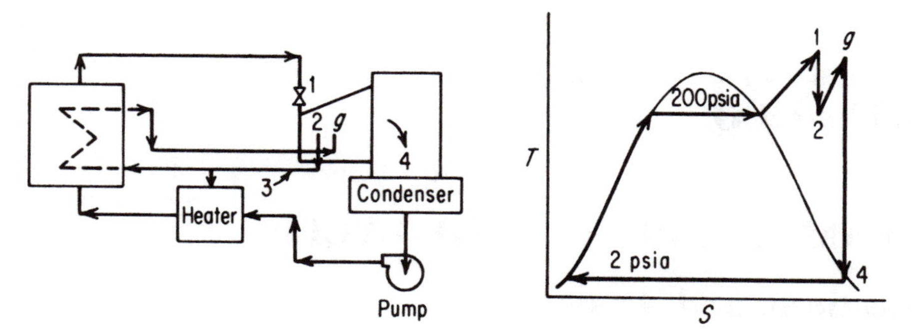

Fig. I-2

From Mollier diagram,

$$h_1 = 1{,}322.1 \text{ Btu per lb}$$
$$h_2 = 1{,}134.4 \text{ Btu per lb}$$
$$h_3 = 1{,}134.4 \text{ Btu per lb}$$
$$h_4 = 974 \text{ Btu per lb}$$

Turbine power needed to generate, assuming generator efficiency = 95 per cent:

$$19{,}200 \times 3{,}413 \times 1/0.95 = 6.90 \times 10^7 \text{ Btuh}$$

If turbine efficiency is assumed to be 70 per cent, then the isentropic power required is

$$6.90 \times 10^7/0.7 = 9.86 \times 10^7 \text{ Btuh}$$

Steam to process = 40,000 lb per hr
Steam for feedwater heating = 30,000 lb per hr
Total = 70,000 lb per hr

Set up an energy balance in this way:

$$(70{,}000 + \text{lb per hr through turbine})(1{,}322.1 - 1{,}134.1) + (\text{lb per hr through turbine})(1{,}134.4 - 974) = 9.86 \times 10^7 \text{ Btuh}$$

From which lb per hr through turbine = 216,982 lb per hr.

Steam to process = 40,000 lb per hr
Steam to feedwater heating = 30,000 lb per hr
Steam through turbine = 216,892 lb per hr
Boiler capacity needed = sum of above = **286,892 lb per hr**

Supplement **J**

STEAM TURBINES AND CYCLES

Questions and Answers

J-1. A steam turbine carrying full load of 50,000 kw uses 569,000 lb of steam per hr. Engine efficiency is 75 per cent and its exhaust steam is at 1 in. Hg abs and has an enthalpy of 950 Btu per lb. What is the temperature and pressure of the steam at the throttle?

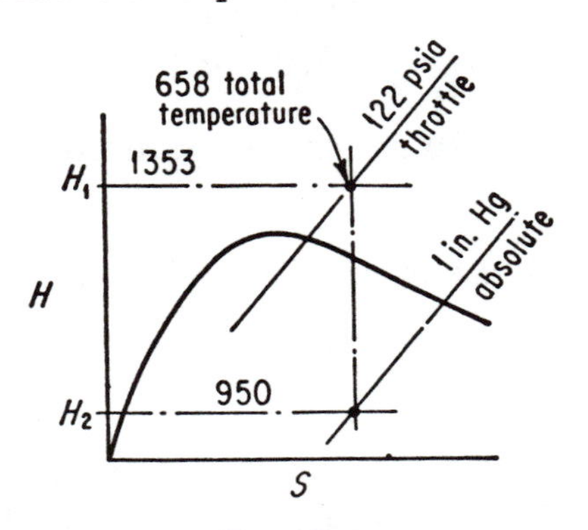

Fig. J-1

ANSWER. See Fig. J-1. Engine efficiency is given in accordance with

$$E_e = \frac{3{,}413}{w_s(H_1 - H_2)}$$

First we see that w_s may be found by a simple ratio from given data.

$$w_s = 569{,}000/50{,}000 = 11.38 \text{ lb per kwhr}$$
$$H_1 - H_2 = 3{,}413/(11.38 \times 0.75) = 403 \text{ Btu}$$
$$H_1 = 403 + 950 = 1{,}353 \text{ Btu per lb}$$

Referring to the skeleton Mollier diagram (Fig. J-1) and following the isentropic path upward, we see that the vertical path intersects the enthalpy line of 1353 Btu per lb at **122 psia** and **658°F total steam temperature.**

J-2. By actual test after erection how would you check the manufacturer's claimed operating conditions and state what you would expect the efficiency to be for a steam turbine.

ANSWER. An acceptance test should be made of a steam turbine to determine whether the guaranteed value of the steam consumption has been met. The test should be made under the operating conditions which were specified for the machine guarantee conditions. The steam supply (assumed 100,000 lb per hr) should be noted from a steam flow meter. The system steam pressure (assumed 450 psig) is noted by pressure gauge (calibrated). The steam temperature is noted from a calibrated thermometer or indicator. The exhaust or back pressure (29 in. Hg) is noted by vacuum gauge (calibrated). The load (12,000 kw-hr) is noted from voltmeter and ammeter. The actual steam consumption is about 8.35 lb per kwhr, while the ideal steam consumption would be about 6.82 lb per kwhr, which would give an efficiency of about 82 per cent. Note it would not be remiss to repeat that all flow-measuring and temperature- and pressure-indicating instruments must be carefully calibrated before a test is conducted.

J-3. A single-stage turbine of the DeLaval type has a rotor 18 in. in diameter which revolves at 1,200 rpm. The steam jet strikes the buckets at an angle of 20° with the plane of the rotor. It is desired to operate the rotor at 40 per cent of the jet velocity. If exhaust pressure is 14.7 psia and nozzle pressure is 100 psia, what steam temperature is necessary at admission to the nozzle?

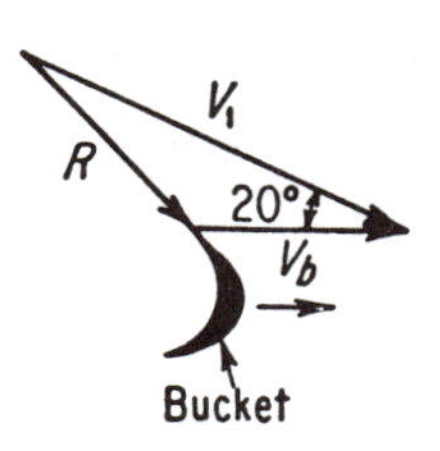

Fig. J-2

ANSWER. Refer to the discussion of the flow of steam through nozzles in Chap. 6. See also Fig. J-2. Let v_b be rotor velocity in feet per second. Then

$$v_b = (2\pi)(^{18}\!/_2)(^1\!/_{12})(12{,}000/60) = 942 \text{ fps}$$
$$v_1 = (1/0.4)(942) = 2{,}355 \text{ fps}$$

Drop through blading is isentropic across terminal conditions. Proceeding,

$$\frac{v_1^2}{2gJ} = \text{ideal heat drop} \times 0.80 = \text{actual heat drop}$$

Ideal heat drop $= (2{,}355)^2/(64.4 \times 778 \times 0.80) = 138.5$ Btu per lb. Following the path of the expansion on the Mollier diagram between 100 psia and 14.7 psia, we obtain close to 138.5 Btu per lb. This gives us a temperature of **327.8°F.**

J-4. Steam enters the fixed blade of a turbine with a relative velocity of 300 fps, a pressure of 70 psia and 60° superheat. If the steam leaves the blade with a relative velocity of 800 fps and a pressure of 60 psia, find (*a*) its specific volume and entropy, (*b*) is the blade loss in per cent of available heat drop, and (*c*) if better blades were used so that the present loss is only 5 per cent of available heat drop, what will be the relative velocity at the exit?

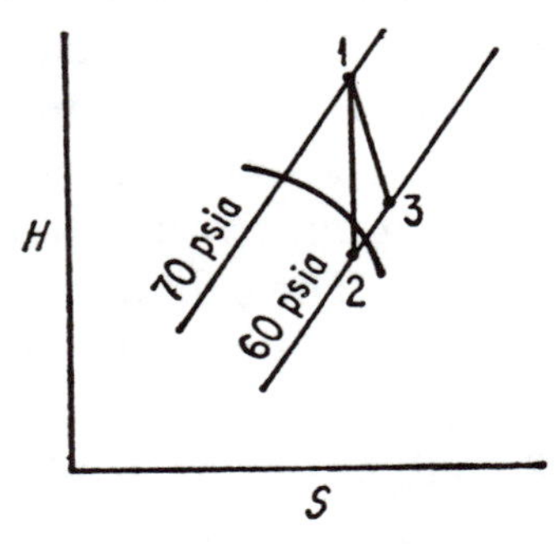

Fig. J-3

ANSWER. Using a form of the general energy equation and referring to the skeleton Mollier diagram (Fig. J-3).

$$(a) \qquad v_1^2 - v_2^2 = 50{,}000(H_1 - H_2)(1 - y)$$

where $H_1 - H_2$ is the ideal heat drop and y is an expression of turbine losses as a decimal less than one. Then for the actual heat drop

$$v_1^2 - v_2^2 = 50{,}000(H_1 - H_3)$$
$$300^2 - 800^2 = 50{,}000(1{,}212 - H_3)$$

From this H_3 is found to be equal to 1201.5 Btu per lb. Referring to Fig. J-3 and knowing H_3 and the pressure condition of 60 psia, we locate this point on the Mollier diagram. The final entropy is 1.6750 and the specific volume may be determined by the method of mixtures, page 101. Thus, the specific volume is **7.689 cu ft per lb.**

(*b*) Blade loss is $(H_3 - H_2)/(H_1 - H_2)$.

$(1{,}201.5 - 1{,}199.3)/(1{,}212.5 - 1{,}199.3) = 0.1675$, or **16.75 per cent** (*c*) For better blades we substitute for the y the value 0.05, insert in the first equation given containing the term y, and solve for v_2. This is found to be **867 fps.** Note $S_1 = S_2 = 1.6721$.

J-5. A 20,000-kw turbogenerator is supplied with steam at 300 psia and 650°F. Back pressure is 1 in. Hg abs. At best efficiency the combined steam rate is 10. Find (*a*) combined thermal efficiency, (*b*) combined engine efficiency, and (*c*) ideal steam rate.

ANSWER.

(*a*) Combined thermal efficiency (see Fig. J-4):

$$\frac{3{,}413}{w_e(H_1 - H_3)} = \frac{3{,}413}{10(1{,}340.6 - 47.06)} = 0.264, \text{ or } \mathbf{26.4\ per\ cent}$$

(*b*) Combined engine efficiency: ideal steam rate/actual steam rate.

$$\text{Ideal steam rate} = 3{,}413/\text{ideal heat drop}$$
$$= 3{,}413/(1{,}340.6 - 885.06) = w_i$$
$$w_i = 7.48 \text{ lb steam per kwhr}$$
$$7.48/10 = 0.748, \text{ or } \mathbf{74.8\ per\ cent}$$

(*c*) Ideal steam rate is **7.48 lb steam per kwhr.** Ideal heat drop is $H_1 - H_2$ on the Mollier diagram.

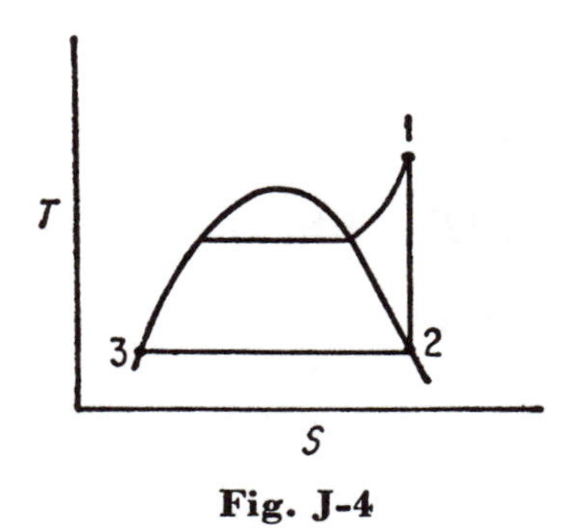

Fig. J-4

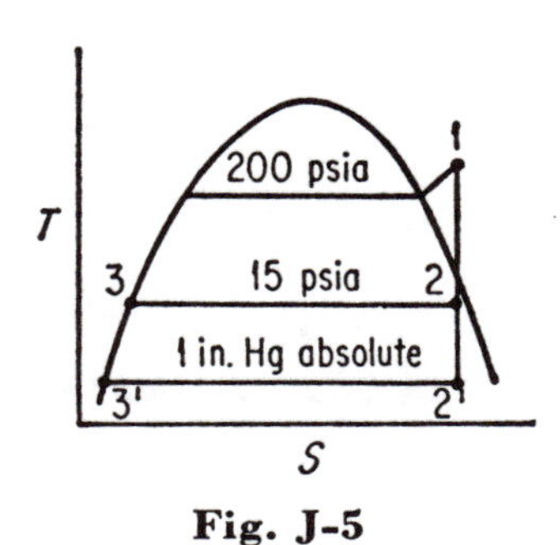

Fig. J-5

J-6. If steam at a pressure of 200 psia and 500°F is supplied to a steam turbine, what per cent increase in the efficiency of the Rankine engine cycle would result from lowering the back pressure from 15 psia (atmospheric) to 1 in. Hg?

ANSWER. See Fig. J-5. Rankine efficiency is $(H_1 - H_2)/(H_1 - H_3)$. At 15 psia back pressure:

$(1{,}267.9 - 1{,}062)/(1{,}267.9 - 181.04) = 0.1892$, or **18.92 per cent**

At 1 in. Hg back pressure:

$$(1{,}267.9 - 874)/(1{,}267.9 - 47.06) = 0.3226, \text{ or } \mathbf{32.26\ per\ cent}$$

Per cent increase:

$$(32.26 - 18.92)/(18.92) = 0.70.51, \text{ or } \mathbf{70.51\ per\ cent}$$

J-7. Steam expands in a nozzle of a steam turbine from an initial pressure of 200 psia and a temperature of 450°F to a back pressure of 2 in. Hg abs. Weight of steam discharged is 7,200 lb per hr. Friction loss in nozzle is 15 per cent of theoretical (ideal) heat drop and radiation loss is 1 per cent of ideal heat drop. Find quality of exhaust at nozzle exit.

ANSWER. Review the questions on nozzle flow in Chap. 6. Use steam tables and Mollier diagram to determine values of enthalpy. Refer also to Fig. J-6.

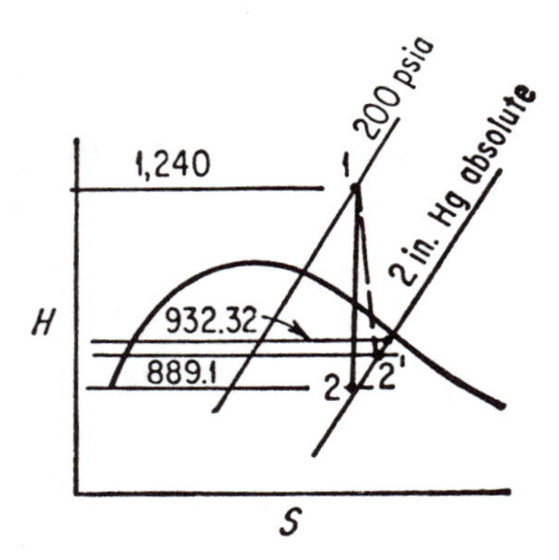

Fig. J-6

$$\begin{aligned} H_{2'} &= H_2 + [0.15(H_1 - H_2)] \\ &= 889.1 + [0.15(1{,}240 - 889.1)] \\ H_{2'} &= 889.1 + 52.64 \\ &= 941.74 \text{ Btu per lb} \end{aligned}$$

Heat loss by radiation $= 0.01(1{,}240 - 889.1) = 9.42$ Btu per lb

$$H_{2'} - H_r = 941.74 - 9.42 = 932.32 \text{ Btu per lb}$$

Refer to the Mollier diagram. The point of intersect between enthalpy of 932.32 Btu per lb and 2 in. Hg abs pressure line shows 18.2 per cent wetness. Therefore, quality is 100 − 18.2 = **81.8 per cent.**

J-8. An industrial plant operates a 5,000-kw turbine having an engine efficiency of 73 per cent. The initial steam conditions are 600 psia, 600°F and the back pressure is 1 in. Hg abs.

(*a*) Find the turbine steam rate.

(*b*) Find the pound per hour to the turbine throttle.

(*c*) Find the turbine heat rate if the turbine auxiliaries require 4,000 lb of steam per hr and their exhaust heats the feed water to 160°F.

(*d*) Find the station heat rate if the boiler efficiency is 80 per cent and the electrically driven boiler auxiliaries require 100 kw.

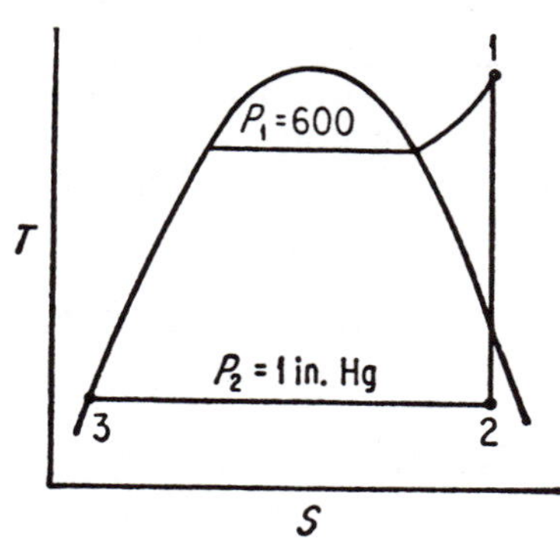

Fig. J-7

(*e*) What is the over-all station efficiency?

ANSWER. Refer to Fig. J-7. From steam tables and/or Mollier diagram the following state points may be determined:

$$H_1 = 1290.9 \text{ Btu}$$
$$S_1 = 1.5334$$
$$H_2 = 823.5 \text{ Btu}$$
$$S_2 = 1.5334$$
$$x_2 = 0.74$$

Ideal heat drop is ideal work $= H_1 - H_2 = 1{,}290.9 - 823.5$
$= 467.4$ Btu per lb

Ideal Rankine steam rate is

$$\frac{3{,}413}{H_1 - H_2} = \frac{3{,}413}{467.4} = 7.3 \text{ lb steam per kwhr}$$

Actual work energy conversion $= 467.4 \times 0.73$ (engine eff)
$= 342$ Btu per lb

(*a*) Turbine steam rate $= 3{,}413/342 =$ **10 lb per kwhr**

(*b*) Lb steam per hr to throttle $= 5{,}000 \text{ kw} \times 10 =$ **50,000 lb per hr**

(*c*) Turbine heat rate = input, Btu/output, kwhr

Input = (steam to throttle + steam to auxiliaries)$(H_1 - H_3)$

Input = $(50{,}000 + 4{,}000)(1{,}290.9 - 127.9)$

$$\text{Turbine heat rate} = \frac{\text{input}}{5{,}000 \text{ kwhr}} = \textbf{12,560 Btu per kwhr}$$

Note enthalpy equal to 127.9 Btu per lb is that of saturated liquid at the turbine back pressure.

(*d*) Station heat rate $= \dfrac{\text{turbine heat rate}}{\text{boiler efficiency}}$

$$\text{Turbine heat rate} = \frac{\text{input}}{5{,}000 - 100 \text{ (for aux.)}} = 12{,}800 \text{ Btu per kwhr}$$

$$\text{Station heat rate} = \frac{12{,}800}{0.80} = \mathbf{16{,}000\ Btu\ per\ kwhr}$$

(*e*) Station efficiency $= \dfrac{3{,}413}{\text{station heat rate}} = \dfrac{3{,}413}{16{,}000} = 0.214$, or

21.4 per cent

J-9. An automatic extraction turbine operates with steam at 400 psia and 700°F at the throttle. Its extraction pressure is 200 psia and it exhausts at 110 psia. At full load 80,000 lb of steam per hr are supplied to the throttle and 20,000 lb per hr are extracted at the bleed point. What is the kilowatt output?

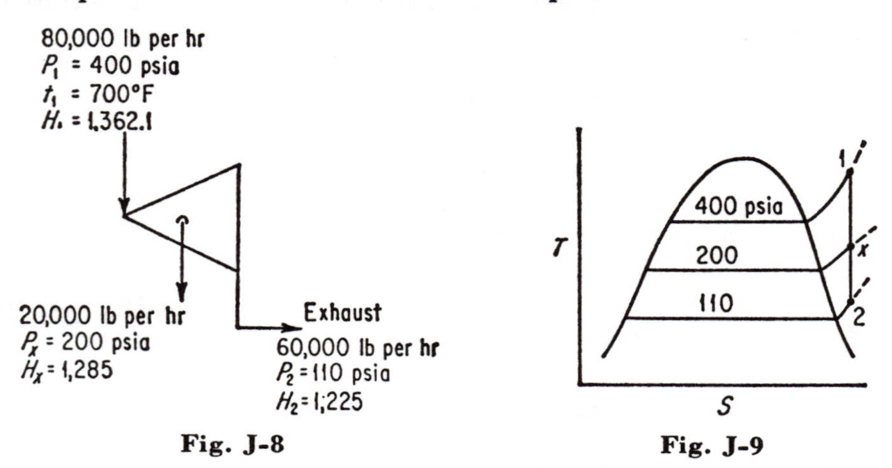

Fig. J-8 Fig. J-9

ANSWER. The ratio of total available energy to Btu per kwhr is that which we are looking for. Refer to Figs. J-8 and J-9. Use the Mollier diagram and steam tables freely. We can then set up the following values:

$H_1 = 1362.1$ Btu	$S_2 = 1.6393$	$H_x = 1285$ Btu
$S_1 = 1.6393$	$H_2 = 1225$ Btu	$P_x = 200$ psia

Total available energy:

For extraction:

$$80{,}000 \times (H_1 - H_x) = 80{,}000 \times (1{,}362 - 1{,}285)$$
$$= 6{,}160{,}000 \text{ Btu per hr}$$

For exhaust:

$$(80{,}000 - 20{,}000)(H_x - H_2) = 60{,}000(1{,}285 - 1{,}225)$$
$$= 3{,}600{,}000 \text{ Btu per hr}$$

$$\text{kw output} = \frac{9{,}760{,}000}{3{,}413} = \mathbf{2{,}860\ kw}$$

Since turbine efficiency is not involved, the 2,860-kw answer is entirely theoretical. The power transfer to the generator shaft would have to take into account the mechanical efficiency of the turbine as a whole.

J-10. A 100-megawatt turbogenerator is supplied with steam at 1,250 psia and 1,000°F and the condenser pressure is 2 in. Hg abs. At rated load the steam supplied per hour is 1,000,000 lb and at zero load, is 50,000 lb. What is the steam rate in pounds per kilowatt hour at 4/4, 3/4, 2/4, and 1/4 load?

ANSWER. For the purposes of this problem the curve of total steam consumption (Willans's line) is practically a straight line. This is a characteristic of such curves for practically all types of steam turbines when operating without overloading. If Willans's line is extended to intercept the Y axis (total steam per hour), this intercept represents the steam required to operate the turbine when delivering no power. It is the amount needed to overcome the friction of the turbine and the windage, that required for driving the governor, oil pumps, etc., and that for meeting the losses due to turbulence, leakage, and radiation under no-load conditions.

The steam rate may be determined from the simple expression

$$\frac{50}{L} + 9.5 = \frac{F}{L} = \frac{50 + (1{,}000 - 50)/100 \times L}{L}$$

where F is expressed in terms of per 1,000 lb of steam per hr and from the conditions of the problem. Then

Load fraction	Load (L), mw	Steam rate, lb per kwhr
1/4	100 × 1/4 = 25	50/25 + 9.5 = **11.5**
2/4	100 × 2/4 = 50	50/50 + 9.5 = **10.5**
3/4	100 × 3/4 = 75	50/75 + 9.5 = **10.17**
4/4	100 × 4/4 = 100	50/100 + 9.5 = **10.00**

Supplement K

GAS TURBINES AND CYCLES

Questions and Answers

K-1. An airplane supercharger consists of a turbine and a rotary compressor. It is receiving exhaust from the engine at 15 psia and 1000°F, and compressing air from 5 to 15 psia, the initial temperature being 60°F. What is the lowest efficiency it can have and still do the job?

ANSWER. The turbine uses the products of combustion to generate work. For lowest efficiency,

$$\frac{\text{Minimum compressor work required}}{\text{Energy supplied to the turbine}}$$

The minimum compressor work required may be developed through isothermal compression. Using the equation for isothermal compression for 1 lb air,

$$Q = \frac{wRT}{J} 2.3 \log \frac{V_2}{V_1}$$

and $$P_1V_1 = P_2V_2$$

where 5 psia $= P_1$
15 psia $= P_2$
60°F $= t_1$

$$\frac{V_2}{V_1} = \frac{P_2}{P_1}$$

$$Q = 1 \times 53.3 \times (460 + 60)/778 \times 2.3 \log 15/5 = 30 \text{ Btu}$$

This is equivalent to 30 × 778 = 23,400 ft-lb. This is minimum work. The energy supplied to the turbine is equivalent to the heat drop from 15 psia and 1000°F to that assumed as 60°F. A good average c_p is 0.26.

$$c_p(t_1 - t_4) = 0.26(1{,}000 - 60)778 \equiv 190{,}000 \text{ ft-lb}$$
$$\text{Efficiency} = 23{,}400/190{,}000 = 0.123, \text{ or } \mathbf{12.3 \text{ per cent}}$$

K-2. (*a*) Sketch the aircraft turbojet cycle on *TS* coordinates indicating the effect of inefficiencies in all apparatus except the diffuser.

(*b*) The diffuser (ram) of a turbojet engine has an efficiency of 100 per cent. Estimate the temperature and pressure of air leaving the ram when it is flying at sea level at a speed of 500 mph.

(*c*) What is the relationship between turbine work and compressor work in a turbojet engine?

(*d*) What air flow is required to produce a thrust of 3,000 lb at an air speed of 500 mph and a jet exit velocity of 1,600 fps?

ANSWER.

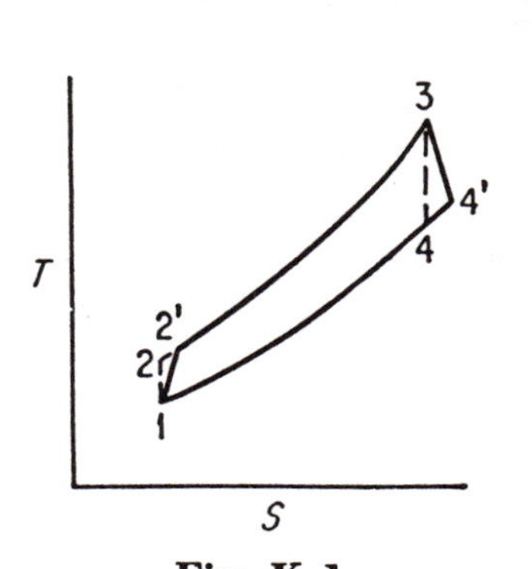

Fig. K-1

(*a*) For the effect of inefficiencies, see Fig. K-1 herewith. The increases in entropy are due to internal friction in the compressor (1–2) and in the turbine during expansion (3–4). Figure 11-3 helps tell the story too.

(*b*) The ramming intake increases both the temperature and the pressure at the compressor inlet. The temperature rise is the full temperature rise equivalent to the forward speed of the gas turbine.

$$\Delta T_r = \frac{v^2}{2gJc_p} = \frac{(500/60 \times 88)^2}{64.4 \times 778 \times 0.26} = 41.5°\text{F}$$
$$\text{Temp. leaving} = 60 + 41.5 = \mathbf{101.5°F}$$

The pressure rise is not as efficient as the temperature rise and only a portion of the pressure rise theoretically available can be obtained. In aircraft literature, the ram efficiency is defined as follows:

$$p_1 - p_{am} = \eta_r q_c$$

where p_1 = pressure at compressor inlet, psi
p_{am} = pressure of ambient, psi

η_r = ram efficiency

q_c = dynamic pressure rise, psi

When stagnation takes place, i.e., when utilizing the kinetic energy of the air stream to compress isentropically the air to a state of higher temperature (stagnation), higher pressure, and zero velocity, the pressure at stagnation may be calculated from

$$\frac{(460 + 101.5)}{520} = \left(\frac{p_0}{p_{am}}\right)^{(k-1)/k} = \left(\frac{p_0}{14.7}\right)^{0.283}$$

$$p_0{}^{0.283} = 1.08 \times 14.7^{0.283} = 1.08 \times 2.14 = 2.31$$

$$0.283 \times \log p_0 = \log 2.31$$

$$\log p_0 = 0.3636/0.283 = 1.285$$

$$p_0 = \mathbf{19.3\ psia}$$

(*c*) Review Chap. 11, pp. 160 to 161. In the turbojet the turbine produces work, all of which is used to drive the air compressor. The forward thrust of the plane is produced wholly from the jet action of the exhaust gases from the turbine. There is no propeller. Again, the diffuser produces part of the necessary air compression.

(*d*) The thrust equation may be represented by the following equation:

$$\text{Thrust} = \text{lb air per sec}/g \times (\text{jet exit velocity} - \text{plane velocity})$$

Then by rearrangement

$$\text{lb per sec} = \frac{3{,}000 \text{ lb thrust} \times 32.2}{1{,}600 - 735} = \mathbf{111.8\ lb\ per\ sec}$$

K-3. At a certain section of an air stream the Mach number is 2.5, the stagnation temperature is 560°R, and the static pressure is 0.5 atm. Assuming that the flow is steady, isentropic, and follows one-dimensional theory, calculate (all at the point where the Mach number is 2.5) (*a*) temperature, (*b*) stagnation pressure, (*c*) velocity, (*d*) specific volume, and (*e*) mass velocity.

ANSWER. The relation between Mach number M, velocity relative to some point v, fps, and acoustic velocity v_c, fps, may be expressed by

$$M = \frac{v}{v_c} = \frac{v}{\sqrt{kgRT}}$$

The Mach number varies with position in the fluid and the compressibility effect likewise varies from point to point. If a body moves through the atmosphere, the over-all compressibility effects are a function of the Mach number.

(*a*) For the problem at hand Marks* gives the following formula:

$$\frac{T}{T_0} = \frac{1}{1 + [(k-1)/2]M^2}$$

Solve for stagnation temperature T_0 by rearrangement of the above equation.

$$T_0 = T\left(1 + \frac{1.4-1}{2} \times 2.5^2\right) = 560°\text{R}$$

$$T = 560/2.25 = \textbf{249°R the static temperature}$$

(*b*) See previous problem and determine stagnation pressure as follows: since process is isentropic

$$p_0 = p\left(\frac{T_0}{T}\right)^{k/(k-1)}$$

$$= 14.7 \times 0.5(560/249)^{3.5} = \textbf{125.5 psia}$$

(*c*) $v = Mv_c = 2.5 \times \sqrt{1.4 \times 32.2 \times 53.3 \times 249} = \textbf{1,936 fps}$

(*d*) From the perfect-gas law

$$\bar{V} = \frac{RT}{p} = \frac{53.3 \times 249}{0.5 \times 14.7 \times 144} = \textbf{12.6 cu ft per lb}$$

(*e*) Mass velocity may be expressed as G lb per sec-ft² and ρ is density in pounds per cubic foot.

$$G = \rho v = (1/12.6)(1{,}936) = \textbf{153.3 lb per sec-ft}^2$$

For a nozzle, the stagnation condition is the state of the fluid at rest in the chamber ahead of the nozzle. For a body moving through the atmosphere, the stagnation state occurs only at the stagnation point at the nose of the body and may be computed by the formulas used in the previous and this problem.

K-4. A jet plane flies at an altitude of 5,000 ft where atmospheric temperature is 40°F. An observer on the ground notes that he hears

* "Mechanical Engineers' Handbook," revised by T. Baumeister, 6th ed., pp. 11-91, McGraw-Hill Book Company, Inc., New York, 1958.

the sound of the plane exactly 3 sec after the plane has passed directly overhead. Assuming that the velocity of sound remains constant at its value corresponding to 40°F, estimate the speed of the jet plane.

ANSWER. A plane traveling slower than the speed of sound would be heard before it passed directly overhead; if at the same speed as

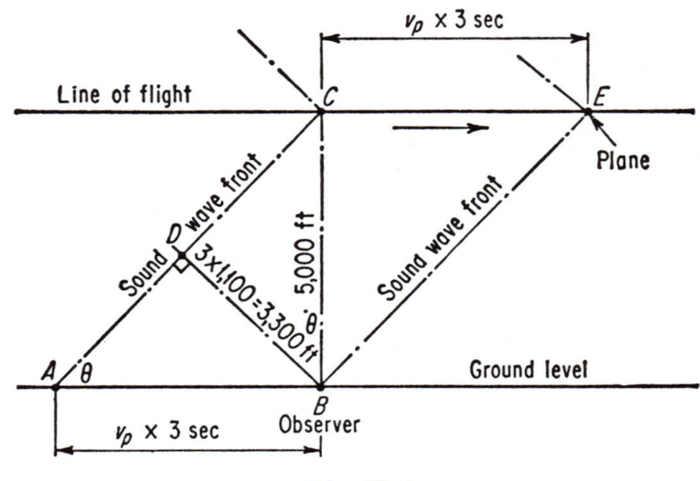

Fig. K-2

sound (Mach 1), then it will be heard at the same time it passed directly overhead; if at a speed greater than sound, it would be heard after it had passed overhead. From the conditions set by the problem make up sketch Fig. K-2.

$$v_c = \sqrt{1.4 \times 32.2 \times 53.3 \times (460 + 40)} = 1,100 \text{ fps}$$

The speed of the plane must be greater than the acoustic velocity. When plane is at point C, wave front AC develops and sound is heard by observer 3 sec later at point B. By this time plane has traveled distance $v_p \times 3$, where v_p is velocity of plane in fps. Sound-wave fronts run parallel and are conical in shape. Likewise, the distance AB is $v_p \times 3$. Angle CAB is equal to angle CBD. Distance $CD = \sqrt{(5,000^2 - 3,300^2)} = 3,760$ ft. From geometry $AB/BD = BC/BD$. The Mach number at which the plane is traveling is AB/BD, since they represent the distances traveled by the plane/traveled by the sound. Thus,

$$\text{Mach No.} \frac{AB}{BD}$$

Then by similar triangles

$$\frac{BC}{BD} = \frac{5{,}000}{3{,}760} = 1.33$$

Velocity of plane = 1.33 × 1,100 = 1,465 fps, or **1,000 mph**

K-5. Air with a specific heat ratio equal to 1.4 flows at supersonic velocity in a duct whose cross-sectional area is 2 sq ft. Static temperature in the main body of flow is estimated to be 400°R. Static pressure measurements by means of a mercury manometer indicate a manometer deflection of 10 in. of mercury below atmospheric. Pitot tube measurements indicate a manometer deflection of 90 in. of mercury above atmospheric pressure. The barometric pressure is 30 in. of mercury abs. Determine the mass flow rate of air in the duct.

ANSWER. Review question K-3. The sonic velocity may be calculated.

$$v_c = \sqrt{1.4 \times 32.2 \times 53.3 \times 400} = 980 \text{ fps}$$

The specific volume is

$$\bar{V} = \frac{RT}{p} = \frac{53.3 \times 400}{9.85 \times 144} = 15 \text{ cu ft per lb}$$

Absolute pressure equivalent to 10 in. Hg below atmospheric is given by

$$(^{20}\!/_{30})14.7 = 9.85 \text{ psia}$$

The duct velocity without slight correction for temperature is found to be

$$\sqrt{2 \times 32.2 \times 90 \times 1.136 \times 62.4 \times 15} = 2{,}480 \text{ fps}$$

$$\text{Mass flow rate } G = \rho v$$

$$= (1\!/_{15})(2{,}480)(2) = \mathbf{330 \text{ lb per sec}}$$

$$\text{Mach No.} = 2{,}480/1{,}065 = 2.33$$

K-6. A convergent-divergent nozzle operates with a constant stagnation pressure of 100 psia but under variable back-pressure conditions. It is noted during start-up that when the back pressure reaches 91.8 psia, flow through the nozzle becomes critical; that is, any further reduction in back pressure does not vary the mass flow rate. As the back pressure is lowered beneath 91.8 psia, plane one-dimensional normal shock sets in so as to result in pressures at the exit plane being identical with back pressure. What is the maximum back pressure that can exist and still result in supersonic flow for the entire length of the divergent portion of the nozzle?

ANSWER. For any convergent-divergent nozzle with a fixed value of P_1, the inlet pressure, the flow is at first subsonic at all points along the nozzle for values of P_2 slightly less than P_1; and we have the case of the ordinary venturi tube. As the exit pressure P_3 is reduced, velocity at the throat increases, and the weight discharge also increases. When P_2 reaches the critical value such that

$$\frac{P_2}{P_1} = \left(\frac{2}{k+1}\right)^{k/(k-1)} \quad \text{or} \quad \frac{P_c}{P_1} = \left(\frac{2}{k+1}\right)^{k/(k-1)}$$

and $P_c = P_1 \times 0.528$ when k is 1.4. At critical then sonic velocity exists at the throat and further reduction in P_2 fails to increase flow. Thus flow rate is a maximum when $P_c = 0.528\ P_1 = 52.8$ psia. Here the back pressure P_3 is 91.8 psia and sonic velocity (Mach 1)

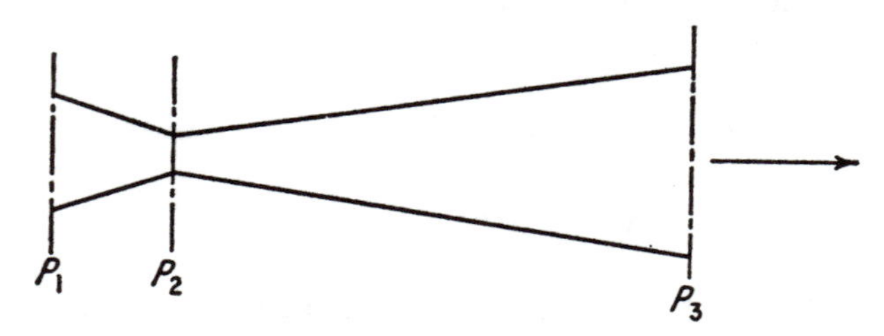

Fig. K-3

occurs only at the throat and no further reductions in P_2/P_1 are possible if the stream fills the passage. When the flow is entirely supersonic (Mach > 1) for the entire length of the divergent portion of the nozzle, results may be calculated by isentropic flow relations and any shock wave will occur outside the nozzle proper. Refer to Fig. K-3.

From isentropic gas flow tables* and normal shock tables*

M	P_2/P_1	P_2 calculated	P_3/P_2	P_3 calculated
1.0	0.52828	52.828	1.0000	52.83
1.3	0.36092	36.092	1.8050	55.15
1.5	0.27240	27.240	2.4583	**66.96**
1.7	0.20259	20.259	3.2050	64.93

Thus, the maximum back pressure that can exist and still have supersonic flow appears to take place at a Mach number equal to 1.5 at a pressure of 66.96 psia. For greater accuracy examine Mach number either side of 1.5 in same manner as above.

K-7. A stationary gas-turbine power plant is to deliver 20,000 hp to an electric generator. The plant is to have no regeneration. The maximum temperature in the turbine is set at 1540°F, and the system is designed for a maximum pressure of 60 psia. The compressor air intake is located outside the building, where the mean temperature is 60°F. For the preliminary design of the plant, determine the air intake required in cubic feet per minute.

ANSWER. Assume the Brayton air standard cycle, Fig. K-4, and neglect pressure drop through the filter. $T_1 = (460 + 60) =$

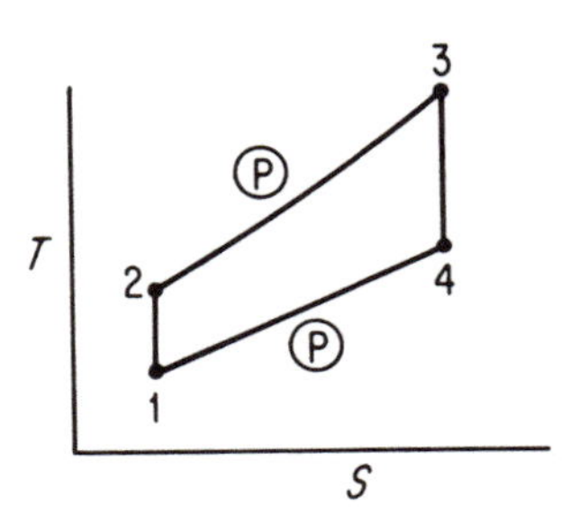

Fig. K-4

* Ascher H. Shapiro, "The Dynamics and Thermodynamics of Compressible Flow," Parts I and II from Volume I, The Ronald Press Company, New York, 1958.

520°R; $P_1 = 14.7$ psia $= P_4$; $P_2 = P_3 = 60$ psia; and $T_3 = (1540 + 460) = 2000$°R. Then from thermodynamics,

$$T_2 = T_1 \left(\frac{P_2}{P_1}\right)^{(k-1)/k} = 520 \left(\frac{60}{14.7}\right)^{0.286} = 778°\text{R}$$

And likewise

$$T_4 = \frac{T_3}{(P_2/P_1)^{(k-1)/k}} = 1339°\text{R}$$

Heat pickup between 1 and 2 is given by

$$Q_{1-2} = c_p(T_2 - T_1) = 0.24(778 - 520) = 62 \text{ Btu per lb}$$
$$Q_{3-4} = c_p(T_3 - T_4) = 0.24(2{,}000 - 1{,}339) = 159 \text{ Btu per lb}$$

Then the difference between the two states is 159 − 62 = 97 Btu per lb. Then the air weight rate is

$$20{,}000 \times 42.42/97 = 8{,}750 \text{ lb per min}$$

This represents a volume of (8,750 × 53.34 × 520)/(14.7 × 144) **= 114,000 cu ft per min.**

Supplement L

INTERNAL-COMBUSTION ENGINES AND CYCLES

Questions and Answers

L-1. A gas engine is supplied with 1,100 cfh of gas at 80°F and 5 in. WG. Barometer reads 30.2 in. Hg. Heating value of gas is 540 Btu per cu ft at 60°F and 30 in. Hg. Engine develops 52 hp. Calculate the thermal efficiency of the engine.

ANSWER. Efficiency is equal to ratio of output to input. The input is

$$\frac{1{,}100 \text{ cfh} \times \text{actual heating value under operating conditions}}{2{,}545}$$

Actual heating value corrected for temperature and pressure

$$540 \times \frac{460 + 80}{460 + 60} \times \frac{30}{30.2 + 5/13.6} = 550 \text{ Btu per cu ft at firing}$$

$$\text{Input} = 1{,}100 \times 550/2{,}545 = 238 \text{ hp}$$

$$\text{Efficiency} = {}^{52}\!/_{238} = 0.219, \text{ or } \mathbf{21.9 \text{ per cent}}$$

Be sure in gas-combustion problems that heating value at standard conditions is corrected to actual firing conditions.

L-2. Calculate the temperature and pressure at the beginning of the expansion stroke in a gasoline engine, assuming no heat losses and combustion completed. The air to fuel ratio is 15, compression ratio is 6.2, the compression curve exponent is 1.3, and the pressure at the beginning of the compression stroke is 13 psia with a temperature of 150°F. Assume the heating value of the gasoline to be 20,000 Btu per lb and the specific heat during expansion is 0.3. Assume gases to be 1500°F and 17 psia when remaining in the clearance volume from previous explosion cycle.

ANSWER. Assume a four-cycle engine and refer to pressure-volume diagram (Fig. L-1). Solve for pressure and temperature at point 3. $P_1 = 13$ psia $= P_5$. $T_1 = 460 + 150 = 610°R$. Compression ratio $= r_c = V_1/V_2 = 6.2$. Clearance volume $= V_2$. Then

$$P_2 = P_1 r_c^n = 13 \times 6.2^{1.3} = 165 \text{ psia}$$
$$T_2 = T_1 r_c^{n-1} = 610 \times 6.2^{0.3} = 1045°R, \text{ or } \mathbf{585°F}$$

This is the temperature-pressure condition of the mixture after compression and just before ignition. The addition of the heat of combustion will increase temperature and pressure. Then the heat

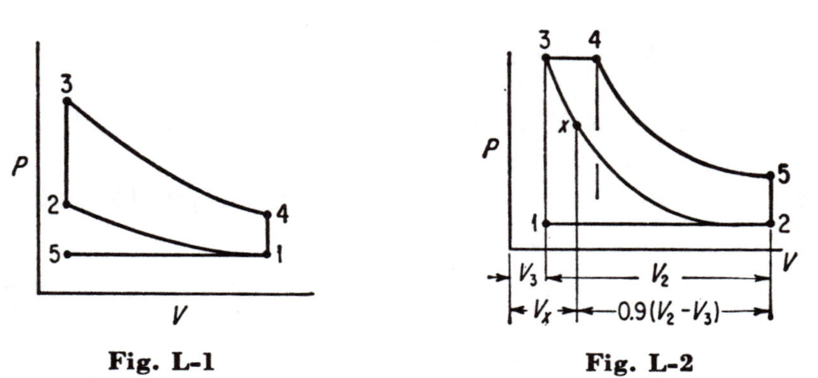

Fig. L-1 **Fig. L-2**

added to this mixture, since there are no heat losses, is that given up on combustion of 1 lb of gasoline. Since this is a constant-volume process the temperature rise $(T_3 - T_2)$ is given by

$$T_3 - T_2 = \frac{\text{Btu transferred by combustion}}{\text{wt of mixture} \times \text{sp ht}}$$

$$= \frac{20{,}000}{15 + 1} \times \frac{1}{0.3} = 4160°F$$

From this $T_3 = 5205°R$, or $4745°F$. If we assume the validity of the perfect-gas law and V_1 is fresh ($1/15$) mixture

$$P_3 = P_2 \frac{T_3}{T_2} = 165 \frac{5{,}205}{1{,}045} = \mathbf{825\ psia}$$

L-3. Calculate the pressure and temperature at 90 per cent of the completion of the compression stroke in a four-cycle diesel engine having a compression ratio of 14. The air temperature at beginning of stroke is 150°F and the pressure is 13.7 psia.

ANSWER. Refer to Fig. L-2. Assume n for air equal to 1.3.

$$r_c = \frac{V_2}{V_3} \qquad t_2 = 150°\text{F} \qquad P_2 = 13.7 \text{ psia}$$

$$P_x = P_2\left(\frac{V_2}{V_x}\right)^n \qquad T_x = T_2\left(\frac{V_2}{V_x}\right)^{n-1}$$

Obviously, the problem is to find V_2/V_x.

$$V_x = V_2 - 0.90(V_2 - V_3) = V_2 - 0.90\left(V_2 - \frac{V_2}{14}\right)$$

$$V_x = (2.3/14)(V_2)$$

By rearrangement

$$\frac{V_2}{V_x} = 6.08$$

$$P_x = (13.7)(6.08)^{1.3} = \mathbf{142.5\ psia}$$

$$T_x = (150 + 460)(6.08)^{0.3} = 1048°\text{R, or } \mathbf{588°F}$$

L-4. The volume in the clearance space of a 6- by 10-in. Otto gas engine is 0.06 cu ft. Find the ideal thermal efficiency (ITE) of the engine on the air standard basis, if the exponent of the expansion and compression lines is 1.35.

ANSWER. Refer to Fig. L-3. $V_1 = V_2 = 0.06$ cu ft. In the usual manner piston displacement $V_4 - V_1$ is found to be equal to 0.163

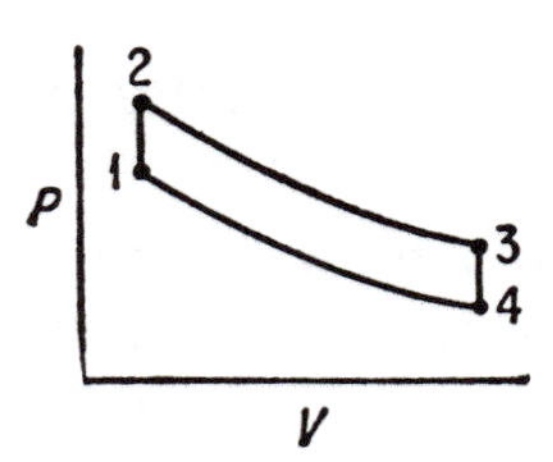

Fig. L-3

cu ft. First find V_4 by use of polytropic process and n equal to 1.35.

$$V_4 = 0.163 + 0.06 = 0.223 \text{ cu ft}$$

The ideal thermal efficiency is

$$\text{ITE} = 1 - \left(\frac{V_1}{V_4}\right)^{n-1} = 1 - \left(\frac{0.06}{0.223}\right)^{1.35-1} = 0.369, \text{ or } \mathbf{36.9\ per\ cent}$$

L-5. A two-stroke-cycle internal-combustion engine operating at 3,000 rpm has cylinders $3\frac{1}{2}$ in. in diameter with a 5-in. stroke. Inlet valves are $1\frac{1}{4}$ in. in diameter, open to a mean height of $\frac{3}{4}$ in., and are open for 120 degrees of engine rotation. The air-fuel mixture is at a pressure of about 14.7 psia and a temperature of 140°F. If a discharge coefficient of 0.70 is assumed for the inlet ports, what pressure differential is required to transfer a charge, equal to the engine displacement, from the inlet manifold to the cylinder?

ANSWER. The velocity of a fluid passing through an opening or aperture is

$$v = c\sqrt{2g\,\Delta H}$$

Since pressure differential is the object of the solution, ΔH is easily converted: $\Delta H = [(P_1 - P_2)/w] \times 144$.

In this equation we know c, and g (32.2 gravitational constant). We can determinw w, the air density under the conditions, and also calculate v.

$$w = \frac{29}{379} \times \frac{520}{600} \times \frac{14.7}{14.7} = 0.0662 \text{ lb per cu ft}$$

To find velocity, we shall calculate piston displacement and then the flow rate in cubic feet per second. Then from the port area calculate velocity from $v = Q/A$. All calculations are based on inlet valve being open. Then

$$PD = \frac{\pi D^2 L}{4} = \frac{\pi(3.5/12)^2(3.75/12)}{4} = 0.0209 \text{ cu ft}$$

Note that the full stroke of 5 in. was not used, but as required by the problem the stroke for 120° of revolution $= \frac{5}{2} + \frac{5}{2}\sin 30° = 3.75$ in. For the time element involved, i.e., for the piston to move 3.75 in., the time for 1 rev. is $t = 1/(3{,}000/60)$, or $\frac{1}{50}$ sec. This is for a total of 360°. For 120°: $t = (\frac{120}{360})(\frac{1}{50}) = \frac{1}{150}$ sec. The flow rate Q then becomes

$$Q = 0.0209/(1/150) = 3.13 \text{ cfs}$$

The area of the valve port is $A = 0.0085$ sq ft for a diameter of 1.25 in. When valve is in the raised position, a larger opening is available, i.e.,

$$\pi \times 1.25 \times 3/4/144 = 0.02045 \text{ sq ft}$$

Since the constriction occurs at the port itself, it would be more reasonable to use 0.0085 sq ft to find the velocity (higher and will require a greater pressure differential). Thus, the average velocity $v = Q/A = 3.13/0.0085$, or 368 fps. Substitute the proper values in the first equation for velocity.

$$368 = 0.70\sqrt{64.4 \times \frac{\Delta P}{0.0662} \times 144}$$

$$\Delta P = P_1 - P_2 = \mathbf{1.97\ psi}$$

L-6. Calculate from the following data the number of pounds of fuel used by an automobile engine: air temperature 70°F, barometer 30.2 in. Hg, air entering 60 cfm, measured gasoline 30 pints per hr, specific gravity of gasoline 0.735.

ANSWER. Air density at conditions given is first determined.

$$(29/379)(520/530)(30.2/30.0) = 0.0755 \text{ lb per cu ft}$$

Weight of air drawn into the cylinders of the engine is

$$0.0755 \times 60 \times 60 = 272 \text{ lb per hr}$$

Weight of gasoline consumed in same period

$$(30/8)(1/7.48)(62.4)(0.735) = \mathbf{23\ lb\ per\ hr}$$

Pounds of air per pound of fuel: $272/23 = \mathbf{11.85}$

L-7. A 2-stroke-cycle engine has a 4- by 6-in. cylinder with a connecting rod 10 in. long. The intake port is 1 in. high. Determine the mean port opening during the time required to transport the required charge of fuel and air.

ANSWER. The mean port opening is its area divided by height of opening. The mean port area is

$$\frac{\text{Displacement, cfs} \times 144}{\text{Charge velocity, fps}} = \text{sq in.}$$

In order to get displacement we need to know speed at which machine is operating and the velocity of charge. The problem does not give these data so that it must be assumed in accordance with good practice.

Assume 1,800 rpm and 200 fps charge velocity. Then

$$\text{Displacement} = \frac{30 \times (\pi 4^2/4) \times 6}{1{,}728} = 1.31 \text{ cfs}$$

where 30 is suction strokes per sec = rpm/60 = 1,800/60. The mean port area is $1.31 \times 144/200 = 0.942$ sq in. Finally, the mean port opening is $0.942/1 =$ **0.942 in.**

L-8. An automobile engine has a rating of 250 bhp at 4,500 rpm. The engine torque peaks at 2,000 rpm, and the horsepower peaks at 4,500 rpm. The fuel has a heating value of 20,000 Btu per lb, and the overall efficiency is 28 per cent. Find the fuel consumption in pounds per hour when the engine is running at 4,500 rpm.

ANSWER.

$$\text{Brake energy converted to heat} = 250 \text{ hp} \times 2{,}545 = 6.3625 \times 10^5 \text{ Btuh}$$

$$\text{Fuel conversion} = 6.3625 \times 10^5/20{,}000 = 31.81 \text{ lb per hr}$$

Actual fuel required $= 31.81/0.28 =$ **113.6 lb per hr at peak load**

$$\text{Brake hp} = (2\pi NT)/33{,}000 \qquad \text{and} \qquad T = \text{bhp} \times 33{,}000/2\pi N$$

At peak load,

$$\text{Torque} = 250 \times 33{,}000/(2\pi \times 4{,}500) = 292 \text{ lb-ft}$$

L-9. A Carnot engine and air-standard Otto, diesel, and gas-turbine engines are each operating with a heat-addition rate of 10,000 Btu/h. The Carnot engine is operating heat reservoirs at 1200 and 300°F. The other engines have compression ratios of 8:1, and the diesel engine has an expansion ratio of 2:1.

Determine the net horsepower produced by

(*a*) Carnot engine
(*b*) Otto engine
(*c*) Diesel engine
(*d*) Gas-turbine engine

ANSWER.

(a) Carnot engine

$$\eta = \frac{T_1 - T_2}{T_1} = \frac{1200 - 300}{1200 + 460} = 0.5422$$

$$W = Q_1\eta = (10{,}000)(0.5422) = 5422 \text{ Btu/h}$$

Converting to horsepower

$$\text{hp} = \frac{5422 \times 778 \text{ ft-lb/Btu}}{550 \times 3600} = 2.13 \text{ hp} \qquad \textbf{Ans.}$$

(b) Otto engine, $r = 8/1 = 8$

$$\text{Efficiency } \eta = 1 - (1/r)^{r-1} = 1 - (1/8)^{0.4}$$
$$= 0.5647 \text{ or } 56.47 \text{ percent}$$
$$W = Q_1\eta = 10{,}000 \times 0.5647 = 5647 \text{ Btu/h}$$

Converting to hp in the same manner as above, we obtain 2.22 hp **Ans.**

(c) Diesel engine, $r = 8$ and $r_e = 2$

$$\eta = 1 - \frac{r}{\text{r}}\,\frac{r_e^{-r} - r^{r}}{r/r_e - 1} = 1 - \frac{8}{1.4} \times \frac{2^{-1.4} - 8^{-1.4}}{8/2 - 1}$$
$$= 0.3819 \text{ or } 38.19 \text{ percent}$$

$$W = Q_1\eta = 10{,}000 \times 0.3819$$
$$= 3819 \text{ Btu/h or the equivalent of 1.5 hp} \qquad \textbf{Ans.}$$

(d) Gas-turbine engine, $r = 8$

$$\eta = 1 - (1/r)^{r-1/r} = 1 - (1/8)^{0.4/1.4} = 0.4480$$
$$W = Q_1\eta = 4480 \text{ Btu/h or 1.76 hp} \qquad \textbf{Ans.}$$

Supplement M

PUMPS AND PUMPING

Questions and Answers

M-1. A centrifugal pump delivers 300 gpm of water at 3,000 ft total dynamic head when operating at 3,500 rpm.

(*a*) At what speed must a geometrically similar pump operate to deliver 200 gpm at the same total dynamic head?

(*b*) What must be the diameter of the impeller of this new pump if the diameter of the 3,500-rpm pump is 6 in.?

ANSWER.

(*a*) This is an application of the specific speed of the pumps. First determine the specific speed of the existing pump, and then use this work to obtain the speed sought.

$$N_s = \frac{N_1\sqrt{Q_1}}{H^{3/4}} = \frac{3{,}500\sqrt{300}}{3{,}000^{3/4}} = \frac{3{,}500 \times 17.3}{420} = 144$$

For the new pump

$$144 = \frac{N_2\sqrt{200}}{420} \qquad N_2 = \frac{144 \times 420}{\sqrt{200}} = 4{,}277 \textbf{ rpm}$$

(*b*) Two geometrically similar centrifugal pumps will have similar flow conditions if the ratio of the fluid velocities to the velocities of the rotating parts is the same, i.e., if

$$\frac{Q_1}{N_1D_1^3} = \frac{Q_2}{N_2D_2^3}$$

Thus, for the problem at hand insert the proper values in the above relation.

$$\frac{300}{3{,}500 \times 6^3} = \frac{200}{4{,}250 \times D_2^3}$$

$$D_2^3 = \frac{200 \times 3{,}500 \times 6^3}{4{,}250 \times 300} = 118.5 \qquad D_2 = \textbf{5 in. even}$$

Note that for two geometrically similar pumps the following conditions also exist, where Q = cfs, H = head ft, P = power, N = rpm, D = impeller diameter, in.

$$\frac{H_1}{(Q_1/D_1^2)^2} = \frac{H_2}{(Q_2/D_2^2)^2} \qquad \frac{H_1}{(N_1D_1)^2} = \frac{H_2}{(N_2D_2)^2}$$

And for the same fluid

$$\frac{P_1}{Q_1H_1} = \frac{P_2}{Q_2H_2} \qquad \frac{P_1}{N_1^3D_1^5} = \frac{P_2}{N_2^3D_2^5}$$

M-2. A boiler-feed pump is driven by a variable-speed steam turbine. The characteristic curve of the pump at a speed of 6,000 rpm

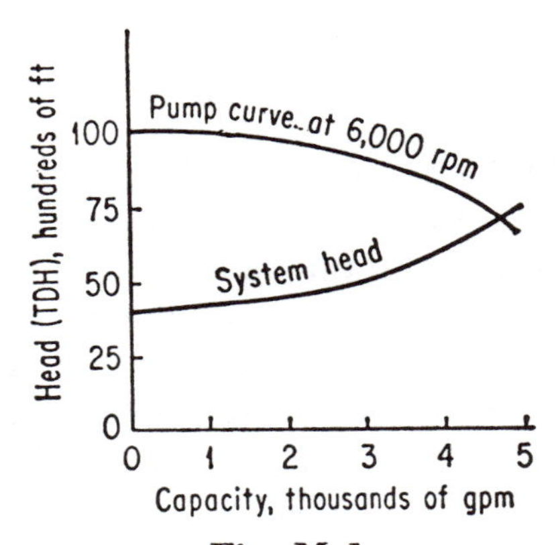

Fig. M-1

and the system-head curve of the boiler–feed-water system are shown in Fig. M-1. Determine the speed at which the pump should be operated to match a system requirement of 3,000 gpm.

ANSWER. Refer to Fig. M-1. Since capacity is directly proportional to speed and since system curve holds, select a gpm at 6,000 rpm and set up a proportion.

At 3,000 gpm and 6,000 rpm, head is 92 ft
At 3,000 gpm and x rpm, head is 50 ft

$$\frac{92}{50} = \left(\frac{6{,}000}{x}\right)^2 = 1.84 = 36 \times \frac{10^6}{x^2} \qquad x^2 = \frac{36 \times 10^6}{1.84} = 19.6 \times 10^6$$

x = **4,320 rpm**

M-3. A model centrifugal pump with a 3-in. diameter impeller delivers 600 gpm of 60°F water at a total head of 350 ft when operating at 1,750 rpm. Find the diameter of a geometrically similar pump that will deliver 1,000 gpm when operating at 3,500 rpm. What will be the total head of the 3,500-rpm pump when it is delivering 1,000 gpm?

ANSWER. Review Prob. M-1 entirely. Note the additional equations.

Diameter:

$$D_2 = \frac{Q_2 N_1 D_1^3}{Q_1 N_2} = \frac{1{,}000 \times 1{,}750 \times 3^3}{600 \times 3{,}500} = \mathbf{2.25\ in.}$$

Total head:

$$H_2 = H_1 \frac{N_2^2 D_2^2}{N_1^2 D_1^2} = 350 \times \frac{3{,}500^2 \times 2.25^2}{1{,}750^2 \times 3^2} = \mathbf{785\ ft}$$

M-4. A centrifugal pump handles 1,500 gpm water at 150°F. Suction is from an open pit 5 ft below the pump center line and discharge is to an open tank 50 ft above the pump center line. Suction and discharge lines are 8 and 6 in. nominal size, respectively. Pipe is Schedule 40 (standard). The discharge piping system consists of 500 ft of pipe, 1 swing check valve, 3 gate valves (full open), 10 standard tees, and 20 standard elbows. What motor size is indicated?

ANSWER. Refer to Fig. M-2. First, will the pump "pump" the hot water? The ability of a pump to lift hot water decreases as the temperature increases. This is due to the fact that the vapor pressure of the water increases with temperature and opposes the atmospheric pressure tending to force the liquid into the pump suction. But once the liquid is picked up by the impeller vanes, it can be pumped. For the situation at hand let us assume the equivalent suction lift is within the design limits of the particular pump. For details as to its calculation see Buffalo-Forge *Centrifugal Pump Application Manual*, 2nd Edition (1959).

From Bernoulli's theorem the total dynamic head (TDH) may be found to be

$$\begin{aligned} \text{TDH} &= (X_B - X_A) + \left(\frac{v_B^2}{2g} - \frac{v_A^2}{2g}\right) + \left(\frac{P_B}{w_B} - \frac{P_A}{w_A}\right) + h_f \\ &= (55 - 0) + (0 - 1.41) + (0 - 0) + h_f \\ &= 55 - 1.41 + h_f \end{aligned}$$

The friction head h_f must be determined for 150°F water which has a lower viscosity than water at 60°F., in this case 0.5 centistoke instead of 1.13 centistokes at 60°F. Also the total equivalent length

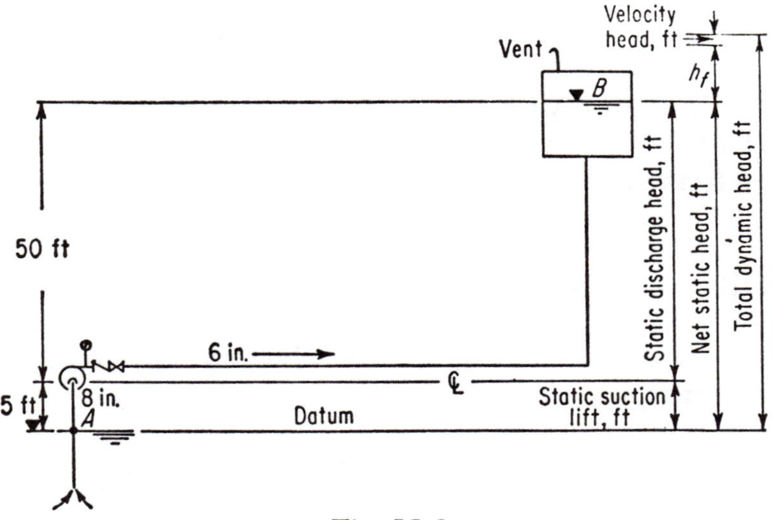

Fig. M-2

of pipe (straight plus that for fittings) must be determined. Use Fig. 5-15.

Length of straight pipe....................	500 ft
1 8-in. long radius elbow..................	14
1 check valve............................	39
3 gate valves (wide open) at 3.5 = 10.5......	11
10 std. tees at 33..........................	330
20 std. ells at 16 = 320....................	320
Total....................................	1,214 ft

Assume 2 ft for suction pipe and entrance loss to pump so that this must be added to the head loss determined from tables or by calculation. Now for discharge pipe loss, first find Reynolds number $Dv\rho/\mu = (6/12)(17)(62.4 \times 0.98)/(0.49 \times 0.000672) = 1{,}575{,}000$. This indicates turbulent flow. Viscosity in centipoise is equal to sp gr $\times$ viscosity in centistokes. For ordinary steel pipe the Fanning friction factor may be read to be 0.0048. Then the head loss is

$$h_f = \frac{2f_2Lv^2}{gD} = 2 \times 0.0048 \times 1{,}214 \times 17.0^2/(32.2 \times 6/12)$$
$$= 2{,}090 \text{ ft}$$

Check. From "Cameron Hydraulic Data Book" for standard weight (Schedule 40) steel pipe and 1,500 gpm water at 60°F, friction loss reads 24.3 ft per 100 ft of 6-in. pipe. Thus,

$$24.3 \times (1{,}214/100) = 2{,}950 \text{ ft}$$

Compare this figure with the 2,090 ft calculated for 180°F water. This latter friction loss is lower because of lower viscosity effects and density effects and is a reasonable value. Thus, the total dynamic head is

$$\text{TDH} = 55 - 1.41 + 2{,}090 + 2 = 2{,}145.59, \text{ say } 2{,}145 \text{ ft}$$

Refer to Chap. 13, p. 176. Efficiency of pump may be assumed to be 70 per cent.

$$\text{bhp} = \frac{1{,}500 \times 8.33 \times 0.98 \times 2{,}145}{33{,}000 \times 0.70} = \mathbf{1{,}140}$$

Criticism. The brake horsepower indicated is not a practical one. In actual design work the discharge line size would be selected to be either 10- or 12-in. diameter, and suction line (one pipeline size larger) 12 or 14 in. This would reduce motor size considerably. For good practical design for water (or liquid of equal viscosity) a pressure drop of 10 to 30 psi per 1,000 ft of pipe is often used.

M-5. A d-c motor driven pump running at 100 rpm delivers 500 gpm of water against a total pumping head of 90 ft with a pump efficiency of 60 per cent. (*a*) What motor horsepower is required? (*b*) What speed and capacity would result if the pump rpm was increased to produce a pumping head of 120 ft, assuming no change in pump efficiency? (*c*) Can a 25-hp motor be used under conditions indicated in (*b*) above?

ANSWER.

(*a*)
$$\frac{500 \times 8.33 \times 90}{33{,}000 \times 0.60} = 18.9$$

Use 20-hp motor.

(*b*) By application of the laws of affinity

$$\frac{90}{120} = \frac{100^2}{x^2} \qquad x^2 = 100^2 \times \frac{120}{90} = 1.33 \times 100^2 \qquad x = 1.151 \times 100$$

The new rpm will be **115.** The new capacity will be

$$\frac{500}{y} = \frac{100}{115} \qquad y = 500 \times 115/100 = \mathbf{575\ gpm}$$

(*c*) Again by application of the laws of affinity

$$\frac{18.9}{z} = \frac{100^3}{115^3} \qquad z = 18.9 \times 115^3/100^3 = 18.9 \times 1.52 = 28.7 \text{ bhp}$$

A 25-hp motor cannot be used. Use a 30-hp motor.

M-6 A client purchased and installed a new centrifugal pump for which the manufacturer estimated the following performance based on design calculations: speed 250 rpm, total dynamic head 25 ft of water, efficiency 84.5 per cent, brake horsepower of driving motor 750, all at rated capacity of 100,000 gpm. A test yielded the information shown in Fig. M-3. All data shown were rechecked for

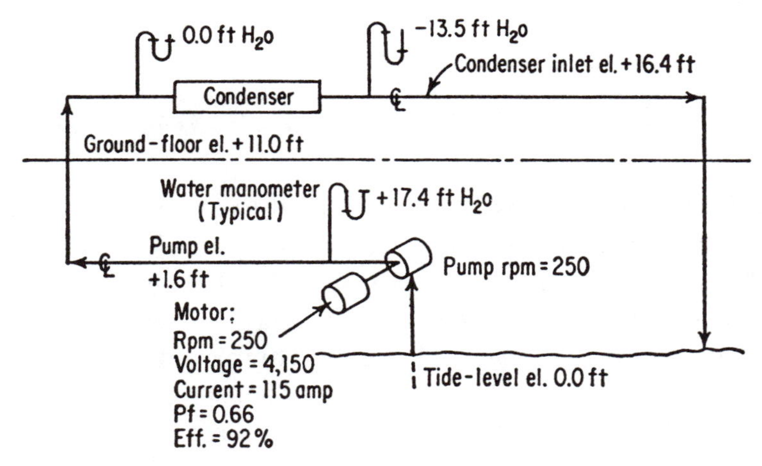

Fig. M-3

accuracy and found correct. Flow rate could not be measured due to size of pump. On the basis of the above, what would you tell your client? For a more complete answer, what additional test data would you require? How would you propose to obtain it?

ANSWER. A quick check of pipe friction shows this to be in order. This was done using the Darcy equation with a friction factor of 0.02 for clean, new steel pipe. Next check TDH of 25 ft, using Bernoulli's equation. Velocity and pressure heads cancel out. From which TDH is found to be 17.2 ft. This figure would actually be greater because of suction pipe loss, although it would be too small to be included. Now check brake horsepower from calculated figures.

$$\text{Bhp} = \frac{\text{gpm} \times 8.33 \times 17.2}{33{,}000 \times 0.845} = \frac{100{,}000 \times 8.33 \times 17.2}{33{,}000 \times 0.845} = 515$$

Electrical horsepower is now checked, assuming 3-phase a-c power.

$$\frac{1.73 \times I \times E \times \text{eff} \times \cos \Phi}{746} = \frac{1.73 \times 115 \times 4{,}150 \times 0.92 \times 0.66}{746}$$

Result is 673 hp. The motor horsepower was selected originally on the basis of

$$\text{Bhp} = \frac{100{,}000 \times 8.33 \times 25}{33{,}000 \times 0.845} = 746. \quad \textbf{This is acceptable.}$$

Although all pressure has been consumed to the point of pipe entry to the condenser, the losses through condenser and in the line back to tidewater are more than compensated for by the hydrostatic leg from elevation plus 16.4 to elevation 0.00 ft. I would tell the client that the system is in order. However, there are further checks that would help confirm these findings. Condenser operation should be checked for vacuum so that water flow is as required. Pump curve and system curve superimposed thereon would give the final check for performance point. If this lines up with the system curve, this clinches it as performing to expectations.

M-7. Discuss the selection of centrifugal pumps for handling viscous liquids.

ANSWER. To select a centrifugal pump for handling viscous liquids, it is necessary to make certain corrections to head, capacity, and efficiency. Below is a working table showing correction factors for the above mentioned items vs. viscosity expressed in seconds Saybolt Universal (SSU), with water at 30 SSU. It should be understood that the table below is based on limited test data taken on pumps of many types and makes tested at different times by several investigators and therefore will not give results equally accurate for all pumps. It is believed, however, it is the best that can be done when taking into account the very complicated nature of pump behavior when handling viscous liquids.

SSU	Head factor	Capacity factor
30	100	100
40	99.5	99
50	99.4	98.5
60	99.2	98.2
100	98.5	98
200	96	95
300	94	92
400	91.2	89
500	89	86
600	87	84
700	85.8	82
800	84	80
900	82.3	78
1,000	81	77
2,000	70	63

Thus, a pump is wanted for pumping 100 gpm of 600 SSU viscosity oil at 100°F and 0.925 specific gravity at a total dynamic head of 70 ft. From table, the capacity correction factor is 84 per cent and the head correction factor is 87 per cent. Then

$$\text{Corrected gpm } 100/0.84 = 119 \text{ gpm}$$
$$\text{Corrected head } 70/0.87 = 80.6 \text{ ft}$$

These are the water-pumping conditions for which the pump is to be

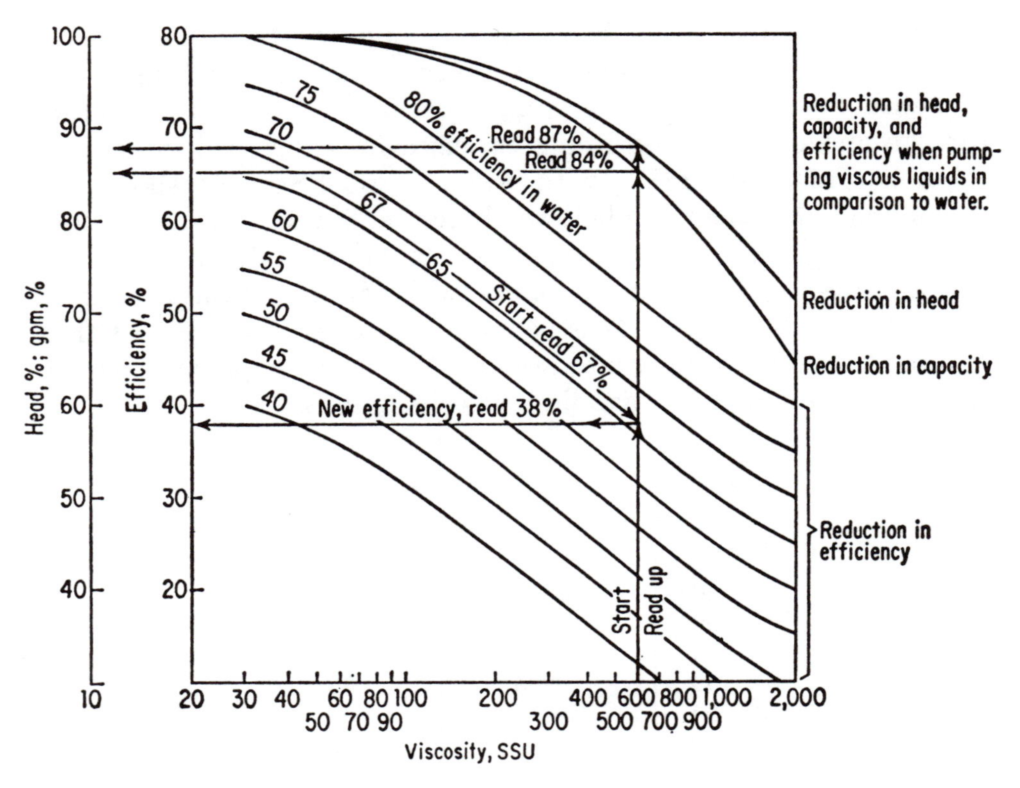

Fig. M-4

selected. Assume pump efficiency of 67 per cent. From Fig. M-4, new efficiency is 38 per cent when pumping oil of 600 SSU viscosity.

$$\text{Brake hp} = \frac{100 \times 70 \times 0.925}{3960 \times 0.38} = 4.3. \quad \textbf{Use a 5-hp motor.}$$

As we have seen, the effect of viscosity is to decrease the head capacity and increase the power that is markedly due to increase in disc friction within the pump. An impeller with a flat characteristic will be more efficient when pumping viscous liquids than an impeller with a steep characteristic. The head-capacity characteristics and the efficiency of a centrifugal pump as applied to viscous oils improve with an increase in pump size. In general, pumps up to 3 in. inclusive will handle oils of a viscosity as high as 800 SSU. Larger-sized pumps can handle oils of viscosities up to 6,000 SSU. The percentage of power absorbed by disc friction is a reasonable portion of the total power consumed by the pump. How much the efficiency of a centrifugal pump will decrease depends on the type of pump, its hydraulic design, and the relation of its operation to the conditions at best efficiency as well as to the liquid's viscosity. Brake horsepower curves when pumping viscous oils arrange themselves approximately parallel to the water-brake horsepower curves. Because disc friction varies as the fifth power of the impeller diameter and only as the speed cubed, it is desirable to use the highest speed (and resulting smaller impeller diameter) consistent with design conditions. Head loss due to viscosity increases as operating conditions approach the best efficiency point of the pump because of increased velocity through the impeller. Therefore, there is an advantage in selecting a relatively oversized pump for a highly viscous liquid so as to place the operating conditions well to the left of capacity at best efficiency on low viscosity service.

M-8. A fire breaks out in a three-story warehouse. A pressure pump engine has a 2½-in. fire hose, 300 ft long, with a nozzle 50 ft above the pump. If the discharge is 250 gpm and $C = 100$ (William and Hazen) or $n = 0.009$ (Kutter and Manning), the pressure behind the nozzle is 28 psi and minor losses are neglected.

(*a*) Compute pressure at the pump.

(*b*) Draw the hydraulic gradient line.

(*c*) If the system is 80 per cent efficient and there is a pressure of 10 psi at the pump inlet, find the horsepower required.

ANSWER

(*a*) Use Bernoulli's theorem between A and B shown in Fig. M-5, with the line through B as the datum.

$$\frac{V_A^2}{2g} + \frac{P_A}{W} + Z_A = \frac{V_B^2}{2g} + \frac{P_B}{W} + Z_B + H_f$$

Velocity heads are equal and thus cancel out.

$$\frac{P_A}{W} + Z_A = \frac{P_B}{W} + Z_B + H_f$$

$$\frac{P_A}{W} + 0 = \frac{28}{0.433} + 50 + H_f$$

$$P_A = 62.4(65 + 50 + 197) = 62.4(312) \text{ psf}$$

$$= 62.4(312/144) = \mathbf{135\ psi} \text{ at the pump}$$

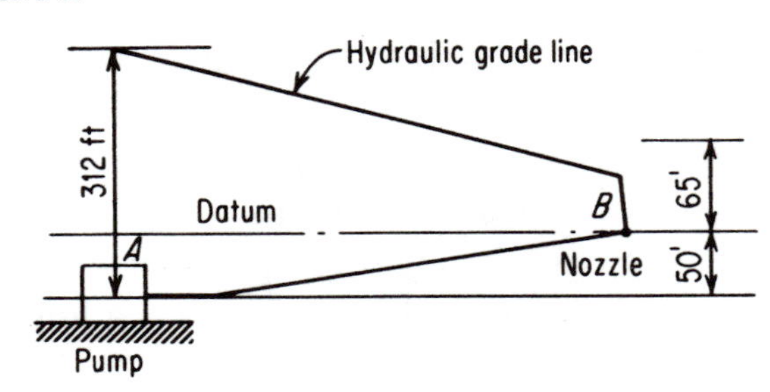

Fig. M-5

(*b*) Hydraulic gradient: see Fig. M-5

(*c*)

$$\text{Hp of pump} = \frac{\text{lb water per minute} \times \text{head}}{(33{,}000 \times \text{Efficiency})}$$

$$= \frac{8.34 \times 250 \times (135 - 10) \times 2.307}{33{,}000 \times 0.80}$$

$$= \textbf{22.8 brake hp}$$

Use a 25-hp motor.

M-9. Vacuum pumps are often used at altitudes high enough (Denver or Mexico City, for example) to appreciably affect their performance characteristics. This altitude factor is often overlooked by design and plant engineers, but if ignored, it can spell trouble for air-operated equipment.

(*a*) A vacuum pump's capability can be measured in terms of the percentage of atmosphere it can exhaust from a closed system. If a barometer at sea level registers air pressure of 30 in. Hg under standard conditions, what would be the maximum vacuum rating of a vacuum pump that has a maximum capability of 24 in. Hg?

(*b*) What will a vacuum pump rate at 5,000 ft altitude in air flow production and maximum vacuum Hg if at sea level it is rated to provide a flow of 10 cfm with 25 in. Hg maximum suction?

ANSWER.

(*a*) Maximum vacuum rating = $^{24}/_{30}$ = 0.8 or **80 per cent.** This is the percentage of atmosphere it can exhaust from a closed system.

(*b*) A 10-cfm vacuum pump with a 25-in. Hg maximum rating at sea level will produce an air flow of only 10 × 0.8 = **8 cfm** at 5000 ft altitude and will result in 25 × 0.8 = **20 in. Hg** vacuum. This pump would not provide sufficient capacity. A model with a sea-level airflow of 12 cfm would be required, since $12 \times 0.83 = 10$ cfm.

Maximum vacuum possible decreases with increasing altitude. Figure M-6 graphically illustrates this relationship. Figure M-7 illustrates altitude vs. open flow requirements. For Fig. M-7, the relationship between altitude and maximum vacuum may be determined.

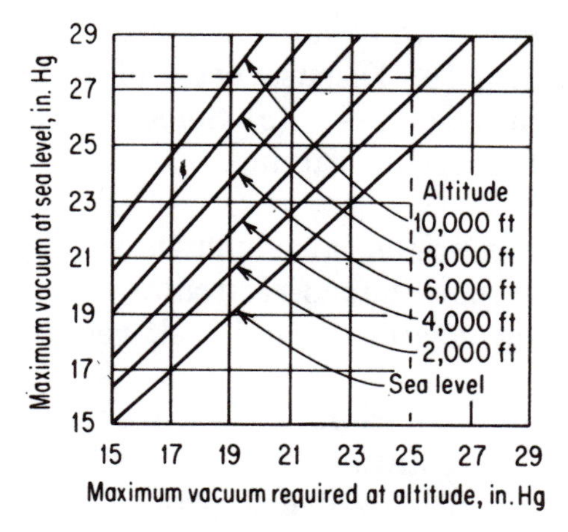

Fig. M-6

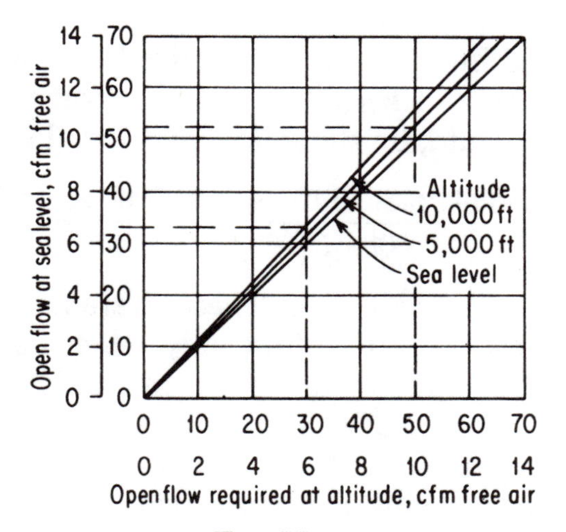

Fig. M-7

M-10. At the higher end of its capacity range, a centrifugal pump is conveniently self-cooling; the volume of fluid moving through the pump is sufficient to cool off the pump by carrying away the heat generated by the pump's rotating parts and bearings. However, at lower flow rates, particularly when a pump is operating at less that 25 percent capacity, the volume of fluid being discharged can become inadequate for cooling, increasing the risk of overheating and seizure of bearings.

(*a*) To protect centrifugal pumps from overheating damage and instability under low-flow conditions, what protective measures must be instituted? (*b*) Discuss each of these measures with advantages and disadvantages. (*c*) Select one of the systems and sketch the pump, piping, and valving.

ANSWER.

(*a*) Recirculation systems, also known as bypass systems, are necessary, such as continuous-recirculation systems, flow-controlled recirculation systems, and self-regulated recirculation systems.

(*b*) Continuous recirculation: Because of increasing electrical rates, continuous-recirculation systems, which recirculate fluid regardless of whether it is needed to protect the pump, are becoming cost-prohibitive. At 3 cents per kWh, for example, the electrical power costs for recirculating 700 gal/min at a discharge head of 500 ft are $27,000 annually. In addition, because this system bypasses fluid even when process demand is at a maximum, capital costs are increased by the need to oversize the pump and its prime mover at the outset. Check the above operating costs roughly, using water as the fluid, by the following formula:

$$\frac{700 \times 8.33 \times 500 \times 8640 \text{ h/year} \times 746 \times 0.03}{33{,}000 \times 0.6} = \$28{,}472$$

Flow-controlled recirculation: These systems reduce electrical consumption by recirculating fluid only when flow approaches a minimum safe rate. However, while energy and pump size are reduced, these systems are both complex and costly. A multiplicity of components increases the problem of breakdowns, and the need for instrument loops increases installation, operating, and maintenance costs.

Self-regulated recirculation: These systems are finding wider use and acceptance in both utility and process applications. They offer economies in the areas of operation, installation, and maintenance and combine all the functions of flow sensing, reverse-flow protection, pressure breakdown, and on-off or modulated bypass flow in a single unit. Capable of handling high recirculation pressure drops, self-powered units employ a rising-disk type of check valve that acts as the flow-sensing element. The system requires only three connections.

(*c*) Refer to Figs. M-8, M-9, and M-10.

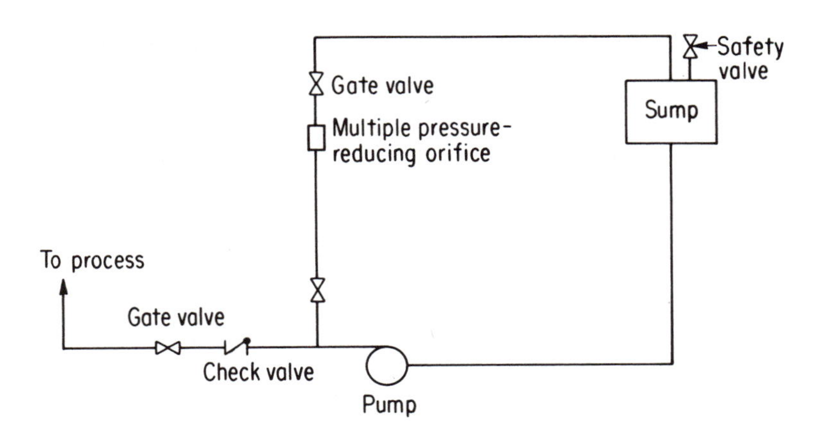

Fig. M-8. Continuous recirculation system.

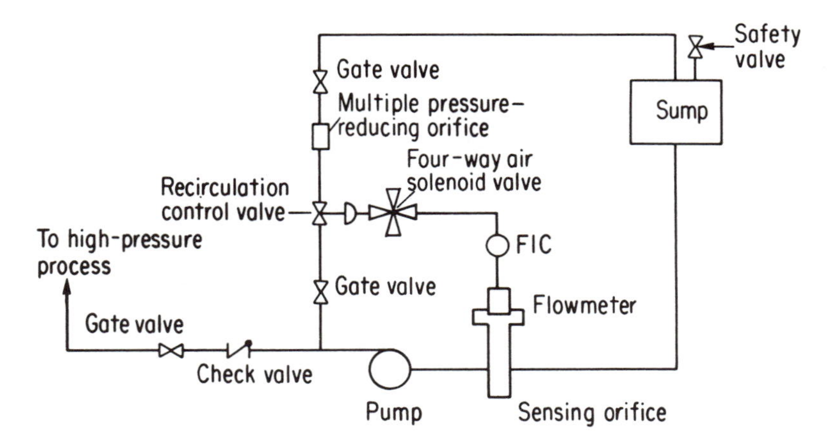

Fig. M-9. Flow-controlled recirculation system.

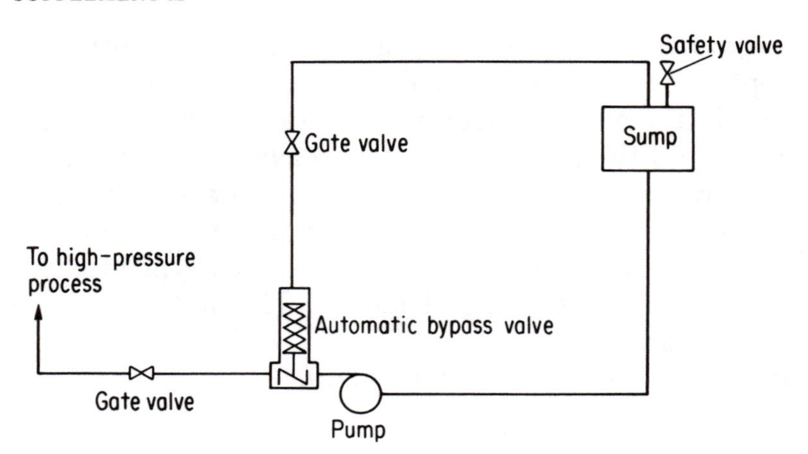

Fig. M-10. Self-regulated recirculation system.

Supplement **N**

FANS, BLOWERS, AND COMPRESSORS

Questions and Answers

N-1. The characteristic and horsepower curves shown in Fig. N-1 indicate the performance of a double-inlet forced-draft fan operating at sea level with a speed of 1,180 rpm when handling air at 40°F. Determine the static pressure developed by the fan and the power required to drive it when it handles 514,000 lb per hr of 80°F air at

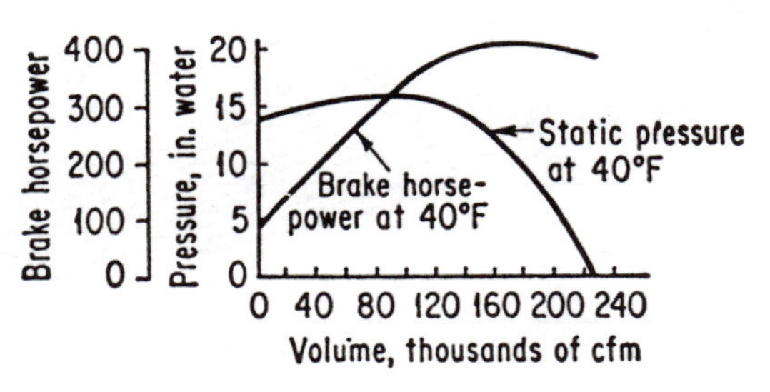

Fig. N-1

an altitude of 5,000 ft, where the barometric pressure is approximately 24.9 in. Hg abs.

ANSWER. The fan will handle the same cfm no matter where located and so long as the rpm and fan configuration does not change. However, because air in the actual location is less dense than the curve characteristic indicates, both new static pressure and bhp developed will be less than at sea level. The cfm handled under all conditions may be obtained in the usual manner.

$$\frac{514{,}000}{60} \times \frac{379}{29} \times \frac{460 + 80}{460 + 60} \times \frac{29.9}{24.9} = 139{,}550 \text{ cfm}$$

At sea level from the characteristic curve at 40°F the static pressure reads 16 in. WG. The brake horsepower reads to be equal to 325. Since brake horsepower and static pressure will vary directly with relative density, use Fig. N-2 to find this factor. Correction factor reads 0.90, the relative density factor.

$$\text{New static pressure} = 16 \times 0.90 = \mathbf{14.4\ in.\ WG}$$
$$\text{New bhp} = 325 \times 0.90 = \mathbf{292.5\ bhp}$$

N-2. (*a*) Draw an approximate head vs. capacity curve for a fan-handling flue gas of any standard type for a design pressure of 10 in. water and 150,000 cfm at 300°F.

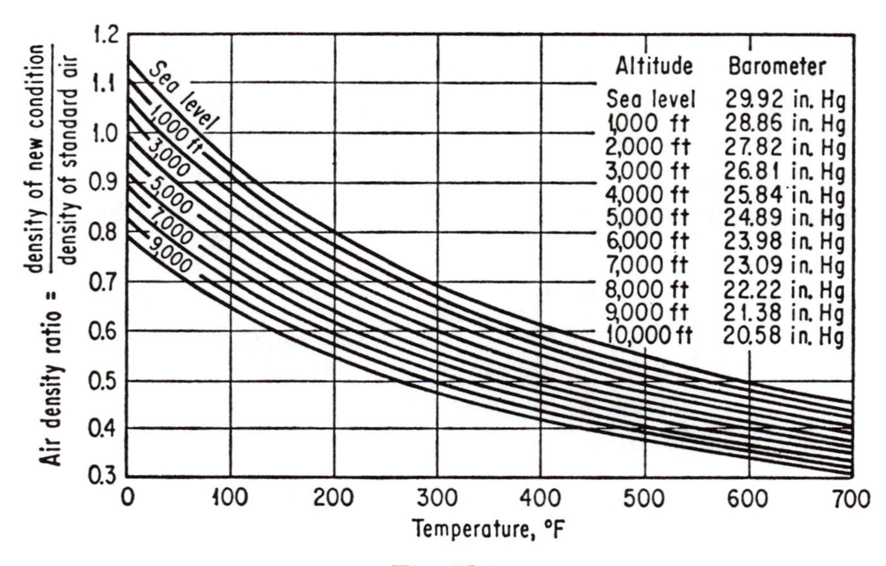

Fig. N-2

(*b*) Assuming the fan to be driven by a constant-speed motor, explain how to determine the head curve with gas at 400°F.

(*c*) If a 30 per cent pressure drop is added to the system resistance, what will be the output of the fan in cfm with the gas at 300°F?

ANSWER.

(*a*) See Fig. N-3. Fan manufacturer will guarantee design point only. When purchasing a fan (or pump), the fan characteristic curve should be made a part of the specification.

(*b*) As indicated previously in Prob. N-1, the cfm will always remain practically the same so long as the rpm does not change. However, the static pressure will decrease with decrease in density, and with increase in temperature (from 300 to 400°F) the air density ratio will become less than unity. Assume fan at sea level, and refer to Fig. N-2. The density ratio is read off to be 0.62 at 400°F and 0.70 at 300°F. The correction factor to be applied to the 10 in. is 0.62/0.70 = 0.888. The new static pressure for the 400°F condition is 8.88 in WG. Locate this point along the 150,000 cfm line and

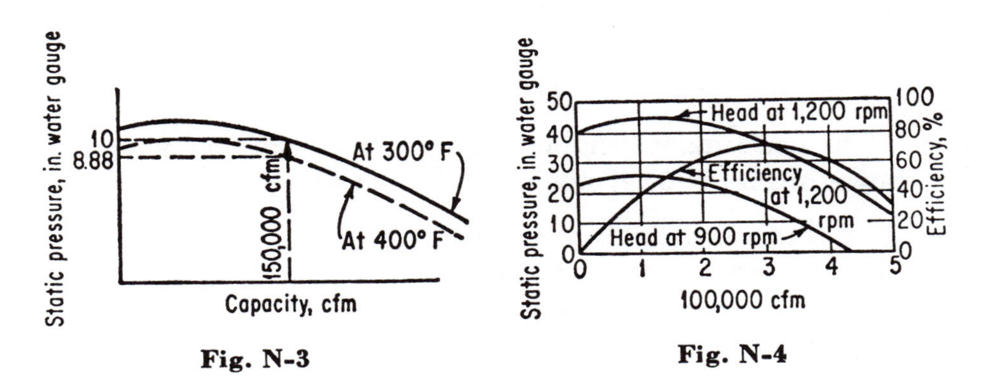

Fig. N-3

Fig. N-4

where it crosses the 8.88-in. line from the left side of the figure. This locates the locus of points and runs parallel to the 300°F curve and establishes the new head vs. capacity curve.

(*c*) Assume constant speed. Then there will be a reduction in flow in accordance with

$$\text{New cfm at } 300°\text{F} = 150{,}000 \times \sqrt{\frac{1}{1 + 0.3}}$$

$$150{,}000 \times 0.877 = \mathbf{131{,}550\ cfm}$$

N-3. Characteristic curves in Fig. N-4 are for a forced-draft fan operated at speeds of 1,200 and 900 rpm. Also shown is the efficiency curve obtained at an operating speed of 1,200 rpm. Estimate the horsepower required to drive the fan when operated at a speed of 900 rpm and at a flow rate of 300,000 cfm at 15 in. water static pressure.

ANSWER. Refer to Fig. N-4. At 900 rpm read curve left for 15 in. static pressure and 300,000 cfm. At 1,200 rpm the new static pressure developed is

$$\frac{15}{y} = \frac{1{,}200^2}{900^2} \qquad y = 15 \times \frac{1{,}200^2}{900^2} = 26.7 \text{ in. WG}$$

The new cfm is

$$\frac{300{,}000}{x} = \frac{900}{1{,}200} \qquad x = 300{,}000 \times \frac{1{,}200}{900} = 400{,}000 \text{ cfm}$$

From Fig. N-4 for 400,000 cfm and 1,200 rpm the fan efficiency is read to be equal to 62 per cent. Assuming the laws of affinity apply and efficiency remains constant at 62 per cent the horsepower required is by use of a recognized formula

$$\text{bhp} = 0.000160 \times 300{,}000 \times 15 \times 1/0.62 = \mathbf{1{,}160}$$

N-4. The performance characteristics of a centrifugal compressor are shown in Fig. N-5; symbols are defined as follows:

P_1 = compressor inlet pressure, absolute
P_2 = compressor discharge pressure, absolute
T_1 = compressor inlet temperature, °R
M = mass flow rate, lb per sec
N = compressor speed, rps

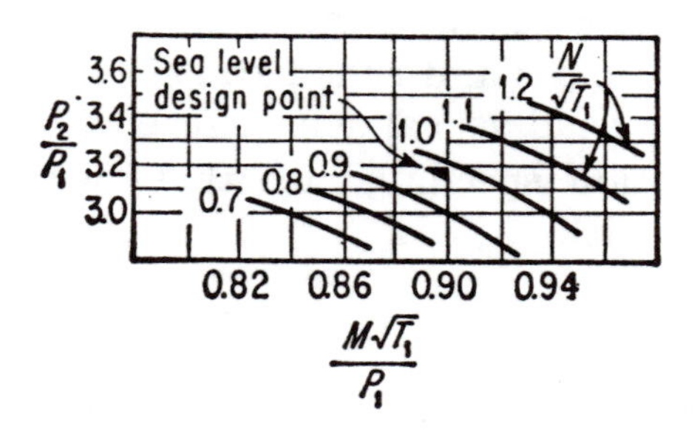

Fig. N-5

The compressor is designed to operate at sea level at 14.7 psia and 60°F. Estimate in psia the compressor exit pressure if the compressor is operated at the same speed and at the same volume flow rate in Denver, Colorado, on a day when the atmospheric pressure and temperature are 12 psia and 0°F, respectively.

ANSWER. For sea-level compressor operation the following apply:

$$P_1 = 14.7 \text{ psia} \qquad \text{from curve, } \frac{M\sqrt{T_1}}{P_1} = 0.9 \qquad \frac{P_2}{P_1} = 3.2$$

$$T_1 = 520°\text{R} \qquad \text{from curve, } \frac{N}{\sqrt{T_1}} = 1$$

The mass flow rate at sea level is first calculated, converted to cfm. Since the compressor is a constant-volume machine, the new weight rate can be determined for Denver conditions.

$$M = 0.9 \times 14.7/520^{1/2} = 30.2 \text{ lb per sec}$$
$$\text{cfs} = {}^{379}\!/_{29} \times 30.2 \times (14.7/14.7) \times ({}^{520}\!/_{520}) = 395 \text{ cfs}$$

The new M in Denver is equal to

$$395 \times {}^{29}\!/_{379} \times 12/14.7 \times {}^{520}\!/_{460} = 27.8 \text{ lb per sec}$$

Since we are asked to estimate the exit pressure we can assume that the work in both locations is the same. Thus,

$$20.2 \times c_p \times 520[1 - 3.2^{0.26}] = 27.8 \times c_p \times 460 \left[1 - \left(\frac{P_2}{P_1}\right)^{0.26}\right]$$

where $0.26 = (n - 1)/n = (1.35 - 1)/1.35$. Then since the c_p's cancel out, it may be shown that $(P_2/P_1)^{0.26} = 1.43$. To find P_2/P_1, proceed as follows:

$$0.26 \log \frac{P_2}{P_1} = \log 1.43 \qquad \log \frac{P_2}{P_1} = \frac{0.1553}{0.26} = 0.598$$
$$\frac{P_2}{P_1} = 3.96 \qquad P_2 = 12 \times 3.96 = \mathbf{47.5\ psia}$$

N-5. The delivery pressure in a reciprocating compressor having isentropic compression and handling 0.1 lb of air per cycle is 50 psia. The temperature of the air drawn in is 60°F and the average atmospheric pressure where the compressor is located may be taken as 14.7 psia. Determine (*a*) the net work of the compressor cycle if the compressor has no clearance; (*b*) the net work of the compressor cycle if the compressor has 5 per cent clearance; (*c*) if the compressor has 5 per cent clearance, find the weight of the air in the cylinder during the compression; (*d*) what per cent increase in the required piston displacement results from the 5 per cent clearance when compared to the compressor of the same capacity with no clearance?

ANSWER.

(*a*) Refer to Fig. N-6 (*PV* diagram) and Fig. N-7 (*TS* diagram) for no clearance.

$$\text{Net work} = \frac{k}{k-1}\, wRT_{2,}\left[\left(\frac{P_3}{P_2}\right)^{(k-1)/k} - 1\right]$$
$$= \frac{1.4}{0.4} \times 0.1 \times 53.3 \times 520 \left[\left(\frac{50}{14.7}\right)^{0.286} - 1\right]$$
$$= \mathbf{4{,}055\ ft\text{-}lb}$$

(*b*) The clearance does not affect the net work delivered by the compressor when handling a given volume of air or a given weight of air. Therefore, the net work for 5 per cent clearance is **4,055 ft-lb.**

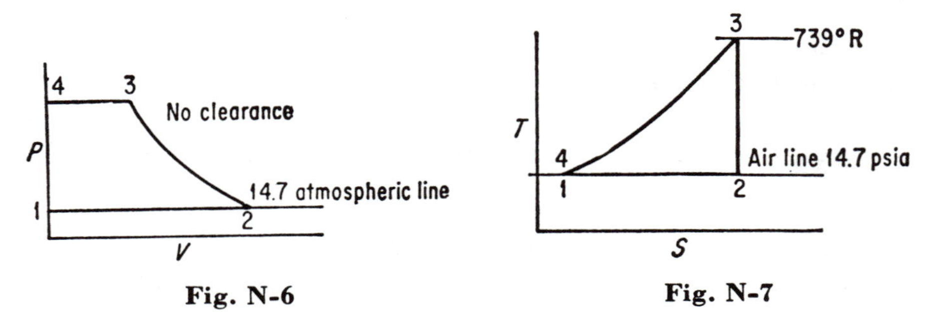

Fig. N-6 **Fig. N-7**

(*c*) Per cent clearance is equal to

$$m = \frac{\text{clearance volume}}{\text{piston displacement}} = 0.05$$

$$V_1 = 1 + m - m\left(\frac{P_3}{P_2}\right)^{1/n} = 1 + 0.05 - 0.05\left(\frac{50}{14.7}\right)^{1/1.4} = 0.93$$

$V_3 = 1.24/0.93 = 1.332 \qquad 1.332 \times 0.0807 = \mathbf{0.113\ lb}$

(*d*) $$\frac{1.332 - 1.24}{1.24} \times 100 = \mathbf{7.4\ per\ cent}$$

N-6. Determine the capacity of either compressor (no clearance and with clearance) given in the above problem in terms of the volume of free air handled per minute if the compressor operates at a speed of 180 rpm and is double-acting.

ANSWER.

$$P_1V_1 = wRT_1 \qquad V_1 = \frac{0.1 \times 53.3 \times 520}{14.7 \times 144} = 1.315\ \text{cu ft/rpm/cylinder}$$
$$1.315 \times 180 \times 2 = \mathbf{471\ cu\ ft}$$

N-7. Assume three compressor cylinders without clearance and with a piston displacement of 1 cu ft, one cylinder so arranged as to give adiabatic compression; one arranged to give isothermal compression; and one arranged to give a compression curve with an exponent equal to 1.25. (*a*) Determine the work during the compression process alone in each cylinder and the final temperature in

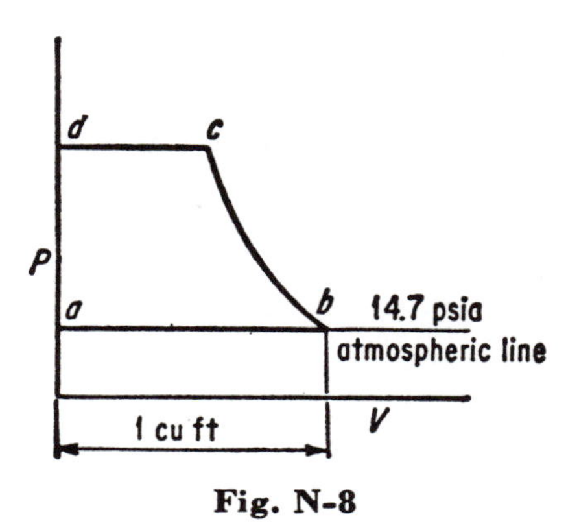

Fig. N-8

each case if air with an initial temperature of 60°F and an initial pressure of 14.7 psia is compressed to 100 psia. (*b*) Express the second and third cases as a percentage of the compression work in the case of the adiabatic process. (*c*) Determine the work per cycle in each case, assuming discharge to occur at the constant pressure of 100 psia. (*d*) Express results as percentages as in (*b*).

ANSWER. Refer to Fig. N-8.

(*a*) Adiabatic compression cylinder:

$$\frac{T_1}{T_2} = \left(\frac{P_1}{P_2}\right)^{(k-1)/k} = \left(\frac{14.7}{100}\right)^{0.286} = 0.578 \qquad T_2 = \frac{520}{0.578} = 900°\text{R},$$

or **440°F**

$$\text{Work of compression} = \frac{P_3V_3 - P_2V_2}{k-1}$$

$$\frac{V_3}{V_2} = \left(\frac{P_4}{P_3}\right)^{1/k} = 0.254 \qquad V_3 = 0.254 \times 1 = 0.254 \text{ cu ft}$$

$$\text{Work} = \frac{(100 \times 25.4) - (14.7 \times 1)}{1.4 - 1} \times 144 = \frac{10.7}{0.4} \times 144$$

$$= \mathbf{3{,}860 \text{ ft-lb}}$$

Isothermal compression cylinder:

$$T_3 = T_2 = 520 \text{ R, or } \mathbf{60°F}$$

$$\text{Work} = P_2V_2 \ln \frac{V_2}{V_3} = 14.7 \times 144 \times 1 \times \ln \frac{1}{0.254}$$

$$= 14.7 \times 144 \times 1 \times 1.3712$$

$$= \mathbf{2{,}900 \text{ ft-lb}}$$

Polytropic compression cylinder:

$$T_3 = T_2 \left(\frac{100}{14.7}\right)^{(k-1)/k} = 520 \times 1.468 = 763°\text{R, or } \mathbf{303°F}$$

$$\text{Work} = \frac{P_3V_3 - P_2V_2}{n-1} = \frac{(100 \times 144 \times 0.209) - 2{,}120}{0.25}$$

$$= \frac{3{,}010 - 2{,}120}{0.25}$$

$$= \mathbf{3{,}560 \text{ ft-lb}}$$

(*b*)

$$\frac{2{,}900}{3{,}850} \times 100 = \mathbf{75.5 \text{ per cent}} \qquad \frac{3{,}560}{3{,}850} \times 100 = \mathbf{92.5 \text{ per cent}}$$

(*c*)

$$\text{Net work adiabatic} = \frac{k}{k-1} P_2V_2 \left[\left(\frac{P_3}{P_2}\right)^{(k-1)/k} - 1\right] = \frac{1.4}{0.4} \times 14.7 \times 144$$

$$\times 1 \left[\left(\frac{100}{14.7}\right)^{0.4/1.4} - 1\right]$$

$$= 3.5 \times 14.7 \times 144 \times 1(1.728 - 1) = \mathbf{5{,}380 \text{ ft-lb}}$$

$$\text{Net work isothermal} = P_2V_2 \ln \frac{P_3}{P_2} = 14.7 \times 144 \times 1 \times 1.9184 = \mathbf{4{,}055 \text{ ft-lb}}$$

$$\text{Net work polytropic} = \frac{n}{n-1} P_2V_2 \left[\left(\frac{P_3}{P_2}\right)^{(n-1)/n} - 1\right]$$

$$\frac{1.25}{0.25} \times 14.7 \times 144 \times 1 \left[\left(\frac{100}{14.7}\right)^{0.25/1.25} - 1\right]$$

$$5 \times 14.7 \times 144 \times (1.468 - 1) = \mathbf{4{,}950 \text{ ft-lb}}$$

(*d*)

$$\frac{4{,}055}{5{,}380} \times 100 = \mathbf{75.3 \text{ per cent}}$$

$$\frac{4{,}950}{5{,}380} \times 100 = \mathbf{92 \text{ per cent}}$$

N-8. What is the volumetric efficiency of a compressor having 5 per cent clearance and operating as in question N-5?

ANSWER.

$$\frac{1.24}{1.33} \times 100 = \textbf{93.2 per cent}$$

N-9. A centrifugal compressor handling air draws in 12,000 cu ft of air per minute at a pressure of 14 psia and a temperature of 60°F. This air is delivered from the compressor at a pressure of 70 psia and a temperature of 164°F. The area of the suction pipe is 2.1 sq ft, the area of the discharge pipe is 0.4 sq ft, and the discharge pipe is located 20 ft above the suction pipe. The weight of the jacket water, which enters at 60°F and leaves at 110°F, is 677 lb per min. Find the horsepower required to drive this compressor assuming no loss from radiation.

ANSWER.

$$\text{hp} = \frac{w}{0.707}\left[c_p(t_2 - t_1) + \frac{V_2^2 - V_1^2}{50{,}000} + \frac{Z_2 - Z_1}{778}\right] + \left[\frac{w_j(t_0 - t_i) + R_c}{0.707}\right]$$

$$12{,}000 \text{ cfm} = 200 \text{ cfs} \qquad P_1 = 14.0 \text{ psia} \qquad T_1 = 60 + 460 = 520°\text{R}$$

$$P_1V_1 = wRT \qquad w = \frac{P_1V_1}{RT_1} = \frac{14.0 \times 144 \times 200}{53.3 \times 520} = 14.55 \text{ lb.}$$

Velocity at entrance = 200/2.1 = 95.3 fps
Velocity at discharge = 200/0.4 = 500 fps

$$\text{hp} = \frac{14.55}{0.707}\left[0.24(624 - 520) + \frac{500^2 - 95.3^2}{50{,}000} + \frac{20}{778}\right] + [677/60 \times (110 - 60)]/0.707$$

$$= 20.6(24.95 + 4.8 + 0.0256) + 797 = \textbf{1,409 hp.}$$

Radiation loss $R_c = 0$.

N-10. Air is admitted to the cylinder of a 12- by 16-in. air engine under pressure of 100 psia until one-fourth of the stroke is reached and then is allowed to expand to the end of the stroke following the law $PV = C$. Find the work done between cutoff and end of stroke.

ANSWER. Refer to Fig. N-9. One-fourth of the stroke is 4 in. Cylinder diameter is 12 in. The equation for work is given by

$$\text{Work} = P_1V_1 \ln \frac{V_2}{V_1}$$

$$P_1 = 100 \times 144 = 14{,}400 \text{ psfa} \qquad V_1 = 0.785\,\frac{12^2}{144} \times \frac{16}{12} \times \frac{1}{4}$$

$$= 0.263 \text{ cu ft}$$

$$\text{Work after cutoff} = 100 \times 144 \times 0.263 \ln 4 = 5{,}260 \text{ ft-lb}$$

$$\text{Work up to cutoff} = 100 \times 144\,(0.263 - 0) = 3{,}780 \text{ ft-lb}$$

$$\text{Total} = \mathbf{9{,}040 \text{ ft-lb}}$$

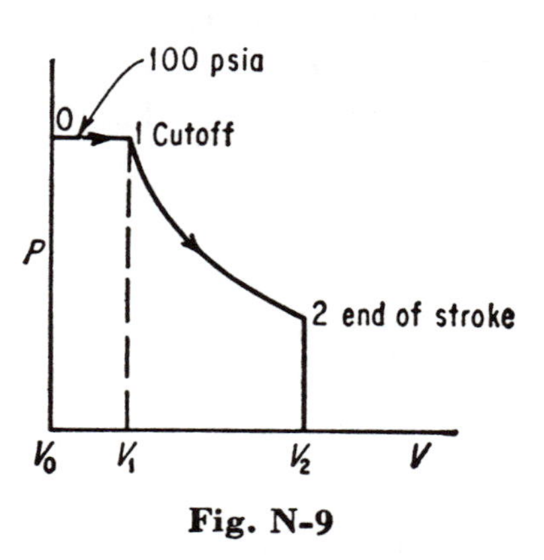

Fig. N-9

N-11. An ammonia oxidation unit will use 16,000 cfm of air (60°F and 1 atm), compressing it from standard conditions of pressure to 125 psig in a centrifugal compressor with nine stages and no intercoolers. Part of the power to drive the compressor will be supplied by a turboexpander which is fed with the waste nitrogen from the nitric acid absorption towers at 75 psig. It may be assumed that this waste gas is pure nitrogen and that all of it was derived from the air. The nitrogen from the absorbers is preheated to 650°C by the effluent gases from the converter and enters the expander at this temperature. The remainder of the power required for air compression is to be supplied by an electric motor. Estimate this motor in horsepower.

ANSWER. Assume efficiency of compressor as 80 per cent, that of expander as 70 per cent. Now refer to Fig. N-10. Work of compression

$$W_c = \frac{kP_1V_1}{k-1}\left[1 - \left(\frac{P_2}{P_1}\right)^{(k-1)/k}\right]$$

$$W_c = \frac{1.4 \times 14.7 \times 144 \times 16,000}{(1.4-1) \times 33,000}[1 - 9.5^{0.40/1.4}] = 4,100 \text{ hp}$$

Assume k is constant. Use 1.4 at 650°C. Gas is ideal at 75 psig.

$V_1 = 12,640 \times (15/90) \times (1,660/520) = 6,700$ cfm at 650°C and 75 psig. Work of expansion

$$W_e = \frac{1.4 \times 90 \times 144 \times 6,700}{0.4 \times 33,000}[1 - (15/90)^{0.4/1.4}] = 2,590 \text{ bhp}$$

Work to be supplied by motor is $-4,100 + 2,590 = -1,510$ bhp. And estimated motor size is **1,750 hp.**

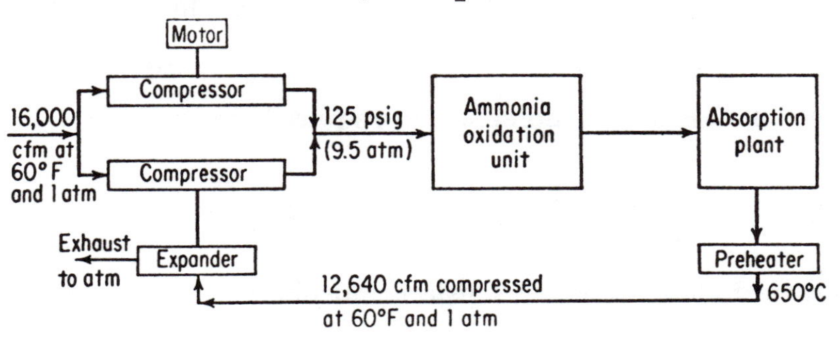

Fig. N-10

N-12. A single-stage compressor is designed to compress air at standard conditions of pressure but at 80°F to a pressure of 80 psig. The clearance is estimated to be 8 per cent. Estimate the percentage change in the capacity of the same compressor (expressed as per cent) and in the horsepower needed to drive it if it were located at an elevation where the air pressure is only 0.5 atm absolute. Discharge gauge pressure remains the same. Estimate the increase in the maximum temperature of the air in the cylinder.

ANSWER.

$$V_1 = V_d\left[1 + C - C\left(\frac{P_2}{P_1}\right)^{1/k}\right]$$

$$\frac{V_1}{V_d} = 1 + 0.08 - 0.08(94.7/14.7)^{1/1.4} = 0.777$$

$$\frac{V_1'}{V_d} = 1.08 - 0.08(87.35/7.35)^{1/1.4} = 0.61$$

Capacity decrease:

$$1 - \{(0.61 \times 7.35)/(0.777 \times 14.7)\} = 0.617 \text{ or } \mathbf{61.7\ per\ cent}$$

Horsepower change:

$$W = \frac{kP_1V_1}{k-1}\left[1 - \left(\frac{P_2}{P_1}\right)^{(k-1)/k}\right]$$

$$W = \frac{1.4 \times 14.7 \times 0.777V_d}{0.4}[1 - (94.7/14.7)^{0.286}]$$

$$W = 40.0V_d(1 - 1.706) = -28.2V_d$$

$$W' = \frac{kP_1V_1'}{k-1}[1 - (87.35/7.35)^{0.286}]$$

$$W' = (1.4 \times 7.35)/0.4 \times 0.61V_d(1 - 2.024) = -16.1V_d$$

Decrease in theoretical horsepower required per unit volume is

$$1 - (16.1/28.2) = 0.43 \text{ or } \mathbf{43\ per\ cent}$$

Increase in theoretical horsepower required per unit weight flow rate is

$$\frac{-16.1V_d}{-28.2V_d \times 0.383} = 1.49 \text{ or } \mathbf{49\ per\ cent}$$

temperature increase for 14.7 psia atmospheric pressure case is

$$T_2 = T_1\left(\frac{94.7}{14.7}\right)^{(k-1)/k} = 540 \times 1.706 = 920 \text{ R or } \mathbf{460°F}$$

and for 7.35 psia atmospheric pressure case is

$$T_2 = T_1\left(\frac{87.35}{7.35}\right)^{(k-1)/k} = 540 \times 2.024 = 1.090 \text{ R or } \mathbf{630°F}$$

N-13. A centrifugal compressor driven by a 2,000-hp steam turbine has been used to compress methane from 60°F and 14.7 psia to 150 psia. (*a*) It is desired to change over to air compression with same intake conditions and discharge pressure. Will the compressor perform satisfactorily? (*b*) A need has arisen for 1,500 lb of saturated steam per hr at 800 psia but it is available only as saturated steam at 200 psia. It is proposed to compress the given steam to the desired pressure in a continuous reciprocating compressor. Estimate the work required in kwhr per 100 lb of saturated steam at 800 psia.

ANSWER. (*a*) Use k for air as 1.4; for methane (CH_4), 1.3.

$$\text{For methane power } W_c = \frac{1.3P_1V_1}{0.3}[1 - 10^{2.3}] = -3.03P_1V_1$$

$$\text{For air } W_a = \frac{1.4P_1V_1}{0.4}[1 - (259/15)^{0.286}] = -4.4P_1V_1$$

$(4.4 - 3.03)/3.03 = 0.45$ or **45 per cent.** At constant rpm and suction cfm, 45 per cent more power is required for the air service. **Compressor not satisfactory.**

(*b*) Steam 200 psia saturated has an enthalpy of 1,199 Btu.

Steam at 800 psia saturated under constant entropy compression from 200 psia has an enthalpy of 1,340 Btu. Theoretically $(1{,}340 - 1{,}199)/1{,}340 = 0.041$ kw hr per lb of superheated steam at an enthalpy of 1,340 Btu. Now let X = lb superheated steam required per lb saturated steam. Then $1 - X$ = lb saturated condensate required per lb saturated steam. Then

$$1{,}199 = (1 - X)689 + (1{,}340X)$$

Solving for X. This is found to be 0.783, from which

$$0.041 \times 0.783 \times 100 = \textbf{3.21 kwhr} \text{ per 100 lb}$$
800 psia saturated steam produced

N-14. A 5-stage centrifugal compressor running at 4,300 rpm is tested with air at 120°F and standard atmospheric pressure. At a capacity of 16,000 cfm at inlet conditions, the discharge pressure was 32.5 psig and the brake hp was 2,100. Calculate the following quantities: (*a*) adiabatic head; (*b*) adiabatic power efficiency; (*c*) polytropic exponent; (*d*) polytropic efficiency; (*e*) fluid or polytropic head; (*f*) pressure which would be developed if the compressor were fed carbon dioxide at 60°F and 3 psig; (*g*) final temperature of the carbon dioxide; (*h*) brake horsepower required when the feed is carbon dioxide. Note compressor has no intercoolers.

ANSWER.

$$(a) \quad h_a = \frac{kRT_1}{(k-1)M}\left[\left(\frac{P_2}{P_1}\right)^{(k-1)/k} - 1\right]$$

$$h_a = \frac{1.4 \times 1544 \times 580}{0.4 \times 29}[(47.2/14.7)^{0.286} - 1] = \mathbf{43{,}400\ ft}$$

(b) $W_{ad} = \dfrac{1.4 \times 14.7 \times 144 \times 16{,}000}{0.4 \times 33{,}000} [1 - (47.2/14.7)^{0.286}]$

$= -1{,}435$

$E_{ad} = (1{,}435/2{,}100)100 =$ **68.5 per cent**

(c) $$\frac{n-1}{n} = \ln \frac{\left\{1 - \left[\left(\frac{P_2}{P_1}\right)^{0.286} - 1\right]/0.685\right\}}{\ln (P_2/P_1)}$$

$$\ln 1.585/\ln 3.21 = 0.2/0.5065 = 0.395$$

From which n is found to be **1.65.**

(d) $E_p = \dfrac{n/n-1}{k/k-1} = \dfrac{2.54}{3.5} 100 =$ **72 per cent**

(e) $h_f = \dfrac{E_p \times h_a}{E_a} = \dfrac{0.72 \times 43{,}400}{0.685} =$ **45,600 ft**

(f) For CO_2

$$h_f = \frac{nRT_1}{(n-1)M}\left[\left(\frac{P_2}{P_1}\right)^{(n-1)/n} - 1\right]$$

$T_1 = 60 + 460 = 520$ R. $P_1 = 17.7$ psia Use $n = 1.65$

$$\left(\frac{P_2}{P_1}\right)^{0.395} - 1 = \frac{45{,}600 \times 0.65 \times 44}{1.65 \times 1544 \times 520} = 0.981$$

$$\left(\frac{P_2}{P_1}\right)^{0.385} = 1 + 0.981 = 1.981$$

Thus, $P_2 = 17.7 \times 5.65 =$ **100 psia**

(g) $T_2 = T_1 \left(\dfrac{P_2}{P_1}\right)^{(n-1)/n} = 520 \times 1.981 = 1030$ R or **570 F**

(h) $W = \dfrac{1.65 \times 17.7 \times 144 \times 16{,}000}{0.65 \times 33{,}000 \times 0.72} [1 - 1.981] =$ **−4,280 bhp**

N-15. For a 40-in.-diameter fan delivering 160,000 cfm of standard air with 16 in. static pressure (SP) at 1,200 rpm, what is the diameter of a geometrically similar fan running at 60,000 rpm that will deliver 1.0 cfm and 1 in. SP?

ANSWER.

Specific speed $N_s = \dfrac{\text{rpm} \times \text{cfm}^{1/2}}{\text{SP}^{3/4}} = \dfrac{1{,}200 \times 160{,}000^{1/2}}{16.0^{3/4}} = 60{,}000$

Diameter $d_s = \dfrac{d \times \text{SP}^{1/4}}{\text{cfm}^{1/2}} = \dfrac{40 \times 16^{1/4}}{160{,}000^{1/2}} =$ **0.20 in.**

N-16. A compressor is used to supply 50 lb/in^2 gauge air to an air-lift pump that is raising 20 gal/min of a liquid with specific gravity of 1.5 to a height of 50 ft. Pump efficiency is 30 percent, and it may be assumed that the compressor operates isentropically ($C_p/C_v = 1.4$). Determine the horsepower required for the compressor.

ANSWER. $W_p = (20)(8.33)(1.5)(50) = 1.25 \times 10^4$ ft-lb/min

Actual work $= (1.25 \times 10^4/0.3) = 4.17 \times 10^4$ ft-lb/min

Volume of air required at ambient temperature assuming that the efficiency of the pump is based on reversible isothermal expansion is

$W_{\min} = NRT \ln (P_1/P_2) = P_1 V_1 \ln (P_1/P_2)$

$1.25 \times 10^4 = (14.7)(144)(V_1) \ln (64.7/14.7)$, from which V_1 is found to be equal to 3.98 ft^3/min. Therefore, the actual volume is (3.98/0.3) or 13.27 ft^3/min.

Isentropic work for compressor:

$$-W = \frac{NRT_1^r}{r^{-1}} \left[\left(\frac{P_2}{P_1} \right)^{r-1/r} - 1 \right]$$

$$= \frac{(14.7)(144)(13.27)(1.4)}{0.4} \left[\left(\frac{64.7}{14.7} \right)^{0.4/1.4} - 1 \right]$$

$$-W = 5.2 \times 10^4 \text{ ft-lb/min}$$

$$\text{hp} = \frac{5.2 \times 10^4}{33{,}000} = \mathbf{1.57}$$

Supplement O

HEAT TRANSMISSION

Questions and Answers

O-1. A 4,530-sq ft heating surface counterflow economizer is used in conjunction with a 150,000-lb per hr boiler. The inlet and outlet water temperatures are 210 and 310°F. The inlet and outlet gas temperatures are 640 and 375°F. Find the over-all heat transfer coefficient in Btu per hr per sq ft per °F.

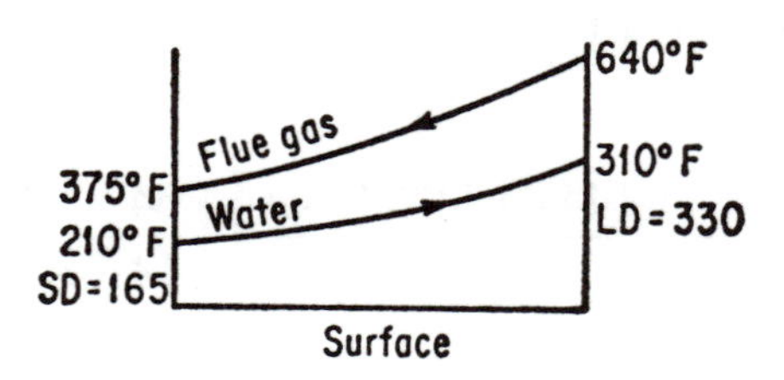

Fig. O-1

ANSWER. Refer to Fig. O-1. Make use of the basic heat transfer equation: $Q = UA\ \Delta t_m$, $U = Q/(A\ \Delta t_m)$. The amount of heat transferred Q is obtained from the familiar expression $Q = wc_p\ \Delta t$.

$$Q = 150{,}000 \times 1 \times (310 - 210) = 15{,}000{,}000 \text{ Btu per hr}$$

From the formula for log mean temperature difference determine Δt_m.

$$\Delta t_m = \frac{330 - 165}{\ln (330/165)} = 238$$

We now know all factors in the basic equation so that

$$U = \frac{Q}{A\ \Delta t_m} = \frac{15 \times 10^6}{4{,}530 \times 238} = \mathbf{13.9\ Btu/(hr)(sq\ ft)(°F)}$$

O-2. A surface feed-water heater is to be designed to heat 500,000 lb of water per hr from 200 to 390°F; the heater is to be straight condensing with no desuperheat or subcooling zones. Saturated steam at 400°F is to be used as the heating medium; drains leave as saturated liquid at 400°F. The tubes are 7/8 in. outside diameter with 1/16-in. walls. The heater is to have two water passes and it has been estimated that the over-all coefficient of heat transfer is 700 Btu/(hr)(°F)(sq ft) of outside tube surface. Specify the number and length of the tubes.

ANSWER. Refer to Fig. O-2 for the heat-transfer process. The log mean temperature difference may be determined by direct

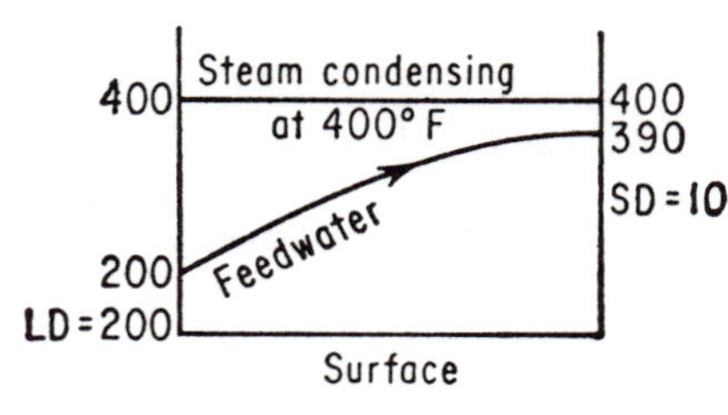

Fig. O-2

calculation; is found to be 64°F. The total amount of heat transferred may be found from the water side, assuming complete insulation of the shell.

$$Q = 500{,}000 \times 1 \times (390 - 200) = 95 \times 10^6 \text{ Btu per hr}$$

The surface on the heater may also be found to be equal to

$$A = \frac{Q}{U\,\Delta t_m} = \frac{95 \times 10^6}{700 \times 64} = 2{,}120 \text{ sq ft}$$

The number of tubes per pass is simply determined by first finding the flow rate in cubic feet per second, then the total cross-sectional area of the tubes and finally dividing by the internal diameter of each tube.

$$\text{Tubes per pass} = \frac{\text{cu ft per sec}}{6 \text{ fps}} \times \frac{1}{0.00307 \text{ sq ft}}$$

$$\text{Tubes per pass} = \frac{500{,}000/(62.4 \times 3{,}600)}{6} \times \frac{1}{0.00307} = 121 \text{ tubes}$$

The velocity of 6 fps for water flow speed is an acceptable value and is often used in heat-exchanger design. Now for a two-pass heater the number of tubes total is 2 × 121 = **242 tubes.** The effective length of tubes is

$$\frac{2{,}120}{A_0 \times 242} = \frac{2{,}120}{0.2291 \times 242} = \mathbf{38.2\ ft}$$

The outside area per foot length of tube A_0 is 0.2291 sq ft.

O-3. A company is heating a gas by passing it through a pipe with steam condensing on the outside. It is proposed to triple the capacity of the heater by changing its length while maintaining the same terminal conditions as regards temperature. What percentage change in length is necessary?

ANSWER. Present capacity is $Q_1 = UA_1\Delta t_m$. New capacity will be

$$Q_2 = U3A_1\,\Delta t_m = 3Q_1$$

Therefore, increase length **200 per cent.** Note area is equal to diameter × length.

O-4. A properly designed steam-heated tubular preheater is heating 45,000 lb per hr of air from 70 to 170°F when using steam at

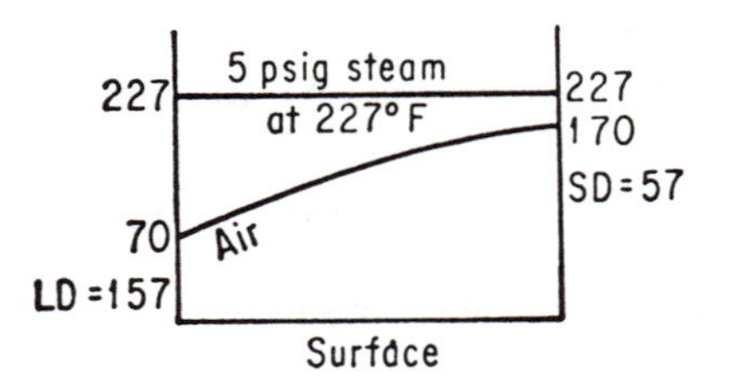

Fig. O-3

5 psig. It is proposed to double the rate of air flow through the heater and yet heat the air from 70 to 170°F; this to be accomplished by increasing the steam pressure. Calculate the new steam pressure required to meet the changed condition, expressed as psig.

ANSWER. Refer to Fig. O-3. Under present conditions the heat transferred is

$$Q_1 = 45{,}000 \times 0.25 \times (170 - 70) = 1{,}125{,}000 \text{ Btu per hr}$$

Under the new conditions the heat to be transferred is

$$Q_2 = 2 \times Q_1 = 2 \times 1{,}125{,}000 = 2{,}250{,}000 \text{ Btu per hr}$$

The only change is an increase in Δt_m, with U and A remaining the same. From Fig. O-3 the log mean temperature difference is 98, say 100°F. Under the new conditions the Δt_m would be 200°F. The net effect will be to raise the horizontal steam line in Fig. O-3. As this happens we can safely say that the arithmetic mean is permissible. Then

$$\frac{LD + SD}{2} = 200 \qquad LD + SD = 400$$

$$LD = x - 70 \qquad SD = x - 170 \qquad (x - 70) + (x - 170) = 400$$

$$x - 70 + x - 170 = 400 \qquad x = \frac{400 + 70 + 170}{2} = \frac{640}{2} = 320°\text{F}$$

The saturated steam pressure corresponding to 320°F may be obtained from the steam tables as 89.65 psia, or

$$89.65 - 14.7 = \mathbf{75\ psig}$$

O-5. A counterflow bank of boiler tubes has a total area of 900 sq ft and its U is 13 Btu/(hr)(sq ft)(°F). The boiler tubes generate

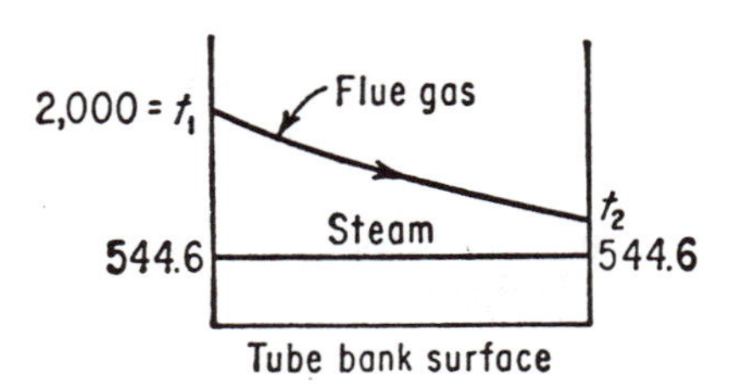

Fig. O-4

steam at a pressure of 1,000 psia. The tube bank is heated by flue gas which enters at a temperature of 2000°F and at a rate of 450,000 lb per hr. Assume an average specific heat of 0.25 Btu/(lb)(°F) for the gas and calculate the temperature of the gas that leaves the bank of boiler tubes. Also calculate the rate at which steam is being generated in the tube bank.

ANSWER. The basic heat-transfer equations as indicated below apply:

$$Q = w_g c_p \, \Delta t_g = UA \, \Delta t_m$$

All terms are familiar to the student. Now refer to Fig. O-4, steam being generated in the tube bank and steam line is below gas line. From the steam tables the steam temperature corresponding to a saturation pressure of 1,000 psia is 544.56°F. At this condition the enthalpy of the liquid H_f is 542.4, of vaporization H_{fg} is 649.5 and steam enthalpy H_g is equal to 1,191.9. We know all terms to be inserted into the basic heat transfer equations except the leaving gas temperature. Let us call this t_2 (see Fig. O-4). Then

$$Q = 450{,}000 \times 0.25 \times (2{,}000 - t_2) = 13 \times 900 \times \Delta t_m$$

$$\Delta t_m = \frac{2{,}000 - t_2}{\ln [(2{,}000 - 544.6)/(t_2 - 544.6)]}$$

By rearrangement then

$$\ln \frac{2{,}000 - 544.6}{t_2 - 544.6} = \frac{13 \times 900}{450{,}000 \times 0.25} = 0.104$$

The antilog of 0.104 = 1.11.

(*a*) Leaving gas temperature:

$$\frac{2{,}000 - 544.6}{t_2 - 544.6} = 1.11$$

from which t_2 = **1850°F.**

(*b*) Steam-generation rate: by simple heat balance the heat absorbed by the water to generate steam is equal to the heat lost by the flue gas.

$$Q = w_g c_p(2{,}000 - t_2) = 450{,}000 \times 0.25 \times (2{,}000 - 1{,}850) = 16.9 \times 10^6$$

If we assume water enters tube bank saturated at 544.6°F, then

$$\text{Steaming rate} \times H_{fg} = 16{,}900{,}000 \text{ Btu per hr}$$

Finally,

Steaming rate = 16,900,000/649.5 = **26,000 lb steam per hr**

O-6. A fuel-oil heater is to be purchased to heat 10,000 gal per hr of an oil initially at 60°F. Exhaust steam at atmospheric pressure from a reciprocating engine is to be used as the heating medium. The general design of the heater has already been decided and it is believed that the over-all coefficient of heat transfer will be 12.5 Btu/(hr)(sq ft)(°F) arithmetic mean temperature difference. Total installed cost of the heater will be $3 per sq ft. Annual fixed charges are to be figured at 40 per cent; maintenance and repairs are to be considered negligible. Heat added to the oil is valued at 40 cents per million Btu. No charge is to be placed against the heating steam because it would be wasted to atmosphere if not used in the heater. The heater is to operate 3,000 hr per yr. The oil has a constant specific heat of 0.5. Its average specific gravity during passage through the heater is to be taken as 0.9.

(*a*) To what economical temperature should the oil be heated? No differential in pumping charge is to be assumed.

(*b*) What heating surface is to be associated with that economical temperature?

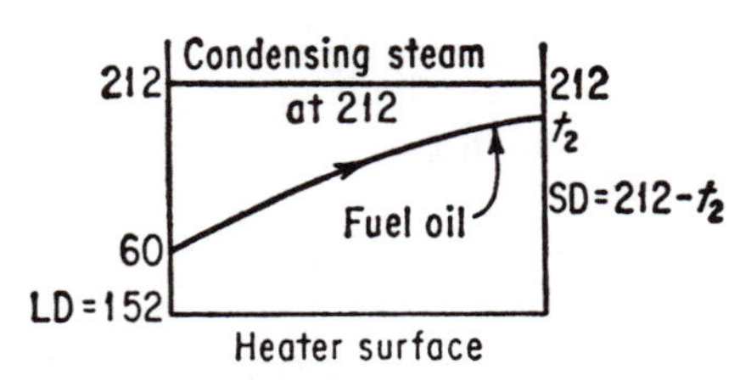

Fig. O-5

ANSWER. Refer to Fig. 0-5. The weight of oil to be heated is simply

$$w_0 = 10{,}000 \times 8.33 \times 0.90 = 75{,}000 \text{ lb per hr}$$

The heat transferred to the oil to raise it from 60°F to t_2 is also simply

$$Q = w_0 c_p (t_2 - t_1) = 75{,}000 \times 0.5(t_2 - 60) \text{ Btu per hr}$$

Annual fixed charges (FC) based on the heater surface A in square feet are

$$\text{FC} = \$3 \times A \times 0.40 = \$1.20\,A$$

Ordinarily the steam would be wasted to atmosphere, but it is used to heat the oil and is thus saved. This saving expressed in terms of leaving-oil temperature is so many dollars worth.

$$\begin{aligned} S &= Q \times 3{,}000 \times 0.40 \text{ dollars per million Btu} \\ &= 75{,}000 \times 0.5(t_2 - 60) \times 3{,}000 \times 0.40 = (45t_2 - 2{,}700) \end{aligned}$$

The fixed charges are to balance out the saving. We now have two equations with two unknowns. Let us solve for the heater surface A. We have been given the over-all coefficient U as 12.5. For the conditions of the problem setup we can safely assume the arithmetic mean temperature difference. Thus,

$$U = \frac{Q}{A \times \Delta t_a} \quad A = \frac{Q}{U \times At_a} = \frac{37{,}500(t_2 - 60)}{12.5 \times \Delta t_a}$$

$$\Delta t_a = \frac{LD + SD}{2} = \frac{(152 - 60) + (212 - t_2)}{2} = \frac{364 - t_2}{2}$$

$$\text{FC} = 1.2\ \text{A} = 1.2 \times \frac{37{,}500(t_2 - 60)}{12.5\ \Delta t_a} = \frac{3{,}600(t_2 - 60)}{\Delta t_a}$$

$$= \frac{3{,}600(t_2 - 60)}{\Delta t_a} = \frac{7{,}200(t_2 - 60)}{364 - t_2}$$

Equating S equal to FC, we have

$$45t_2 - 2{,}700 = \frac{7{,}200(t_2 - 60)}{364 - t_2}$$

$$45t_2 - 2{,}700 = \frac{7{,}200t_2 - 432{,}000}{364 - t_2}$$

This may be simplified to the binomial equation

$$t_2^2 - 264t_2 - 12{,}280 = 0$$

(*a*) from which t_2 = **206°F**

(*b*) heating surface $A = \dfrac{37{,}500(206 - 60)}{12.5[(364 - 206)/2]}$ = **5,520 sq ft**

O-7. A water-to-air heat exchanger operates under the following conditions:

Air mass flow rate	701,400 lb per hr
Air specific heat, constant	0.240
Air entering temperature	60°F
Air leaving temperature	171°F
Water mass flow rate	286,000 lb per hr
Water specific heat	1.0
Water entering temperature	203°F

The exchanger is to be used to heat the same air quantity from minus 10°F to 121°F with water entering at 203°F. Based on the assumption that the U in Btu/(hr)(sq ft)(°F) log mean temperature difference is unchanged, determine the required mass flow rate of water for these new conditions.

ANSWER. First determine the over-all coefficient U. This will remain unchanged for both loading conditions. Also the product of $U \times A$ will remain constant. Let us denote by subscript 1 the first set of conditions and by subscript 2 the second set of conditions. Also subscript w is for water.

$$Q_1 = 701{,}400 \times 0.24 \times (171 - 60) = 286{,}000 \times 1 \times \Delta t_w$$
$$\Delta t_w = \frac{701{,}400 \times 0.24 \times 111}{286{,}000 \times 1} = 65.5°\text{F, say } 66°\text{F}$$

Now set up Fig. O-6 for the first condition. Note water leaving temperature is 203 − 66 = 137°F. By calculation in the usual

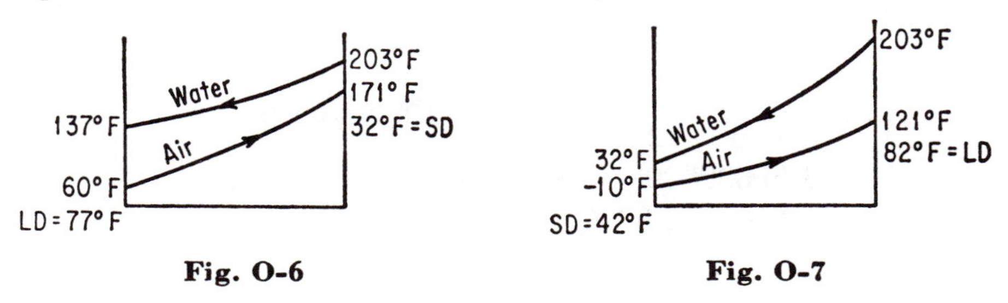

Fig. O-6

Fig. O-7

manner the log mean temperature difference for this set of conditions is 51°F. If Q_2 is the new air loading (−10°F to 121°F), then we can say

$$Q_1 = UA\,\Delta t_{m_1} = C\,\Delta t_{m_1}$$
$$Q_2 = UA\,\Delta t_{m_2} = C\,\Delta t_{m_2}$$
$$Q_1 = 701{,}400 \times 0.24 \times (171 - 60) = 18.6 \times 10^6 \text{ Btu per hr}$$
$$Q_2 = 701{,}400 \times 0.24 \times (121 + 10) = 22.0 \times 10^6 \text{ Btu per hr}$$

also
$$\frac{Q_1}{Q_2} = \frac{C\,\Delta t_{m_1}}{C\,\Delta t_{m_2}} = \frac{18.6}{22.0} = 0.85 = \frac{51}{\Delta t_{m_2}}$$
$$\Delta t_{m_2} = 51/0.85 = 60°\text{F}$$

Now set up Fig. O-7 for the second condition. Again by calculation log mean temperature difference of 60°F, and large difference of 203 − 121 = 82°F, the small difference is found to be 42°F. This makes the leaving water temperature 42 − 10 = 32°F. This is the freezing temperature of water, indeed, but no freeze-up will occur so long as the water is flowing. For best practical operation of the exchanger a control temperature of 35°F should be used. Finally,

$$Q_2 = 22 \times 10^6 = \text{lb water per hr} \times c_p \times \Delta t_{w_2}$$
$$\text{lb water per hr} = \frac{22 \times 10^6}{1 \times 171} = \mathbf{128{,}500}$$

or the equivalent of

$$128{,}500/(60 \times 8.33) = \mathbf{257\ gpm}$$

O-8. A hot-water heater consists of a 20-ft length of copper pipe of ½-in. average diameter and 1⁄16-in. thickness. The outer surface of the pipe is maintained at 212°F. What is the capacity of the coil in gpm if water is fed into the coil at 40°F and is expected to emerge heated to 150°F? Assume the conductivity of the copper to be 2100 Btu/(sq ft)(°F)(hr) per in. of thickness.

ANSWER. Assuming clean pipe and no fouling,

$$\frac{1}{U} = R\text{, steam film} + R\text{, pipe wall} + R\text{, water film}$$

$$\frac{1}{U} = \frac{1}{1{,}000} + \frac{0.0625}{2{,}100} + \frac{1}{550} = 0.0028298$$

$$U = 1/0.0028298 = 354 \text{ Btu/(hr)(sq ft)(°F)}$$

$$A = 20 \times \pi \times 0.5/12 = 2.618 \text{ sq ft active surface}$$

$$\begin{array}{rcl} & 212 \rightarrow 212 & \\ & 150 \leftarrow \ \ 40 & \\ \text{SD} = & 62 \qquad 172 & = \text{LD} \end{array}$$

log MTD = 108°F. This is uncorrected for the type of flow. Further refinement is not required. Note also that U was determined for thin-wall tubing with no correction for wall thickness. The resistance due to wall is negligible. Now,

$$Q = UA\ \Delta t_m = 354 \times 2.618 \times 108 = 100{,}091 \text{ Btu per hr}$$

Assuming that all heat goes into heating up the water (complete insulation),

$$Q = wc_p\ \Delta t = 100{,}091 = w \times 1 \times (150 - 40)$$

$$w = 100{,}091/110 = 910 \text{ lb per hr, or } 910/500 = \mathbf{1.82\ gpm}$$

O-9. A steam condenser receives steam at 1.5 in. Hg absolute and a quality of 0.96. Water "on" at 65°F, "off" at 82°F. Condensate leaves at 86°F, condensing surface 40,000 sq ft, condensing rate 30,000 lb per hr. Determine (*a*) over-all coefficient of heat transfer and (*b*) circulating water required in gpm.

ANSWER. Use Mollier diagram and steam tables. Assume no radiation losses. Also refer to Fig. O-8. Let the following nomen-

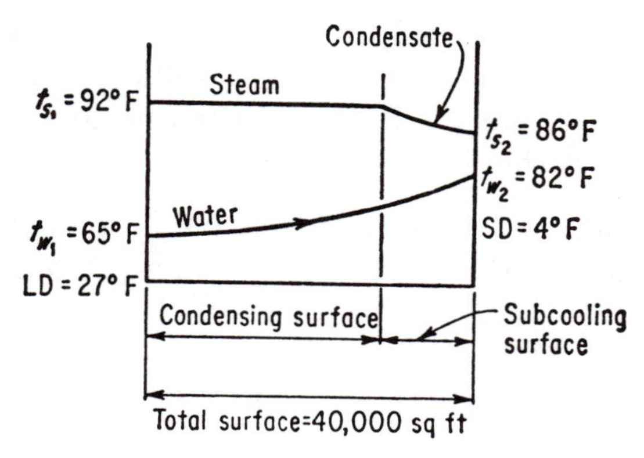

Fig. O-8

clature apply:

Q = heat added to condenser water, Btu per lb
= heat given up by steam, Btu per lb
c_p = specific heat of water, Btu/(hr)(°F)
t_{w_1} = entering water temperature, 65°F
t_{w_2} = leaving water temperature, 82°F
t_{s_1} = entering steam temperature, 92°F
t_{s_2} = leaving condensate temperature, 86°F
t_m = log mean temperature difference, °F
H_2 = enthalpy of entering steam, 1060 Btu per lb
H_3 = enthalpy of condensate, 54 Btu per lb
A = condensing surface, 40,000 sq ft
U = over-all heat-transfer coefficient, Btu/(hr)(sq ft)(°F)

(*a*) Over-all coefficient of heat transfer:

$$Q = 30{,}000(H_2 - H_3) = 30{,}000(1{,}060 - 54) = 30.1 \times 10^6 \text{ Btu per hr}$$

Refer to Fig. O-8 and determine log MTD to be 12°F. Then,

$$U = \frac{Q}{A\,\Delta t_m} = \frac{30.1 \times 10^6}{40{,}000 \times 12} = \mathbf{63\ Btu/(hr)(sq\ ft)(°F)}$$

(*b*) Cooling water:

$$w = \frac{30.1 \times 10^6}{82 - 65} = 1.772 \times 10^6 \text{ lb per hr}$$

$$\frac{1.772 \times 10^6}{500 \times \text{sp gr}} = \frac{1.772 \times 10^6}{500 \times 1} \doteq \mathbf{3{,}540\ gpm}$$

In certain problems presented by several state boards conductivity of metal wall has been presented as $27 - t/180$ Btu/(hr)(sq ft)(°F)(ft) thickness for steel. Here t is the arithmetic average of the two temperatures across the wall having a flat surface (not rounded).

O-10. Repairs are to be made on a valve in a steam line. When the valve is isolated, the insulation is removed from the valve and

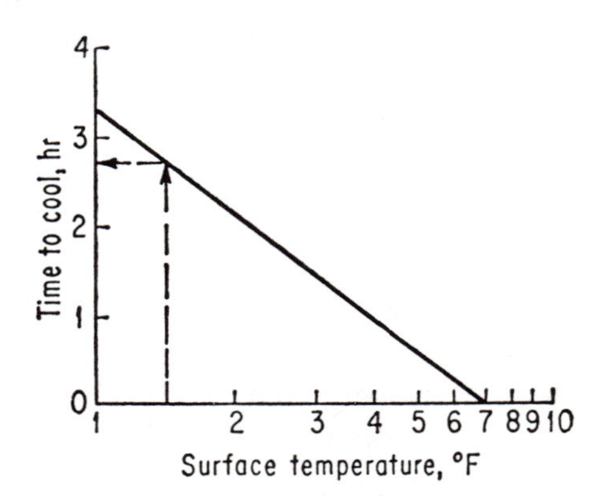

Fig. O-9

the valve surface temperature is measured and found to be 700°F with an ambient temperature of 60°F. Thirty minutes later the valve surface temperature is again measured and found to be 500°F with an ambient temperature still 60°F. Estimate the time required for the valve surface temperature to reach 150°F, assuming that the ambient temperature remains constant.

ANSWER. For more accurate results three initial surface temperatures should be taken and then plotted so that temperature is taken along the abscissa in logarithm form and time is plotted along the ordinate in cartesian form as in Fig. O-9.

From the plot in Fig. O-9 the time is approximately 2 hr, 75 min, say **3 hr.** The greatest change takes place after the first reading then linear thereafter.

O-11. A concrete storage tank in the shape of a cube and lined with steel plate with no air space between steel and concrete is to hold 7,500 gal of a sodium hydroxide solution. Walls are 12 in. thick. (*a*) How much heat in Btuh must be added to the solution to prevent it from dropping below 25°F if the outside temperature is 0°F? (*b*) How many feet of 1-in. steel pipe would you install as a heating coil, using 5 psig saturated steam to maintain the proper temperature?

ANSWER. (*a*) Using outside area of cube and the temperature difference of 25°F and the over-all coefficient which we shall calculate, we can find the heat loss under steady state conditions. By familiar method we find the cube to be 10 ft on a side. If we assume the tank is sitting on the ground, the exposed surface totals 500 sq ft. Determine the coefficient U by getting the summation of all resistances: air film, concrete, steel plate, and sodium hydroxide liquid film and dividing into unity. $R_t = R_1 + R_c + R_s + R_2$. Where

$$R_1 = 1/6 = 0.167 \qquad R_c = \frac{L_c}{k_c} = \frac{12}{8.3} = 1.45$$

$$R_s = \frac{L_s}{k_s} = \frac{1}{35 \times 12} = 0.00238 \qquad \text{Note 1-in. steel plate.}$$

$$R_2 = 1/500 = 0.0020 \qquad R_t = 0.167 + 1.45 + 0.00238 + 0.0020$$

From which R_t is 1.621. Now $U = 1/R_t = 1/1.621 = 0.618$ Btuh/(sq ft)(°F).

$$Q = 0.618 \times 500 \times 25 = \mathbf{7{,}720\ Btuh}$$

(*b*) Heat transfer coefficient for pipe in sodium hydroxide solution

$$U_p = \frac{1}{1/600 + 0 + 0.133/(35 \times 12) + 0(0.344/0.275) + 1/(1{,}000 \times 0.275/0.344)}$$

$$U_p = \frac{1}{0.00167 + 0 + 0.0003165 + 0 + 0.00125} = 308.5 \text{ Btuh/(sq ft)(°F)}$$

$$\text{Btuh/ft of pipe} = U_p \times A_0 \times (227 - 25) = 308.5 \times 0.344 \times 202 = 21{,}400$$

Length of pipe required 7,720/21,400 = **0.36 ft.** For all practical purposes this is an unwieldly length. The best way is to run pipe around perimeter 1 ft away from sides of tank 6 in. off bottom. This will require about **36 ft** of pipe.

O-12. A certain clean, metal, double-pipe heat exchanger heats turbulently flowing air to a particular temperature by means of condensing steam. How much greater an air rate should we be able to process with two such heat exchangers in series, if steam pressure and terminal temperatures remain the same?

ANSWER. In the exchanger described, virtually all the resistance to heat transfer would be in the air film. We can say, therefore, $U = h$. Substituting this in the usual $Q = UA\ \Delta t_m$ we have

$$Wc_p\ \Delta t = h(\pi DL)\ \Delta t_m$$

Since c_p, Δt, D and Δt_m are the same for one exchanger as for the two together, we can simplify above equation to: $W \alpha hL$. Now, the Dittus-Boelter equation states: $Nu = 0.023\ \mathrm{Re}^{0.8}\ \mathrm{Pr}^{0.4}$. From the Dittus-Boelter equation, we can say $h \alpha W^{0.8}$ since diameter and physical properties are also constant in our problem. Then by substitution from equations substituting $h \alpha W^{0.8}$ in $W \alpha hL$, we get $W \alpha W^{0.8}L$ or $W^{0.2} \alpha L$. Also $W \alpha L^5$ or $W_2/W_1 = (L_2/L_1)^5$. Thus, the flow rate ratio equals the length ratio raised to the fifth power. For the numbers in our problem

$$\frac{W_2}{W_1} = \left(\frac{2}{1}\right)^5 = 32$$

The two exchangers in series can process **32 times as much air** as the single exchanger. Of course, before actually trying this doubling up of heat exchangers, we should calculate the increase in pressure drop. Taking as an approximate friction factor $f \alpha W^{-1/5}$, we have from the Fanning equation

$$\Delta P = \alpha W^{-1/5} W^2 L \text{ or } \Delta P \alpha L^{-1} L^{10} L \text{ or } \Delta P \alpha L^{10}$$

For a doubled length

$$\frac{\Delta P_2}{\Delta P_1} = \left(\frac{2}{1}\right)^{10} = 1{,}024$$

This is more than a thousandfold increase in the exchanger pressure drop. However, if we only wish to double the air rate through the exchanger, rather than multiply it by 32, then

$$\frac{L_2}{L_1} = \left(\frac{W_2}{W_1}\right)^{0.2} = 2^{0.2} = 1.15$$

Finally

$$\frac{\Delta P_2}{\Delta P_1} = \left(\frac{L_2}{L_1}\right)^{10} = \left(\frac{W_2}{W_1}\right)^2 = \left(\frac{2}{1}\right)^2 = 4$$

Thus, to double the air rate we need 15 per cent greater length for our heat exchanger and must suffer four times the pressure drop through it.

O-13. Stack gases from a chemical processing unit operation are composed primarily of air, with small amounts of noxious vapors that must be condensed out to prevent a community air pollution problem. The gases are flowing at a rate of 80,000 cu ft per hr, leaving the process at a temperature of 600°F, and are to be cooled down to 100°F. Cooling water is available at 50°F to cool the tube side of a shell-and-tube heat exchanger of cross-flow design. Gases will flow on the shell side of the exchanger. Estimate the surface required to do the job.

ANSWER. Since the surface area is to be estimated, certain assumptions will be made to simplify the solution.

Assumptions: Cooling water temperature rise = 25°F
Specific heat of stack gases (mostly air) = 0.24
Stack gases are at atmospheric pressure, 14.7 psia
Average stack gas temperature is arithmetic
Molecular weight of stack gases = 29
No fouling factor on shell or tube sides of heat exchanger
Tubes considered thin-walled

Average gas temperature = (600 + 100)/2 = 350°F or 810°R
Gas density is found by

$$(29/379) \times (460 + 60)/(460 + 350) = 0.049 \text{ lb per cu ft}$$

Cooling load to be charged to exchanger:

$$80{,}000 \times 0.049 \times 0.24 \times (600 - 100) = 470{,}400 \text{ Btuh} = Q$$

Refer to Fig. O-10. Then since $Q = UA\,\Delta t_m$, we must obtain Δt_m and U, then solve for A, surface area in square feet.

$$\text{Log mean temperature difference } \Delta t_m = \frac{LD - SD}{2.3 \log (LD/SD)}$$

$$\Delta t_m = \frac{525 - 50}{2.3\ (525/50)} = 203°\text{F}$$

$$U = \frac{1}{(1/h_o) + (1/h_i)} = \frac{1}{(1/250) + (1/500)} = 167 \text{ Btuh/(°F)(hr)}$$

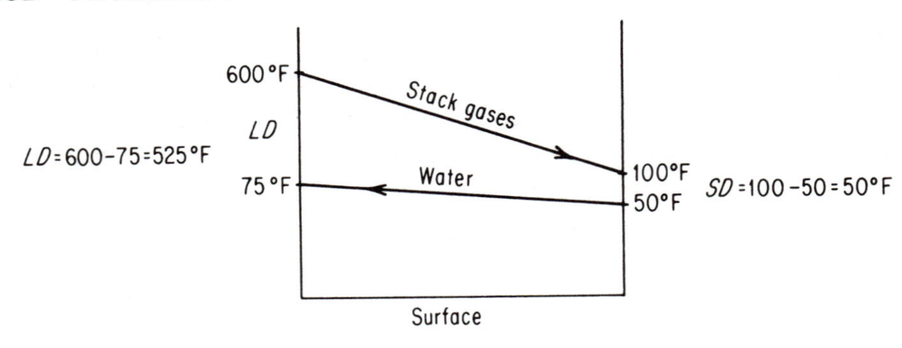

Fig. O-10

Compare U calculated with Table 15-4, gas (atmospheric pressure) to liquid. **Note that h_o = 250 for air flow across staggered tubes outside tubes' film coefficient, and h_i = 500 for water inside tubes of 1 in. diameter flowing at 2 fps.**

$$A = 470{,}400/(167 \times 203) = \mathbf{14\ sq\ ft}$$

O-14. If 10,000 lb per hr of distilled water is required from untreated water, steam is available at 300°F, and the condenser will vent to the atmosphere, how much surface is required? Assume a pressure drop through the condenser and lines of about 5 psi. The saturation temperature in the evaporator shell will be 19.7 psia or 226°F. Take the overall coefficient = 605.

ANSWER. Heat balance is

$$Q_{\text{evap}} = 10{,}000 \times 961 = 9{,}610{,}000 \text{ Btuh}$$
$$Q_{300\text{F}} = 10{,}550 \times 910 = 9{,}610{,}000 \text{ Btuh}$$
$$\text{Temperature head } \Delta t = 300 - 226 = 74°\text{F}$$

$$\text{Surface area} = \frac{Q}{U\,\Delta t} = \frac{9{,}610{,}000}{605 \times 74} = \mathbf{215\ sq\ ft}$$

O-15. Find the area of housing required to maintain the film temperature of a 3-in.-diameter by 4.5-in.-long, 120° central partial bearing at 165°F; rpm = 900 and r/c = 1,000. Room temperature is 100°F. Assume the housing temperature rises one-half as much as that of the oil film. Cooling rate is 2 Btuh/(sq ft)/(°F). SAE oil is used. The load is 100 psi of projected journal area. Viscosity of the oil at 165°F is 0.00000118 lb-sec per sq in.

ANSWER. A lubricant is used to reduce the friction of bearings and sliding surfaces in machines and thus diminish the wear, heat, and possibility of seizure of parts. Although a layer of oil will eliminate the excessive friction of metal-to-metal contact, the friction within the oil film itself must be taken into account. The study of lubrication and the design of bearings is therefore concerned mainly with phenomena related to the oil film between the moving parts.

$$\text{Tangential velocity } U = \pi dn/60$$
$$= \pi \times 3 \times {}^{900}\!/_{60} = 141.4 \text{ in. per sec}$$
$$\Delta T = (165 - 100)/2 = 32.5°\text{F}$$

The relationship that will now be determined relates to the load and friction characteristics of the bearing in question.

$$\frac{W_1}{\mu U} \times \left(\frac{c}{r}\right)^2 = \frac{300}{0.00000118 \times 141.4 \times 1{,}000^2} = 1.79$$

From Fig. O-11, $\epsilon = 0.512$ and $(F_1/\mu U)(c/r) = 3.69$.

$$F_1 = 3.69\mu U \frac{r}{c}$$
$$= 3.69 \times 0.00000118 \times 141.7 \times 1{,}000 = 0.6176 \text{ lb/in.}$$
$$c_2 = \frac{F_1 n}{1.0805 c_1 \Delta T} = \frac{0.6176 \times 900}{1.0805 \times 2 \times 32.5} = 7.91$$
$$A_c = \pi dZ \times 7.91 = 336 \text{ in.}^2 \text{ or } \mathbf{2.33\ ft^2}$$

where A_c = cooling area of bearing housing
c = radial clearance of journal bearing
d = diameter of journal bearing
F_1 = tangential friction force per axial inch
ϵ = eccentricity ratio
μ = viscosity, lb-sec per sq ft
n = revolutions per minute
r = radius of journal
ΔT = rise in temperature of bearing housing
U = tangential velocity
Z = viscosity, centipoises

Reference: by M. C. Shaw and E. F. Macks, "Analysis and Lubrication of Bearings," McGraw-Hill Book Company, New York, 1949.

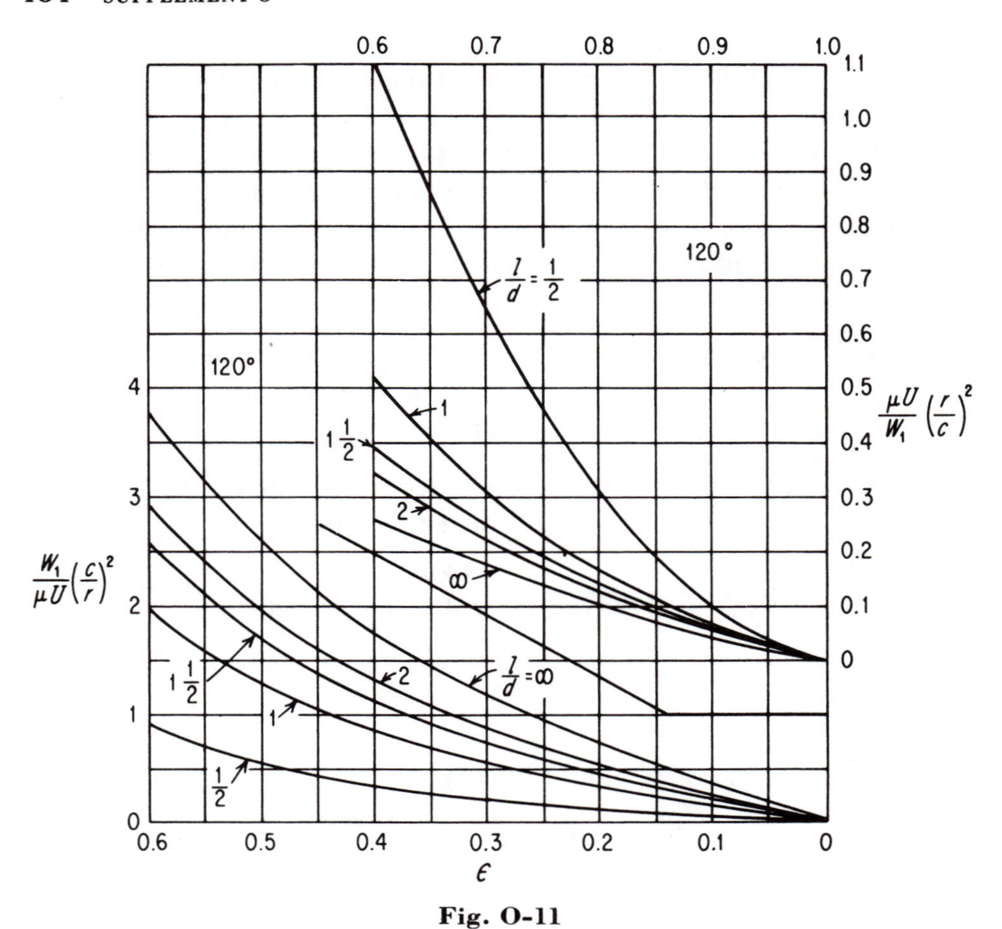

Fig. O-11

O-16. A fixed-displacement pump supplies fluid to two hydraulic motors driving a rolling mill. The speed of the motor is accurately controlled by two pressure-compensated flow-control valves used in a meter in the circuit. The average controlled flow is 20 gpm, the flow from the pump is 25 gpm, and the system pressure is 2,500 psi. Heat is generated by the pumping of the superfluous 5 gpm through the relief valve, but the extra capacity this represents must be maintained to handle the peak loads. The following data apply:

1. The oil temperature rise by natural convection from the pump reservoir is 137°F.

2. It is desired to hold the reservoir temperature to 90°F; a 40°F rise in fluid temperature (to 130°F) is permissible.

3. Cooling water from the spray towers has a summer temperature of 85°F.

4. A water flow of 10 gpm is available and seems reasonable.

5. The over-all coefficient of heat transfer oil-to-water is 100 Btu/(hr)(sq ft)(°F).

Find the surface required for the oil cooler to hold the reservoir temperature at 90°F.

ANSWER. Calculate heat load Q from the following equation:

$$Q = 1.484 \times R \times P$$

Thus,

$$Q = 1.484 \times 5 \text{ gpm} \times 2{,}500 \text{ psi} = 18{,}550 \text{ Btu per hr}$$

Temperature rise of cooling water ΔT_w is given by

$$\Delta T_w = 18{,}550/(60 \times 10 \times 8.33) = 3.725°\text{F}$$

Thus, cooling water can be assumed to enter at 85°F and leave at 89°F. The log mean temperature difference LMTD is now found as follows:

$$\text{LMTD} = \frac{\Delta T_{\text{max}} - \Delta T_{\text{min}}}{2.3026 \log_{10} (\Delta T_{\text{max}}/\Delta T_{\text{min}})}$$

$$= \frac{(130 - 85) - (90 - 89)}{2.3026 \log_{10} (130 - 85)/(90 - 89)}$$

$$\text{LMTD} = 11.6°\text{F}$$

Finally, the surface required is found.

$$A = Q/(U \times \text{LMTD}) = 18{,}550/(100 \times 11.6) = \mathbf{16\ sq\ ft}$$

If an exchanger having this heat transfer surface is not available, the next larger size should be selected. It is advisable to check the water pressure drop to make certain it is within acceptable limits.

O-17. **A kiln is to be designed for our plant to withstand a temperature of 2000°F while limiting the heat loss to 250 Btu/(h)(ft^2)**

with an outside temperature of 100°F. We have available the following types of brick:

Fireclay	4.5 in thick		$k = 0.90$
Insulating	3.0 in thick	1800°F max allowable temp	$k = 0.12$
Building	4.0 in thick	300°F max allowable temp	$k = 0.40$

(*a*) What will be the minimum wall thickness?

(*b*) Determine the actual heat loss through the wall.

ANSWER. Refer to Fig. O-12. x_1, x_2, and x_3 must be multiples of 4.5, 3, and 4 in, respectively.

(*a*) Basis: 1 ft² of surface area. Let us use the series formula

$$Q = \Delta t/\Sigma R, \; R = x/kA, \; A = 1 \text{ ft}^2$$
$$Q = 250 \text{ Btu/(h)(ft}^2\text{)(1 ft}^2\text{)} = 250 \text{ Btu/h}$$

For fireclay, $t_a \geq (2000 - 1800) = 200°\text{F}$.

$$R = (x_1/k_1A) = \Delta t_1/Q \qquad \text{Also } x_1 = k_1A(\Delta t_1/Q)$$
$$x_1 = 0.9 \text{ (1 ft}^2\text{)(200/250)} = 0.72 \text{ ft or 8.64 in}$$

Thus we need two layers of firebrick. $x_1 = (2)(4.5) = 9$ in or 0.75 ft

Then $\quad \Delta t_1 = (x_1)/(k_1A)(Q) = (0.75)(250)/(0.9) = 208°\text{F}$

Now $\quad \Delta t_2 \geq 2000°\text{F} - 208°\text{F} - 300°\text{F} = 1492°\text{F}$

$$x_2 = 0.12\text{(1 ft}^2\text{)(1492/250)} = 0.716 \text{ ft or 8.59 in}$$

Thus we need three layers of insulating brick. $x_2 = (3)(3 \text{ in}) = 9$ in or 0.75 ft

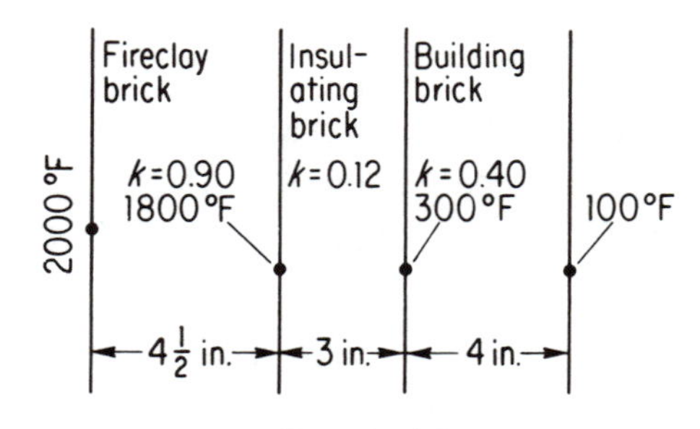

Fig. O-12

Then $\Delta t_2 = (x_2)/(k_2A)(Q) = (0.75)(250)/(0.12) = 1563°F$
Now $\Delta t_3 \geq (2000°F - 208 - 1563 - 100) = 129°F$
$x_3 = 0.4(1)(129/250) = 0.206$ ft or 2.48 in

Thus we need one layer of building brick. $x_3 = 1(4 \text{ in}) = 0.33$ ft
Now the minimum wall thickness is 9 in + 9 in + 4 in = **22 in**

(*b*) The actual heat loss is

$$Q = \frac{\Delta t}{\Sigma R} = \frac{2000 - 100}{(0.75/0.9) + (0.75/0.12) + (0.333/0.4)}$$
$$= (1900/(7.916) = \mathbf{240\ Btu/(h)(ft^2)}$$

O-18. A furnace is constructed with 9 in of firebrick, 4½ in of insulating brick, and 9 in of building brick. The inside surface temperature is 1400°F and the outside air temperature is 85°F. If the furnace wall loses heat at the rate of 200 Btu/(h)(ft²), determine the air heat-transfer coefficient.

ANSWER.

$$\frac{Q}{A} = \frac{k_1(t_1 - t_2)}{l_1} = \frac{k_2(t_2 - t_3)}{l_2} = \frac{k_3(t_3 - t_4)}{l_3} = h(t_4 - t_5)$$

Overall temperature drop across wall $= t_1 - t_4$

$$\Delta t = \frac{Q}{A}\left(\frac{l_1}{k_1} + \frac{l_2}{k_2} + \frac{l_3}{k_3}\right) = 200\left[\frac{9}{12(0.8)} + \frac{4.5}{12(0.15)} + \frac{9}{12(0.4)}\right]$$
$$= \frac{200}{12}(11.25 + 30 + 22.5) = 1063°F$$

$1400 - t_4 = 1063°F \qquad h(337 - 85) = 200$

$t_4 = 337°F \quad \text{and} \quad h = 0.794 \text{ Btu/(h)(ft}^2\text{)(°F)}$ **Ans.**

O-19. List the steps in the design of a heat exchanger.

ANSWER.

1. By heat balance determine the terminal temperatures or heat load as the case may be.
2. Calculate the ΔT_{LMTD}.
3. Assure a design U, including fouling resistance.

4. Select a tube size and length based on available space, etc., and determine the number of tubes needed.

5. Calculate exchanger area required.

6. Select a tube pitch and layout angle based on the kind of fluid, cleaning requirements, etc. Determine shell size for them by looking up standard tables. Allowance should be provided for the tie rods and spacers.

7. Select kind of baffles, their orientation and spacing.

8. Calculate shell-side heat-transfer coefficient.

9. Calculate tube-side heat-transfer coefficient.

10. Calculate the overall design heat-transfer coefficient by taking into account wall resistance and fouling factor.

11. Compare the calculated value in step 10 with the assumed value in step 3. If the former is equal to or greater than the latter, proceed to next step or otherwise repeat step 3 onward.

12. Calculate shell-side pressure drop.

13. Calculate tube-side pressure drop.

14. If either of the calculated pressure drops does not satisfy the allowable, and if the allowable limit cannot be increased, then the configuration of the exchanger has to be modified. Check if the pressure drop in the nozzles or baffle windows is excessive, since these are almost a waste. Various modifications may include changing the baffle spacing, baffle cut, number and length of tubes, and passes. The whole procedure starting from step 3 has to be repeated.

15. An exchanger designed to satisfy the heat-transfer and pressure-drop requirements should also be checked to satisfy that the vibrations will not damage the tube bundle. If they will, the design has to be modified by changing the baffle spacing or providing intermediate tube supports. The shell side should be rerated, since any change in the baffles will change the shell-side heat-transfer coefficient.

16. Finally, stress analysis of critical parts should be made if so required by the various codes, local regulations, or customer request.

O-20. Is the overall heat-transfer coefficient uniform throughout a heat exchanger?

ANSWER. Although for all intents and purposes it is uniform, in actual practice it is not. Its uniformity or lack of uniformity depends on the h_o and h_i film coefficients. These are in turn dependent on fluid properties, viz., heat capacity C_p, thermal capacity k, viscosity u, and density p, which are themselves temperature-dependent. Hence, as the fluid temperature changes along the exchanger length, the properties also change, which affects the h_o and h_i and hence U. One of the future developments is likely to be the design of heat exchangers based on local values of U instead of the overall U. The total heat duty will then be the sum of the heat transferred over small segments of the exchanger.

O-21. What are some rules of thumb in the preliminary design of shell-and-tube heat exchangers?

ANSWER.

1. Tube-side fluid velocity for process fluids: 3 to 6 ft/s.
2. Tube side: for cooling water about 6 ft/s.
3. Shell-side liquid velocity: 2 to 5 ft/s.
4. Vapor velocity to be 20 to 250 ft/s, depending on pressure (high, atmospheric, vacuum) and the molecular weight of the vapor (high or low).
5. More corrosive and/or higher-pressure liquid should be put on the tube side.
6. Odd number of tube passes should be avoided.

O-22. Discuss the use of safety factors in the design of an exchanger.

ANSWER. The safety factor allows for uncertainties in the heat-transfer equations which are normally up to plus or minus 20 to 30 percent in the best design methods. The safety factor is preferably applied to the overall heat-transfer coefficient rather than the individual film coefficient. The resulting increase in heat-transfer area should be accommodated by increasing the length of the tubes rather than the number, since the latter decreases both the shell- and tube-side velocities and the heat-transfer coefficient, thus reducing the effective safety as well.

O-23. What are some other reasons for providing extra heat-transfer surface in the heat exchanger?

ANSWER.

1. To take care of deviations in operating conditions from the design conditions.

2. To provide for emergencies and upset conditions when the flow rates, temperatures, and pressures might change for some time.

3. Excessive fouling might block off some of the tubes.

4. Expansion of plant or process changes with developing technology may require more output.

5. Some tubes may have to be plugged because of leakage.

O-24. A clean heat exchanger has a higher heat-transfer value coefficient than the design value. It will deliver a larger heat load than necessary that might upset the downstream process conditions. How would you tackle such a situation?

ANSWER.

1. Bypass part of the process stream around the exchanger until the design conditions are met in the exchanger; then send the total stream through the exchanger.

2. Install another unit downstream to compensate for the changed conditions of the fluid to bring them back into line. When the main unit reaches the design conditions, take the downstream unit out. However, no one likes to incur this extra cost on an extra unit. Use step 1 preferably.

Supplement **P**

REFRIGERATION

Questions and Answers

P-1. Liquid ammonia expands from a tank at 180.6 psia and 90°F in an evaporating coil and is then compressed. The compressor suction is 4.3 psi below atmospheric pressure. What is the refrigerating effect per pound of ammonia?

ANSWER. Pressure of ammonia leaving coil and entering compressor suction is $14.7 - 4.3 = 10.4$ psia. This assumes no pressure drop in suction line. From ammonia tables for dry and saturated (not superheated) ammonia the enthalpy of saturated ammonia vapor at 10.4 psia is 597.6 Btu per lb. The enthalpy of saturated liquid at 180.6 psia is 143.5 Btu per lb. Note also that the temperature of 90°F given is also the saturation temperature at that pressure. Then, the refrigerating effect per pound circulated is

$$597.6 - 143.5 = \mathbf{454.1\ Btu\ per\ lb\ circulated}$$

P-2. An ammonia refrigeration unit has an 180-rpm, double-acting compressor whose cylinder has a diameter of 8 in. with a 10-in. stroke and whose clearance is 5 per cent. It is desired to calculate the refrigeration capacity in tons of refrigeration at an evaporation temperature of 12°F and a condensing temperature of 105°F. Assume that the 12°F vapor is saturated as it enters the compressor. Also assume that the liquid ammonia is not subcooled at the entrance to the expansion valve.

ANSWER. The following table lists the various thermodynamic data for both the evaporation and condensing temperatures:

Condition	12°F	105°F
Pressure, psia	40.31	228.9
Sp vol liquid, cu ft per lb	0.02451	0.02769
Sp vol vapor, cu ft per lb	6.996	1.313
Enthalpy of liquid, Btu per lb	56.0	161.1
Enthalpy of vaporization, Btu per lb	559.5	472.3
Enthalpy of vapor, Btu per lb	615.5	633.4
Entropy units liquid	0.1254	0.3269
Entropy vaporization units	1.1864	0.8366
Entropy units vapor	1.3118	1.1635

Refer to Fig. P-1 for *TS* diagram and Fig. P-2 for diagram of refrigeration-cycle mechanical presentation. Tons of refrigeration TR is given by

$$\mathrm{TR} = \frac{VDV_e(H_1 - H_4)}{200 \times \text{sp vol sat vapor}}$$

Refrigerating effect with no liquid subcooling is $H_1 - H_3$ or $H_1 - H_4$

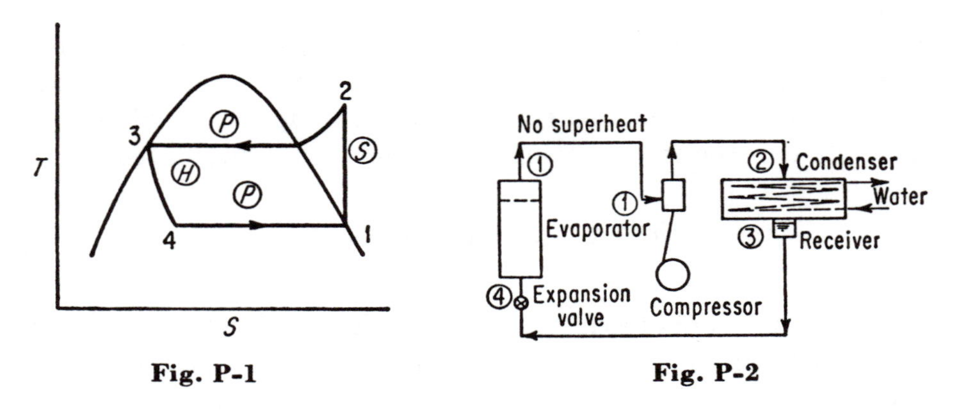

Fig. P-1 **Fig. P-2**

with only slight liquid flashing. *VD* is volumetric displacement of compressor at 100 per cent volumetric efficiency $V_e = 1$. If we neglect the area of the piston rod to simplify calculations,

$$VD = \text{rpm} \times 2 \times \frac{\pi d^2 L}{4 \times 1{,}728} = 180 \times 2 \times \frac{\pi \times 8^2 \times 10}{4 \times 1{,}728} = 105 \text{ cfm}$$

The equation for volumetric efficiency with *c* as clearance (decimal) and the familiar polytropic exponent n is

$$V_e = 1 + c - c\left(\frac{P_2}{P_1}\right)^{1/n}$$

Thus, the volumetric efficiency of this machine is accordingly

$$V_e = 1 + 0.05 - 0.05\left(\frac{228.9}{40.31}\right)^{1/1.3} = 1.05 - 0.19 = 0.86$$

$$\text{TR} = \frac{105 \times 0.86 \times (615.5 - 161.1)}{200 \times 6.996} = \mathbf{29.5}$$

P-3. Ammonia enters the cooler of an ammonia refrigerating machine at 0°F and leaves it at 15°F. When operating at 10 tons of refrigeration (ice melting capacity) and 15 indicated horsepower at the compressor, how many gallons per minute of condensing water must be supplied for a temperature rise from 50°F at entrance to 70°F at condenser exit. Assuming the liquid ammonia to leave the condenser at 50°F, calculate the ideal coefficient of performance. What is the actual coefficient of performance?

ANSWER. Heat pickup in evaporator is

$$10 \times 200 = 2{,}000 \text{ Btu per min}$$

Heat added in compressor cylinder is 15 hp $\times$ 2,545/60 = 634 Btu per min. Thus the total heat pickup is the sum of the two items 2,000 + 634 = 2634 Btu per min, which is to be absorbed by the condensing water through the 20°F range. The flow of condensing water is

$$\text{gpm} = \frac{2{,}634}{20 \times 8.33} = \mathbf{15.9\ gpm}$$

The ideal coefficient of performance (COP) is

$$(460 - 0)/(70 - 0) = \mathbf{6.57}$$

The actual COP is

$$\frac{12{,}000 \text{ Btu per hr per TR} \times 10}{15 \times 2{,}545} = \mathbf{3.14}$$

P-4. In a test on an ammonia refrigerating machine 30 gpm of water are collected at the condenser outlet. For the same period the water enters the condenser at 70°F and leaves at 85°F. Assume "28 gal-deg per ton of refrigeration," what is the system capacity?

ANSWER. The figure of 28 gal-deg is a common factor used to test refrigeration systems with water-cooled condensers. This means that 28 gpm will be raised 1° or 1 gpm will be raised 28° to remove the equivalent of 1 ton of refrigeration from the system. This takes into account both sensible and superheat added to the refrigerant when passing through the compressor. Now for the problem at hand,

$$\frac{30 \times (85 - 70)}{28} = \mathbf{16\ TR}$$

P-5. In an ammonia ice machine, how many pounds of ice at 24°F are produced by the evaporization of 1 lb of ammonia, the liquid ammonia entering the cooler at 60°F and leaving as a vapor at 20°F? The pressure in the expansion coils is 15 psig. Cooler efficiency is 75 per cent. Water is placed in the cooler at 50°F. What is the ideal COP in the above case?

ANSWER. Assume no liquid subcooling in condenser and liquid expands through float valve directly into the cooler. There is no pressure drop assumed to take place in the piping. Temperature leaving condenser and entering float valve is 60°F, and its corresponding saturation pressure is 107.6 psia. The pressure in the cooler after expansion is 15 + 14.7 = 29.7 psia, say 30 psia. The temperature therein is, therefore, 0°F at saturation.

Suction temperature is 20°F. Suction pressure is same as in coils, assuming no pressure drop, and is equal to 30 psia. The compressor discharge pressure is same as condenser, or 107.6 psia. At isentropic compression temperature is 182°F.

Heat (enthalpy) of liquid before expansion at 60°F is 109.4 Btu per lb. Enthalpy of saturated vapor at compressor suction (no superheat) is 623.5 Btu per lb. Enthalpy of compressor discharge vapor is 702 Btu per lb.

(*a*) Ice formed per lb ammonia:

Heat absorbed by ammonia is 623.5 − 109.4 = 514.1 Btu per lb
Heat removed from water per lb is:

From 50 to 32°F..............	18 Btu
During freezing...............	144 Btu
Ice* cooling 32 to 24°F........	4 Btu
Grand total...................	166 Btu per lb

* 1 lb × 0.5 sp ht × (32 − 24) = 4 Btu.

Ice formed = 514.1 × 0.75/166 = **2.2 lb ice per lb ammonia**

(*b*) COP.

$$\text{Actual COP} = \frac{514.1}{702 - 623.5} = 6.55$$

$$\text{Ideal COP} = (460 - 0)/(60 - 0) = \mathbf{7.7}$$

P-6. (*a*) What characteristics are necessary in a refrigerant to produce a high COP? (*b*) Assuming an actual COP of 3, calculate the indicated horsepower required to produce 1 TR. (*c*) With a temperature of 15°F in the cooler and condensing water available at 50°F, calculate the ideal COP.

ANSWER.

(*a*) COP is an indication of *refrigerating effect* in the evaporator for the work of compression. Therefore, the characteristics of a refrigerant must give high refrigerating effect for the low work of compression. Thus, the necessary characteristics are:

1. Large latent heat of vaporization at moderately low pressure.
2. High critical point to prevent excessive power consumption.
3. Evaporation pressures above atmospheric to avoid leakage into the system and thus keep head pressures low to reduce power consumption and increase refrigerating effect.
4. Boiling point at atmospheric pressure lower than the lowest temperature to be produced under ordinary conditions and thus keep head pressures low. High head pressures cause excessive power consumption.
5. Low density so as to reduce pressure required to raise the head of liquid when condensing unit and evaporator are not at same level.
6. Low piston displacement to reduce friction loss and power.
7. Low viscosity to keep pressure drop small and thus power requirements low.

(*b*) Indicated horsepower is, by dimensional analysis,

$$\text{ihp per TR} = (\text{Btu per hr per TR})(\text{hr per day})/(\text{ihp} \times 2{,}545 \times \text{hr per day})$$
$$= 12{,}000 \times 24/(\text{ihp} \times 2{,}545 \times 24) = 3$$
$$\text{ihp} = 12{,}000 \div (3 \times 2{,}545) = \mathbf{1.58\ ihp}$$

(*c*) Ideal COP is

$$\text{Ideal COP} = (\text{evap. temp.})/(\text{water temp.} - \text{evap. temp.})$$
$$= (460 + 15)/(50 - 15) = \mathbf{13.5}$$

P-7. An ammonia compressor works between a pressure of 40 psia, 30°F and 100 psia.

(*a*) What is the final temperature if the compression is adiabatic?

(*b*) What work is required per pound of ammonia if the compression is adiabatic?

(*c*) If the compressor is water-jacketed so that the final compression temperature is 180°F, how many foot-pounds of work are required per lb of ammonia circulated?

(*d*) What per cent saving in energy to drive the compressor is made by the use of the water jacket?

(*e*) How many gpm of water must be circulated through the water jacket per pound of ammonia if the temperature rise in the water jacket is 10°F?

(*f*) If expansion takes place in the expansion coils from liquid ammonia at 75°F to ammonia vapor at 40 psi and 25°F, how much refrigerating effect is obtained per pound of ammonia?

(*g*) What is the COP?

(*h*) For a capacity of 200 TR, what horsepower is required in each case to drive the compressor if the mechanical efficiency of the compressor is 80 per cent?

(*i*) For a capacity of 200 TR, how many pounds of ammonia must be circulated per minute in each case?

(*j*) How many gpm circulating water must be used in the compressor jacket?

(*k*) If cooling water enters the condenser at 60°F and leaves at a temperature 40°F below that of the entering ammonia, how many gpm of circulating water are required in the condenser in each case?

(*l*) If the compressor clearance is 5 percent and the piston speed is 600 fpm what is the cylinder diameter if the compressor is single-acting?

(*m*) If the length of stroke is 1.25 times the diameter of the cylinder and the compressor is to be belt-driven from a motor making 900 rpm and having a pulley 14 in. in diameter, what diameter pulley should be provided on the compressor?

ANSWER. With the appropriate charts and thermodynamic tables for other refrigerants the solution to this problem could apply

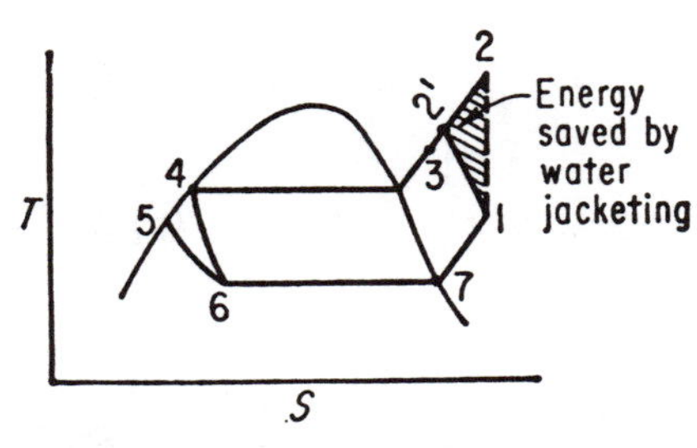

Fig. P-3

to Freon-12, Freon-22, methyl chloride, sulphur dioxide, etc. Refer to standard handbooks for charts and tables. See also Fig. P-3.

(*a*) Final temperature is **240°F.**

(*b*) Adiabatic compression work is

$$H_2 - H_1 = 731 - 626 = \mathbf{105\ Btu\ per\ lb}$$

(*c*) Polytropic compression work when water jacketed is $H_{2'} - H_1$

$$693 - 626 = 67 \text{ Btu per lb}$$

(*d*) Per cent saving polytropic over adiabatic compression:

$$\frac{105 - 67.0}{105} \times 100 = \mathbf{36.2\ per\ cent}$$

(*e*) Gallons circulated:

$$(Q_1 - Q_2)\left(T_1 + \frac{T_{2'} - T_1}{2}\right)$$

$$= (1.335 - 1.280)\left(490 + \frac{640 - 490}{2}\right) \text{ Btu per lb}$$

31.1 Btu per lb, or 31.1/10 = 3.11 lb water

$$\text{gal} = \frac{3.11}{8.33} = \mathbf{0.374\ gal\ jacket\ water\ per\ lb\ ammonia}$$

(*f*) Refrigerating effect per pound of ammonia:

$H_4 = 623$ Btu per lb enthalpy of liquid = 127 Btu per lb

623 − 127 = **496 Btu per lb ammonia**

(*g*) COP:

$$\text{COP adiabatic} = \frac{496}{105} = \mathbf{4.72}$$

$$\text{COP polytropic} = \frac{496}{67.0} = \mathbf{7.40}$$

(*h*)

$$\text{Adiabatic hp} = \frac{4.71}{472} \times 200 \times \frac{1}{0.80} = \mathbf{250\ hp}$$

$$\text{Polytropic hp} = \frac{4.71}{7.40} \times 200 \times \frac{1}{0.80} = \mathbf{159.5\ hp}$$

(*i*) Ammonia circulated:

$$\frac{200\ \text{TR} \times 200\ \text{Btu per min per TR}}{496} = \mathbf{80.8\ lb\ ammonia}$$

(*j*) Compressor jacket cooling water:

$$80.8 \times 0.374 = \mathbf{30.2\ gpm}$$

(*k*) Adiabatic cooling water: final temperature is 240°F. Heat removed from condenser per minute is

$$200 \times 200 + (105 \times 80.8) = 48{,}490\ \text{Btu per min}$$

With temperature of exit water at 200°F, we can set up the following heat balance:

$$8.33 \times \text{gpm}\ (200 - 60) = 48{,}490$$
$$\text{gpm} = 48{,}490/(8.33 \times 140) = 41.5\ \text{gpm}$$

Polytropic cooling water:

$$200 \times 200 + (67 \times 80.8) = 45{,}200\ \text{Btu per min}$$
$$\text{gpm} = 45{,}200/(8.33 \times 80) = \mathbf{67.8\ gpm}$$

(*l*) Cylinder diameter. We must first determine volumetric efficiency of compressor and then the piston displacement. Assume compression exponent of 1.3.

$$E_v = 1 + 0.05 - 0.05\left(\frac{190}{40}\right)^{1/1.3} = 1.05 - (0.05 \times 3.32) = 0.884$$

The volumetric efficiency is, therefore, 88.4 per cent. Then the piston displacement is 80.8/0.884 = 91.5 lb per min. This is equivalent to 7.5 cu ft per lb × 91.5 = 686 cfm. Cylinder area is simply $^{686}/_{600}$ = 1.143 sq ft. Now

$$0.7854d^2 = 1.143 \times 144 \qquad d^2 = 210 \qquad d = \mathbf{14.5\ in.}$$

(*m*) Pulley diameter:

$$\text{Length of stroke} = \frac{14.5 \times 1.25}{12} = 1.51 \text{ ft}$$

$$\text{Number of strokes} = 600/1.51 = 398 \text{ strokes per min}$$

$$\frac{398 \text{ strokes per min}}{2 \text{ strokes per rev}} = 199 \text{ rpm for compressor pulley}$$

Then

$$\pi \times 14 \times 900, \text{ for motor pulley} = \pi \times \text{diameter compressor pulley} \times 198$$

$$\text{Diameter of compressor pulley} = \frac{14 \times 900}{198} = \mathbf{63.2\ in.}$$

P-8. An ice plant is freezing 1,000 gal of water per hr from 65°F into ice at 20°F. Find the output of the refrigeration machine in tons, assuming 20 per cent additional load due to losses in heat leakage, etc.

ANSWER.

Heat of liquid water at 65°F........	33 Btu per lb
Heat of fusion.....................	144 Btu per lb water
Subcooling solid ice (32 − 20)0.5....	6 Btu per lb ice
Total...........................	183 Btu per lb of water (ice)

Convert gallons of water to pounds: 1,000 × 8.33 = 8,330 lb. Finally,

$$\frac{8{,}333 \times 183 \times 1.20}{12{,}000} = \mathbf{152.5\ tons\ refrigeration}$$

P-9. In a simple single-stage ammonia refrigeration system, the cooler is supplied with saturated liquid from the condenser at 180 psig, and the compressor is drawing saturated gas from the cooler at 25 psig. The compressor is a 10- by 12-in. single-cylinder, double-acting machine running at 300 rpm. If the volumetric efficiency is 75 per cent, what is the refrigerating capacity of the compressor under these conditions?

ANSWER. Volume of ammonia pumped is

$$\frac{0.785 \times 10^2}{144} \times \frac{12}{12} \times 2 \times 300 \times 0.75 = 245 \text{ cfm}$$

Ammonia vapor at 25 psig and saturated is 7.0 cu ft per lb and the weight of ammonia is simply 245/7.0 = 35 lb. From ammonia tables the enthalpy of saturated vapor at 25 psig is 615 Btu per lb while the enthalpy of the liquid at 180 psig and saturated is 149 Btu per lb. The difference of refrigerating effect is 466 Btu per lb. And the capacity of the system is

$$\frac{466 \times 35}{200} = \mathbf{81.5 \text{ tons of refrigeration}}$$

P-10. In an industrial plant it is planned to insulate a large steel reservoir in a 15°F brine refrigeration system with molded cork slabs. Refrigeration costs $1 per ton. Insulation is available in 1- to 5-in. thicknesses with 1-in. variations at installed costs of 22.5 cents, 35 cents, 45 cents, 52 cents, and 75 cents per sq ft. What thickness of insulation should be applied to realize a minimum of 25 per cent return on the cost of insulation.

ANSWER. Set up the following table:

Thickness of insulation, in.	1	2	3	4	5
Installed cost per sq ft......	$0.225	$0.35	$0.45	$0.52	$0.75
25 per cent return per sq ft	$0.056	$0.0875	$0.1125	$0.13	$0.188

Assume steel reservoir ½ in. thick, 50°F room temperature, 24 hr per day, and 365 days per year operation. With no insulation heat gain in steel shell is

$$Q = UA\,\Delta t = 15 \times 1 \times (50 - 15) = 2{,}100 \text{ Btu/(sq ft)(hr)}$$

Cost of heat loss with no insulation is 2,100/12,000 × $1 = $0.175. Conductance with 1-in. cork insulation, assuming no surface coefficients, is

$$Q = 0.30 \times 1 \times (50 - 15) = 10.5 \text{ Btu/(sq ft)(hr)}$$

10.5/12,000 × $1 = $0.000875. Saving = 0.175 − 0.000875 = $0.174125 per sq ft. Since $0.174125 is over $0.056, use **1-in. thickness as minimum.** Note that many factors were assumed to speed up development of answer. This is acceptable if examiners do not require complete calculations.

P-11. A single stage propane compressor driven by an electric motor in a refrigeration system will be used under a variety of suction conditions but at a constant speed and constant discharge pressure. Cooling water is available at 70°F and will be allowed to rise to 90°F. Minimum temperature differential in the condenser is to be 10°F. Evaporator temperature on the propane side will vary from −30 to −10°F. Saturated propane vapor will enter the suction side of the compressor. Assume propane to behave like a perfect gas. Neglect variations in volumetric efficiency and power efficiency due to change in suction conditions. Estimate the approximate maximum horsepower necessary to drive the compressor per 100 cu ft of intake volume per minute. Estimate k equal to 1.13.

ANSWER. Try $T_2 = 100°\text{F}$ $\qquad \left(\frac{P_2}{P_1}\right)^{0.115} = \frac{560}{430} = 1.3$

$$P_1 = 20.2 \qquad \frac{P_2}{20.2} = 9.8 \qquad P_2 = 198$$

$$T_1 = -30°\text{F}$$

Assumed T_2 of 100°F is slightly lower, but for all practical purposes call $P_2 = 198$. For

$$P_1 = 45.5 \qquad T_2 = (198/45.5)^{0.115} \times 470 = 560 \text{ R}$$

$$T_1 = 10°\text{F} \qquad \text{This is acceptable.}$$

$$P_2 = 198$$

For the maximum Δt condition

$$W_1 = \frac{1.13 \times 20.2 \times 144 \times 100}{0.13 \times 33{,}000} \times 0.3$$

$$= -22.8 \text{ hp per 100 cu ft suction}$$

For the minimum Δt condition

$$W_2 = \frac{1.13 \times 45.5 \times 144 \times 100}{0.13 \times 33{,}000} (1 - {}^{560}\!/_{470})$$

$$= \mathbf{-32.5\ hp} \text{ per 100 cu ft suction}$$

P-12. Ammonia is to be liquefied by compression and cooling. Cooling water is available at 60°F and since it is limited in quantity, it is desired that the temperature rise be 30°F in the condensing section of the cooler. A log mean over-all Δt of 20°F in the condensing section is also desired. Estimate the work in kilowatthour required per 100 lb ammonia liquefied if the compression is adiabatic and ammonia enters as a saturated vapor at 30 psia. First assume ammonia is an ideal gas and then repeat the calculation without making the assumption. Assume that any superheat is removed in a separate cooler ahead of the condenser.

ANSWER. Saturated ammonia vapor at 30 psia has a temperature of 0°F. Determine the temperature of saturated ammonia vapor entering condenser.

$$\text{log mean } \Delta t = 20 = \frac{30}{2.3 \log[(t - 60)/(t - 90)]}$$

From which t is found to be 98.5°F. The vapor pressure of ammonia at 98.5°F is 206 psia from ammonia tables.

(*a*) Ideal gas adiabatic case power calculated

$$W = \frac{kRT_1}{k - 1}\left[1 - \left(\frac{P_2}{P_1}\right)^{(k-1)/k}\right]$$

$$= \frac{1.31 \times 10987 \times 558.5}{0.31 \times 3415}[1 - (206/30)^{0.236}]$$

$W = -0.656$ kwhr per lb mol

For 100 lb ammonia, $W_{\text{theo}} = -0.656 \times (100/17) = \mathbf{-3.84\ kwhr}$

(*b*) Nonideal using enthalpy chart

$$H_1 \text{ at 30 psia and 0°F} = 611.0 \text{ Btu per lb}$$

$$H_2 \text{ at 206 psia (constant temperature)} = 735.0 \text{ Btu per lb}$$

$$W = -\Delta H = -124 \times 100 \times 1/3415 = \mathbf{-3.63\ kwhr}$$

P-13. A Freon-12 refrigeration unit has a double-acting compressor 6 by 10 in. with 6 per cent clearance and runs at 200 rpm. What is the rated capacity of the unit in standard tons? How many tons of refrigeration will it furnish if the Freon is evaporated at −12°F and condensed at 110°F? Assume k of 1.14.

ANSWER. A standard ton of refrigeration is based on an evaporator temperature of 5°F and a discharge temperature of 110°F.

At 5°F: $P_1 = 26.51$ psia	$H_1 = 78.79$ Btu per lb (sat)
At 14°F: $P_1 = 26.51$ psia	$\bar{V}_1 = 1.52$ cf per lb
At 86°F: $P_2 = 107.9$ psia	$H_2 = 27.72$ Btu per lb (sat liq)
$H_1 - H_2 = 53.7$ Btu per lb	$H_2 = 25.1$ Btu per lb at 77°F (sat liq cooled)

$$\text{Cylinder displacement} = (2 \times 0.785 \times 36 \times 10 \times 200)/1{,}728 = 65.5 \text{ cfm}$$

$$\text{Actual displacement} = 65.5[1 + 0.06 - 0.06(107.9/26.51)^{0.88}] = 56 \text{ cfm}$$

$56/1.52 \times 53.7 = 1{,}976$ Btu per min. This is equivalent to 1,976/200 **9.88 standard tons of refrigeration.** It would be safe to say 10.

At an evaporator temperature of −12°F

At −12°F: $P_1 = 18.37$ psia	$H_1 = 76.81$ Btu per lb (sat)
At 3°F: $P_1 = 18.37$ psia	$\bar{V}_1 = 2.0$ cf per lb
At 110°F: $P_2 = 150.7$ psia	$H_2 = 33.65$ Btu per lb (sat liq)

$H_1 - H_2 = 76.81 - 33.65 = 43.16$ Btu per lb

$$\text{Actual displacement} = 65.5[1 + 0.06 - 0.06(150.7/18.37)^{0.88}] = 44.3 \text{ cfm}$$

$44.3/2.0 \times 43.16/200 =$ **4.75 tons of refrigeration**

P-14. A propane refrigeration system in a processing plant supplies 80,000 Btuh of refrigeration at 20°F (temperature of the refrigerant). Cooling water may be assumed to be 80°F, condensing the propane at 100°F. Refer to Fig. P-4 for *T-S* relations.

(*a*) Determine conditions at exit of condenser.

(*b*) After expansion at constant enthalpy, −20°F propane is saturated at 25.05 psia and enthalpys are equal. What is the new enthalpy?

(*c*) After evaporation, vapor at −20°F is saturated at the back pressure of 25.05 psia. What is the enthalpy?

(*d*) Determine the refrigeration effect.

(*e*) Determine the propane circulation rate to maintain the refrigeration effect.

(*f*) Determine the ideal work of compression under adiabatic no-work conditions, $\Delta S = 0$. Follow the constant entropy line from −20°F saturated propane to 188.7 psia in the superheat region of the *T-S* curve.

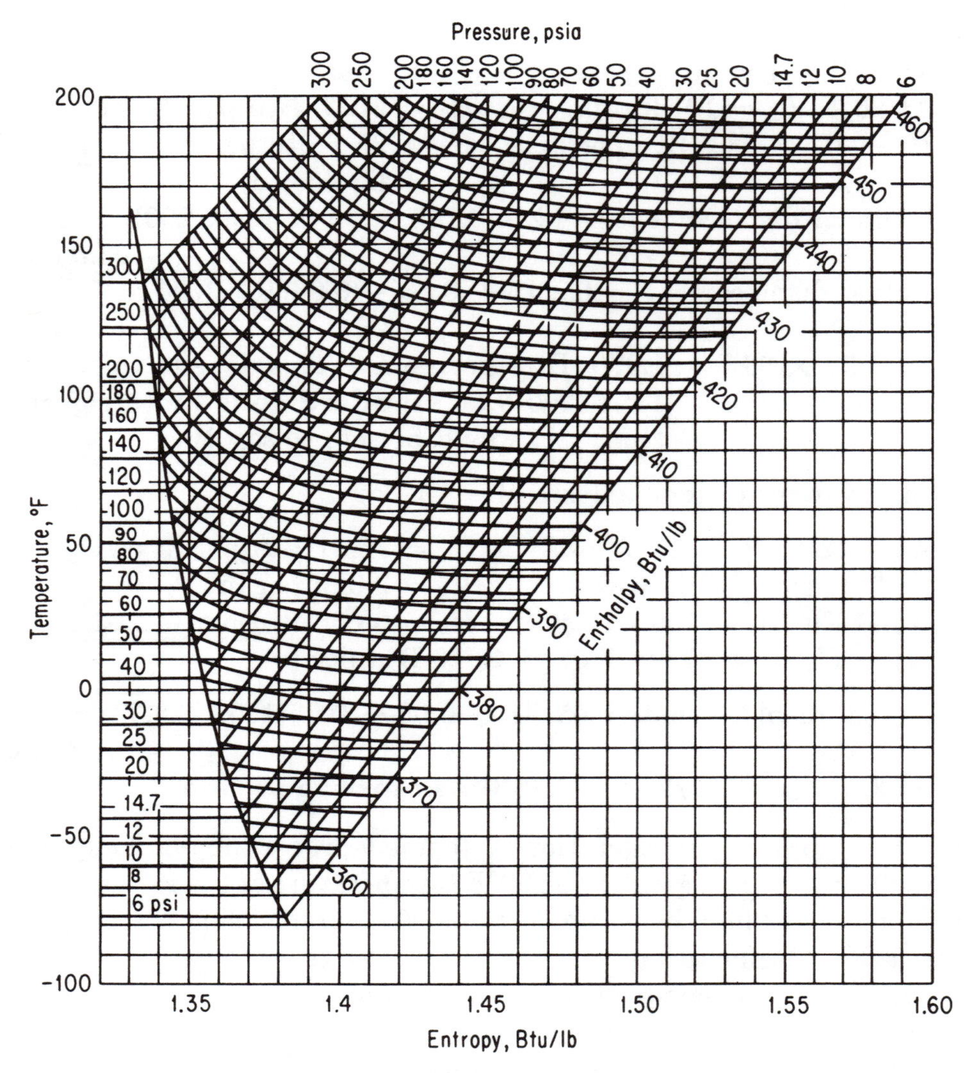

Fig. P-4. *T-S* diagram for propane.

(*g*) Determine the actual work of compression, assuming compressor efficiency of 75 per cent.

(*h*) Determine the brake horsepower required to compress the propane circulated.

(*i*) Determine the heat to be removed by the cooling water.

(*j*) Determine the amount of cooling water required for this heat removal.

ANSWER.

(*a*) Since 100°F liquid propane is saturated at 188.7 psia, enthalpy h_1 = **264.6 Btu per lb** from propane tables.

(*b*) $h_1 = h_2$ = **264.6 Btu per lb.** No heat removed or work performed.

(*c*) h_3 = **371.5 Btu per lb** from propane tables.

(*d*) $h_3 - h_2 = 371.5 - 264.6 =$ **106.9 Btu per lb.** This is the evaporation load.

(*e*) 80,000/106.9 = **748 lb per hr.**

(*f*) h_5 = 413 Btu per lb is enthalpy at the end of ideal compression.

$$\text{Ideal work} = 413 - 371.5 = \mathbf{41.5\ Btu\ per\ lb}$$

(*g*) The actual work of compression is 41.5/0.75 = **55.3 Btu per lb**

(*h*) The propane circulation rate is 748 lb per hr, and thus

$$(55.3 \times 748)/2{,}545 = \mathbf{16.25\ brake\ hp}$$

(*i*) Conditions at compressor discharge:

Enthalpy = enthalpy at saturation on entry to compressor suction plus work of compression

Thus, $371.5 + 55.3 = 426.8$ Btu per lb

This corresponds to a temperature of 151°F if no heat is removed by compressor jacket cooling. Then the heat to be removed is 426.8 − 264.6 = 162.2 Btu per lb. Finally, heat to be removed is 426.8 − 748 = 121,326 **Btuh.**

(*j*) $Q = Wc_p\,\Delta t$, and by rearrangement, $W = Q/(c_p\,\Delta t) =$ 121,326 (1 × 10) or 12,133 lb per hr of water. Flow in gpm is 12,133/500 = **24.2 gpm.**

P-15. **In air-conditioning refrigeration systems certain design and operating criteria have been found useful to provide quick checks for load calculations and to check whether or not a system is performing as designed and installed. Listed below are a number of these checks together with values of temperatures, pressures, airflows, and flow rates. These have been mismatched purposely.**

1. **Condensing-water inlet temperature, °F**
2. **Condensing-water outlet temperature, °F**

3. Water-temperature rise in condenser, °F.
4. Freon-12 head pressure, lb/in^2 gauge
5. Freon-12 suction pressure, lb/in^2 gauge
6. Freon-22 head pressure, lb/in^2 gauge
7. Freon-22 suction pressure, lb/in^2 gauge
8. Temperature of air entering cooling coil, °F
9. Temperature drop of air through cooling coil, °F
10. Temperature of air leaving cooling coil, °F
11. Total air circulated per ton of refrigeration, ft^3/min
12. Room temperature for comfort cooling, °F
13. Condensing water, gal/min per ton of refrigeration
14. Temperature of chilled water "on" coolers, °F
15. Temperature of chilled water "off" coolers, °F
16. Condensing cooling water, gal-deg per ton refrigeration
17. Chilled water, gal-deg per ton refrigeration

Match up the correct values with each description.

Cover up the right side column while working up the left side. Merely list the number you select beside the values.

Correct or mismatch	*Correct values*
2. 85–115°F	1. 60–95°F
1. 60–95°F	2. 85–115°F
9. 12–25°F	3. 15–25°F
6. 170–250 lb/in^2 gauge	4. 10–150 lb/in^2 gauge
4. 100–150 lb/in^2 gauge	5. 25–50 lb/in^2 gauge
5. 25–50 lb/in^2 gauge	6. 170–250 lb/in^2 gauge
7. 50–80 lb/in^2 gauge	7. 50–80 lb/in^2 gauge
12. 70–86°F	8. 75–90°F
3. 15–25°F	9. 12–25°F
14. 50–65°F	10. 50–70°F
11. 300–600 ft^3/min	11. 300–600 ft^3/min
8. 75–90°F	12. 70–86°F
13. 3–5 gal/min per ton	13. 3–5 gal/min per ton
10. 50–70°F	14. 50–65°F
15. 40–50°F	15. 40–50°F
17. 28 gal-deg per ton	16. 30 gal-deg per ton
16. 30 gal-deg per ton	17. 28 gal-deg per ton

P-16. An ammonia refrigeration unit has a 180-r/min double-acting compressor whose cylinder has a diameter of 8 in with a 10-in stroke and whose clearance is 5 percent.

(*a*) Sketch the T-S diagram of the cycle.

(*b*) Sketch the mechanical hookup of the cycle.

(*c*) Calculate the refrigeration capacity of the unit in tons of refrigeration.

Data and Assumptions:

Evaporation temperature, 12°F

Condensing temperature, 105°F

Vapor is saturated as it enters the compressor suction.

Liquid ammonia is not subcooled at the entrance to the expansion valve.

	At 12°F	At 105°F
P, lb/in^2 abs	40.31	228.9
v_f, ft^3/lb	0.02451	0.02769
v_g, ft^3/lb	6.996	1.313
h_f, Btu/lb	56.0	161.1
h_{fg}, Btu/lb	559.5	472.3
h_g, Btu/lb	615.5	633.4
S_f, entropy units	0.1254	0.3269
S_{fg}, entropy units	1.1864	0.8366
S_g, entropy units	1.3118	1.1635

ANSWER.

(*a*) See Fig. P-5.

(*b*) See Fig. P-6.

(*c*) Calculation of unit capacity:

$$\text{Refrigeration effect} = Q_A = h_1 - h_4 = h_1 - h_3$$

$$\text{Tons of refrigeration} = N = \frac{V_D \eta V (h_1 - h_4)}{200 V_1}$$

where V_D = compressor displacement at 100 percent volumetric efficiency

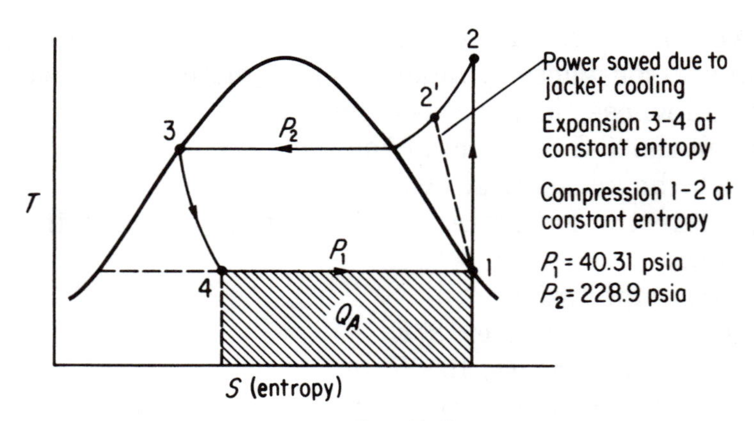

Fig. P-5

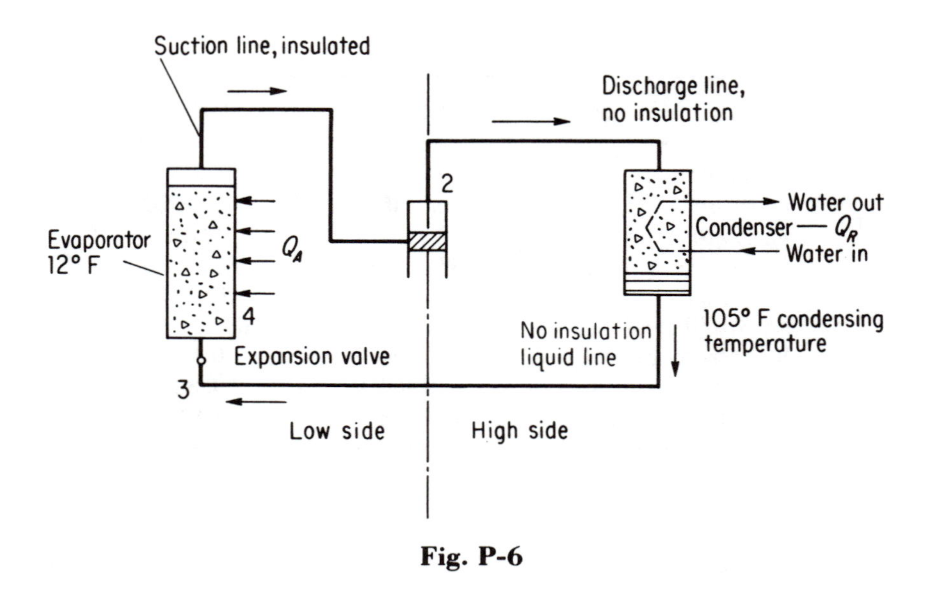

Fig. P-6

For a double-acting compressor and neglecting the area of the piston rod:

$$V_D = \frac{(2)(\text{r/min})(\pi d^2 L)}{4 \times 1728} = \frac{(2)(180)(\pi^2 \times 10)}{4 \times 1728} = 105 \text{ ft}^3/\text{min}$$

$$\eta V = \text{volumetric efficiency} = 1 + c - c\left(\frac{P_2}{P_1}\right)^{1/n}$$

where c = percent clearance as a decimal = 0.05
n = polytropic exponent = 1.3 for ammonia vapor

$$\eta V = 1.05 - 0.05\left(\frac{228.9}{40.31}\right)^{1/1.3} = 1.05 - 0.19 = 0.86$$

h_1 = enthalpy of saturated vapor at 12°F = 615.5 Btu/lb
h_4 = enthalpy of saturated liquid at 105°F = 161.1 Btu/lb
v_1 = specific volume of saturated vapor = 6.996 ft³/lb
1 ton refrigeration = 200 Btu/min heat removal

$$N = \frac{(105)(0.86)(615.5 - 161.1)}{(200)(6.996)} = \mathbf{29.4\ tons}$$

Supplement Q

HEATING AND VENTILATING

Questions and Answers

Q-1. The following data were obtained by a heating engineer during a survey of a client's office in New York City. The only heat loss is through one wall facing north. The exposed wall measures 20 by 10 ft. The wall has one single pane, ⅛-in. thick window, 20 sq ft in area. The wall is constructed of 4-in. face brick, 4-in. common brick, 1-in. air space, ¾ in. wood fiber insulating board (density 15 lb per cu ft), ⅜-in. gypsum lath on 2- by 4-in. studs, and ½-in. plaster. The outside design temperature is 0°F and the inside temperature is to be 70°F. Infiltration is assumed to be 10 cfm.

(*a*) Calculate the room heat loss.

(*b*) How many pounds of saturated steam at 2 psig are required to maintain temperature of 70°F?

(*c*) How much air at 100°F is required to heat the room?

(*d*) What is the temperature in the 3⅝-in. air space (4-in. stud space)?

ANSWER. From the Guide or any other source the conductivities of the various resistances are found and set up.

U_g (single glass) . . . . 1.13 Btu/(hr)(sq ft)(°F)
k for face brick . . . . 9.2 Btu/(hr)(sq ft)(°F) per in. thickness
k for common brick 5.0
k for fiberboard 0.34
k for plaster 8.00
k for gypsum lath 3.30
Coefficient air film inside 1.1
Coefficient air film outside (15-mph wind) 6.00
Coefficient air film inside room 1.65

Since U = summation of reciprocal of resistances in series, calculate each and then add.

$$U = \frac{1}{6.2015} = 0.161$$

$$\begin{aligned}
\text{Outside air film} &= 1/6.00 &&= 0.167 \\
\text{Face brick} &= 4/9.2 &&= 0.434 \\
\text{Common brick} &= 4/5 &&= 0.800 \\
\text{Air space} &= 1/1.1 &&= 0.909 \\
\text{Fiberboard} &= 0.75/0.34 &&= 2.200 \\
\text{Air space} &= 1/1.1 &&= 0.909 \\
\text{Gypsum lath} &= 0.375/3.3 &&= 0.114 \\
\text{Cement plaster} &= 0.5/8.00 &&= 0.0625 \\
\text{Inside air film} &= 1/1.65 &&= 0.606 \\
&\text{Summation} &&= 6.2015 \\
\text{Wall area (gross)} &= 20 \times 10 &&= 200 \text{ sq ft} \\
&\text{Glass area} &&= 20 \text{ sq ft} \\
&\text{Net wall area} &&= 180 \text{ sq ft}
\end{aligned}$$

Heat required to raise infiltration air = $10 \times 1.08 \times (70 - 0)$ = 756 Btu per hr

(*a*) Room heat loss:

$$\begin{aligned}
\text{Loss through wall} &= 0.161 \times 180 \times 70 &&= 2100 \text{ Btu per hr} \\
\text{Loss through glass} &= 1.13 \times 20 \times 70 &&= 1580 \\
&\text{Infiltration loss} &&= 756 \\
&\text{Total loss} &&= \mathbf{4436\ Btu\ per\ hr}
\end{aligned}$$

From steam tables heat of condensation (H_{fg}) of steam at 2 psig is 966.1 Btu per lb. The steam rate is

(*b*) $$\text{lb steam per hr} = \frac{\text{total loss}}{966.1} = \frac{4{,}436}{966.1} = 4.64, \text{ say } \mathbf{5\ lb\ per\ hr}$$

(*c*) $$\text{cfm air at 100°F} = \frac{4{,}436}{1.08(100 - 70)} = \mathbf{137\ cfm}$$

(*d*) Since summation of resistances is previously found to be 6.2015 and the sum of the resistances measured from the inside to the air space is found to be 0.7825(0.114 + 0.0625 + 0.606), the temperature within the 3⅝-in. air space is taken from the Guide to be

$$t_i - (t_i - t_o)\,\frac{0.7825}{6.2015} = 70 - (70 - 0)\,\frac{0.7825}{6.2015} = 70 - 8.18 = \mathbf{61.8°F}$$

Q-2. A pitched roof construction of asbestos shingles on roofing felt plus wood sheathing has an over-all coefficient of 0.01 Btu/(hr)(sq ft)(°F). The roof area is 1,000 sq ft. The ceiling, with an area of 600 sq ft, is made up of ¾ in. of gypsum lath and plaster having a conductance of 1.80 Btu/(hr)(sq ft)(°F) for thickness stated,

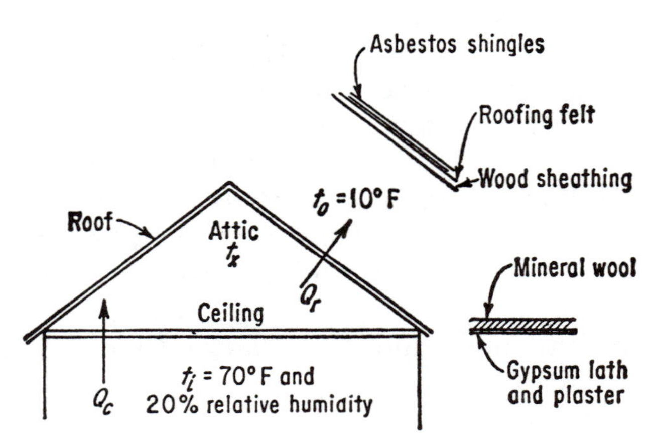

Fig. Q-1

plus 2 in. of mineral wool having a conductivity of 0.27 Btu/(hr)(sq ft)(°F) for 1-in. thickness. Inside and outside film conductances are 1.65 and 6.0, respectively. Outdoor temperature is plus 10°F. Indoor air is at 70°F and 20 per cent relative humidity. Attic space is not ventilated. Will condensation form on the attic side of the wood sheathing? Support your answer by numerical calculations.

ANSWER. See Fig. Q-1. Over-all coefficient for roof is given as 0.61. Since attic temperature is unknown and will be useful in determining dew point let this be denoted by t_x. When equilibrium sets, i.e., when heat gain through ceiling into attic *equals* heat loss from attic through the roof to the outside, the temperature t_x may be determined. Heat gain through ceiling into attic is

$$Q_c = U_c A_c (t_i - t_x)$$

Loss through the roof to outside is

$$Q_r = U_r A_r (t_x - t_0)$$

$$U_c = \frac{1}{\frac{1}{1.65} + \frac{1}{1.80} + \frac{2}{0.27} + \frac{1}{1.65}} = 0.109 \text{ Btu/(hr)(sq ft)(°F)}$$

$$Q_c = 0.109 \times 600(70 - t_x)$$
$$Q_r = 0.61 \times 1{,}000(t_x - 10)$$
$$Q_c = Q_r = 0.109 \times 600(70 - t_x) = 0.61 \times 1{,}000(t_x - 10)$$

from which t_x is calculated to be equal to 15.9°F, say 16°F. If this temperature is below the dew point corresponding to the room conditions, then condensation will occur on the attic side of the sheathing, provided there is no vapor seal on the attic side of the mineral wool insulation. If the temperature is above the dew point, then condensation will not occur. From the psychrometric chart, dew point corresponding to a condition of 70°F dry bulb and 20 per cent relative humidity is 26°F. **Without vapor seals condensation will occur.**

Q-3. How many pounds of moisture must be added per hr to air entering a building at 32°F and 60 per cent relative humidity to produce an inside relative humidity of 30 per cent at 70°F? Building volume is 500,000 cu ft and there are three air changes per hour. Take specific volume of mixture of air and water vapor as 13.8.

ANSWER. Pressures are so low that the perfect-gas law may be considered valid. With the use of steam tables we see that:

At 32°F vapor pressure of moisture = 0.180 in. Hg. The partial pressure of moisture in the air at 60 per cent relative humidity is $0.60 \times 0.180 = 0.108$ in. Hg. The humidity ratio becomes

$$\frac{0.108}{29.92 - 0.108} \times 0.622 = 0.00226 \text{ lb water per lb dry air}$$

This is also equivalent to

$$\frac{0.00226}{1 + 0.00226} = 0.00224 \text{ lb water per lb wet mixture}$$

At 70°F vapor pressure of water is 0.739 in. Hg. The partial pressure of moisture in air at 30 per cent relative humidity is

$$0.30 \times 0.739 = 0.221 \text{ in. Hg}$$

The new humidity ratio becomes

$$\frac{0.221}{29.92 - 0.221} \times 0.622 = 0.00463 \text{ lb water per lb dry air}$$

This is now also equivalent to

$$\frac{0.00463}{1 + 0.00463} = 0.00462 \text{ lb water per lb wet mixture}$$

Water to be added is, therefore,

$$500{,}000 \times 3 \times (0.00462 - 0.00224)/13.8 = \mathbf{258\ lb\ per\ hr}$$

or

$$258/500 = \mathbf{0.515\ gpm}$$

Q-4. Find the season fuel cost of heating a building whose heat loss calculated on the basis of 70 and 0°F is 240,000 Btu per hr, for a season of 7,000 degree days using coal at $15 per ton and 12,000 Btu per lb, burned with an efficiency of 60 per cent.

ANSWER. The total heat loss for the season is given by the formula

$$H' = \frac{H(t_r - t_m)N}{t_r - t_0} = \frac{24DH}{t_r - t_0}$$

where H' = total loss, Btu for season
H = calculated heat loss, Btu per hr for estimated temperature difference $(t_r - t_0)$
N = number of hours that heating is required and for which t_m is average temperature
D = number of degree-days in the heating season
t_r = desired room temperature, °F
t_0 = outside temperature assumed for heat calculation H, °F
t_m = mean outside temperature for season of N hr, °F

The question implies the use of the second portion of the above equation.

$$H' = \frac{24 \times 7{,}000 \times 240{,}000}{70 - 0} = 576 \times 10^6 \text{ Btu}$$

The amount of fuel used for the season is given by

$$F = \frac{H'}{eh} = \frac{576 \times 10^6}{0.60 \times 12{,}000} = 80{,}000 \text{ lb coal}$$

$$\text{Cost} = 80{,}000/2{,}000 \times 15 = \mathbf{\$600}$$

Q-5. In order to render worker environment healthful for the normal 8-hr period of work maximum allowable concentrations (MAC) have been established by industrial hygienists for various solvents used in industry. Discuss dilution air volumes required to render vapor-laden atmospheres healthful for workers.

ANSWER. Maximum allowable concentrations have been determined and are used by industrial hygienists to calculate the volume of air required to dilute the contaminant to safe limits. They are expressed in parts per million (ppm) for the normal eight-hour working day in industry. The air dilution volume for various solvents may be calculated using the following formulas:

$$\text{cu ft air per pint evaporated} = \frac{403 \times \text{sp gr liquid} \times 1 \times 10^6 \times K}{\text{mol weight liquid} \times \text{MAC}}$$

$$\text{cu ft air per lb evaporated} = \frac{387 \times 1 \times 10^6 \times K}{\text{mol weight liquid} \times \text{MAC}}$$

K may be defined as a constant, depending on the toxicity of the solvent involved, uniformity of distribution, dilution of vapors in air, location of fan and its proximity to evolved vapors, population of area, etc., and may vary from 3 to 10.

The following table of values reflects the use of the above formulas The MAC's are in parentheses; MAC values are subject to change and revision if and when further research or experience indicates the need. If the MAC values change, the dilution air requirements should be calculated from the above formulas.

Q-6. Technical literature is replete with data charts and nomographs for readily determining fluid friction (air, gases, water, etc.) in circular or rectangular conduits of constant cross section. There are few data, however, for estimating such losses when the cross section is nonstandard (triangular, trapezoidal, flat oval, elliptical, etc.). Here we will present a method for determining such systems.

Table Q-1. Dilution Air Volumes for Vapors*

Liquid	Cu ft air required for dilution to MAC	
	Per pint evaporation	Per pound evaporation
Acetone (500)	11,000	13,300
Amyl acetate (200)	13,600	14,900
Isoamyl alcohol (100)	37,200	43,900
Benzol (50)	Not recommended	
n-Butanol (butyl alcohol) (50)	Not recommended	
n-Butyl acetate (200)	15,300	16,600
Butyl cellosolve (200)	15,400	16,400
Carbon disulfide (20)	Not recommended	
Carbon tetrachloride (50)	Not recommended	
Cellosolve (200)	20,800	21,500
Cellosolve acetate (100)	29,700	29,300
Chloroform (100)	50,200	32,400
1-2 Dichloroethane (75) (ethylene dichloride)	Not recommended	
1-2 Dichloroethylene (200)	26,900	20,000
Dioxane (100)	47,300	43,900
Ethyl acetate (400)	10,300	11,000
Ethyl alcohol (1000)	6,900	8,400
Ethyl ether (400)	9,630	13,100
Gasoline (500)	6,000–7,000	8,000–10,000
Methyl alcohol (200)	49,100	60,500
Methyl acetate (200)	25,000	26,100
Methyl butyl ketone (100)	33,500	38,700
Methyl cellosolve (25)	Not recommended	
Methyl cellosolve acetate (25)	Not recommended	
Methyl ethyl ketone (200)	22,500	26,800
Methyl isobutyl ketone (100)	32,300	38,700
Methyl propyl ketone (200)	19,000	22,400
Naphtha (coal tar) (200)	15,000–19,000	20,000–25,000
Naphtha (petroleum) (500)	6,000–7,000	8,000–10,000
Nitrobenzene (1)	Not recommended	
Isopropyl alcohol (400)	13,200	16,100
Propyl acetate (200)	17,500	18,900
Isopropyl ether (400)	7,150	9,460
Stoddard solvent (500)	6,000–7,000	8,000–10,000
Tetrachloroethane (5)	Not recommended	
Tetrachloroethylene (100)	39,600	23,400
Toluol (toluene) (200)	19,000	21,000
Trichloroethylene (100)	45,000	29,400
Xylol (xylene) (200)	16,500	18,200

* The tabulated dilution air quantities must be multiplied by the selected K value.

Table Q-1 has been abstracted by permission from *Industrial Ventilation*, American Conference of Governmental Industrial Hygienists, January, 1951.

MAC is now replaced by TLV (threshold limit value).

An equilateral triangular duct of constant cross section is to carry 1000 ft^3/min of air at atmospheric conditions, not over 100°F. Unit friction is to be 0.1 in WG per 100 ft of straight duct.

Size the duct, using sheet metal, using chart graph from ASHRAE Fundamentals volume.

ANSWER. From the ASHRAE graph the circular equivalent is 14 in. The hydraulic radius $= D/4 = 14/4 = 3.5$ in. Also the hydraulic radius may be given as

$$r = \frac{\text{cross-sectional area}}{\text{wetted perimeter}} = \frac{A}{P}$$

First trial: 18-in equilateral triangle

$$P = 54 \text{ in} \qquad A = 1/2ah = 1/4a^2\sqrt{3}$$
$$A = 1/4 \times 18^2\sqrt{3} = 140 \text{ in}^2$$
$$A/P = 140/54 = 2.6$$

And the equivalent diameter $= 4r = 4 \times 2.6 = 10.4$ in. Too small.

Second trial: 22-in equilateral triangle

$$P = 66 \text{ in} \qquad A = 1/4 \times 22^2\sqrt{3} = 209 \text{ in}^2$$
$$A/P = 419/66 = 3.17$$

And the equivalent diameter $= 4 \times 3.17 = 12.68$ in. Getting close.

Third trial: 24-in equilateral triangle

$$P = 72 \text{ in} \qquad A = 249 \text{ in}^2 \; A/P = 249/72 = 3.47$$

And the equivalent diameter $= 4 \times 3.47 = 13.88$ in. **Ans.**

Q-7. It is frequently stated that gases and vapors heavier than air should be exhausted at or near floor level and those lighter than air should be exhausted at or near the ceiling. Experience indicates that it is important to consider the distribution, concentration, and temperature of the gas or vapor in the air before drawing any conclusions as to the relative merits of which system to use. This is especially true of general ventilation where the forces influencing the rate of gas or vapor diffusion are complex or uncertain. In the situation that follows, these considerations are to be taken into account.

A processing room 20 ft wide by 40 ft long by 15 ft high is to be provided with general ventilation also known as dilution ventilation, at the rate of 15 air changes per hour. The room contains 10,000 ppm of a vapor having a specific gravity (compared with air) of 5. The vapor is well mixed with room air because of drafts, plant activity, and convection currents within the room.

(*a*) Determine the effective specific gravity of the room air. Discuss the implications and room effects.

(*b*) Determine the cubic feet per minute of room air to be exhausted.

(*c*) At what level or levels should exhaust points be located, at floor, midway, or at ceiling?

(*d*) Determine the concentration of vapor in the room air within the exhaust system.

(*e*) Estimate fan brake horsepower if air hp requirements are given by

$$\text{Air hp} = \frac{0.16 \times Ap \times \text{ft}^3/\text{min}}{1000}$$

Assume a reasonable fan efficiency and a system duct resistance from most remote exhaust intake point to outdoor discharge to be 1.5 in WG.

(*f*) What would be the makeup air heating load if design conditions were + 10°F outdoors and air leaving the duct heater is to be 80°F?

ANSWER.

(*a*) Room air constitutes 99 parts having a specific gravity of 1.0, and vapor constitutes 1 part having a specific gravity of 5.0. Then

$$\begin{aligned} 0.99 \times 1.0 &= 0.99 \\ 0.01 \times 5.0 &= \underline{0.05} \\ & 1.04 \end{aligned}$$

This is the effective specific gravity of the room air mixture of air and vapor. Therefore, the above room air mixture, compared with incoming outside air, would tend to move downward to the floor ever so slightly, expressed by the ratio of 104:100 and not by the ratio of 5:1, as is so frequently implied. This means that, in industry, the effects of drafts, window leakage and ventilation, plant traffic disturbances, or convection currents set up by process heat can easily dwarf into insignificance the effect of pure vapor specific gravity.

(*b*) Exhaust $ft^3/min = (20 \times 40 \times 15)(15/60) = 3000\ ft^3/min$ **Ans.**

(*c*) Locate exhaust points at floor level to avoid drawing room air across the workers' breathing zone.

(*d*) Since this is general (dilution) ventilation, room air-vapor concentration is the same as within the duct air stream.

(*e*) $\text{Bhp} = \dfrac{\text{air hp}}{\text{fan efficiency}} = \dfrac{0.16 \times 1.5 \times 3000}{1000 \times 0.6} = 1.2\ \text{bhp}$ Use a 2-hp motor.

(*f*) Makeup air heating load is obtained from

$$3000 \times 1.08 \times (80 - 10) = 226{,}800\ \text{Btu/h} \qquad \textbf{Ans.}$$

This is equivalent to the consumption of 230 lb steam per hour or 230 ft^3/h of a gaseous fuel having a gross heating value of 1000 Btu/ft^3.

Q-8. Ethyl ether is being evaporated from a product in a room-temperature drying hood operating at atmospheric pressure. Solvent vapor must be removed at the rate of 46 lb/h and the solvent vapor concentration must be maintained below the 1.8 percent lower explosive limit (LEL). Calculate the air volume which must be supplied to the dryer. Room air is at 75°F.

ANSWER. This is a gas-dilution problem, the solution being straightforward if you remember that gas analyses are based on percent by volume; i.e., for ideal gas mixtures, volume percent is proportional to the mols of each gas. Molecular weight of ethyl ether is 74 ($C_2H_5-O-C_2H_5$). Then the mols of ether removed per hour is 46/74 = 0.622.

Since the LEL is given as 1.85 percent ether by volume, this is the same as saying 1.85 mol percent ethyl ether in the exhaust air leaving the dryer.

For each 1.85 mols ether exhausted, 98.15 mols air must dilute it. Then, the leaving air with 0.622 mol ethyl ether is

$$\frac{0.622}{0.0185} = x \qquad \text{and} \qquad x = 33\ \text{mols air/h}$$

At standard conditions of 60°F and 14.7 lb/in^2 abs the molal volume is 379 ft^3/mol. Then at a room temperature of 75°F the volume per mol of air is

$$\frac{379(460 + 75)}{460 + 60} = 390 \text{ ft}^3\text{/mol}$$

Finally, the volume of air being exhausted is found to be

$$(33)(390) = 12{,}870 \text{ ft}^3\text{/h} \qquad \textbf{Ans.}$$

And this is the same as the volume being supplied. This amount of air is the least permissible to prevent the explosive limit from being attained in the main bulk of the exhaust system. However, local concentrations within the dryer itself may be explosive, so that a sparkproof design is a requirement.

Q-9. In the manufacture of pharmaceuticals and chemicals there are a large number of different contaminants which are deleterious to personnel, products, building, and equipment. There are a wide variety of mists and dusts; vapors such as acetone, benzol, and methyl alcohol; gases such as carbon dioxide, hydrogen, and ammonia. Although some gases and vapors as well as dusts are nontoxic, it is advisable to control them at their point of generation to provide an acceptable environment for the worker.

A chemical plant is considering upgrading its dust-collection system in its processing area consisting of five rooms containing open drums, kettles, and coating pans. Each room is presently air-conditioned, and each piece of processing equipment is provided with a local exhaust system under suction to a central exhaust system and dust collector of the bag type. Because of the nature of new processing techniques, increased capture velocities are required with the attendant increase in exhaust-air requirements. Room air conditioning is to be maintained at present levels.

Outline a procedure to evaluate the needs of the required changes to upgrade the system, including effects on air-conditioning equipment and room pressure differentials.

ANSWER.

1. Obtain the facts by making a site visit to review the "as built" installation drawings. Look for

Design conditions
Room configuration
Exhaust-system layout
Air-conditioning system: supply and return or once-through
Refrigeration system: air-cooled condensers, water-cooled condensers, chilled water, direct expansion
Equipment nameplate data: supply fans, exhaust fans, HVAC units, air handlers, chilled water units
Air quantities: supply to each room, return from each room, exhaust to collector for each room, airflow pattern for each room
Airflow control between rooms
Pressurization needs for each room
Negative pressure needs for each room
Room with most hazardous contaminant
Room with least hazardous contaminant
Visual tests with titanium tetrachloride smoke bombs
Diversity factors in plant operations
Need to shut down certain rooms or have systems run continuously
Alarm systems for safe operation
Air-filtration-plant data from nameplates
Safety devices

2. Evaluation of data obtained:
Airflow quantities by measurement: supply to each room, return from each room, exhaust air from each piece of equipment, excess of supply over exhaust, excess of exhaust over supply (negative room)
Effluent-air-treatment efficiency
Pressure differentials between rooms and adjacent nonprocess areas and outdoors to prevent air movement from contaminated rooms to clean rooms
Capacity of present air-conditioning system
Capacity of existing air-filtration system
Capacity of existing supply and return air system
Capacity of existing exhaust system

3. Program to upgrade system:
Selection of new capture velocities for each unit
Increases in exhaust airflow thereby

Increases in supply and return air

Compare present duct systems with new airflow increses as to accommodation of new flows

Determine new friction losses and their effect on fan units and other factors

Compare new flows and system resistance with existing equipment. Can fans be speeded up within safe limits or are new units needed?

Give same treatment as above to all air handlers

4. Specifications for upgraded system:
   Duct systems, fans and air handlers, air conditioners, etc.

Q-10. Sulfur dioxide is to be recovered from a dry gas containing (by volume) 13 percent SO_2 and 87 percent air, using a tower at atmospheric pressure with water as the solvent. By means of lead cooling coils, the temperature is maintained at 40°C throughout the operation. The raw gas, entering at the rate of 100 ft^3/min at 40°C, is to be stripped of 99 percent of its SO_2 content. Data: See Fig. Q-2. Determine the minimum amount of water that can be used.

ANSWER. Enter chart at 98.8 mm Hg. At the 40°C curve turn down and read 8.8 SO_2 per 1000 g water and

$$(8.8/1000)(2.205/2.205) = 19.4 \text{ lb } SO_2/2205 \text{ lb } H_2O$$
$$= 0.0088 \text{ lb/lb or } 0.88 \text{ lb } SO_2/100 \text{ lb } H_2O$$

In 100 ft^3 "rich" gas, there are

$$(100/459)(0.87)(273/313)(29) = 4.79 \text{ lb air}$$
$$(100/359)(0.13)(273/313)(64) = 2.02 \text{ lb } SO_2$$

On an hourly basis,

$$(4.79)(60) = 287.4 \text{ lb air/h} = G$$
$$(2.02)(60) = 121.2 \text{ lb } SO_2\text{/h}$$

Partial pressure SO_2 in the entering gas stream is (0.13)(760) = 98.8 mm Hg.

Unabsorbed SO_2 in the waste gas leaving the absorber is (0.01)(121.2) = 1.21 lb SO_2 per hour.

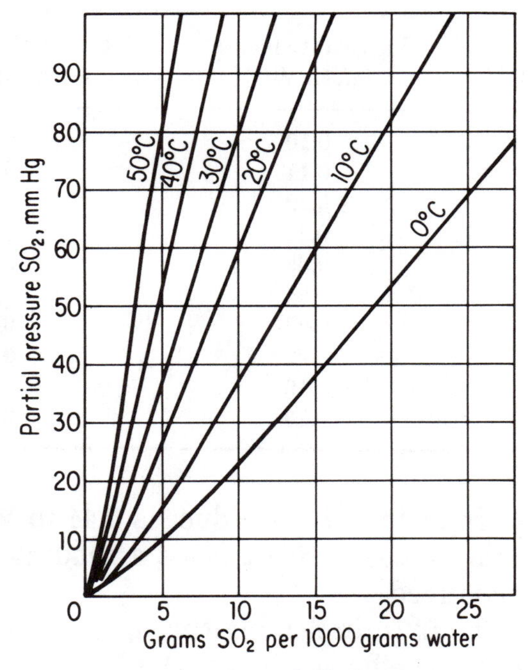

Fig. Q-2

From Fig. Q-2, it has been found that the concentration of an aqueous solution of SO_2 at a partial pressure of SO_2 of 98.8 mm Hg at 40°C is 0.88 lb SO_2 per 100 lb water. This is the maximum strength of solution that can be produced from the entering gas. Then,

$$L(X_1 - X_0) = G(Y_1 - Y_0) = L(0.0088 - 0)$$

$$= 287.4(121.2/287.4 - 1.2/287.4)$$

$$0.0088\ L = 121.2 - 1.2 = 120 \text{ and } L$$

$$= 13{,}600 \text{ lb water/h} \qquad \textbf{Ans.}$$

Q-11. A 10-point pitot tube traverse for air at 70°F and normal barometric pressure of 14.7 lb/in² abs flowing in a 12-in circular duct gives the following data:

Traverse point, in from wall	Vertical traverse, *ah* in WG	Horizontal traverse, *Ah* in WG
⅜	0.10	0.14
1	0.11	0.18
1¾	0.19	0.22
2¾	0.27	0.30
4⅛	0.35	0.35
7⅞	0.34	0.36
9¼	0.29	0.24
10¼	0.22	0.19
10⅞	0.16	0.14
11⅝	0.13	0.11

Throat suction at inlet to the 12-in duct is 0.44 in WG. A subsequent throat suction measurement gives a reading of 0.28 in WG with the inlet unchanged.

(*a*) Determine the ft³/min flowing through the pipe.

(*b*) Determine the coefficient of entry C_e of the inlet. Throat suction is 0.44 in WG.

(*c*) Determine the quantity of air flowing through the pipe under the above conditions.

ANSWER.

(*a*) Set up the following tabulation:

ah	$V = 4{,}005\sqrt{h}$, ft/min	*Ah*	$V = 4{,}005\sqrt{h}$, ft/min
0.10	1,266	0.14	1,498
0.11	1,328	0.18	1,699
0.19	1,746	0.22	1,879
0.27	2,081	0.30	2,193
0.35	2,369	0.35	2,369
0.34	2,335	0.36	2,403
0.29	2,157	0.24	1,962
0.22	1,879	0.19	1,746
0.16	1,602	0.14	1,498
0.13	1,444	0.11	1,328
Totals	18,207		18,575

$$\text{Average} = (18{,}207 + 18{,}575)/20 \text{ readings} = 1839 \text{ ft/min}$$

$$\text{ft}^3/\text{min flow} = VA = 1839 \times 0.7854\,(12/12)^2 = \mathbf{1444\ ft^3/min}$$

(*b*) Finding C_e, knowing the velocity and the throat suction at inlet:

$$V = 4005 C_e \sqrt{h}$$

from which $C_e = (V)/(4005\sqrt{h}) = (1839)/(4005 \times \sqrt{0.44}) = \mathbf{0.69}$

(*c*) New airflow:

$$\text{ft}^3/\text{min} = 4005 C_e \sqrt{h} A$$

$$= 4005(0.69)(0.53)(0.7854) = \mathbf{1150\ ft^3/min}$$

Supplement **R**

AIR CONDITIONING

Questions and Answers

R-1. One hundred cu ft of air at 70°F dry bulb and 20 per cent relative humidity are passed through a spray chamber where it comes in contact with water at the adiabatic-saturation temperature. It is cooled adiabatically and leaves at 53°F. What is the final percentage saturation of the air and how much moisture does it pick up?

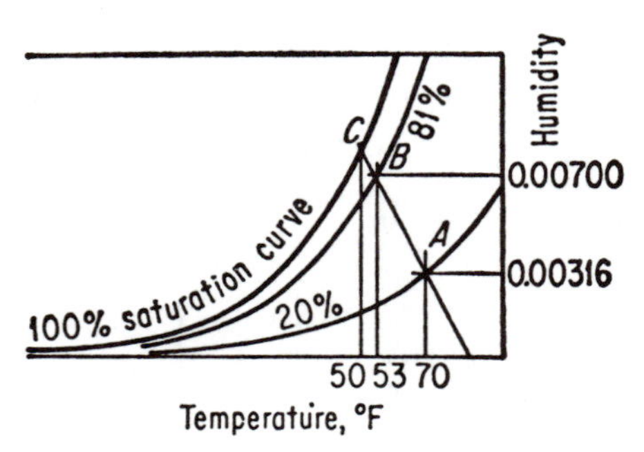

Fig. R-1

ANSWER. See Fig. R-1 skeleton psychrometric chart. Point *A* gives the value of 0.00316 lb of water per lb of dry air, or

$$0.00316 \times 7{,}000 = 22.12 \text{ grains}$$

By following along the adiabatic-saturation line to a dry-bulb temperature of 53°F (point *B*), a final humidity of 0.00700 and a percentage saturation (relative humidity) of **81** are obtained. The adiabatic-saturation temperature is about 50°F.

From the psychrometric chart by interpolation the volume of air at 70°F and 20 per cent relative humidity is found to be 13.45 cu ft per lb of dry air. Therefore, total weight of dry air is

$$100/13.45 = 7.44 \text{ lb}$$

And the weight of water picked up by the air is

$$7.44(0.00700 - 0.00316) = \mathbf{0.0286\ lb}$$

R-2. In a drying process air at 70°F and 60 per cent relative humidity enters the dryer and is discharged at a temperature of 90°F and 90 per cent relative humidity. How many cubic feet of air must be circulated through the dryer to remove 10 lb of water per min from the material to be dried? Note that a cubic foot of entering air has a greater volume on leaving.

ANSWER.

$$0.60 = \frac{\text{wt vapor per cu ft actual}}{\text{wt vapor per cu ft in air saturated with vapor}}$$

$$0.60 = \frac{x}{1/869} \qquad x = \frac{0.60}{869} = 0.691 \times 10^{-3} \text{ lb at entrance}$$

For leaving condition if volume is same as at entrance condition,

$$0.90 = \frac{y}{1/468.5} \qquad y = \frac{0.90}{468.5} = 1.92 \times 10^{-3} \text{ lb}$$

For every cubic foot entering there is

$$1 \times \frac{460 + 90}{460 + 70} = 1.039 \text{ cu ft leaving}$$

That is to say, for every cubic foot entering 1.039 cu ft leaves due to increase in temperature. Then the weight of vapor leaving is found by

$$0.00192 \times 1.039 = 0.00199468 \text{ lb}$$

Thus, there is removed for each cubic foot of air passing through dryer: $0.001995 - 0.000691 = 0.001304$ lb vapor. Finally, by dimensional analysis,

$$\frac{\text{lb water}}{\text{per min}} \times \frac{1}{\text{lb per cu ft removed}} = \frac{10}{0.001304} = \mathbf{7{,}660\ cfm}$$

R-3. An air-conditioning engineer has been given the following data on a room to be air conditioned:

Room sensible heat gain.	40,000 Btu per hr
Room latent heat gain.	8,000 Btu per hr
Outside conditions.	95°F dry bulb and 75°F wet bulb
Inside conditions.	80°F dry bulb and 50 per cent relative humidity

(*a*) Calculate the refrigeration tonnage required with 10 per cent safety factor.

(*b*) Calculate the dry- and wet-bulb temperature of a mixture of 25 per cent outside air and 75 per cent return air.

(*c*) Calculate the room sensible heat factor and apparatus dew point.

(*d*) Calculate the cfm supply of air to satisfy room requirements.

ANSWER.

$$(a) \qquad \frac{(40{,}000 + 8{,}000)1.10}{12{,}000} = \mathbf{4.4\ TR}$$

(*b*) Dry-bulb temperature:

$$\begin{aligned} \text{Outside air} &= 0.25 \times 95 = 23.8°\text{F} \\ \text{Return air} &= 0.75 \times 80 = \underline{60.0°\text{F}} \\ \text{Mixture temperature} &= \mathbf{83.8°F} \end{aligned}$$

Wet-bulb temperature: On psychrometric chart draw a straight line connecting the outside air condition and the inside air condition

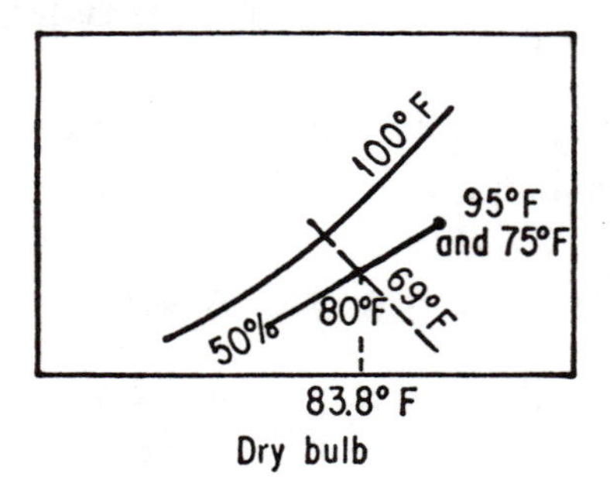

Fig. R-2

(see Fig. R-2). Locate the calculated dry-bulb temperature of 83.8°F along the base of the chart and move up in a straight line until you hit the straight line just drawn. Then read up and to the left along constant wet-bulb line the wet-bulb temperature of **69°F.**

(*c*) Room sensible heat factor (SHF):

$$\text{SHF} = \frac{\text{room sensible heat}}{\text{room sensible heat} + \text{room latent heat}}$$

$$= \frac{40{,}000}{40{,}000 + 8{,}000} = 0.833$$

From the Carrier System Design Manual, Part 1, Load Estimating, pp. 1–147, the apparatus dew point may be calculated from the following equation:

$$\text{SHF} = \frac{0.244(t_{rm} - t_{adp})}{0.244(t_{rm} - t_{adp}) + 1{,}076/7{,}000(W_{rm} - W_{adp})}$$

where

W_{rm} = room moisture content, grains per lb dry air

W_{adp} = moisture content at apparatus dew point, grains per lb dry air

t_{rm} = room dry-bulb temperature, °F

t_{adp} = apparatus dew point, °F

0.244 = specific heat of moist air at 55°F dew point, Btu/(°F)(lb) of dry air

1,076 = average heat removal required to condense 1 lb water vapor from room air

$0.244(t_{rm} - t_{adp})$ = total sensible load

$\frac{1{,}076}{7{,}000}(W_{rm} - W_{adp})$ = total latent load

Of course, the Carrier Manual above-mentioned will give you the apparatus dew point (ADP) at once. But it is required that you calculate it. To do this, assume a value of ADP from the psychrometric chart, using the slope for the SHF directly. This assumption is found to be between 56 and 58°F. Try 57°F. Also from the psychrometric chart W_{rm} and W_{adp} are found. Now insert these in the SHF equation, in its simplified form below

$$\text{SHF} = \frac{1}{1 + 0.628\,\dfrac{W_{rm} - W_{adp}}{t_{rm} - t_{adp}}}$$

$$= \frac{1}{1 + 0.628\,\dfrac{77 - 69.5}{80 - 57}} = 0.831$$

Thus, the ADP assumed is close enough. Use **57°F.**

(*d*) Air supply, cfm. If we assume all air passing through the conditioner will come in contact with the cooling surface (Carrier zero bypass factor), then the cfm is given by

$$\text{cfm} = \frac{\text{RSH}}{1.08(t_{rm} - t_{adp})} = \frac{40{,}000}{1.08(80 - 57)} = \mathbf{1{,}610\ cfm}$$

R-4. A conditioner cooling surface fin section has a total capacity of 8.99 TR with an SHF of 0.67. With an entering air condition of 80°F dry-bulb temperature and 50 per cent relative humidity and air flow rate of 200 lb per min, find the leaving air conditions of *(a)* dry-bulb temperature, *(b)* dew point, *(c)* wet-bulb temperature, and *(d)* relative humidity.

ANSWER. Set up the table of *entering conditions* as follows:

Dry-bulb temperature	80°F
Wet-bulb temperature	67°F
Dew-point temperature	60°F
Relative humidity	50 per cent
Total heat (enthalpy)	31.15 Btu per lb
Grains moisture per lb dry air	77.3
Sp vol, cu ft per lb	13.84
Air flow rate, lb per min	200
Sensible heat factor	0.67

(a)
$$0.67 = \frac{\text{total sensible heat}}{\text{sensible heat} + \text{latent}}$$

Total sensible heat = 0.67 × 8.99 = 6.03 TR. The temperature drop through the cooling surface is

$$\text{Temp. difference} = \frac{6.03 \times 200}{220 \times 0.24} = 22.8°\text{F}$$

Therefore, the leaving dry-bulb temperature is 80 − 22.8 = **57.2°F**

(b) Total load minus sensible load equals moisture (latent) load, i.e.,

$$(8.99 \times 200) - (6.03 \times 200) = 592 \text{ Btu per min}$$

$$\frac{592 \times 7{,}000}{1{,}076 \times 220} = 17.4 \text{ grains per lb dry air removed by coil}$$

At 80°F dry-bulb temperature and 50 per cent relative humidity the moisture content is 77.3 grains per lb dry air. Then the moisture content for the leaving condition is 77.3 − 17.4 = 59.9 grains per lb dry air. This corresponds approximately to a dew point of **53°F.**

(*c*) Wet-bulb temperature. From psychrometric chart with use of leaving conditions of dry-bulb and wet-bulb temperatures read **54.9°F** wet-bulb temperature.

(*d*) Relative humidity. Also from the chart, read **85 per cent** relative humidity.

R-5. It is desired to condition 10,000 cfm of air to a leaving dew-point temperature corresponding to 20 grains per lb dry air by means of silica gel adsorption beds. Air enters the beds at 65°F dry-bulb temperature and 55° dew-point temperature. The heat of adsorption is plus 200 Btu per lb of vapor adsorbed. The following information is desired:

(*a*) What is the outlet dry-bulb temperature?

(*b*) If reactivation reduces the moisture content of the silica gel to 5 per cent by weight, what is the weight of silica gel that is required in the beds to permit extracting the required moisture on a 4-hr cycle and that limits the total moisture gain in the beds to 20 per cent by weight for each cycle?

(*c*) The beds are heated to 150°F by reactivation and are cooled back to 80°F by circulating, dry air through them; air enters at 60°F and leaves the beds at 70°F. How much cooling air is necessary if the cooling cycle is to be restricted to 1 hr? The specific heat of silica gel may be taken as 0.22 Btu/(lb)(°F).

ANSWER. This is considered chemical drying and is represented by a straight line along the wet-bulb temperature between the limits of the drying process *AB* only in case the drying is purely by adsorption (the drying agent does not dissolve in the water extracted from the air) and only in case the drying agent does not retain an appreciable amount of the heat of vaporization liberated when the water is condensed on the surface of the adsorber. In case an appreciable amount of heat of vaporization is retained by the adsorber, the process takes place along a line below the wet-bulb temperature *AB′*. If the drying agent is soluble in water (such as calcium chloride), the drying process takes place along a line which lies above *AB″* or below *AB′*, depending on whether heat is liberated or absorbed when the agent is dissolved in water (see Fig. R-3).

(*a*) From psychrometric chart at inlet air conditions of 65° dry-bulb temperature and 55° dew-point temperature the specific volume is 13.4 cu ft per lb and the moisture content is found to be 64 grains per lb. The change in temperature will depend on the moisture pickup converted to Btu. Knowing the cfm being handled and the Btu pickup the temperature rise can be calculated. The moisture pickup is 64 − 20 = 44 grains per lb of dry air. This converted to pounds per hour is given as

$$\frac{10{,}000 \times 60 \text{ min}}{13.4} = 44{,}800 \text{ lb per hr of dry air}$$

The temperature rise is

$$\frac{44{,}800 \times 44 \times 200}{7{,}000} \times \frac{1}{1.08} \times \frac{1}{10{,}000} = 5.22°\text{F}$$

The leaving air temperature is 65 + 5.22 = **70.22°F.** Examination of Fig. R-3 and the psychrometric chart indicates that an

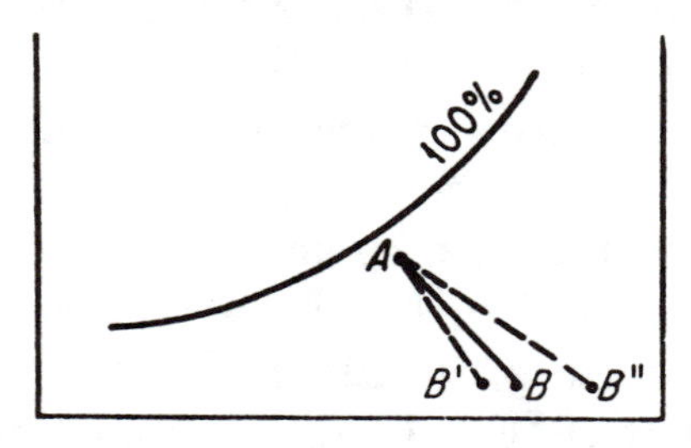

Fig. R-3

appreciable amount of heat of vaporization has been retained by the silica gel.

(*b*) For the 4-hr cycle operation the total moisture removed is simply

$$\frac{44{,}800 \times 44 \times 4}{7{,}000} = 1{,}125 \text{ lb in 4 hr}$$

Problem says that the moisture gained percentage wise is

$$20 - 5 = 15 \text{ per cent}$$

Thus, the weight of silica gel required is 1,125/0.15 = **7,500 lb.**

(*c*) The heat picked up by the reactivation air is represented by the temperature drop.

$$7{,}500 \times 0.22 \times (150 - 80) = 115{,}000 \text{ Btu per hr}$$

And the air required to accomplish this is $Q/[1.08(t_2 - t_1)]$

$$\text{cfm} = 115{,}000/(1.08 \times 10) = \mathbf{10{,}650\ cfm}$$

R-6. Outside air at 88°F dbt and 75 per cent RH and at 29.92 in. Hg barometric pressure is passed through an air washer supplied with cooled water. It is then reheated to 68°F dbt and it is desired that the RH should be 40 per cent. Determine the temperature of the air leaving the washer. Also determine, per 1,000 cf of outside air, the pounds of water and the Btu of heat removed in the washer and the Btu added in the reheater.

ANSWER. Refer to Fig. R-4. Now list all data from psychrometric chart at points 1, 2, and 3. At point 2 air is assumed saturated, so its RH is 100 percent. At point 3, dpt is 43, which is also

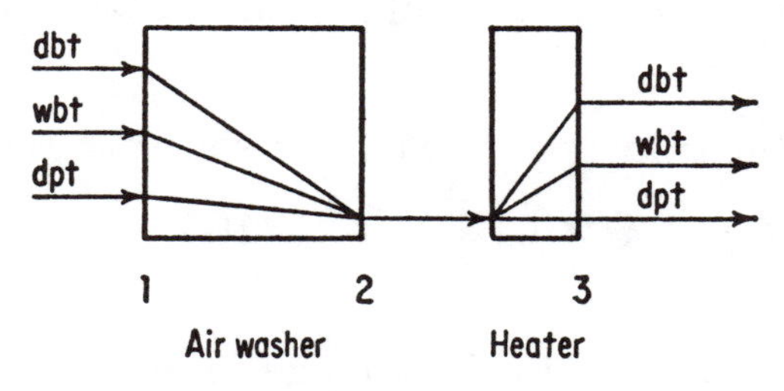

Fig. R-4

true at point 2, since no moisture is added in reheater (no dpt change). Also, since at point 2, RH is 100 per cent, then dbt and wbt are also 43°F. Therefore, temperature of air leaving washer is **43°F.**

	1	2	3
dbt, °F	88	43	68
wbt, °F	81.2	43	54.3
dpt, °F	79.1	43	43
RH, %	75	100	40
S, gr per lb	151	41	41 dry air
H, Btu per lb	44.8	16.7	22.9 dry air
$\bar{V}$, cf per lb	14.28	12.78	13.42

$(151 - 41)/7{,}000 = 0.0157$ lb vapor per lb dry air removed in washer

$(0.0157/14.28) \times 1{,}000 =$ **1.1 lb** moisture removed per 1,000 cf outside air

44.8 − 16.7 = 28.1 Btu per lb dry air removed in washer
(28.1/14.28) × 1,000 = **1,960 Btu removed per 1,000 cf outside air**
22.9 − 16.7 = 6.2 Btu per lb dry air added in reheater
(6.2/14.28) × 1,000 = **433 Btu** added per 1,000 cf outside air

R-7. A cooling tower installation is to be made near the top of a high mountain where the average atmospheric pressure is 13 psia. Four cfs of water is to be cooled from 105 to 82°F. Air enters the tower at 96°F at a partial vapor pressure of 0.75 psia and leaves as saturated air at 102°F. Calculate (*a*) the weight of air required per minute, and (*b*) the make-up water in pounds per hour. Note that the standard psychrometric chart cannot be used in the solution of this problem.

ANSWER. Heat transferred from the water to the air:

$$Q = wC_p(T_1 - T_2) = 4 \times 62.5 \times 1 \times (105 - 82) = 5{,}750 \text{ Btu per sec}$$

Heat absorbed by the air in the change in enthalpy of the air mixture from inlet to outlet:

$$Q = H_0 - H_i = w_a(h_0 - h_i) \qquad \text{where } h_i = \text{enthalpy of air ``in''}$$

$$p_r = p_v + p_a = 13.0 \text{ psia}$$

$$p_a = 13 - 0.75 = 12.25 \text{ psia}$$

If we assume the vapor and air act as perfect gases, the humidity ratio is

$$w_i = \frac{p_v R_a}{p_a R_v} = \frac{0.75 \times 53.3}{12.25 \times 85.7} = 0.0381 \text{ lb vapor per lb dry air}$$

Total weight of inlet air mixture w_r

$$= 1.0000 + 0.0381 = 1.0381 \text{ lb}$$

From Keenan and Keyes' Steam Tables, for an inlet temperature $t_i = 96°F$, $h_g = 1103.5$ Btu/lb. Now the enthalpy of the vapor is given by

$$h_g = w_v h_g = 0.0381 \times 1103.5 = 42.05$$

Enthalpy of dry air is $h_a = C_p t_i = 0.24 \times 96 = 23.05$.

$$h_g + h_a = 42.05 + 23.05 = 65.10$$

Enthalpy of air in:

$$h_i = (h_v + h_a)/1.0381 = 65.10/1.0381 = 63.0 \text{ Btu per lb}$$

Enthalpy of air out h_0. Since the outlet air is saturated at 102°F,

$$p_v = 1.0078 \text{ psia}$$
$$p_a = p_r - p_v = 13 - 1.0078 = 11.9922 \text{ psia}$$

Humidity ratio:

$$w_0 = \frac{p_v R_a}{p_a R_v} = \frac{1.0078 \times 53.3}{11.9922 \times 85.7} = 0.0522 \text{ lb vapor per lb dry air}$$

From Keenan and Keyes' Steam Tables, for an outlet temperature $t_0 = 102°F$,

$$h_g = 1106.1 \text{ Btu per lb}$$

Enthalpy of vapor:

$$h_v = w_v h_g = 0.0522 \times 1106.1 = 57.5$$

Enthalpy of dry air:

$$h_a = C_p t_0 = 0.24 \times 102 = 24.5$$

Then $h_v + h_a = 57.7 + 24.5 = 82.2$. The enthalpy of air out is

$$h_0 = (h_v + h_a)/1.0522 = 82.2/1.0522 = 78.0 \text{ Btu per lb}$$

(*a*) Weight of air (w_a) required per minute is

$$w_a = \frac{Q \times 60}{h_0 - h_i} = \frac{5{,}750 \times 60}{78 - 63} = \mathbf{23{,}000 \text{ lb per min}}$$

(*b*) Pounds per hour of makeup water. Each pound of air picks up an amount of water vapor equal to

$$\Delta w = w_a - w_i = 0.0522 - 0.0381 = 0.0141 \text{ lb vapor per lb air}$$
$$w_{H_2O} = 23{,}000 \times 60 \times 0.0141 = \mathbf{19{,}500 \text{ lb/hr}} \text{ or}$$
$$19{,}500/500 = 39 \text{ gpm}$$

R-8. How many cubic feet per minute of air can be cooled with a 10-ton Freon-12 air-conditioning system from an initial temperature of 90°F dbt and 75 percent relative humidity to a final condition of 75°F dbt and 45 per cent relative humidity? High-side temperature is 110°F and low-side is 40°F.

ANSWER. Cooling load is 10 × 200 = 2000 Btum. From the psychrometric chart, enthalpy per pound of air at entering conditions is 46.9 Btu; at leaving conditions it is 27.1 Btu per lb. Thus, heat removed per lb air = 46.9 − 27.1 = 19.8 Btu per lb. Weight of air flow per min = 2,000/19.8 = 101 lb. From the psychrometric chart, specific volume of air = 14.36 cu ft per lb. Finally, cfm = 101 × 14.36 = **1,450.** Note that specific volume is based on outside air conditions.

R-9. A high-pressure air-conditioning system is undergoing analysis. Fan ft^3/min = 28,000; static pressure = 5.5 in WG; fan outlet velocity = 2800 ft/min. System is a draw-through configuration.

(*a*) What is the temperature rise across the fan?

(*b*) How many tons of refrigeration are needed to offset this temperature rise?

ANSWER.

(*a*) Temperature rise across fan is given by

$$\frac{5.19 \times (TP_2 - TP_1)}{778 \times C_p \times d_1 \times FTE}$$

where TP_2 = downstream total pressure, in WG
TP_1 = upstream total pressure, in WG
FTE = fan total efficiency
C_p = specific heat of air = 0.24
d_1 = upstream air density, lb/ft^3

Now continuing, fan outlet velocity = 2800 ft/min and the fan velocity pressure = $(2800/4005)^2$ = 0.49 in WG. Now assume casing velocity pressure is negligible. Normally, it would be about 500 ft/min, which is equal to a velocity pressure of $(500/4005)^2$ = 0.0156 in WG.

Downstream total pressure is 0.49 + 5.5 = 5.99 in WG, say 6.00 in WG = TP_2. Also assume a fan efficiency of 0.84, or 84 percent.

The total pressure on the upstream side of the fan is 0 in WG, i.e., $TP_1 = 0$. *Note:* On any system where you have louvers, filters, and coils, TP_1 would be a negative number, as the pressure when related to atmospheric is negative. However, we have already calculated the required fan static pressure. Therefore, fan static plus velocity pressure at the an outlet gives us $TP_2 - TP_1$, neglecting the velocity pressure in the inlet plenum. Let $d_1 = 0.075$ lb/ft^3 air density.

Finally, temperature rise across fan is

$$\frac{5.19 \times 6}{778 \times 0.24 \times 0.075 \times 0.84} = 2.65°\text{F}$$

(*b*) Tonnage required over and above system heat load

$$\frac{28{,}000 \times 1.08 \times 2.65}{12{,}000} = 6.7 \text{ tons} \qquad \textbf{Ans.}$$

R-10. The case for industrial comfort air conditioning has been proved over and over again. If a plant is located where summers are hot and sticky, production will suffer during summer, no matter how you slice profit and loss statements. In production areas where there is high internal heat gain from motors and machinery, in the drafting room, in engineering, planning, and research, there is bound to be a summer's slump, a slow-down.

You are the plant manager of a manufacturing complex and are considering air conditioning your plant but want to keep cooling costs down. List a number of practical ways you could reduce cooling costs. Limit the number to 10.

ANSWER.

1. Use existing ductwork and air-handling equipment, with minor modifications, to handle air conditioning.

2. Keep plant closed up and run cooling equipment at night to store cooling capacity in walls, machinery, etc., and so reduce size of cooling equipment needed. Also power rates are generally lower at night.

3. For operations where air conditioning may not be practical but heat is a problem as in steel mills, foundries, etc., consider air conditioning "rest areas" where personnel can recuperate from heat exposure.

4. Check water consumption in your plant. Volume may be sufficient to use in water-cooled condenser first before going to the plant for process use, thus eliminating a cooling tower.

5. It is not necessary to bring temperature and humidity down to levels found in offices. Dry-bulb temperatures between 80 and

85°F and 50 percent relative humidity are good enough in the main plant area when it is 95°F dry-bulb temperature in the shade outside and relative humidity is 65 percent or higher.

6. For areas where sedentary work is being done, it is important to reduce relative humidity as well as dry-bulb temperature. Reduced humidity permits bodily heat-removing functions to work at greater efficiency.

7. Install needed equipment in hottest "problem" spots first and add more later when budget permits.

8. Consider leasing arrangements for packaged air-conditioning equipment.

9. Exhaust cooled conditioned air out through the "hot" spots of your plant, such as foundry and paint spray room, where air conditioning may not be practical.

10. Consider pinpointing air conditioning at isolated "hot" spots instead of cooling the entire plant to cool a few people. Package units are ideal for such applications.

11. Forget about 20- to 30-ft ceilings. Supply conditioned air at 8- to 10-ft levels. Figure cooling load as if ceiling were all glass.

12. If packaged units are being considered, avoid long duct runs and deliver air into room direct from unit. Air can be delivered farther with higher velocity than in commercial jobs because noise is not so great a factor.

13. Ventilate or condition only partitioned areas of extreme heat from rest of shop. Segregate the heat-producing areas so you can cool only personnel.

14. Check actual operating time and heat output of machinery and motors and use these findings to size cooling-load equipment.

15. Provide automatic door closures on all outside doors or in segregated areas. You may also wish to consider door alarms with timers to draw attention to doors left open over extended periods of time.

R-11. The outside design temperature is 100°F and relative humidity is 15 percent. It is required to cool 1500 ft^3/min of the air to the lowest possible temperature by evaporative cooling with the relative humidity of the air increasing to 60 percent. Determine how many pounds of water per minute are required and what is the final temperature.

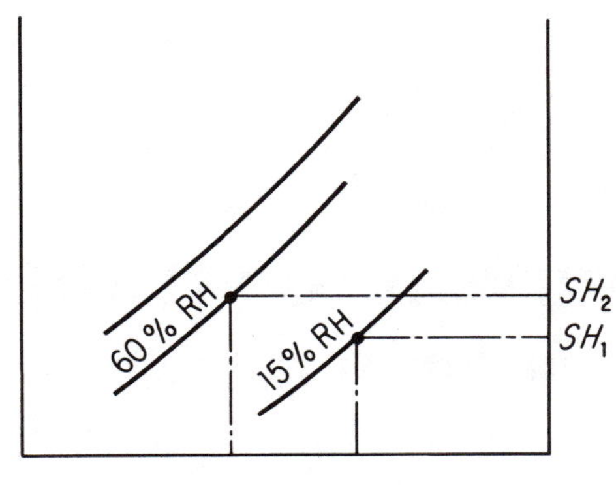

Fig. R-5

ANSWER. Use the psychrometric chart for normal atmospheric conditions. Refer to Fig. R-5 and note point 1:

$$T_{db1} = 100°\text{F} \qquad \text{RH}_1 = 15 \text{ percent}$$

$$T_{wb} = 66°\text{F} \qquad \text{SH}_1 = 42 \text{ gr/lb dry air}$$

Assume water supply is at 66°F or equal to T_{wb}. Then follow the constant wet-bulb line at 66°F from RH_1 = 15 percent to RH_2 = 60 percent and read T_{db2} = 75.7°F final temperature. **Ans.**

Also read SH_2 = 80 gr/lb dry air and V_2 = 13.74 ft^3/lb dry air.

Mass of airflow in = 1500/14.24 = 105.2 lb/min

Amount of water in incoming air = 105.2 × 42 = 4418 gr/min

Amount of water in air at final condition = 105.2 × 80 = 8416 gr/min

Amount of water added to airstream = 8416 − 4418 = 3998 gr/min

Finally, rate of water added = 3998/7000 = 0.571 lb min **Ans.**

Supplement S

ENVIRONMENTAL CONTROL
Questions and Answers

S-1. A 2-lb cylinder of chlorine gas fell off a lab table and broke, permitting the gas to escape into a closed room of 50 by 27 by 15 ft.

(*a*) Calculate the concentration in parts per million. The barometric pressure is 760 mm Hg, and the temperature is 22°C.

(*b*) On the basis of the current TLVs (threshold limit values), would it be dangerous to remain in this room for 3 hr?

ANSWER.

(*a*)

$$\text{ppm} = \frac{(\text{weight})(22.4L) \times (760/P) \times [(273 + T°\text{C})/273] \times 10^6}{(\text{mw})(\text{gm-mol})(\text{vol})}$$

$$= (2 \text{ lb})(454 \text{ gm})(22.4L) \times \frac{(760/760)(295/273)(10^6)}{(50)(27)(15)(1L/0.035 \text{ cu ft})}$$

$$= \mathbf{536\ ppm}$$

(*b*) Using a TLV of 1 ppm, it would be very dangerous.

S-2. (*a*) Calculate a worker's daily 8-hr exposure to a solvent with a current threshold limit value (TLV) of 100 ppm if the worker spends 1 hr at an operation where the concentration is 250 ppm, 4 hr at 200 ppm, and 3 hr at 100 ppm.

(*b*) If you returned to this same plant two weeks later and found similar exposures at the same operation, would you recommend that control measures be instituted? If so, why?

ANSWER. (*a*) Calculate 8-hr exposure to a solvent with a current TLV of 100 ppm if:

1 hour at 250 ppm

4 hours at 200 ppm
3 hours at 100 ppm

$$\frac{(1 \times 250) + (4 \times 200) + (3 \times 100)}{8} = 1{,}350/8 = \mathbf{169\ ppm}$$

(*b*) Control is necessary, since the 8-hr exposure exceeds the TLV of 100 ppm.

S-3. The acoustical treatment of a room has reduced the over-all noise level by 3 dB. By what factor has the acoustical power been reduced?

ANSWER. Reducing the over-all noise level 3 dB reduces the acoustical power by **one-half.** Thus

$$W = 10 \log_{10} \frac{P_1}{P_2}$$

$$3 \text{ dB} = 10 \log_{10} \frac{P_1}{P_2}$$

$$3/10 = \log_{10} \frac{P_1}{P_2}$$

$$\frac{P_1}{P_2} = 10^{0.3}$$

$$\frac{P_1}{P_2} = 2 \qquad \text{or} \qquad P_2 = \frac{P_1}{2}$$

S-4. Estimate the sound pressure level at 10 ft for the following vents:

Vent No. 1

Saturated steam
Valve size 4 in.
Temperature 400°F
Pressure 250 psig
Molecular weight 18

Vent No. 2

Saturated steam
Valve size 6 in.
Temperature 200°F
Pressure 200 psig
Molecular weight 18

Vent No. 3

Air

Valve size 2 in.

Temperature 75°F

Pressure 250 psig

Molecular weight 29

Vent noise correction factor for saturated steam = −5.0, for air = 0.

Vent sound pressure correction factor for 4-in.-diameter outlet = 12, for 6-in.-diameter outlet = 15, for 2-in.-diameter outlet = 6.

ANSWER. The sound pressure level is determined from $L_p = L_p(1) + L_p(2) + L_p(3)$.

Vent No. 1

$L_p(1) = 124$

This is the base sound pressure level at 10 ft

$L_p(2) = 12$

$L_p(3) = -5$

Total = **131 dB sound pressure level at 10 ft**

The frequency of the maximum sound energy is

$$F_0 = \frac{52.8}{d}\sqrt{\frac{T}{MW}} = \frac{52.8}{4/12}\sqrt{\frac{860}{18}} = 1000 \text{ Hz acoustic frequency}$$

Vent No. 2

$L_p(1) = 119$

This is the sound pressure level at 10 ft at pressure.

$L_p(2) = 15$

$L_p(3) = -5$

Total = **129 dB** sound pressure level at 10 ft

The frequency of the maximum sound energy is

$$F_0 = \frac{52.8}{6/12}\sqrt{\frac{660}{18}} = 500 \text{ Hz acoustic frequency}$$

Vent No. 3

$L_p(1) = 120$

This is the sound pressure level at 10 ft at pressure.

$L_p(2) = 6$
$L_p(3) = 0$
Total = **125 dB** sound pressure level at 10 ft

The frequency of the maximum sound energy is

$$F_0 = \frac{52.8}{2/12}\sqrt{\frac{535}{29}} = 1000 \text{ Hz acoustic frequency}$$

Base sound pressure level at 10 ft from an air vent at pressures shown:

250 psig	124 (steam)
200 psig	119 (steam)
250 psig	120 (air)

Reference: A. Thumann and R. K. Miller, "Secrets of Noise Control," Fairmont Press (1974).

S-5. Given an octave-band analysis based on center frequencies:

Octave band	1	2	3	4	5	6	7	8
SPL (sound pressure level)	88	88	94	96	96	92	89	76
Band correction	−26	−16	−9	−3	0	+1	+1	−1

Determine equivalent dB(A) level overall.

ANSWER. Proceed as follows:

Octave Band	*Band SPL*	*Band Correction*	*Corrected SPL*	*dB/10*	*Antilog*
1	88	−26	62	6.2	0.02×10^8
2	88	−16	72	7.2	0.16×10^8
3	94	−9	85	8.5	3.16×10^8
4	96	−3	93	9.3	19.99×10^8
5	96	0	96	9.6	39.81×10^8
6	92	+1	93	9.3	19.95×10^8
7	89	+1	90	9.0	10.00×10^8
8	76	−1	75	7.5	0.32×10^8
				Total	93.37×10^8

Then

$$\log 93.37 \times 10^8 = 10(9.97) = 99.7 \text{ or } \mathbf{100\ dB(A)}$$

S-6. A noise consultant is reviewing a residential ordinance which is as follows:

Old Octave Bands, H	*Sound Pressure Level, dB*
37–75	64
75–150	64
150–300	59
300–600	54
600–1200	46
1200–2400	41
2400–4800	34
4800–9600	36

Convert this ordinance to new octave bands, using the following table:

TABLE S-1
CONVERSION OF OLD OCTAVE BAND LEVELS TO NEW OCTAVE BAND LEVELS

$L_2 - L_1$	*Correction to L_1 to get new octave data*
0	0
1	0.2
2	0.5
3	0.7
4	1.0
5	1.2
6	1.4
7	1.7
8	1.9
9	2.1
10	2.4
11	2.6
12	2.8

where L_1 = level in any old octave band
L_2 = level in next higher old octave band

Note: If $L_2 - L_1$ is negative, subtract the correction factor from L_1. If $L_2 - L_1$ is positive, add the correction factor to L_1.

ANSWER.

$L_2 - L_1$	*New Octave Band Center Frequency, Hz*	*Correction*	*Sound Pressure Level in New Octave Band, dB*
0	63	0	64
−5	125	−1.2	62.8
−5	250	−1.2	57.8
−8	500	−1.9	52.1
−5	1,000	−1.2	44.8
−7	2,000	−1.7	39.3
+2	4,000	+0.5	34.5
+2	8,000	+0.5	36.5

S-7. A fabric dust collector handles 400 cfm with a dust loading of 10 gr per cu ft. Its initial resistance is 1 in. water gage. At the end of 6 hr operation it reaches the maximum permissible resistance of 5 in. water gage. How soon would 5 in. water gage be reached for (*a*) a dust loading of 20 gr per cu ft and a flow of 400 cfm; (*b*) a dust loading of 10 gr per cu ft and a flow of 800 cfm?

ANSWER.

(*a*) Resistance builds up at a rate of 4 in. per 6 hr, or ⅔ in. per operating hour. Since flow through the bags is in the streamline range, resistance is directly proportional to flow rate. If there is twice as much dust, the given resistance will be reached in one-half the time, or 6 hr × ½ = **3 hr.**

(*b*) Initial resistance is doubled to 2 in., leaving 5 − 2 = 3 in. for buildup. Twice the dust is caught per minute, doubling buildup rate. Twice the flow doubles the buildup rate:

$$3 \text{ in. WG}/(\tfrac{2}{3} \text{ in. per hr} \times 2 \times 2) = \tfrac{9}{8} = \mathbf{1.13\ hr}$$

S-8. In a hammermill dust collector the air leaving the mill and entering the collector has the following conditions:

Total dry solids = 25 std (short tons per day)
Maximum moisture = 4 per cent
Dry gas flow = 729 std or 13,197 scfm

Water vapor = 31 std or 904 scfm
Dry gas plus moisture = 760 std or 14,101 scfm at 60°F dbt
Temperature of moisture is 130°F dbt
Actual cfm (acfm) = 16,000 at 130°F
Operating conditions = sea level or 14.7 psia
Collector sits in open air at 50°F dbt design conditions

(*a*) Find the dew point of the mixture entering the collector.

(*b*) Will condensation take place inside the steel plate collector walls and cause clogging?

(*c*) Will jacketing the collector with insulation be necessary to prevent condensation?

(*d*) What thickness of fiberglass insulation is necessary? The collector wall thickness is ⅛ in. steel.

ANSWER. Please refer to Fig. S-1, and note that all data given are needed to solve this problem.

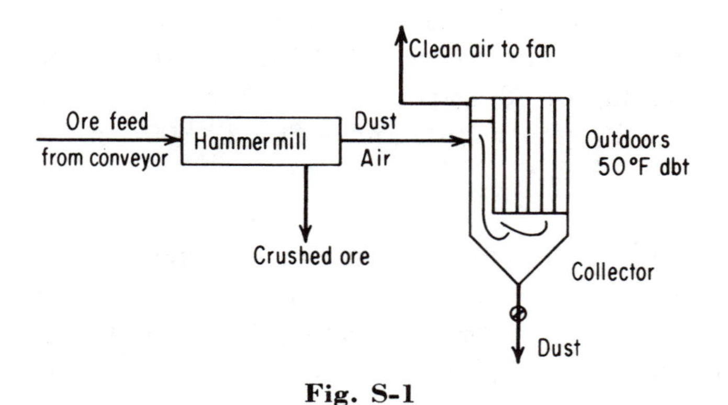

Fig. S-1

(*a*) Apply the humidity ratio equation, letting pp = partial pressure of water vapor, psia. Set up the equation and solve for pp. Then refer to standard steam tables (saturated vapor); opposite pp read the corresponding saturation temperature. Molecular weight of water = 18; molecular weight of air = 29.

$$\frac{\text{pp}}{14.7 - \text{pp}} \times \frac{18}{29} = \frac{31}{729} = 0.042 \text{ lb water per lb dry air}$$

from which pp = 0.93 psia. The corresponding saturation temperature = **100°F** approximately. This is the dew point of the air-dust mixture.

(*b*) Since 100°F is greater than 50°F, condensation **will** take place inside the collector and will cause clogging.

(*c*) An insulation jacket is necessary to increase the wall temperature above 100°F. Increase the collector shell temperature to 110°F. From tables, ⅛-in.-thick steel plate with 1-in. rigid insulation has an overall coefficient of heat transfer U equal to 0.25 for 15-mph wind outside the shell. Assume the inside shell coefficient = 1.65, and the over-all resistance is then found to be

$$\frac{1}{U} = \frac{1}{0.25} = 4$$

Then (1/1.65)/4 = 0.606/4 = 0.151.

Now let T_i = 130°F; T_{wi} = inside wall surface temperature, °F. Also, T_0 = 50°F.

$$\frac{T_i - T_{wi}}{T_i - T_0} = 0.151$$

$$T_i - T_{wi} = 0.151(130 - 50) = 12$$

and it follows that

$$T_{wi} = 130 - 12 = 118°\text{F}$$

This is the new inside wall temperature and is above 100°F dpt. Since the wall temperature is above 100°F dpt, vapor will **not** condense out inside the collector.

(*d*) Use **1-in.** fiberglass rigid insulation. Common practice is to use 2-in. insulation as a minimum with an aluminum or stainless jacket overlapped at the joints and sealed.

S-9. Calculate the diameter a constant-flow orifice must have to provide an air flow of 1.0 cfm for a membrane filter sample. Assume the following design data:

1. Resistance through filter 2 cm Hg at 1.0 cfm
2. Source of suction capable of delivering 28 in. Hg vacuum
3. Barometric reading 30 in. Hg

Given the following:

W = mass of air in lb per sec (density of air at 70°F is 0.074 lb per cu ft)
C_v = constant = 1
A_2 = orifice area in sq in. = to be found
T_2 = absolute temperature in °R at 70°F = 530°R

P_1 = upstream pressure in lb per sq in. (1 in. Hg = 0.4912 lb per sq in.)

ANSWER. Constant-flow-orifice calculation for a flow of 0.5 cfm is given by

$$W = \frac{0.533 C_v A_2 P_1}{\sqrt{T_2}}$$

where W = lb per sec of air at STP (standard temperature and pressure)
C_v = constant = 1
A_2 = orifice area, in.^{2}
T_2 = 350°R (70°F)
P_1 = upstream pressure, psi

Then

$$W = 1 \text{ cfm} \times 0.074/60 = 0.00123 \text{ lb per sec}$$

$$A_2 = \frac{0.00123 \times \sqrt{530}}{0.533 \times 1 \times (14.696 - 0.387)} = 0.00371 \text{ in.}^2$$

$$D_2 = \sqrt{\frac{4A_2}{\pi}} = \sqrt{\frac{4 \times 0.00371}{\pi}} = \sqrt{0.00473} = \mathbf{0.069\ in.}$$

S-10. You have collected three samples of a given fume using a portable electrostatic precipitator which you have assumed was operating at 3 cfm. Each sample was collected for 30 min and on this basis the laboratory reported the following results:

No. 1: 0.085 mg/cu m
No. 2: 0.070 mg/cu m
No. 3: 0.040 mg/cu m

However, upon recalibrating the precipitator just after you returned from the field trip, you found it was operating at 2.75 cfm. What are the correct values of these three samples (1 cu ft = 0.0283 cu m)?

ANSWER.

$$\left.\begin{aligned} &\text{No. 1: } 0.085 \times 3/2.75 = 0.093 \text{ mg/cu m} \\ &\text{No. 2: } 0.070 \times 3/2.75 = 0.076 \text{ mg/cu m} \\ &\text{No. 3: } 0.040 \times 3/2.75 = 0.044 \text{ mg/cu m} \end{aligned}\right\} \textbf{Answer}$$

S-11. A glove box is being designed for a plutonium facility and is to be protected against fire by the injection of Halon 1301. The glove box volume is 25,000 cu ft at 80°F. The volume of Halon 1301 injected into the room through suitable nozzles at constant temperature is 1,800 cu ft for adequate protection. Assume a homogeneous mixture after injection, and also assume constant temperature due to expansion cooling of Halon 1301, although there would be a tendency toward a drop in pressure. Molecular weight of Halon 1301 is 150. Assume no effect of decomposition products: hydrogen fluoride, hydrogen bromide, bromine. Phosgene, generally feared as a byproduct of decomposed halogenated hydrocarbons, is not formed because no chlorine is present.

Determine the pressure buildup inside the glove box due to Halon 1301. Is this pressure excessive for the gloves to withstand? Assume the box is airtight.

ANSWER. The basic equation is $PV = WRT$. Both temperature t and volume V are constant. Further, P = psfa; V = volume, cu ft; W = weight, lb; R = gas constant for the mixture; and T = temperature, °R.

Weight of atmospheric air in box originally is found:

$$25{,}000/13.8 = 1{,}800 \text{ lb}$$

Weight of added Halon 1301:

$$379/150 \times (460 + 80)/520 = 2.6 \text{ cu ft per lb}$$
$$1{,}800/2.6 = 690 \text{ lb Halon 1301}$$

Total weight of gas mixture (air plus Halon 1301) in glove box after Halon injection is 1,800 + 690 = 2,490 lb.

Determine the gas mixture constant $R = 1544/M_m$. But first determine the average molecular weight of the gas mixture M_m as follows:

$$\text{Moles of air} = 1{,}800/29 = 63$$
$$\text{Moles Halon 1301} = 690/150 = 4.65$$
$$\text{Total moles} = 63 + 4.65 = 67.65$$
$$M_m = \text{total weight/total moles} = 2{,}490/67.65 = 36.8 = M_m$$
$$R_m = 1{,}544/36.8 = 42$$

Now proceed to determine the resulting pressure due to injection, using $P = WR_mT/V$.

$$P = \frac{2{,}490 \times 42 \times (460 + 80)}{25{,}000} = 2.250 \quad \text{psfa}$$

Expressed in psia, 2,250/144 = 15.6 psia or 15.6 − 14.7 = 0.9 psi. This expressed in inches water gage manometer = 2.31 × 12 × 0.9 = **25 in.** Since normal glove pressure resistance is from 2 to 4 in. water gage, the pressure buildup is excessive and could cause rupture and escape of radioactive materials. **Explore and develop more sturdy glove material.**

S-12. There is a release of neutron-activated air from a beam tube into a reactor area. It is considered that the primary hazard is due to ^{41}A. The argon concentration in the area at the time of the release is such that a worker would receive a tolerance dose in 5 min. If a ventilation system capable of six complete changes of air per hour is turned on, how long will it be until a worker can enter and work indefinitely without exceeding the tolerance dose due to that source?

ANSWER. First let us find the maximum allowable dose in 5 min. Assume N_0 = original number of atoms per cubic centimeters of ^{41}A in the enclosure to be ventilated.

$$\text{For } {}^{41}\text{A} \rightarrow t/2 = 109.2 \text{ min}$$
$$\lambda^{41}A = 0.693/(t/2) = 6.346 \times 10^{-3} \text{ per min}$$

The activity of ^{41}A = λN: $\rightarrow \lambda N(t) = \lambda N_0 e^{-\lambda t}$. Therefore, the total dose in 5 min

$$D_t = \int_0^5 \lambda N_0 e^{-\lambda t}\, dt = -N_0 e^{-\lambda t}\Big|_0^5$$
$$= N_0[1 - e^{-25}] = 0.0313N_0 \qquad \text{dosage per worker}$$

We can now approximate the decay of ^{41}A with an exponential function with decay constant λ_A. To find λ_A we need to know the time when one-half the amount of ^{41}A has been removed from the room by the ventilating system.

Using the standard assumption that 1/10 of ^{41}A is removed every minute, we can set up a time table as shown below.

Time, min	*Concentration of* ^{41}A
0	1
1	0.9
2	0.81
3	0.729
4	0.656
5	0.590
6	0.531
7	0.478
8	.
9	.
10	.

By interpolation we find that $t/2 = 6.5$ min. And therefore

$$\lambda_A = 0.1067 \text{ min}^{-1}$$

If a graph was to be prepared, the decay would most likely be exponential, according to our assumptions. To find the time when the ventilated area will again be safe, we must go back to the exponential and integrate.

$$\int_t^\alpha \lambda N_0 e^{-\lambda_A t} e^{-\lambda t}\,dt \le 0.0313 N_0$$

$$\int_t^\alpha \lambda N_0 e^{-(\lambda_A+\lambda)t}\,dt \le 0.0313 N_0$$

Solving,

$$\lambda N_0 \left[\frac{-e^{-(\lambda_A+\lambda)t}}{\lambda_A + t}\right]\Bigg|_t^\alpha = \lambda N_0 \frac{e^{-(\lambda_A+\lambda)t}}{\lambda_A+\lambda} = N_0[5.618 \times 10^{-2} e^{-0.113t}] \le 0.0303\, N_0$$

Therefore,

$$e^{-0.113t} \le 5.57 \times 10^{-1} \qquad \text{and} \qquad t = 5.17 \text{ min}$$

Now, adding a safety margin of ≮20 per cent, the safe time for reentry should be in approximately $t =$ **6.5 to 7 min.**

S-13. A steam line in the primary coolant loop is shielded by a 4-in. concrete wall in a nuclear power plant. The steam is radioactive

(N^{16} from O^{16} n.p.), with an emission rate of 2.5×10^9 6-Mev gamma rays per cm-sec. The piping is made of steel. Calculate the dose rate just outside 6-in.-diameter penetration. Assume no self-absorption in the steam or in the penetration. Refer to Fig. S-2.

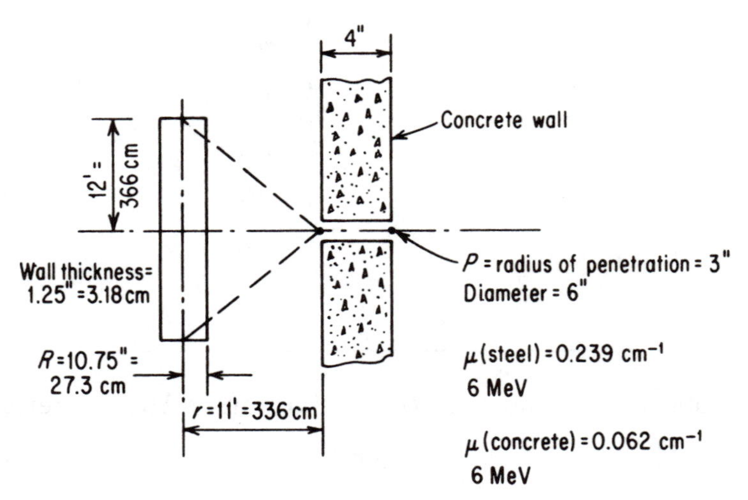

Fig. S-2

ANSWER. Assume no self-absorption in the steam or in the penetration. Find the dose rate outside penetration. The reaction of interest is

$$O^{16} + {}_1n^1 \rightarrow N^{16} + {}_1P^1$$

As stated, radioactive N^{16} emits at a rate of

$$2.5 \times 10^9 \text{ gammas per cm-sec}$$

with 6-Mev X-rays. Therefore, we have 1.5×10^{10} Mev per cm-sec with no self-absorption. The first step is to look at the dose rate on the outer edge of the steam pipe with a wall thickness of 3.18 cm (steel). For coefficients:

$$\mu \text{ for 6-Mev } \gamma \text{ in steel} \equiv 0.239 \text{ cm}^{-1}$$
$$\mu \text{ for 6-Mev } \gamma \text{ in concrete} \equiv 0.062 \text{ cm}^{-1}$$

Using the line source formula,

$$\Phi = \frac{\beta S_i}{4\pi a} - [F(\theta_2,b_1) - F(\theta_1,b_1)]$$

In this case $\theta_1 = \theta_2$. Therefore, $\Phi = \frac{\beta S_i}{4\pi a} F(\theta, b)$. Solving for the angle θ,

$$\tan\theta = 366/336 = 1.0892 \qquad \text{and} \qquad \theta = 47.4°$$

For buildings in steel,

$$A_1 = e^{-\alpha_1 b} + A_2 e^{-\alpha_2 b}$$
$$A_2 = 1 - A_1$$
$$A_1 = 3.2$$
$$A_2 = 2.2$$
$$\alpha_1 = 0.08$$
$$\alpha_2 = 0.058$$

Solving for μ_x,

$$\mu_x = (0.239)(3.18) = 0.76$$

Looking at Φ,

$$\Phi = \frac{15 \times 10^{10}}{2\pi(336)} F(47.4°, 0.76)[3.2e^{0.062} - 2.2e^{-0.0441}]$$
$$\Phi = (7.105 \times 10^6)\{0.3[3.2(1.062) - 2.2(0.96)]\}$$
$$\Phi = 3.214 \times 10^6 \text{ Mev per cm}^2\text{-sec}$$

Now looking at the duct, we have an incident

$$\Phi = 3.214 \times 10^6 \text{ Mev per cm}^2\text{-sec}$$

Assuming that absorption by the source does not contribute because of spherical distribution, we can consider the gamma flux to be distributed uniformly in every direction. We can also assume for simplicity that there are no fringe effects, no reflection, and no albedos (structural defects). We were given a cylindrical duct with uniform distribution and no attenuation within the duct. Because of this we can assume a spherical distribution incident at the duct exit. From Rockwell we know that only $1/2\pi$ of Φ is headed out the exit of the duct. An angle of escape is therefore going to be related by $2\pi \sin\theta\, d\theta$. We integrated from $\theta = 0$ to θ_1, when $\theta_1 = \arcsin\sqrt{\frac{r^2}{r^2 + L^2}}$. Assuming $r^2 \ll L^2$, we can find the total number of gammas escaping.

$$\Phi_T = \int_0^{\sin^{-1} r/L} \Phi_0 \times \frac{\pi r^2}{2\pi} \times \frac{2\pi \sin \theta \, d\theta L^2}{L^2}$$

$$\Phi_T = \Phi_0 \frac{\pi r^4}{2L^2} \text{ Mev per sec flux.}$$

Answer. Note that Φ_0 = Initial flux

Rockwell—*Shielding Design Manual,* T. Rockwell, Van Nostrand (1956).

S-14. An analytical procedure requires a minium of 5 mg of offending material in a total sample in order to obtain satisfactory analytical accuracy. The TLV (threshold limit value) of this material is 75 ppm. It is suspected that the air concentration is three times greater than the TLV. Molecular weight of the offending material is 138. The sampling rate for the collecting device is 0.1 ft³/min. The temperature is 22°C and the barometric pressure is 750 mm Hg. Determine the minimum time necessary to collect one air sample.

ANSWER.

$$t = \frac{(\text{minimum amount of material})(22.4 \times \text{L/g-mol})}{(\text{sampling rate})(\text{TLV})(\text{mol wt})} \times \frac{(10^6)(760/P)(273 + C/273)}{28.3 \text{ L/ft}^3}$$

$$t = \frac{(5 \text{ mg})(0.001 \text{ g/mg})(22.4 \text{ L/g-mol})}{(0.1 \text{ ft}^3)(75)(138 \text{ g/g-mol})} \times \frac{(10^6)(760/750)(295/273)}{28.3 \text{ L/ft}^3}$$

$= 4.19$ min **Ans.**

S-15. A room of 100,000 ft³ volume initially having a concentration of benzene equal to 1000 ppm is to have this concentration reduced to 25 ppm, using a 5 ppm benzene-air supply. The room has a ventilation system capable of delivering 50,000 ft³/min. Assume a 1/10 mixing factor.

Notation and Data. Perfect mixing is rarely accomplished; so a mixture factor should normally be applied to the number of room air changes. This mixing factor will depend upon the gas or vapor toxicity, the uniformity of contaminant distribution within the room, location of fans, construction of the room, and the room's population. This factor may vary from ⅓ to ⅒ and is used in conjunction with actual air changes to yield number of effective air changes. Thus,

$$n = \frac{Qt}{R}$$

where n = actual air changes (dimensionless)
Q = airflow rate, ft^3/min
t = time elapsed, min
R = space volume, ft^3
kn = number of effective air changes

Thus, if the mixing factor for a particular room was ⅓ and the room had 9 actual air changes, the room would have only $9 \times \frac{1}{3} = 3$ effective air changes. Then, if the efficiency of air mixing within the room is not known, the recommended procedure would be to use $k = \frac{1}{10}$ to be on the safe side.

Requirement. Determine the time necessary to dilute the room air from 1000 to 25 ppm benzene.

ANSWER.

$$C = C_s(1 - e^{-kn}) + C_i e^{-kn} \tag{S-1}$$

$$C - C_s = (C_i - C_s)e^{-kn} = \text{diluted concentration} \tag{S-2}$$

where C = concentration at time t
C_i = initial concentration at $t = 0$
e = naperian logarithm base, 2.7182818

Substituting values for the problem (situation) in Eq. (S-2)

$$25 - 5 = (1000 - 5)e^{-n/10}$$
$$20 = 995e^{-n/10}$$

from which $n/10 = 3.9$ effective air changes

$$t = \frac{nR}{Q} = \frac{kn/k \times R}{Q} = \frac{39 \times 100{,}000}{50{,}000} = 78 \text{ min} \qquad \textbf{Ans.}$$

S-16. The gamma radiation intensity 1 h after detonation of a nuclear weapon is equal to 3450 roentgens per hour. Determine the intensity (*a*) 8 h, (*b*) 16 h, (*c*) 24 h, (*d*) 10 days, (*e*) 30 days after detonation.

ANSWER. Use $d = d_1 t^{-1.2}$, where $d_1 = 3450$.

(*a*) $d_8 = 3450\,(8)^{-1.2} = 284$ r/h

(*b*) $d_{16} = 3450\,(16)^{-1.2} = 124$ r/h

(*c*) $d_{24} = 3450\,(24)^{-1.2} = 76.2$ r/h

(*d*) d_{10} days $= d_{240} = 3450\,(240)^{-1.2} = 4.8$ r/h

(*e*) d_{30} days $= d_{720} = 3450\,(720)^{-1.2} = 1.3$ r/h

S-17. A client plans to purchase a portable kerosene heater of the convection type for his family room. The heater is rated at 14,000 Btu/h based on the gross heating value of the fuel. Size of the room is 8 by 10 by 20 ft.

Data:

Specific gravity of the kerosene (60/60°F) = 0.825

Gross heating value of the kerosene = 137,000 Btu/gal

Fuel analysis, percent by weight:

Carbon	86.5
Hydrogen	13.2
Nitrogen	0.3
Sulfur	None

Room is airtight because of weatherstripping, caulking, and insulation, and doors will be shut so that no outside air will enter or leave the airtight room.

Requirement

(*a*) Determine the reduction in room oxygen content if the heater is run continuously for 4 h in the room.

(*b*) What advice would you offer the client to provide the safe use of the heater?

ANSWER.

(*a*) Density ρ of kerosene $= 0.825 \dfrac{1}{0.016035} = 51.45\ \text{lb/ft}^3$

$$\text{Gross heating value} = \frac{137{,}000}{(231/1728)(51.45)} = 19.919\ \text{Btu/lb}$$

$$\text{Kerosene burned per h} = \frac{14{,}000}{19.919} = 0.7 \text{ lb/h}$$

Basis: 100 lb kerosene and lb mol O_2 required is 86.5/12 + 13.2/2, or 13.8 lb mol

Gross room volume = $8 \times 10 \times 20 = 1600$ ft^3

Assume room pressure is atmospheric at 14.7 lb/in^2 abs and neglect presence of water vapor. Then since air by volume is 21 percent O_2 and 79 percent N_2, the mols of room air is given by

$$\frac{pV}{RT} = \frac{14.7 \times 144 \times 1600}{1545 \times 520} = 4.3 \text{ mol room air}$$

lb mol O_2 initially in room = $4.3 \times 0.21 = 0.885$

lb mol consumed in 1 h = $13.8/100 \times 0.407 = 0.06$ mol O_2

Consumption of O_2 in 4 h = $0.06 \times 4 = 0.24$ mol O_2

Reduction of room oxygen in 4 h = $0.24/0.885 \times 100 = 27$ percent **Ans.**

(*b*) Advise client that the use of a kerosene heater presents a danger to life and property if certain steps are not taken to prevent asphyxiation of the room's occupants. There are three principal dangers inherent in the use of kerosene heaters: (1) Kerosene heaters are portable and contain a flammable liquid. If tipped over they could cause a fire. Many new models have a tip-over flame extinguisher to eliminate this problem. Nevertheless, heaters should always be placed on a level surface, away from woodwork and other combustible materials. Safer yet, provide a leakproof sheet-metal pan of heavy-gauge soldered joints and turned-up edges around the full perimeter, at least 2 in high, to contain the liquid if it spills. (2) For safe filling of the heater, purchase a heater that has a removable fuel tank. Always extinguish the heater flame before filling, let cool, and fill outdoors. (3) Your kerosene heater can consume as much oxygen as 30 people in a room. While most modern kerosene heaters burn efficiently and produce little or no smoke or odor (except at startup and shutdown) with the proper fuel, they can suffocate.

For greatest safety, purchase heaters wtih oxygen-depletion systems. These heaters turn off automatically when the oxygen level in the room approaches minimum breathing levels. Never use kerosene heaters in unventilated places.

A good rule is to have 1 in^2 of outside air opening for each 1000 Btu/h heater rating. Use water-clear kerosene (classified as 1-K kerosene). This is the purest form of kerosene, and it is sometimes difficult to find. If impure, kerosene can cause bad problems—offensive odors, smoking or gummed-up wicks, and even carbon monoxide gas. And *never* use kerosene that has been adulterated with gasoline or any other highly volatile combustible liquid.

Keep the heater clear of all furniture, drapes, children, and pets. The temperature can reach 500°F. As for local restrictions, check your municipality's building codes on the legality of using kerosene heaters in your home. Very few models have been approved by the Underwriters Laboratories (UL); so look for this mark of safety carefully.

Before purchasing a kerosene heater, review your need. Be certain your intent is to heat a small, well-ventilated area or room, and make sure the model you are considering has the features mentioned above. Determine if a $150 to $300 investment is a sound one—you could save more money by simply investing in energy-efficient improvements such as caulking or weatherstripping.

S-18. A survival shelter is designated for 1000 occupants such that 10,000 ft^2 of floor area is available with an average ceiling height of 8 ft. Normal outside air has a carbon dioxide concentration of 0.04 percent by volume before entering the ventilation system.

(*a*) Determine the time amount the shelter is to be buttoned up without ventilation from the outside before the carbon dioxide concentration reaches 2 percent by volume.

(*b*) If forced mechanical ventilation is initiated after the carbon dioxide concentration reaches 2 percent by volume, how long will it take to drop the concentration to 0.5 percent by volume if the outside air intake rate to the shelter is 3000 ft^3/min?

(*c*) If shelter specifications require that the maximum permissible carbon dioxide concentration is 4 percent by volume and that a time cycle of 3 h for button up and blower on condition is necessary, determine the time the blower must be on during this 3-h cycle if the ventilation rate is 3000 ft^3/min.

(d) At what concentration of carbon dioxide must the blower be turned on in order to adhere to the stated specifications?

ANSWER.

(a) By simple inspection the cubic feet per person = 80. And the button up time before carbon dioxide concentration reaches the requirement of 2 percent by volume is

$$v_t = v_0 + \frac{NPt}{V}$$

$$0.02 = 0.0004 + \frac{1000 \times 0.75t}{80{,}000}$$

from which t is found to be equal to 2.09 h, say, 2 h button up time **Ans.**

(b) Outside air intake = 3000 ft^3/min, or 3 ft^3/min per person

Time to drop carbon dioxide concentration from 2 to 0.5 percent by volume is as follows with outside air at 0.04 percent:

$$Q = \frac{3000 \times 60}{1000} = 180 \text{ ft}^3\text{/h per person}$$

$$V = \frac{80{,}000}{1000} = 80 \text{ ft}^3 \text{ per person}$$

$$t_v = \frac{\overline{V}}{NQ} \ln \frac{V_{\max} - P/Q}{V_{\min} - P/Q}$$

$$= \frac{80}{180} \ln \frac{0.0200 - 0.75/180}{0.005 - 0.75/180} = 0.444 \ln 19.3$$

$$= 0.444 \times 2.960 = 1.31 \text{ h, say 80 min} \qquad \text{Ans.}$$

(c) Cycle of button up time and ventilation time = 3 h

$$\frac{V}{N} = 80 \qquad v_{\max} = 4 \text{ percent} \qquad Q = 180 \qquad v_{\min} = 0.5 \text{ percent}$$

$$t_v = \text{blower on time} = \frac{80}{180} \ln \frac{0.04 - 0.00418}{0.005 - 0.00418}$$

$$= 0.444 \ln \frac{0.03582}{0.00082} = 0.444 \ln 43.7$$

$$= 0.444 \times 2.303 \times 1.640481 = 1.67 \text{ h} = \text{blower on time}$$

Blower off = 3 − 1.67 = 1.33 h **Ans.**

(d) Carbon dioxide concentration needed to turn blower on

$$t_b = \frac{V}{N}\frac{v_{max} - v_{min}}{P} = 1.33 = 80\,\frac{0.04 - v_{min}}{0.75}$$

$$= 1.33 \times 0.75 = 80\,(0.04 - v_{min}) = 80\,\Delta v$$

$$= \frac{0.99}{80} = \Delta v = 0.01$$

$$= 0.04 - v_{min} = 0.01$$

$$= 0.04 - 0.01 = v_{min} = 0.03\text{, or 3 percent carbon dioxide by volume, blower to be turned on} \qquad \textbf{Ans.}$$

S-19. Ventilation, forced or natural, is necessary for certain parts of a building in winter as well as in summer. Buildup of high humidities indoors in cold weather should be avoided by judicious application of exhaust ventilation.

A building has a volume of 30,000 ft^3 and an air infiltration rate of two air changes per hour. Average inside temperature is 70°F at 30 percent relative humidity with outdoors at 0°F. The building is then protected with storm sash, insulation, and vapor barrier and maintained at 45 percent relative humidity.

(*a*) Determine the moisture-removal potential.

(*b*) Under the protected building condition, what will be the necessary exhaust ventilation rate?

(c) What are the infiltration heat losses for both conditions to dispel the water removal under the original conditions?

ANSWER. Using the graph in Fig. S-3,

(*a*) From the graph the air needed to remove 1 lb of water vapor under the original conditions (unprotected) is 3000 ft^3. Then the water-removal rate is

$$\frac{30{,}000 \times 2}{3000} = 20\text{ lb/h moisture-removal potential} \qquad \textbf{Ans.}$$

(*b*) After building is protected and is maintained at the new conditions, the new ventilation rate to remove the same amount of water vapor is

$$2000 \times 20 = 40{,}000\text{ ft}^3\text{/h} \qquad \textbf{Ans.}$$

Note. 2000 is obtained from the graph at 45 percent relative humidity and 70°F room temperature. Thus, the ventilation rate

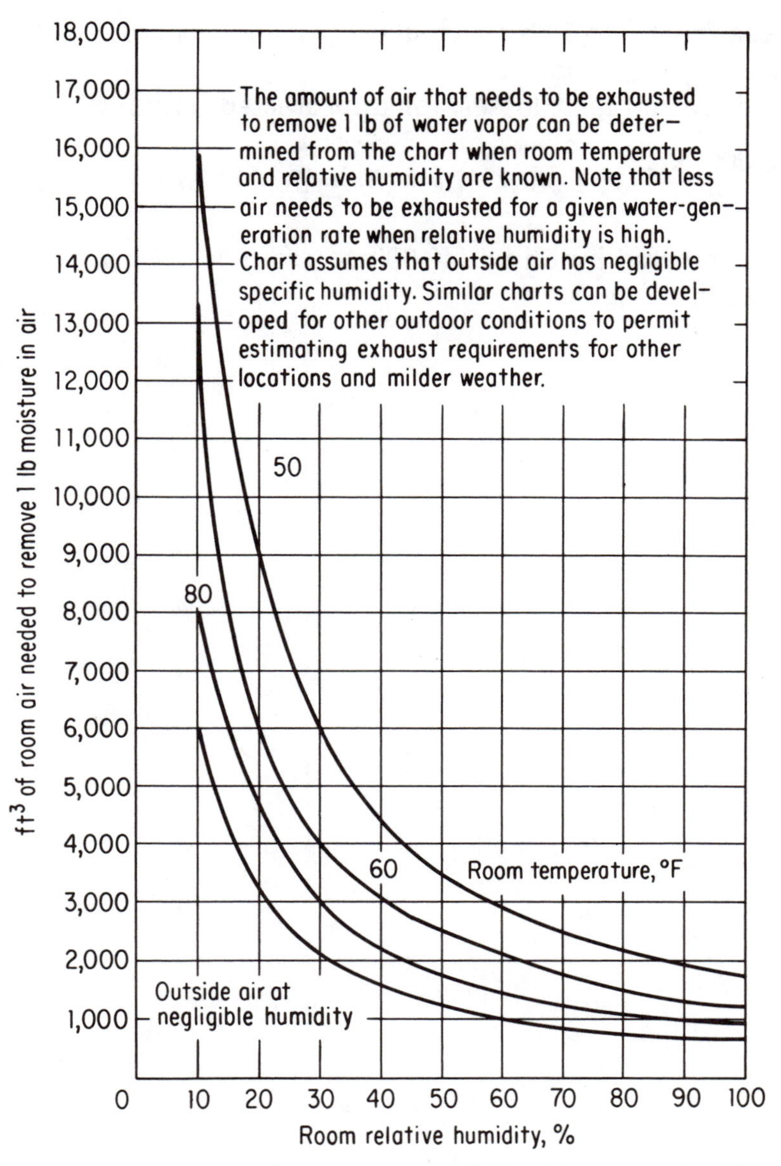

Fig. S-3. Volume of air needed to remove 1 lb of water vapor. [*Reprinted by special permission from Chemical Engineering (May 19, 1980). Copyright 1980 by McGraw-Hill, Inc., New York, N.Y. 10020.*]

is reduced from 2 to 1.33 air changes per hour by (40,000/30,000) = 1.33 air changes per hour.

(*c*) Infiltration heat losses: for unprotected building

$$\frac{30{,}000 \times 2}{60} \times 1.08 \times (70 - 0) = 75{,}600 \text{ Btu/h} \qquad \textbf{Ans.}$$

For the protected building: $(40{,}000 \times 1/60) \times 1.08 \times (70 - 0) =$ 50,500 Btu/h **Ans.**

Supplement T

PRODUCTION ENGINEERING

Questions and Answers

The NCEE *Task Analysis Survey* surfaced areas of professional practice which are now being incorporated into the fabric of the examination for mechanical engineering. This area of the examination, called management or production engineering, includes problem situations that represent professional situations that a mechanical engineer may encounter relating to any of the following: tool and jig design, maintenance management, manufacturing, materials handling, plant design, fire protection, and metal fabrication. This supplement provides a sampling of this area.

T-1. A cutting tool is designed for a nominal 2-in-diameter by 3-in-long round hole. The pitch of the teeth is 0.700 in, and each roughing tooth is 0.004 in larger in diameter than the one that cuts just before it. The tool is to remove a maximum of 0.15 in from the ID of the hole.

(*a*) Determine the number of teeth the cutting tool should have.

(*b*) The cutting tool is placed into service on a hydraulic cutting tool machine with a cylinder pressure of 1000 lb/in^2 when pulling at its rated capacity of 30 tons. The force to move the ram under no load is negligible. A gauge shows a pressure of 700 lb/in^2 in the cylinder when the tool is cutting fully as intended. Determine the specific cutting pressure of this tool in the material being cut.

(*c*) The cutting-tool material has a working strength of 60,000 lb/in^2. Determine the minimum diameter of the space between the teeth of the tool.

ANSWER.

(a) For roughing: 0.15/0.004 = 37.5, say 38 teeth. Add 10 to 15 teeth for finishing. **Ans.**

(b) Cutting force = F_c = (700/1000)(30)(2000) = 42,000 lb
Number of teeth in contact = N_c = (3/0.700) + 1 = 5
Area per tooth = $A_t = 2\pi \times 0.004 = 0.025$ in^2
Specific cutting pressure = $P_B = F_c/(A_t \times N_c)$ = 42,000/(0.025
$\times$ 5) = 336,000 lb/in^2 **Ans.**

(c) Let d = diameter

$$\frac{\pi d^2}{4} = \frac{42{,}000}{60{,}000}$$

Rearranging,

$$d = \sqrt{\frac{42{,}000 \times 4}{60{,}000 \times \pi}} = \sqrt{0.89} = 0.94 \text{ in} \qquad \textbf{Ans.}$$

T-2. A manufacturing process requires that a cast-iron flywheel be stopped periodically. The flywheel OD is 60 in and the rim is 6 in thick by 14 in wide. The rim is connected to the hub by arms, and it may be assumed that the hub and arms contribute one-tenth as much as the rim to the flywheel capacity. For each operation, the flywheel shaft is brought up to a speed of 240 r/min, and it must be brought to rest in a maximum of 20 s. It is suggested that a brake be applied to the rim of the flywheel in order to stop it. Assume density of cast iron is 0.26 lb/in^3. Determine:

(a) Heat generated in stopping the flywheel.

(b) What operating conditions should be considered other than those mentioned in order to select a brake lining material?

(c) What characteristics of the brake lining material might be considered in the selection?

ANSWER.

(a) Kinetic energy = $I_j W_t^2/2$, where $W = 2\pi$ r/min and I_j = $(\pi\rho a/2)(r_o^4 - f_i^4)$.
r_o = 2.5 ft and r_i = 2.0 ft, a = 14/12 = 1.17 ft, and
$\rho = 0.26 \times 1728/g = 449.3/g$
$I_j = (\pi/2)(449.3/32.2)(2.5^4 - 2^4)$ = 591 slug-ft^2

$W = (240/60)(2\pi) = 25.13$ rad/s

$I_{jt} = I_j + 0.1I_j = 591 + 59.1 = 650.1$

Then kinetic energy $= 650.1 \times 25.13^2/2 = 205{,}269$ ft-lb to stop the flywheel.

$$\text{Btu} = \frac{205{,}269}{778} = \mathbf{264\ Btu}$$

(*b*) Because of the heat generated, the frequency of stops should be known. The heat-rate average during the stopping process is 264/20 = 13.2 Btu/s. The conditions of the surroundings should be known. For example, is oil or water in the surrounding atmosphere? Would water cooling be necessary?

(*c*) Material characteristics: (1) The friction coefficient between the material and cast iron should be suitable. (2) The material should perform at high temperature. (3) It should have a high heat-transfer coefficient. (4) It should be wear-resistant. (5) It should be reasonable in cost.

T-3. A 42-in belt conveyor for transporting crushed stone is to carry 400 tons/h. The belt travels 1000 ft along the horizontal and then rises 20 ft in the next 1000 ft.

(*a*) State appropriate assumptions and determine what size motor should be installed.

(*b*) Apart from the foregoing, what would be your recommendations as to the maximum angle for raising mine coal on a belt conveyor?

ANSWER.

(*a*) There are a number of considerations which enter into the horsepower required to drive a belt conveyor:

Friction of the idler pulleys in revolving under the weight of the belt and the material carried.

Bending of the belt over the idlers and end pulleys.

Getting the material up to speed at the loading point.

The lifting and slight disturbances of the load in passing over the idlers.

Friction of those shafts which are driven through the conveyor belt such as foot shafts and snub shafts.

The last four items listed above are, as a rule, relatively small in amount, and it is difficult to determine them. The usual calculations for horsepower assume a value great enough for the first item to include the last four. Then,

$$\text{hp} = \frac{\text{speed, ft/min}}{33{,}000} \times \text{friction force due to first item}$$

This may be reduced to

$$\text{hp} = \frac{CTL}{990}$$

where T = tons/h
L = length of horizontal run, ft
C = constant which depends on the width of the belt, idlers, snubbers, etc. This constant is generally furnished by the conveyor manufacturer.

For lift of load:

$$\text{hp} = \frac{T \times 2000 \times H}{60 \times 33{,}000} = \frac{TH}{990}$$

where H is lift in ft = 20. Combining both equations for total hp

$$\text{Total hp} = \frac{CTL}{990} + \frac{TH}{990}$$

For a 42-in belt one manufacturer gives X = 0.056. Then

$$\text{Total hp} = \frac{(0.056 \times 400 \times 2000) + (400 \times 20)}{1000}$$

$$= 52.8 \text{ hp} \qquad \textbf{Ans.}$$

(*b*) The limit of incline is reached when the material tends to slip on the belt surface. This depends on the coefficient of friction between the material and the belt. For dry coal the maximum angle is between 18 and 20°.

T-4. In cases where dry-bulk materials are being handled in an airstream or where they can be introduced into a vessel in an airstream, consideration should be given to the use of a venturi feeder.

This is a simple device; it is low in cost, light in weight, and can be fabricated by a sheet-metal shop. Design a venturi feeder to handle 600 lb/h of a light dry powder for delivery into a system having 2 in WG static pressure. Neglect temperature effects and any friction loss downstream from the venturi. Use 50 ft^3/min of air per lb of dry material and a discharge velocity of 3000 ft/min at point 3 (see Fig. T-1).

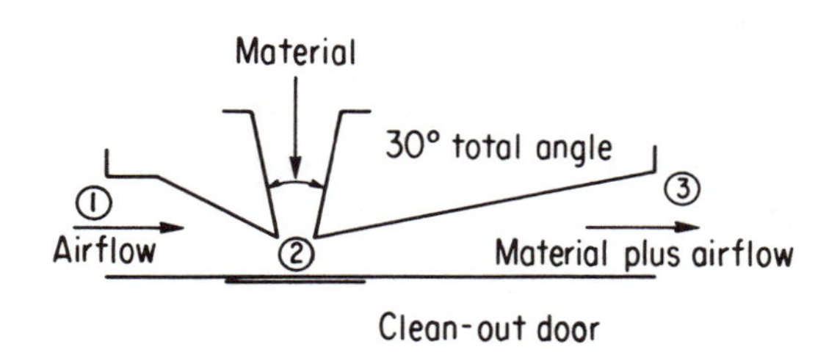

Fig. T-1

(*a*) Determine the air quantities required.
(*b*) Determine the air horsepower required.
(*c*) Determine the size of openings at points 1, 2, and 3.

ANSWER. See Fig. T-1.

Let A = opening area, in^2
Q = air quantity flowing, ft^3/min
V = conveying velocity, ft/min
h_t = total pressure, in WG
h_s = static pressure, in WG
K_1 = loss coefficient = 0.2
K_2 = convergence loss factor for 30° angle entry = 0.075
Points 1, 2, 3 refer to Fig. T-1.

Start design at point 3.

(*a*) $Q = (600/60) \times 50 = 500$ ft^3/min conveying air needed **Ans.**

$A_3 = Q_3/V_3 = 500/3000 = 0.1667$ ft^2 or 24 in^2. Use a 5-in-square opening.

At point 3 the total pressure h_t is

$$h_t = h_{s3} + h_{v3} = 2 + (3000/4005)^2 = 2 + 0.56 = 2.56 \text{ in WG}$$

At point 2

$$h_{v2} = h_{v3} + \frac{h_{s3} - h_{s2}}{1 - K_1} = \left(\frac{3000}{4005}\right)^2 + \frac{2 - 0.05}{1 - 0.2} = 3.12 \text{ in WG}$$

and $V_2 = \sqrt{3.12} \times 4005 = 7075$ ft/min (say 7100) at point 2. Thus $A_2 = (500/7100) \times 144 = 10.14 \text{ in}^2$.

Since one side was established as 5 in, the throat height at point 2 is $10.14/5 = 2.028$ in, say, 2 in throat height.

At point 1 (discharge)

$$h_{t1} = h_{v1} + h_{s2} + (1 + K_2) \times (h_{v2} - h_{v1})$$
$$h_{t1} = (3000/4005)^2 + (-0.05) + (1 + 0.075)\,[3.12 - (3000/4005)^2]$$
$$h_{t1} = 3.27 \text{ in WG}$$

(*b*) Air hp required

$$\frac{62.3 \times \text{ft}^3/\text{min} \times h_{t1}}{12 \times 33{,}000} = \frac{62.3 \times 500 \times 3.27}{12 \times 33{,}000} = 0.26 \qquad \textbf{Ans.}$$

Fan brake hp, assuming fan efficiency of 70 percent, is $0.26/0.7 = 0.37$ bhp.

(*c*) Throat-hole width is made as wide as the venturi body, or 5 in. And its length is made equal to throat height, or 2 in.

T-5. An 8-in-diameter steel shaft is to have a press fit in a 20-in-diameter steel disk. The maximum tangential stress in the disk is not to exceed 10,000 lb/in^2.

(*a*) Determine the required diametral interference of metal.

(*b*) If the disk is 10 in thick in the axial direction, determine the force required to press the parts together.

(*c*) Determine the torque which the joint will transmit because of the force-fit pressure.

ANSWER.

(*a*) Refer to Fig. T-2. $d/D = 8/20 = 0.4$; max $S_{it} = 10{,}000$ lb/in^2.

From J. E. Shigley, *Mechanical Engineering Design,* 3d ed., McGraw-Hill, New York, 1976 (p. 564):

$$S_{it} = p\,\frac{R^2 + r^2}{R^2 - r^2}$$

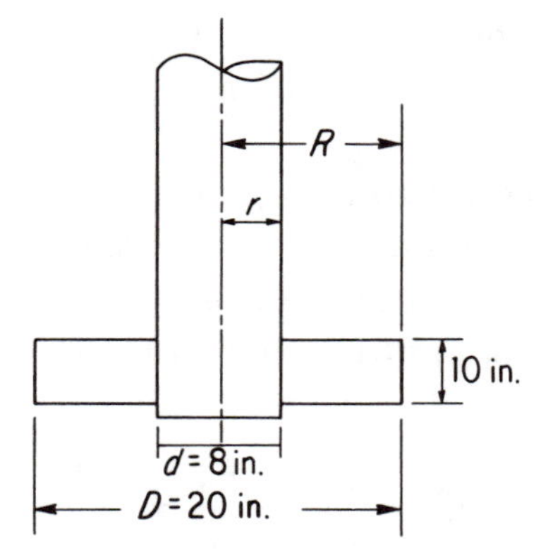

Fig. T-2

$$10{,}000 = p\frac{10^2 + 4^2}{10^2 - 4^2} = 1.38p$$

from which $p = 7246 \text{ lb/in}^2$.

From L. S. Marks, *Mechanical Engineers' Handbook*, 6th ed., McGraw-Hill, New York, 1958 (pp. 8–59): at $d/D = 0.4$ and $p = 7264$ lb/in^2 the diametral interference = 0.00057 in/in of shaft diameter. Then

$$\text{Total diametral interference} = 8 \times 0.00057 = \mathbf{0.00456\ in}$$

(b) $F = \pi fpdl$. Assume $f = 0.12$. Thus, given $p = 7{,}246 \text{ lb/in}^2$; $1 = 10$ in; $d = 8$ in

$$F = \pi(0.12)(7246)(8)(10) = \mathbf{218{,}553\ lb}$$

(c) Torque $= F \times r = 218{,}552 \times 4 =$ **874,210 in-lb.**

T-6. The parameters for the design of a new simple-cycle steam power plant have been established as follows: maximum steam pressure, 1200 lb/in^2 abs; maximum steam temperature, 800°F; maximum moisture in exhaust steam, 15 percent; minimum condenser pressure, 1.5 in Hg abs; and turbine internal (engine) efficiency, 85 percent.

(a) Determine the optimum turbine inlet steam pressure and temperature under these conditions. Indicate how these were obtained.

(*b*) Estimate the overall efficiency and plant heat rate from fuel to switchboard under these conditions.

ANSWER.

(*a*) For the given conditions: P_g = 1200 lb/in^2 abs; t_g = 800°F, maximum moisture at exhaust = 15 percent, exhaust pressure = 1.5 in Hg abs, the corresponding enthalpy = 945 Btu/lb, isentropic efficiency = 85 percent, find the optimum inlet pressure and temperature for the turbine

$$0.85 = \frac{h_1 - h_2}{h_1 - h_{2i}} = \frac{\Delta h_a}{\Delta h_i}$$

Optimum inlet conditions will include the 800°F temperature but with some pressure that will yield the maximum enthalpy drop to point 2 on the H-S diagram, Fig. T-3.

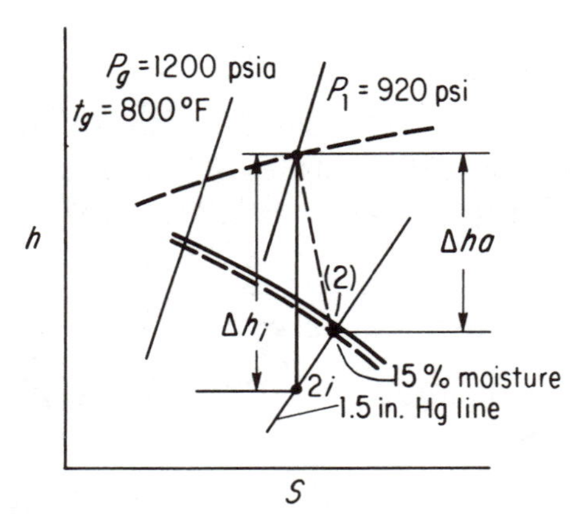

Fig. T-3

By trial and error on the Mollier diagram (H-S), it is possible to find a value for Δh_i such that 85 percent of the value will yield the desired Δh_a. For maximum results, the actual expansion process will end at P = 1.5 in Hg and 15 percent moisture. The initial pressure is found to be P_1 = 920 lb/in^2 abs. Finally,

$$P_1 = \mathbf{920\ lb/in^2\ abs} \qquad t_1 = \mathbf{800°F}$$

(*b*) From Fig. T-3: h_1 = 1394 Btu/lb; h_2 = 945 Btu/lb.

Then: $W = h_1 - h_2 = 1394 - 945 = 449$ Btu/lb
$Q_s = h_1 - h_f = 1394 - 60 = 1334$ Btu/lb
$n_{tc} = W/Q_s = 449/1334 =$ **34 percent**

Assume generator efficiency = 85 percent and electromechanical efficiency = 90 percent. Then: $n_p = 34 \times 0.85 \times 0.9 = 26$ percent and heat rate = 3413/0.26 = 13,127.

Heat rate = **13,127 Btu/kWh**

T-7. When iron- and copper-base alloys are centrifugally cast, an accelerating force of from $40g$ to $60g$ imposed on the material during solidification makes the castings dense without damage to the mold. If the molds in an operation are placed in a 16-in diameter, at what speed should the centrifuge be rotated to impose an average force of $50g$ lb/lb on each casting?

ANSWER. The acceleration on the material is

$a = F/m = v^2/r$ tangential velocity
$v = \pi D_n$ in/min $= \pi D_n/60 \times 12$ ft/s and
$r = D/2.\{$Thus, $a = 2\pi^2 D_n^2/3600 \times 144/12$ ft/s^2
$A = a/g = 2\pi^2 D_n^2/3600 \times 12 \times 32.2 = 1.42 \times 10^{-5} D_n^2$

In this case

$50g/g = 1.42 \times 10^{-5} \times 16 \times n^2$
$n = \sqrt{50 \times 10^4/1.42 \times 1.6} = 468$ r/min **Ans.**

T-8. (*a*) What is the maximum blanking force for an aluminum part if the length of the cut is 30 in, the metal is 0.125 in thick, and the yield strength is 2.5 tons/in^2?

(*b*) How much force is required to draw a 12-in-diameter 0.25-in-thick stainless-steel shell if the yield strength is 15 tons/in^2?

(*c*) What force is required to neck a 0.125-in-thick aluminum shell from a 3- to a 2-in diameter if the necking angle is 30° and the ultimate compressive strength of the material is 14 tons/in^2?

ANSWER.

(*a*) The maximum blanking force for any metal is given by $F = Lts$, where F = blanking force, tons; L = length of cut, in (= circumference of part, in); t = metal thickness, in; s = yield strength of metal, tons/in². Thus, $F = (30)(0.125)(2.5) = 0.375$ ton. **Ans.**

(*b*) Using the same equations as in (*a*) and substituting the drawing edge length or perimeter (circumference of part) for L, $F = (12\pi)(0.25)(15) = 141.5$ tons. **Ans.**

(*c*) The force required to neck a shell is $F = ts_c(d_1 - d_s)/\cos$ necking angle, where F = necking force, tons; t = shell thickness, in; s_c = ultimate compressive strength of the material, tons/in²; d_1 = large diameter of shell, i.e., the diameter *after* necking, in. Thus, for this shell, $F = (0.125)(14)(3.0 - 2.0)/\cos 30° = 2.02$ tons. **Ans.**

Table T-1 presents typical yield strengths of various metals which are blanked or drawn in metalworking operations.

T-9. A 48-in-diameter flywheel of steel having a 12-in wide × 10-in-deep rim rotates at 200 r/min. Determine how long a cut can be stamped in a 1-in-thick aluminum plate if the stamping energy is obtained from this flywheel. The ultimate shearing strength of the aluminum is 40,000 lb/in².

ANSWER. In routine design calculations, the weight of a spoked or disk flywheel is assumed to be concentrated in the rim of the flywheel. The weight of the spokes or disk may be neglected. When computing the energy (kinetic) of the flywheel, the weight of a rectangular, square, or circular rim is assumed to be concentrated at the horizontal centerline. Thus, for the rectangular rim, the

TABLE T-1

Metal	Yield strength, tons/in²
Aluminum, 2S annealed	2.5
Aluminum, 24S heat-treated	23.0
Low brass, ¼ hard	24.5
Yellow brass, annealed	10.0
Cold-rolled steel, ¼ hard	16.0
Stainless steel, 18-8	15.0

weight is concentrated at a radius of (48/2 − 10/2) = 19 in from the centerline of the shaft to which the flywheel is attached.

Then, kinetic energy $K = Wv^2/2g$, where K = kinetic energy of the rotating shaft, ft-lb; W = weight of flywheel rim, lb; v = velocity of the flywheel at the horizontal centerline of the rim, ft/s. The velocity of a rotating rim = $v = 2\pi RD/60$, where R is rotational speed, r/min; D = distance of the rim horizontal centerline from the center of rotation, ft. For this flywheel $v = 2\pi(200)(10/12)/60$ = 33.2 ft/s.

Rim flywheel has a volume of (rim height, in)(rim width, in)(rim circumference measured at a horizontal centerline, in), or $(10)(12)(2\pi)(19) = 14{,}350$ in^3. Now, since the machine steel weighs 0.28 lb/in^3, the weight of the flywheel rim is (14,350)(0.28) = 4010 lb. Finally, $K = (4010)(33.2)^2/(2 \times 32.2) = 68{,}700$ ft-lb.

Dimensions of the hole that can be stamped: A stamping process is a shearing process. The area sheared is the product of the plate thickness and the length of the cut. Each square inch of the sheared area offers a resistance equal to the ultimate shearing strength of the material punched. During the stamping process, the force exerted by the stamp varies from a maximum of F lb at the point of contact to 0 lb when the stamp emerges from the metal. Thus, the average force during stamping is $(F + 0)/2 = F/2$. The work done is the product of $F/2$ and the distance through which this force moves, or the plate thickness t in. Therefore, the maximum length that can be stamped is that which occurs when the full kinetic energy of the flywheel is converted to stamping work.

With a 1-in-thick aluminum plate, the work done W ft-lb = (force, lb) (distance, ft). The work done when all the flywheel energy is used is $W = K$. Thus, $W = K = 68{,}700 = (F/2)(1/12)$ and $F = 1{,}650{,}000$ lb. **Ans.**

T-10. A cut in a steel plate 1 in thick is to be made by hand and by machine. Length of cut is to be 96 in. (*a*) How long will it take by hand? (*b*) How long will it take by machine? (*c*) What is the acetylene and oxygen consumption when hand-cut? (*d*) What is the acetylene and oxygen consumption when cut by machine? Data:

Metal thickness, in	Speed, in/min		Gas consumption, ft^3/h	
	Manual	Machine	Oxygen	Acetylene
¼	16–18	20–26	50–90	8–11
½	12–15	17–22	90–125	10–13
1	8–12	14–18	130–200	13–16
2	5–7	10–13	200–300	16–20
4	4–5	7–9	300–400	21–26
6	3–4	5–7	400–500	26–32
8	3–6	4–6	500–650	28–35
10	2–3	3–4	700–1000	30–38
12	2.5–3.5	3–4	720–880	42–52

ANSWER.

(*a*) For any flame cutting, the cutting time T min is L/C, where L = length of cut, in; C is cutting speed, in/min.

Manual cutting = 96/8 = 12 min.
Use lower manual speed from data.

(*b*) Machine cutting $T = 96/14 = 6.9$ min. Use lower machine speed from data.

(*c*) From data, oxygen consumption is 130 to 200 ft^3/h. Actual consumption = (12 min/60) × 130 = 26 ft^3 at the minimum cutting speed and minimum oxygen consumption.

(*d*) From data, acetylene consumption is 13 to 16 ft^3/h at minimum cutting and (12/60) × 13 = 2.6 ft^3 at lower speed. Use same procedure for higher cutting speeds.

T-11. (*a*) What is the machine unit time to work 25 parts if the setup time is 75 min and the unit standard time is 5.0 min?

(*b*) If one machine tool has a setup standard time of 9 min and a unit standard time of 5.0 min, how many pieces must be handled if a machine with a setup standard of 60 min and a unit standard time of 2.0 min is to be more economical?

(*c*) Determine the minimum lot size for an operation requiring 3 h to set up if the unit standard time is 2.0 min and the maximum increase in the unit standard may not exceed 15 percent.

(*d*) Find the unit time to change a lathe cutting tool if the operator takes 5 min to change the tool and the tool cuts 1.0 min per cycle and has a life of 3 h.

ANSWER.

(*a*) The true unit time for a machine $T_u = (S_u/N) + U_s$, where S_u = setup time, min; N = number of pieces in lot; U_s = unit standard time, min. Thus, for this machine, $T_u = (75/75) + 5.0 = 6.0$ min. **Ans.**

(*b*) Call one machine *X*, the other *Y*. Then, (unit standard time of *X*, min)(number of pieces) + (setup time of *X*, min) = (unit standard time of *Y*, min)(number of pieces) + (setup time of *Y*, min). For these two machines, since the number of pieces *Z* is unknown, $5.0Z + 9 = 2.0Z + 60$. Solving, $Z = 17$ pieces. Thus, machine *Y* will be more economical when 17 or more pieces are made. **Ans.**

(*c*) The minimum lot size $M = S_u/U_sK$, where K = allowable increase in unit standard time, percent. Thus, for this run, $M = (3 \times 60)/(2.0)(0.15) = 600$ pieces. **Ans.**

(*d*) The unit tool-changing time U_t to change from dull to sharp tools is $U_t = T_cC_t/l$, where T_c = total time to change tool, min; C_t = time tool is in use during cutting cycle, min; l = life of tool, min. Thus, for this lathe, $U_t = (5)(1)/(3)(60) = 0.0278$ min. **Ans.**

T-12. A trapezoidal piece shown in Fig. T-4 is to be cut from 2-in-wide strip stock, and the 1-in-diameter hole is to be pierced, all in a progressive die. Two pieces are produced in each stroke. The material is 0.15C steel with a shear strength of 50,000 lb/in^2 and a 40 percent penetration to failure. A 12-ton press is available with a 175-lb flywheel at a radius of gyration of 9 in, mounted on the crankshaft. (*a*) Determine how the punches may be arranged to

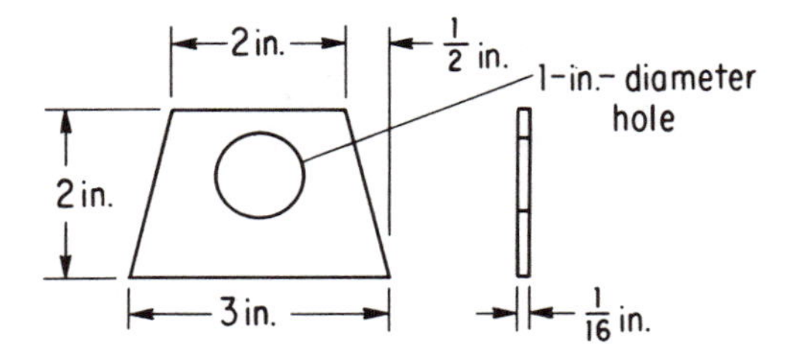

Fig. T-4

strike the work. (*b*) If the flywheel is permitted to slow down 10 percent per stroke for continuous action, at what speed may the press operate?

ANSWER. Two pieces cut at once. Length of one angular side = $(4.25)^{1/2}$ = 2.062 in

Force to cut one side = 2.062 × 1/16 × 50,000 = 6450 lb

Force to pierce one hole = π × 1/16 × 50,000 = 9840 lb

Minimum force without shear = 12,900 lb = 6.45 tons

Energy = 32,580 × 0.40 × 1/16 × 1.16 = 945 in-lb per stroke = 78.7 ft-lb per stroke

If cuts are staggered, minimum force is that for piercing = 4.9 tons, say 5 tons.

(*a*) The maximum force required for two pieces = 16.29 tons. Punches must be staggered. A feasible arrangement is for the two piercing punches to strike with a force of 9.84 tons and the cutoff punches together with a force of 6.45 tons.

(*b*) For 78.7 ft-lb per stroke, the velocity of the flywheel at the radius of gyration should be

$$V = \sqrt{\frac{64.4 \times 78.7}{0.19 \times 175}} = 12.35 \text{ ft/s}$$

$$N = \frac{12.35 \times 60}{1.5\pi} = 157 \text{ strokes/min}$$

This minimum speed is 157 strokes per minute, and speed may be any larger amount up to the capacity of the press in speed and power.

SUGGESTED STUDY REFERENCE LIST

Mechanics and Machine Design

Church, A. H.: *Mechanical Vibrations*, 2d ed., Wiley, New York, 1963.

Creamer, R. H.: *Machine Design*, 2d ed., Addison-Wesley, Reading, Mass., 1976.

Dudley, D. W.: *Gear Handbook*, McGraw-Hill, New York, 1962.

Faires, V. M.: *Design of Machine Elements*, 4th ed., Macmillan, New York, 1965.

Haberman: *Vibration Analysis*, Merrill, Chicago, 1968.

Higdon, A., et al.: *Mechanics of Materials*, 3d ed., Wiley, 1978.

Hinkle, R. T.: *Kinematics of Machines*, 2d ed., Prentice-Hall, Englewood Cliffs, N.J., 1960.

Ford, H., and J. M. Alexander: *Advanced Mechanics of Materials*, Wiley, New York, 1977.

Ohers, E., et al.: *Machinery Handbook*, 20th ed., Industrial Press, New York, 1973.

Olsen, G. A.: *Elements of Mechanics of Materials*, Prentice-Hall, Englewood Cliffs, N.J., 1966.

Singer, F. L.: *Engineering Mechanics*, 3d ed., Harper and Row, New York, 1975.

Spotts, N. F.: *Design of Machine Elements*, 4th ed., Prentice-Hall, Englewood Cliffs, N.J., 1961.

Timoshenko, S.: *Vibration Problems in Engineering*, Wiley, New York, 1974.

Willems, N.: *Strength of Materials*, McGraw-Hill, New York, 1981.

Fluid Mechanics and Hydraulics

Binder, R. C.: *Fluid Mechanics*, 5th ed., Prentice-Hall, Englewood Cliffs, N.J., 1973.

Brater, E. F., and H. W. King: *Handbook of Hydraulics*, 6th ed., McGraw-Hill, New York, 1976.

Centrifugal Pump Application Manual, Buffalo Forge Co., 1959.

Daugherty, R. L., and J. B. Franzinia: *Fluid Mechanics with Engineering Applications*, 7th ed., McGraw-Hill, New York, 1977.

Davis, C. V., and K. E. Sorenson: *Handbook of Applied Hydraulics*, 3d ed., McGraw-Hill, New York, 1969.

Hicks, T. G., and T. Edwards: *Pump Application Engineering*, McGraw-Hill, New York, 1970.

Thompson, P. A.: *Compressible-Fluid Dynamics*, McGraw-Hill, New York, 1971.

Thermodynamics, Heat and Power

ASME *Power Test Codes*, American Society of Mechanical Engineers.

Jones, J. B., and G. A. Hawkins: *Engineering Thermodynamics*, Wiley, New York, 1960.

Potter, P. J.: *Power Plant Theory and Design*, Wiley, New York, 1969.

Zemansky, M. W.: *Heat and Thermodynamics*, 6th ed., McGraw-Hill, New York, 1981.

Engines and Turbines

Lichty, L. C.: *Internal Combustion Engines*, McGraw-Hill, New York, 1951.

Shepherd, D. G.: *Principles of Turbomachinery*, Macmillan, New York, 1961.

Heat Transfer

Kern, D. Q.: *Process Heat Transfer*, McGraw-Hill, New York, 1950.

Holman, J. P.: *Heat Transfer*, 5th ed., McGraw-Hill, New York, 1981.

Welty, J. R.: *Engineering Heat Transfer*, Wiley, New York, 1978.

Refrigeration and Air Conditioning

ASHRAE *Guide and Data Book Applications*, 1980.

ASHRAE *Guide and Data Book Systems*, 1980.

ASHRAE *Handbook of Fundamentals*, 1980.

Carrier Air Conditioning Company, *Handbook of Air Conditioning System Design*, McGraw-Hill, New York, 1966.

Dossat, R. J.: *Principles of Regrigeration*, Wiley, New York, 1981.

Fan Engineering, Buffalo Forge, 1970.

Hemeon, W.: *Plant and Process Ventilation*, 2d ed., Industrial Press, New York, 1963.

McQuiston, F. C., and J. D. Parker, *Heating, Ventilating and Air Conditioning*, Wiley, New York, 1982.

Stoecker, W. F.: *Design of Thermal Systems*, McGraw-Hill, New York, 1980.

Zimmerman and Levine: *Psychrometric Tables and Charts*, Industrial Research Service, 1954.

Environmental and Sound Control

Austin, P. R.: *Design and Operation of Clean Rooms*, rev. ed., Business News, Birmingham, Mich. 1965.

Clayton, G. D., and F. E. Clayton: *Patty's Industrial Hygiene and Toxicology*, 3d ed., rev.; vol. 1: *General Principles*, Wiley Interscience, New York, 1978.

Clayton, G. D., and F. E. Clayton: *Patty's Industrial Hygiene and Toxicology*, 3d ed., rev.; vols. 2A, 2B, and 2C: *Toxocology*, Wiley Interscience, New York, 1982.

Constance, J. A.: "Why Some Dust Control Systems Don't Work," *Pharmaceutical Engineering*, vol. 3, no. 1, January-February 1983).

Constance, J. A.: "Reverse Contamination Caused by Dust Control Exhaust Systems," *Pharmaceutical Engineering*, vol. 3, no. 2, March-April 1983.

Constance, J. D.: *Controlling In-Plant Airborne Contaminants: Systems Design and Calculations*, Marcel Dekker, New York, 1983.

Jennings, B. H.: *Environmental Engineering*, IEP, New York, 1970.

The Industrial Environment—Its Evaluation and Control, U.S. Department of Health, Education and Welfare, 1965.

Industrial Ventilation, American Conference of Governmental Industrial Hygienists, 1985.

Sax, N. I.: *Dangerous Properties of Industrial Materials*, 4th ed., Van Nostrand Reinhold, New York, 1975.

Threlkeld, J. L.: *Thermal Environmental Engineering*, 2d ed., Prentice-Hall, Englewood Cliffs, N.J., 1970.

Nuclear Power

Etherington, H.: *Nuclear Engineering Handbook*, McGraw-Hill, New York, 1958.

Glasstone, S., and A. Sesonski: *Nuclear Reactor Engineering*, Van Nostrand Reinhold, New York, 1967.

Plant and Production Engineering

Apple, J. M.: *Material Handling Systems Design*, Wiley, New York, 1972.

Bolz, H. A., and G. E. Hagemann: *Materials Handling Handbook*, Wiley, New York, 1958.

Chemical Engineering Magazine: *Pneumatic Conveying of Bulk Materials*, McGraw-Hill, New York, 1980.

Higgins, L., and L. Morrow: *Maintenance Engineering Handbook*, McGraw-Hill, New York, 1977.

Society of Manufacturing Engineers: *Tool and Manufacturing Engineers Handbook*, 3d ed., McGraw-Hill, New York, 1976.

Society of Tool and Manufacturing Engineers: *Handbook of Fixture Design*, McGraw-Hill, New York, 1962.

Stanier, W.: *Plant Engineering Handbook*, 2d ed., McGraw-Hill, New York, 1959.

Metrication

Adams, H. F. R.: *SI Metric Units—An Introduction*, McGraw-Hill Paperbacks, 1973, revised 1974.

Caspe, Marc S., and Nicola Scianna: "The Problems of Dual Dimensioning," *Consulting Engineer*, June 1977.

Conversion Factors, Englehard Industries Division of Englehard Minerals and Chemicals Corp., Murray Hill, N.J. 07974.

Feirer, J. L.: *SI Metric Handbook*, The Metric Company, Chas. Scribners Sons, New York, LC Cat. No. 76-40815, ISBN 0-917644.

The International (SI) Metric System and How It Works, Polymetric Services, Inc., Reseda, Calif. 91335, 1973.

Manuals and Handbooks — General

Baumeister, T., et al.: *Marks Standard Handbook for Mechanical Engineers*, 8th ed., McGraw-Hill, New York, 1978.

Cameron Hydraulic Data, Ingersoll-Rand, 1942.

Cameron Pump Operator's Data, Ingersoll-Rand, 1943.

Compressed Air Data, Ingersoll-Rand, 1939.

Keenan, J. H., F. G. Keyes, P. G. Hill, and J. G. Moore: *Steam Tables: Thermodynamic Properties of Water, Including Vapor, Liquid and Solid*, Wiley, New York, 1969 (English units), 1978 (S.I. units).

Kent, R. T.: *Mechanical Engineer's Handbook*, 12th ed., Wiley, New York, 1950.

In addition to the above references, much help can be obtained from the various technical publications: *Chemical Engineering; Heating, Piping and Air Conditioning; Machine Design; Nucleonics; Plant Engineering; Power; Power Engineering; Product Engineering; Research and Development; Specifying Engineer. Note:* Some of the data appearing in various sources are updated periodically by the publishers, so that it behooves interested engineers to contact the publishers and obtain the new data for evaluation and comparison.

PROBLEM INDEX*

* Problem numbers in parentheses.

REFERENCE BOOK
NOT TO BE TAKEN
FROM THE LIBRARY